de Gruyter Expositions in Mathematics 33

de Gruyter Expositions in Mathematics

1 The Analytical and Topological Theory of Semigroups, *K. H. Hofmann, J. D. Lawson, J. S. Pym* (Eds.)

2 Combinatorial Homotopy and 4-Dimensional Complexes, *H. J. Baues*

3 The Stefan Problem, *A. M. Meirmanov*

4 Finite Soluble Groups, *K. Doerk, T. O. Hawkes*

5 The Riemann Zeta-Function, *A. A. Karatsuba, S. M. Voronin*

6 Contact Geometry and Linear Differential Equations, *V. E. Nazaikinskii, V. E. Shatalov, B. Yu. Sternin*

7 Infinite Dimensional Lie Superalgebras, *Yu. A. Bahturin, A. A. Mikhalev, V. M. Petrogradsky, M. V. Zaicev*

8 Nilpotent Groups and their Automorphisms, *E. I. Khukhro*

9 Invariant Distances and Metrics in Complex Analysis, *M. Jarnicki, P. Pflug*

10 The Link Invariants of the Chern-Simons Field Theory, *E. Guadagnini*

11 Global Affine Differential Geometry of Hypersurfaces, *A.-M. Li, U. Simon, G. Zhao*

12 Moduli Spaces of Abelian Surfaces: Compactification, Degenerations, and Theta Functions, *K. Hulek, C. Kahn, S. H. Weintraub*

13 Elliptic Problems in Domains with Piecewise Smooth Boundaries, *S. A. Nazarov, B. A. Plamenevsky*

14 Subgroup Lattices of Groups, *R. Schmidt*

15 Orthogonal Decompositions and Integral Lattices, *A. I. Kostrikin, P. H. Tiep*

16 The Adjunction Theory of Complex Projective Varieties, *M. C. Beltrametti, A. J. Sommese*

17 The Restricted 3-Body Problem: Plane Periodic Orbits, *A. D. Bruno*

18 Unitary Representation Theory of Exponential Lie Groups, *H. Leptin, J. Ludwig*

19 Blow-up in Quasilinear Parabolic Equations, *A. A. Samarskii, V. A. Galaktionov, S. P. Kurdyumov, A. P. Mikhailov*

20 Semigroups in Algebra, Geometry and Analysis, *K. H. Hofmann, J. D. Lawson, E. B. Vinberg* (Eds.)

21 Compact Projective Planes, *H. Salzmann, D. Betten, T. Grundhöfer, H. Hähl, R. Löwen, M. Stroppel*

22 An Introduction to Lorentz Surfaces, *T. Weinstein*

23 Lectures in Real Geometry, *F. Broglia* (Ed.)

24 Evolution Equations and Lagrangian Coordinates, *A. M. Meirmanov, V. V. Pukhnachov, S. I. Shmarev*

25 Character Theory of Finite Groups, *B. Huppert*

26 Positivity in Lie Theory: Open Problems, *J. Hilgert, J. D. Lawson, K.-H. Neeb, E. B. Vinberg* (Eds.)

27 Algebra in the Stone-Čech Compactification, *N. Hindman, D. Strauss*

28 Holomorphy and Convexity in Lie Theory, *K.-H. Neeb*

29 Monoids, Acts and Categories, *M. Kilp, U. Knauer, A. V. Mikhalev*

30 Relative Homological Algebra, Edgar E. Enochs, Overtoun M. G. Jenda

31 Nonlinear Wave Equations Perturbed by Viscous Terms, Viktor P. Maslov, Petr P. Mosolov

32 Conformal Geometry of Discrete Groups and Manifolds, Boris N. Apanasov

Compositions of Quadratic Forms

by

Daniel B. Shapiro

Walter de Gruyter · Berlin · New York 2000

Author

Daniel B. Shapiro
Department of Mathematics
Ohio State University
231 West 18th Avenue
Columbus, Ohio 43210-1328
USA
shapiro@math.ohio-state.edu

Mathematics Subject Classification 2000:
11-02; 11E04, 11E25, 11E39, 11E88, 15A63, 17A75

Key words:
Quadratic Forms, Sums of Squares, Composition Formula, Hurwitz-Radon Theorem, Intercalate Matrix, Hopf map

♾ Printed on acid-free paper which falls within the guidelines of the ANSI to ensure permanence and durability.

Library of Congress — Cataloging-in-Publication Data

Shapiro, Daniel B., 1948–
Compositions of quadratic forms / by Daniel B. Shapiro.
p. cm. — (De Gruyter expositions in mathematics, ISSN 0938-6572 ; 33)
Includes bibliographical references and index.
ISBN 3 11 012629 X (alk. paper)
1. Forms, Quadratic. I. Title. II. Series.
QA243 .S48 2000
512′.74–dc21 00-043016

Die Deutsche Bibliothek — Cataloging-in-Publication Data

Shapiro, Daniel B.:
Compositions of quadratic forms / by Daniel B. Shapiro. — Berlin ; New York : de Gruyter, 2000
(De Gruyter expositions in mathematics ; 33)
ISBN 3-11-012629-X

Typesetting with TEX: I. Zimmermann, Freiburg.
Printing: WB-Druck GmbH & Co., Rieden/Allgäu. Binding: Lüderitz & Bauer-GmbH, Berlin.
Cover design: Thomas Bonnie, Hamburg.

Dedicated to Amanda, Becky and Jacob

Table of Contents

Part II. Compositions of Size $[r, s, n]$

Introduction

This book addresses basic questions about compositions of quadratic forms in the sense of Hurwitz and Radon. The initial question is: For what dimensions can they exist? Subsequent questions involve classification and analysis of the quadratic forms which can occur in a composition.

This topic originated with the "1, 2, 4, 8 Theorem" concerning formulas for a product of two sums of squares. That theorem, proved by Adolf Hurwitz in 1898, was generalized in various ways during the folowing century, leading to the theories discussed here. This area is worth studying because it is so centrally located in mathematics: these compositions have close connections with mathematical history, algebra, combinatorics, geometry, and topology.

Compositions have deep historical roots: the 1, 2, 4, 8 Theorem settled a long standing question about the existence of "n-square identities" and exhibited some of the power of linear algebra. Compositions are also entwined with the nineteenth century development of quaternions, octonions and Clifford algebras.

Another attraction of this subject is its fascinating relationship with Clifford algebras and the algebraic theory of quadratic forms. A general composition formula involves arbitrary quadratic forms over a field, not just the classical sums of squares. Such compositions can be reformulated in terms of Clifford algebras and their involutions. There is also a close connection between the forms involved in compositions and the multiplicative quadratic forms introduced by Pfister in the 1960s.

All the known constructions of composition formulas for sums of squares can be achieved using integer coefficients. A composition formula with integer coefficients can be recast as a combinatorial object: a special sort of matrix of symbols and signs. These "intercalate" matrices have been studied intensively, leading to a classification of the integer compositions which involve at most 16 squares.

Finally this topic is connected with certain deep questions in geometry. For instance, composition formulas provide examples of vector bundles on projective spaces, of independent vector fields on spheres, of immersions of projective spaces into euclidean spaces, and of Hopf maps between euclidean spheres. The topological tools developed to analyze these topics also yield results about real compositions.

Let us now describe the original question with more precision: A composition formula of size $[r, s, n]$ is a sum of squares formula of the type

$$(x_1^2 + x_2^2 + \cdots + x_r^2) \cdot (y_1^2 + y_2^2 + \cdots + y_s^2) = z_1^2 + z_2^2 + \cdots + z_n^2$$

where $X = (x_1, x_2, \ldots, x_r)$ and $Y = (y_1, y_2, \ldots, y_s)$ are systems of indeterminates and each $z_k = z_k(X, Y)$ is a bilinear form in X and Y. Such a formula can be viewed in several different ways, with each version providing different insights and techniques. Hurwitz restated the formula as a system of r different $n \times s$ matrices. More geometrically (assuming that the z_k's have real coefficients), the formula becomes a bilinear pairing

$$f : \mathbb{R}^r \times \mathbb{R}^s \longrightarrow \mathbb{R}^n$$

which satisfies the norm condition: $|f(x, y)| = |x| \cdot |y|$ for $x \in \mathbb{R}^r$ and $y \in \mathbb{R}^s$. For example the usual multiplication of complex numbers provides a formula of size $[2, 2, 2]$. In the original sums-of-squares language, this bilinear pairing becomes the formula:

$$(x_1^2 + x_2^2) \cdot (y_1^2 + y_2^2) = z_1^2 + z_2^2 \quad \text{where } z_1 = x_1y_1 - x_2y_2 \text{ and } z_2 = x_1y_2 + x_2y_1.$$

The quaternion and octonion algebras, discovered in the 1840s, provide similar formulas of sizes $[4, 4, 4]$ and $[8, 8, 8]$. Using his matrix formulation Hurwitz (1898) proved that a formula of size $[n, n, n]$ exists if and only if n is 1, 2, 4 or 8. Hurwitz and Radon used similar techniques to determine exactly when formulas of size $[r, n, n]$ can exist. It is far more difficult to analyze compositions of sizes $[r, s, n]$ when $r, s < n$.

These ideas have been generalized in two main directions, determining the contents of the two parts of this book.

Part I: If the composition involves general quadratic forms over a field in place of the sums of squares, what can be said about those forms? Interesting results have been obtained for the classical sizes $[r, n, n]$.

Part II: What sizes r, s, n are possible in the general case? Does the answer depend on the field of coefficients? Many partial results have been obtained using methods of algebraic topology, combinatorics, linear algebra and geometry.

Further descriptions of the historical background and the contents of this work appear in Chapter 0 and in the Introduction to Part II.

Readers of this work are expected to have knowledge of some abstract algebra. The first two chapters assume familiarity with only the basic properties of linear algebra and inner product spaces. The next five chapters require quadratic forms, Clifford algebras, central simple algebras and involutions, although many of those concepts are developed in the text. For example, Clifford algebras are defined and their basic properties are established in Chapter 3. Later chapters assume further background. For example Chapter 11 uses algebraic number theory and Chapter 12 employs algebraic topology.

Each chapter begins with a brief statement of its content and ends with some exercises, usually involving alternative methods or related results. In fact many related topics and open questions have been converted to exercises. This practice lengthens the exercise sections, but adds some further depth to the book. The Notes at the end of each chapter provide additional comments, historical remarks and references. At the end of the book there is a fairly extensive bibliography, arranged alphabetically by first author.

Most of the material described in this book has already appeared in the mathematical literature, usually in research papers. However there are many items that have not been previously published. These include:

- an improved version of the Eigenspace Lemma (2.10);
- a discussion of anti-commuting skew-symmetric matrices, Exercise 2.13;
- the trace methods used to analyze (2, 2)-families, Chapter 5;
- the treatment of composition algebras, Chapter 1.A (due to Conway);
- the analysis of "minimal" pairs, Chapter 7;
- properties of the topological space of all compositions, Chapter 8;
- monotopies and isotopies, Chapter 8 (due to Conway);
- the matrix approach to Pfaffians, Chapter 10;
- Hasse principle for divisibility, Chapter 11.A (due to Wadsworth);
- general monomial compositions, Chapter 13.B;
- the characterization of all compositions of codimension 2, (14.18);
- nonsingular and surjective bilinear pairings over fields, Exercises 14.16–19.

This book evolved over many years, starting from series of lectures I gave on this subject at the Universität Regensburg (Germany) in 1977, at the Universidad de Chile in 1981, at the University of California-Berkeley in 1983, at the Universität Dortmund (Germany) in 1991, at the Universidad de Talca (Chile) in 1999 and several times at the Ohio State University. I am grateful to these institutions, to the National Science Foundation, to the Alexander von Humboldt Stiftung and to the Fundación Andes for their generous support. It is also a pleasure to thank many friends and colleagues for their interest in this work and their encouragement over the years. Special thanks are due to several colleagues who have made observations directly affecting this book. These include J. Adem, R. Baeza, E. Becker, A. Geramita, J. Hsia, I. Kaplansky, M. Knebusch, K. Y. Lam, T. Y. Lam, D. Leep, T. Smith, M. Szyjewski, J.-P. Tignol, A. Wadsworth, P. Yiu, and S. Yuzvinsky. Extra thanks are due to Adrian Wadsworth for providing great help and support in the early years of my mathematical career.

I am also grateful to those colleagues and students who have proofread sections of this book, finding errors and making worthwhile suggestions. However I take full responsibility for the remaining grammatical and mathematical errors, the incorrect cross references, the inconsistencies of notation and the gaps in understanding.

As mentioned above, this book has been in progress for many years. In fact it is hard for me to believe how long it has been. The writing was finally finished in 1998, barely in time to celebrate the centennial of the Hurwitz 1, 2, 4, 8 Theorem.

Chapter 0

Historical Background

The theory of composition of quadratic forms over fields had its start in the 19th century with the search for n-square identities of the type

$$(x_1^2 + x_2^2 + \cdots + x_n^2) \cdot (y_1^2 + y_2^2 + \cdots + y_n^2) = z_1^2 + z_2^2 + \cdots + z_n^2$$

where $X = (x_1, x_2, \ldots, x_n)$ and $Y = (y_1, y_2, \ldots, y_n)$ are systems of indeterminates and each $z_k = z_k(X, Y)$ is a bilinear form in X and Y. For example when $n = 2$ there is the ancient identity

$$(x_1^2 + x_2^2) \cdot (y_1^2 + y_2^2) = (x_1 y_1 + x_2 y_2)^2 + (x_1 y_2 - x_2 y_1)^2.$$

In this example $z_1 = x_1 y_1 + x_2 y_2$ and $z_2 = x_1 y_2 - x_2 y_1$ are bilinear forms in X, Y with integer coefficients. This formula for $n = 2$ can be interpreted as the "law of moduli" for complex numbers: $|\alpha| \cdot |\beta| = |\alpha\beta|$ where $\alpha = x_1 - i x_2$ and $\beta = y_1 + i y_2$.

A similar 4-square identity was found by Euler (1748) in his attempt to prove Fermat's conjecture that every positive integer is a sum of four integer squares. This identity is often attributed to Lagrange, who used it (1770) in his proof of that conjecture of Fermat. Here is Euler's formula, in our notation:

$$(x_1^2 + x_2^2 + x_3^2 + x_4^2) \cdot (y_1^2 + y_2^2 + y_3^2 + y_4^2) = z_1^2 + z_2^2 + z_3^2 + z_4^2$$

where

$$z_1 = x_1 y_1 + x_2 y_2 + x_3 y_3 + x_4 y_4$$

$$z_2 = x_1 y_2 - x_2 y_1 + x_3 y_4 - x_4 y_3$$

$$z_3 = x_1 y_3 - x_2 y_4 - x_3 y_1 + x_4 y_2$$

$$z_4 = x_1 y_4 + x_2 y_3 - x_3 y_2 - x_4 y_1.$$

After Hamilton's discovery of the quaternions (1843) this 4-square formula was interpreted as the law of moduli for quaternions. Hamilton's discovery came only after he spent years searching for a way to multiply "triplets" (i.e. triples of numbers) so that the law of moduli holds. Such a product would yield a 3-square identity. Already in his *Théorie des Nombres* (1830), Legendre showed the impossibility of such an identity. He noted that 3 and 21 can be expressed as sums of three squares of rational numbers, but that $3 \times 21 = 63$ cannot be represented in this way. It follows that a 3-square identity is impossible (at least when the bilinear forms have rational coefficients). If Hamilton had known of this remark by Legendre he might have given up the search to multiply triplets! Hamilton's great insight was to move on to four dimensions and to allow a non-commutative multiplication.

Hamilton wrote to John Graves about the discovery of quaternions in October 1843 and within two months Graves wrote to Hamilton about his discovery of an algebra of "octaves" having 8 basis elements. The multiplication satisfies the law of moduli, but is neither commutative nor associative. Graves published his discovery in 1848, but Cayley independently discovered this algebra and published his results in 1845. Many authors refer to elements of this algebra as "Cayley numbers". In this book we use the term "octonions".

The multiplication of octonions provides an 8-square identity. Such an identity had already been found in 1818 by Degen in Russia, but his work was not widely read. After the 1840s a number of authors attempted to find 16-square identities with little success. It was soon realized that no 16-square identity with integral coefficients is possible, but the arguments at the time were incomplete. These "proofs" were combinatorial in nature, attempting to insert $+$ and $-$ signs in the entries of a 16×16 Latin square to make the rows orthogonal.

In 1898 Hurwitz published the definitive paper on these identities. He proved that there exists an n-square identity with complex coefficients if and only if $n = 1, 2, 4$ or 8. His proof involves elementary linear algebra, but these uses of matrices and linear independence were not widely known in 1898. At the end of that paper Hurwitz posed the general problem: For which positive integers r, s, n does there exist a "composition formula":

$$(x_1^2 + x_2^2 + \cdots + x_r^2) \cdot (y_1^2 + y_2^2 + \cdots + y_s^2) = z_1^2 + z_2^2 + \cdots + z_n^2$$

where $X = (x_1, x_2, \ldots, x_r)$ and $Y = (y_1, y_2, \ldots, y_s)$ are systems of indeterminates and each $z_k = z_k(X, Y)$ is a bilinear form in X and Y?

Here is an outline of Hurwitz's ideas, given without all the details. Suppose there is a composition formula of size $[r, s, n]$ as above. View X, Y and Z as column vectors. Then, for example, $z_1^2 + z_2^2 + \cdots + z_n^2 = Z^\top \cdot Z$, where the superscript $\top$ denotes the transpose. The bilinearity condition becomes $Z = AY$ where A is an $n \times s$ matrix whose entries are linear forms in X. The given composition formula can then be written as

$$(x_1^2 + x_2^2 + \cdots + x_r^2) Y^\top \cdot Y = Z^\top \cdot Z = Y^\top A^\top A Y.$$

Since Y consists of indeterminates this equation is equivalent to

$$A^\top \cdot A = (x_1^2 + x_2^2 + \cdots + x_r^2) I_s,$$

where A is an $n \times s$ matrix whose entries are linear forms in X.

Of course I_s here denotes the $s \times s$ identity matrix. Since the entries of A are linear forms we can express $A = x_1A_1 + x_2A_2 + \cdots + x_rA_r$ where each A_i is an $n \times s$ matrix with constant entries. After substituting this expression into the equation and canceling like terms, we find:

> There are $n \times s$ matrices $A_1, A_2, \ldots, A_r$ over F satisfying
> $$A_i^\top \cdot A_i = I_s \quad \text{for } 1 \le i \le r,$$
> $$A_i^\top \cdot A_j + A_j^\top \cdot A_i = 0 \quad \text{for } 1 \le i, j \le r \text{ and } i \ne j.$$

This system is known as the "Hurwitz Matrix Equations". Such matrices exist if and only if there is a composition formula of size $[r, s, n]$.

Hurwitz considered these matrices to have complex entries, but his ideas work just as well using any field of coefficients, provided that the characteristic is not 2. Those matrices are square when $s = n$. In that special case the system of equations can be greatly simplified by defining the $n \times n$ matrices $B_i = A_1^{-1}A_i$ for $1 \le i \le r$. Then $B_1, \ldots, B_r$ satisfy the Hurwitz Matrix Equations and $B_1 = I_n$. It follows that:

> There are $n \times n$ matrices $B_2, \ldots, B_r$ over F satisfying:
> $$\left.\begin{aligned} B_i^\top &= -B_i, \\ B_i^2 &= -I_n, \end{aligned}\right\} \quad \text{for } 2 \le i \le r;$$
> $$B_iB_j = -B_jB_i \quad \text{whenever } i \ne j.$$

Such a system of $n \times n$ matrices exists if and only if there is a composition formula of size $[r, n, n]$. Hurwitz proved that the 2^{r-2} matrices $B_{i_1}B_{i_2}\ldots B_{i_k}$ for $2 \le i_1 \le \cdots \le i_k \le r-1$ are linearly independent. This shows that $2^{r-2} \le n^2$ and in the case of n-square identities (when $r = n$) quickly leads to the "1, 2, 4, 8 Theorem".

In 1922 Radon determined the exact conditions on r and n for such a system of matrices to exist over the real field $\mathbb{R}$. This condition had been found independently by Hurwitz for formulas over the complex field $\mathbb{C}$ and was published posthumously in 1923. They proved that:

A formula of size $[r, n, n]$ exists if and only if $r \le \rho(n)$,

where the "Hurwitz–Radon function" $\rho(n)$ is defined as follows: if $n = 2^{4a+b}n_0$ where n_0 is odd and $0 \le b \le 3$, then $\rho(n) = 8a + 2^b$. There are several different ways this function can be described. The following one is the most convenient for our purposes:

$$\text{If } n = 2^m n_0 \text{ where } n_0 \text{ is odd then } \rho(n) = \begin{cases} 2m+1 & \text{if } m \equiv 0, \\ 2m & \text{if } m \equiv 1, \\ 2m & \text{if } m \equiv 2, \\ 2m+2 & \text{if } m \equiv 3 \end{cases} \pmod 4.$$

For example, $\rho(n) = n$ if and only if $n = 1, 2, 4$ or 8, as expected from the earlier theorem of Hurwitz. Also $\rho(16) = 9$, $\rho(32) = 10$, $\rho(64) = 12$ and generally $\rho(16n) = 8 + \rho(n)$. New proofs of the Hurwitz–Radon Theorem for compositions of size $[r, n, n]$ were found in the 1940s. Eckmann (1943b) applied the representation theory of certain finite groups to prove the theorem over $\mathbb{R}$, and Lee (1948) modified Eckmann's ideas to prove the result using representations of Clifford algebras. Independently, Albert (1942a) generalized the 1, 2, 4, 8 Theorem to quadratic forms over arbitrary fields, and Dubisch (1946) used Clifford algebras to prove the Hurwitz–Radon Theorem for quadratic forms over $\mathbb{R}$ (allowing indefinite forms). Motivated by a problem in geometry, Wong (1961) analyzed the Hurwitz–Radon Theorem using matrix methods and classified the types of solutions over $\mathbb{R}$. In the 1970s Shapiro proved the Hurwitz–Radon Theorem for arbitrary (regular) quadratic forms over any field where $2 \neq 0$, and investigated the quadratic forms which admit compositions. One goal of our presentation is to explain the curious periodicity property of the Hurwitz–Radon function $\rho(n)$:

$$\text{Why does } \rho(2^m) \text{ depend only on } m \pmod 4?$$

The explanation comes from the shifting properties of (s, t)-families as explained in Chapter 2.

Here are some of the questions which have motivated much of the work done in Part I of this book. Suppose σ and q are regular quadratic forms over the field F, where $\dim \sigma = s$ and $\dim q = n$. Then σ and q "admit a composition" if there is a formula

$$\sigma(X)q(Y) = q(Z),$$

where as usual $X = (x_1, x_2, \ldots, x_s)$ and $Y = (y_1, y_2, \ldots, y_n)$ are systems of indeterminates and each z_k is a bilinear form in X and Y, with coefficients in F. The quadratic forms involved in these compositions are related to Pfister forms.

In the 1960s Pfister found that for every m there do exist 2^m-square identities, provided some denominators are allowed. He generalized these identities to a wider class: a quadratic form is a *Pfister form* if it expressible as a tensor product of binary quadratic forms of the type $\langle 1, a \rangle$. In particular its dimension is 2^m for some m. Here we use the notation $\langle a_1, \ldots, a_n \rangle$ to stand for the n-dimensional quadratic form $a_1x_1^2 + \cdots + a_nx_n^2$.

Theorem (Pfister). *If φ is a Pfister form and X, Y are systems of indeterminates, then there is a multiplication formula*

$$\varphi(X)\varphi(Y) = \varphi(Z),$$

where each component $z_k = z_k(X, Y)$ is a linear form in Y with coefficients in the rational function field $F(X)$. Conversely if φ is an anisotropic quadratic form over F satisfying such a multiplication formula, then φ must be a Pfister form.

The theory of Pfister forms is described in the textbooks by Lam (1973) and Scharlau (1985). When $\dim \varphi = 1, 2, 4$ or 8, such a multiplication formula exists using no denominators, since the Pfister forms of those sizes are exactly the norm forms of composition algebras. But if $\dim \varphi = 2^m > 8$, Hurwitz's theorem implies that any such formula must involve denominators. Examples of such formulas can be written out explicitly (see Exercise 5).

The quadratic forms appearing in the Hurwitz–Radon composition formulas have a close relationship to Pfister forms. For any Pfister form φ of dimension 2^m there is an explicit construction showing that φ admits a composition with some form σ having the maximal dimension $\rho(2^m)$. The converse is an interesting open question.

Pfister Factor Conjecture. Suppose q is a quadratic form of dimension 2^m, and q admits a composition with some form of the maximal dimension $\rho(2^m)$. Then q is a scalar multiple of a Pfister form.

This conjecture is one of the central themes driving the topics chosen for the first part of the book. In Chapter 9 it is proved true when $m \leq 5$, and for larger values of m over special classes of fields.

The second part of this book focuses on the more general compositions of size $[r, s, n]$. In 1898 Hurwitz already posed the question: Which sizes are possible? The cases where $s = n$ were settled by Hurwitz and Radon in the 1920s. Further progress was made around 1940 when Stiefel and Hopf applied techniques of algebraic topology to the problem, (for compositions over the field of real numbers). In Part II we discuss these topological arguments and their generalizations, as well as considering the question for more general fields of coefficients. Further details are described in the Introduction to Part II.

Exercises for Chapter 0

Note: For the exercises in this book, most of the declarative statements are to be proved. This avoids writing "prove that" in every problem.

1. In any (bilinear) 4-square identity, if $z_1 = x_1y_1 + x_2y_2 + x_3y_3 + x_4y_4$ then z_2, z_3, z_4 must be skew-symmetric. (Compare 4-square identity of Euler above.)

2. **Doubling Lemma.** From an $[r, s, n]$-formula over F construct an $[r+1, 2s, 2n]$-formula.

(Hint. Given the Hurwitz Matrix Equations consider the $2n \times 2s$ matrices

$$C_1 = \begin{pmatrix} A_1 & 0 \\ 0 & A_1 \end{pmatrix}, \ C_j = \begin{pmatrix} A_j & 0 \\ 0 & -A_j \end{pmatrix} \text{ for } 2 \le j \le r \text{ and } C_{r+1} = \begin{pmatrix} 0 & A_1 \\ -A_1 & 0 \end{pmatrix}.)$$

3. Let $x_1, x_2, \dots, x_n$ be indeterminates and consider the set $\mathcal{S}$ of all the vectors $(\pm x_{\alpha(1)}, \pm x_{\alpha(2)}, \dots, \pm x_{\alpha(n)})$ where α is a permutation of the n subscripts and $+/-$ signs are arbitrary. What is the largest number of mutually orthogonal elements of this set of $2^n n!$ vectors, using the usual dot product? Answer: The sharp bound is the Hurwitz–Radon function $\rho(n)$.

(Hint. If $v_1, \dots, v_s$ is a set of mutually orthogonal vectors in $\mathcal{S}$ let A be the $n \times s$ matrix with columns v_j. Then $A^\top \cdot A = \sigma I_s$ where $\sigma = x_1^2 + x_2^2 + \dots + x_n^2$. This provides a composition formula of size $[n, s, n]$.)

4. **Integer compositions.** Suppose a composition formula of size $[r, s, n]$ is given with *integer* coefficients. As above, let $Z = AY$ where A is an $n \times s$ matrix whose entries are linear forms in X, and $A^\top \cdot A = \sigma I_s$ where $\sigma = x_1^2 + x_2^2 + \dots + x_r^2$.

(1) The columns of A are orthogonal vectors, and each column consists of the entries $\pm x_1, \dots, \pm x_r$, placed (in some order) in r of the n positions, with zeros elsewhere. When $r = n$ the columns of A are signed permutations of X.

(2) The matrix A for the 2-square identity is $\begin{pmatrix} 1 & 2 \\ -2 & 1 \end{pmatrix}$, where we write only the subscript of each x_i. Similarly express A for the 4-square identity. Try to construct the 8-square identity directly in this way. (Explicit matrices A were listed by Hurwitz (1898).)

(3) Write down the matrix A for a composition of size $[3, 5, 7]$.

(Hint: (3) Express $(x_1^2 + x_2^2 + x_3^2) \cdot (y_1^2 + y_2^2 + y_3^2 + y_4^2)$ as 4 squares.

5. Define $D_F(n)$ to be the set of those non-zero elements of the field F which are expressible as sums of n squares of elements of F.

Theorem (Pfister). *$D_F(n)$ is a group whenever $n = 2^m$.*

This result motivated Pfister's theorem on multiplicative quadratic forms. Here is a proof.

(1) If $D_F(n)$ is closed under multiplication then it is a group.

(2) **Lemma.** *Suppose $c = c_1^2 + c_2^2 + \dots + c_n^2$ where $n = 2^m$. Then there exists an $n \times n$ matrix C having first row $(c_1, c_2, \dots, c_n)$ and satisfying $C \cdot C^\top = C^\top \cdot C = cI_n$.*

(3) *Proof of Theorem.* Let $c, d \in D_F(n)$, with the corresponding matrices C, D from the lemma. Then $A = CD^\top$ satisfies $AA^\top = cdI_n$, hence $cd \in D_F(n)$.

(4) **Corollary.** *Let $n = 2^m$ and $x_1, , \dots, x_n, y_1, \dots, y_n$ be indeterminates. Then there exist $z_1, \dots, z_n \in F(X)[Y]$ such that*

$$(x_1^2 + \cdots + x_n^2) \cdot (y_1^2 + \cdots + y_n^2) = z_1^2 + \cdots + z_n^2. \qquad (*)$$

In fact we can choose each z_i to be a linear form in Y and $z_1 = x_1y_1 + \cdots + x_ny_n$.

If $n = 16$, there exists such an identity where $z_1, \dots, z_8$ are bilinear forms in X, Y while $z_9, \dots, z_{16}$ are linear forms in Y. What denominators are involved in the terms z_j?

(Hint. (1) Note that $F^{\bullet 2} \subseteq D_F(n)$.

(2) Express $c = a + b$ where a, b are the sums of 2^{m-1} of the c_k^2. Let A, B be the corresponding matrices which exist by induction. If $a \neq 0$ define $C = \begin{pmatrix} A & B \\ \diamond & A^\top \end{pmatrix}$, where the entry $\diamond$ is to be filled in. If $b \neq 0$ use $C = \begin{pmatrix} A & B \\ B^\top & \diamond \end{pmatrix}$. What if $a = b = 0$?

(4) When $n = 16$, start from a bilinear 8-square identity, then build up the matrix C.)

6. **Hurwitz–Radon function.**

(1) If $n = 2^m \cdot (\text{odd})$ then $\rho(n) = \begin{cases} 8a+1 & \text{if } m = 4a, \\ 8a+2 & \text{if } m = 4a+1, \\ 8a+4 & \text{if } m = 4a+2, \\ 8a+8 & \text{if } m = 4a+3. \end{cases}$

(2) Given r define $v(r)$ to be the minimal n for which there exists an $[r, n, n]$ formula. Then $v(r) = 2^{\delta(r)}$ where $\delta(r) = \min\{m : r \leq \rho(2^m)\}$. Then $\rho(2^m) = \max\{r : \delta(r) = m\}$ and

$$\delta(r) = \#\{k : 0 < k < r \text{ and } k \equiv 0, 1, 2 \text{ or } 4 \pmod 8\}.$$

Here are the first few values:

r	1	2	3	4	5	6	7	8	9	10	11	12	13	14	15	16
$\delta(r)$	0	1	2	2	3	3	3	3	4	5	6	6	7	7	7	7

(3) $r \leq \rho(2^m)$ if and only if $\delta(r) \leq m$. Equivalently: $r \leq \rho(n)$ iff $2^{\delta(r)} \mid n$. Note: $r \leq 2^{\delta(r)}$ with equality if and only if $r = 1, 2, 4, 8$.

7. **Vector fields on spheres.** Let S^{n-1} be the unit sphere in the euclidean space $\mathbb{R}^n$.

A (tangent) *vector field* on S^{n-1} is a continuous map $f : S^{n-1} \to \mathbb{R}^n$ for which $f(x)$ is a vector tangent to S^{n-1} at x, for every $x \in S^{n-1}$. Using $\langle x, y\rangle$ for the dot product on $\mathbb{R}^n$, this says: $\langle x, f(x)\rangle = 0$ for every x. Define a vector field to be *linear* if the map f is the restriction of a linear map $\mathbb{R}^n \to \mathbb{R}^n$. A set of vector fields satisfies a property if that property holds at every point x. For instance, a vector field f is non-vanishing if $f(x) \neq 0$ for every $x \in S^n$.

(1) There is a non-vanishing linear vector field on S^{n-1} if and only if n is even.

(2) There exist r independent, linear vector fields on S^{n-1} if and only if $r \leq \rho(n) - 1$. Consequently, S^{n-1} admits $n-1$ independent linear vector fields if and only if $n = 1, 2, 4$ or 8.

(Hint. (1) A linear vector field is given by an $n \times n$ matrix A with $A^{\top} = -A$. It is non-vanishing iff A is nonsingular.

(2) The vector fields, and Gram–Schmidt, provide r mutually orthogonal, unit length, linear vector fields on S^{n-1}. These lead to the Hurwitz matrix equations.)

Remark. With considerable topological work, the hypothesis "linear" can be removed here. For example, the fact that S^2 has no non-vanishing vector field (the "Hairy Ball Theorem") is a result often proved in beginning topology courses. Finding the maximal number of linearly independent tangent vector fields on S^{n-1} is a famous topic in algebraic topology. Adams finally solved this problem in 1962, showing that the maximal number is just $\rho(n) - 1$. In particular, S^{n-1} is "parallelizable" (i.e. it has $n-1$ linearly independent tangent vector fields) if and only if $n = 1, 2, 4$ or 8.

8. **Division algebras.** Let F be a field. An *F-algebra* is defined to be an F-vector space A together with an F-bilinear map $m : A \times A \to A$. (Note that there are no assumptions of associativity, commutativity or identity element here.) Writing $m(x, y)$ as xy we see that the distributive laws hold and the scalars can be moved around freely. For $a \in A$ define the linear maps $L_a, R_a : A \to A$ by $L_a(x) = ax$ and $R_a(x) = xa$. Define A to be a *division algebra* if L_a and R_a are bijective for every $a \neq 0$.

(1) Suppose A is a finite dimensional F-algebra. Then A is a division algebra if and only if it has no zero-divisors.

(2) Suppose D is a division algebra over F, choose non-zero elements $u, v \in D$, and define a new multiplication $\heartsuit$ on D by: $x \heartsuit y = (R_u^{-1}(x))(L_v^{-1}(y))$. Then $(xu) \heartsuit (vy) = xy$ and $(D, \heartsuit)$ is a division algebra over F with identity element vu.

(3) F-algebras $(A, \cdot)$ and $(B, *)$ are defined to be *isotopic* (written $A \sim B$) if there exist bijective linear maps $\alpha, \beta, \gamma : A \to B$ such that $\gamma(x \cdot y) = \alpha(x) * \beta(y)$ for every $x, y \in A$. Isotopy is an equivalence relation on the category of F-algebras, and $A \cong B$ implies $A \sim B$. If A is a division algebra and $A \sim B$ then B is a division algebra. Every F-division algebra is isotopic to an F-division algebra with identity.

Remark. Since we are used to the associative law, great care must be exercised when dealing with such general division algebras. Suppose A is a division algebra with 1 and $0 \neq a \in A$. There exist $b, c \in A$ with $ba = 1$ and $ac = 1$. These left and right "inverses" are unique but they can be unequal. Even if $b = c$ it does not follow that $b \cdot ax = x$ for every x. See Exercise 8.7.

9. **Division algebras and vector fields.** Suppose there is an n-dimensional real division algebra (with no associativity assumptions).

(1) There is such a division algebra D with identity element e. (Use Exercise 8(2).)

(2) If $d \in D$ let $f_d(x)$ be the projection of $d * x$ to $(x)^\perp$. Then f_d is a vector field on S^{n-1} which is non-vanishing if $d \notin \mathbb{R} \cdot e$. A basis of D induces $n-1$ linearly independent vector fields on S^{n-1}.

(3) Deduce the "1, 2, 4, 8 Theorem" for real division algebras from Adams' theorem on vector fields (mentioned in Exercise 7).

10. **Hopf maps.** Suppose there is a real composition formula of size $[r, s, n]$. Then there is a bilinear map $f : \mathbb{R}^r \times \mathbb{R}^s \to \mathbb{R}^n$ satisfying $|f(x, y)| = |x| \cdot |y|$ for every $x \in \mathbb{R}^r$ and $y \in \mathbb{R}^s$. Construct the associated Hopf map $h : \mathbb{R}^r \times \mathbb{R}^s \to \mathbb{R} \times \mathbb{R}^n$ by $h(x, y) = (|x|^2 - |y|^2, 2f(x, y))$.

(1) $|h(x, y)| = |(x, y)|^2$, using the usual Euclidean norms.

(2) h restricts to a map on the unit spheres $h_0 : S^{r+s-1} \to S^n$.

(3) If $x \in S^{r-1}$ and $y \in S^{s-1}$ then $(\cos(\theta) \cdot x, \sin(\theta) \cdot y) \in S^{r+s-1} \subseteq \mathbb{R}^r \times \mathbb{R}^s$. This provides a covering $S^1 \times S^{r-1} \times S^{s-1} \to S^{r+s-1}$. The Hopf map h_0 carries

$$(\cos(\theta) \cdot x, \sin(\theta) \cdot y) \mapsto (\cos(2\theta), \sin(2\theta) \cdot f(x, y)).$$

(4) When $r = s = n = 1$ the map $h_0 : S^1 \to S^1$ wraps the circle around itself twice. When $r = s = n = 2$ the map $h : \mathbb{C} \times \mathbb{C} \to \mathbb{R} \times \mathbb{C}$ induces $h_0 : S^3 \to S^2$. This map is surjective and each fiber is a circle. Further properties of Hopf maps are described in Chapter 15.

Notes on Chapter 0

Further details on the history of n-square identities appear in Dickson (1919), van der Blij (1961), Curtis (1963), Halberstam and Ingham (1967), Taussky (1970), (1981) and van der Waerden (1976), (1985). The interesting history of Hamilton and his quaternions is described in Crowe (1967). Veldkamp (1991) also has some historical remarks on octonions.

The terminology for the 8-dimensional composition algebra has changed over the years. The earliest name was "octaves" used by Graves and others in the 19th century. This term remained in use in German articles (e.g. Freudenthal in the 1950s), and some authors followed that tradition in English. Many papers in English used "Cayley numbers", the "Cayley algebra" or the "Cayley–Dickson algebra". The term "octonions" came into use in the late 1960s, and may have first appeared in Jacobson's book (1968). Several of Jacobson's students were using "octonions" shortly after that and this terminology has become fairly standard. Jacobson himself says that this term was motivated directly from "the master of terminology" J. J. Sylvester, who introduced quinions, nonions and sedenions. See Sylvester (1884).

In proving his 1, 2, 4, 8 Theorem, Hurwitz deduced the linear independence of the 2^{r-2} matrices formed by products of subsets of $\{B_2, \dots, B_{r-1}\}$. His argument used the skew-symmetry of the B_j. The more general independence result, proved in

(1.11), first appeared in Robert's thesis (1912), written under the direction of Hurwitz. It also appears in the posthumous paper of Hurwitz (1923). Compare Exercise 1.12.

It is interesting to note that the law of multiplication of quaternions had been discovered, but not published, by Gauss as early as 1820. See the reference in van der Waerden (1985), p. 183.

The theory of Pfister forms appears in a number of books, including Lam (1973) and Scharlau (1985) and in Knebusch and Scharlau (1980). The basic multiplicative properties of Pfister forms are derived in Chapter 5.

Basic topological results (Hairy Ball Theorem, vector fields on spheres, etc.) appear in many topology texts, including Spanier (1966) and Husemoller (1975). Further information on these topics is given in Chapter 12.

Exercise 5. This property of $D_F(2^m)$ was foreshadowed by the 8 and 16 square identities of Taussky (1966) and Zassenhaus and Eichhorn (1966). The multiplicative properties of $D_F(n)$ were discovered by Pfister (1965a) and generalized in (1965b). The simple matrix proof here is due to Pfister and appears in Lam (1973), pp. 296–298 and in Knebusch and Scharlau (1980), pp. 12–14. Pfister's analysis of the more general products $D_F(r) \cdot D_F(s)$ is presented in Chapter 14.

Exercise 6. The function $\delta(r)$ arises later in (2.15) and Exercise 2.3, and comes up in K-theory as mentioned in (12.17) and (12.19).

Exercise 7. Further discussion of Adams' work in K-theory is mentioned in Chapter 12.

Exercise 8. (2) This trick to get an identity element goes back to Albert (1942b) and was also used by Kaplansky (1953).

(3) Isotopies were probably first introduced by Albert (1942b). He mentions that this equivalence relation on algebras was motivated by some topological results of Steenrod.

Exercise 9. The statement of the 1, 2, 4, 8 Theorem on real division algebras is purely algebraic, but the only known proofs involve non-trivial topology or analysis. The original proofs (due to Milnor and Bott, and Kervaire in 1958) were simplified using the machinery of K-theory. Gilkey (1987) found an analytic proof. Real division algebras are also mentioned in Chapter 8 and Chapter 12. Hirzebruch (1991) provides a well-written outline of the geometric ideas behind these applications of algebraic topology.

Exercise 10. The Hopf map $S^3 \to S^2$ is an important example in topology. Hopf (1931) proved that this map is not homotopic to a constant, providing early impetus for the study of homotopy theory and fiber bundles. Further information on Hopf maps appears in Gluck, Warner and Ziller (1986), and Yiu (1986). Also see Chapter 15.

Part I

Classical Compositions and Quadratic Forms

Chapter 1

Spaces of Similarities

In order to understand the Hurwitz matrix equations we formulate them in the more general context of bilinear forms and quadratic spaces over a field F. The first step is to use linear transformations in place of matrices, and adjoint involutions in place of transposes. Then we will see that a composition formula becomes a linear subspace of similarities. This more abstract approach leads to more general results and simpler proofs, but of course the price is that readers must be familiar with more notations and terminology.

Rather than beginning with the Hurwitz problem here, we introduce some standard notations from quadratic form theory, discuss spaces of similarities, and then remark in (1.9) that these subspaces correspond with composition formulas. The chapter ends with a quick proof of the Hurwitz 1, 2, 4, 8 Theorem.

We follow most of the standard notations in the subject as given in the books by T. Y. Lam (1973) and W. Scharlau (1985). Throughout this book we work with a field F having characteristic not 2. (For if $2 = 0$ then every sum of squares is itself a square, and the original question about composition formulas becomes trivial.)

Suppose (V, q) is a *quadratic space* over F. This means that V is a (finite dimensional) F-vector space $q : V \to F$ is a regular quadratic form. To explain this, we define

$$B = B_q : V \times V \to F \quad \text{by} \quad 2B(x, y) = q(x + y) - q(x) - q(y).$$

Then q is a *quadratic form* if this B_q is a bilinear map and if $q(ax) = a^2 q(x)$ for every $a \in F$ and $x \in V$. The form q (or the bilinear form B) is *regular* if $V^{\perp} = 0$, that is:

$$\text{if } x \in V \text{ and } B(x, y) = 0 \text{ for every } y \in V, \text{ then } x = 0.$$

If q is not regular it is called *singular*. (Regular forms are sometimes called nonsingular or nondegenerate.) Since 2 is invertible in F, the bilinear form B can be recovered from the associated quadratic form q by:

$$q(x) = B(x, x) \quad \text{for all } x \in V.$$

Depending on the context we use several notations to refer to such a quadratic space. It could be called (V, q) or (V, B), or sometimes just V or just q.

To get the matrix interpretation of a quadratic space (V, q), choose a basis $\{e_1, \ldots, e_n\}$ of V. The form q can then be regarded as a homogeneous degree 2 polynomial by setting $q(X) = q(x_1, \ldots, x_n) = q(x_1e_1 + \cdots + x_ne_n)$. The *Gram matrix* of q is $M = (B(e_i, e_j))$, an $n \times n$ symmetric nonsingular matrix. Then $q(X) = X^\top \cdot M \cdot X$, where X is viewed as a column vector and $X^\top$ denotes the transpose. Another basis for V furnishes a matrix M' congruent to M, that is: $M' = P^\top \cdot M \cdot P$ where P is the basis-change matrix.

Here is a list of some of the terminology used throughout this book. Further explanations appear in the texts by Lam and Scharlau.

1_V denotes the identity map on a vector space V. Then $1_V \in \mathrm{End}(V)$.

$F^\bullet = F - \{0\}$, the multiplicative group of F.

$F^{\bullet 2} = \{a^2 : a \in F^\bullet\}$, the group of squares.

$\langle c\rangle = cF^{\bullet 2}$ is the coset of c in the group of square classes $F^\bullet/F^{\bullet 2}$. We also use $\langle c\rangle$ to denote the 1-dimensional quadratic form with a basis element of length c.

$\langle a_1, \ldots, a_n\rangle = \langle a_1\rangle \perp \cdots \perp \langle a_n\rangle$ is the quadratic space (V, q) where V has a basis whose corresponding Gram matrix is $\mathrm{diag}(a_1, \ldots, a_n)$. Interpreted as a polynomial this form is $q(X) = a_1x_1^2 + \cdots + a_nx_n^2$.

$\langle\langle a_1, a_2, \ldots, a_n\rangle\rangle = \langle 1, a_1\rangle \otimes \langle 1, a_2\rangle \otimes \cdots \otimes \langle 1, a_n\rangle$ is the n-fold Pfister form. It is a quadratic form of dimension 2^n. For example $\langle\langle a, b\rangle\rangle = \langle 1, a, b, ab\rangle$.

$\det(q) = \langle d\rangle$ if the determinant of a Gram matrix M_q equals d. Choosing a different basis alters $\det(M_q)$ by a square, so $\det(q)$ is a well-defined element in $F^\bullet/F^{\bullet 2}$.

$dq = (-1)^{n(n-1)/2}\det(q)$ when $\dim q = n$.

$q_1 \perp q_2$ and $q_1 \otimes q_2$ are the orthogonal direct sum and tensor product of the quadratic forms q_1 and q_2.

$q_1 \simeq q_2$ means that q_1 and q_2 are isometric.

$q_1 \subset q_2$ means that q_1 is isometric to a subform of q_2.

$nq = q \perp \cdots \perp q$ (n terms). In particular, $n\langle 1\rangle = \langle 1, 1, \ldots, 1\rangle$.

$\langle a\rangle q = \langle a\rangle \otimes q$ is a form *similar* to q.

$D_F(q) = \{a \in F^\bullet : q \text{ represents } a\}$ is the value set.

$G_F(q) = \{a \in F^\bullet : \langle a\rangle q \simeq q\}$ is the group of "norms", or "similarity factors" of q.

1.1 Definition. Let (V, q) be a quadratic space over F with associated bilinear form B. If $c \in F$, a linear map $f : V \to V$ is a *c-similarity* if

$$B(f(x), f(y)) = cB(x, y)$$

for every $x, y \in V$. The scalar $c = \mu(f)$ is the *norm* (also called the multiplier, similarity factor or ratio) of f. A map is a *similarity* if it is a c-similarity for some scalar c. An *isometry* is a 1-similarity.

Let $\mathrm{Sim}(V, q)$ be the set of all such similarities. Sometimes it is denoted by $\mathrm{Sim}(V)$, $\mathrm{Sim}(V, B)$ or $\mathrm{Sim}(q)$. It is easy to check that the composition of similarities is again a similarity and the norms multiply: $\mu(fg) = \mu(f)\mu(g)$. If f is a similarity and b is a scalar: $\mu(bf) = b^2\mu(f)$.

Suppose $f \in \mathrm{Sim}(V)$ and $\mu(f) = c \neq 0$. Then f must be bijective (using the hypothesis that V is regular to conclude that f is injective), so that f induces an isometry from $(V, \langle c\rangle q)$ to (V, q). Then $\langle c\rangle q \simeq q$ and the norm $c = \mu(f)$ lies in the group $G_F(q)$.

Define $\mathrm{Sim}^\bullet(V, q) = \{f \in \mathrm{Sim}(V, q) : \mu(f) \neq 0\}$, the set of invertible elements in $\mathrm{Sim}(V, q)$. Then $\mathrm{Sim}^\bullet(V, q)$ is a group containing the non-zero scalar maps and the orthogonal group $\mathrm{O}(V, q)$ as subgroups.[1] The norm map $\mu : \mathrm{Sim}^\bullet(V, q) \to F^\bullet$ is a group homomorphism yielding the exact sequence:

$$1 \longrightarrow \mathrm{O}(V, q) \longrightarrow \mathrm{Sim}^\bullet(V, q) \longrightarrow G_F(q) \longrightarrow 1.$$

The subgroup $F^\bullet\,\mathrm{O}(V, q)$ consists of all elements $f \in \mathrm{Sim}^\bullet(V, q)$ where $\mu(f) \in F^{\bullet 2}$. For instance if F is algebraically closed, everything in $\mathrm{Sim}^\bullet(V, q)$ is a scalar multiple of an isometry. The notation "$\mathrm{Sim}(V, q)$" is an analog of "$\mathrm{End}(V)$", emphasizing the *additive* structure of the similarities, so it is important to include 0-similarities.

The adjoint involution I_q is essential to our analysis of similarities. This map I_q is a generalization of the transpose of matrices.[2]

1.2 Definition. Let (V, q) be a quadratic space with associated bilinear form B. The adjoint map $I_q : \mathrm{End}(V) \to \mathrm{End}(V)$ is characterized by the formula:

$$B(v, I_q(f)(w)) = B(f(v), w)$$

for $f \in \mathrm{End}(V)$ and $v, w \in V$. When convenient we write $\tilde{f}$ instead of $I_q(f)$.

It follows from the regularity of the form that I_q is a well-defined map which is an F-linear anti-automorphism of $\mathrm{End}(V)$ whose square is the identity. If a basis of V is chosen we obtain the Gram matrix M of the form q, and the matrix A of the map $f \in \mathrm{End}(V)$. Then the matrix of $\tilde{f} = I_q(f)$ is just $M^{-1}A^\top M$, a conjugate of the transpose matrix $A^\top$. If $q \simeq n\langle 1\rangle = \langle 1, \dots 1\rangle$ and an orthonormal basis is chosen, then I_q coincides with the transpose of matrices.

1.3 Lemma. *Let (V, q) be a quadratic space.*

(1) *If $f \in \mathrm{End}(V)$ then f is a c-similarity if and only if $\tilde{f}f = c1_V$.*

---

1 There seems to be no standard notation for this group. Some authors call it $\mathrm{GO}(V, q)$.

2 The notation I_q should not cause confusion with the $n \times n$ identity matrix I_n (we hope).

(2) *If* $f, g \in \mathrm{Sim}(V)$ *the following statements are equivalent:*
   (i) $f + g \in \mathrm{Sim}(V)$.
   (ii) $af + bg \in \mathrm{Sim}(V)$ *for every* $a, b \in F$.
   (iii) $\tilde{f}g + \tilde{g}f = c1_V$ *for some* $c \in F$.

Proof. Part (1) follows easily from the definitions and (2) is a consequence of (1). □

Similarities $f, g \in \mathrm{Sim}(V)$ are called *comparable* if $f+g \in \mathrm{Sim}(V)$ as well. The lemma implies that if $f_1, f_2, \dots, f_r$ are pairwise comparable similarities in $\mathrm{Sim}(V)$ then the whole vector space $S = \mathrm{span}\{f_1, f_2, \dots, f_r\}$ is inside $\mathrm{Sim}(V)$. For if $g \in S$ then g is a linear combination of the maps f_i, and $\tilde{g}g$ is a linear combination of the terms $\tilde{f}_i f_i$ and $\tilde{f}_i f_j + \tilde{f}_j f_i$. Since these terms are scalars by hypothesis, $\tilde{g}g$ is also a scalar and $g \in \mathrm{Sim}(V)$.

Such a subspace $S \subseteq \mathrm{Sim}(V)$ is more than just a linear space. The map μ restricted to S induces a quadratic form σ on S. (To avoid notatonal confusion we have not used the letter μ again for this form on S.) Generally if $f, g \in \mathrm{Sim}(V)$ are comparable, define $B_\mu(f, g) \in F$ by:

$$\tilde{f}g + \tilde{g}f = 2B_\mu(f, g)1_V.$$

Then B_μ is bilinear whenever it is defined, and $\mu(f) = B_\mu(f, f)$. This map $\mu : \mathrm{Sim}(V) \to F$ has all the properties of a quadratic form except that its domain is not necessarily closed under addition.

The induced quadratic form σ on a subspace $S \subseteq \mathrm{Sim}(V)$ could be singular. Of course, this can occur only when there are non-trivial 0-similarities of V. A map $f \in \mathrm{End}(V)$ is a 0-similarity if and only if the image $f(V)$ is a totally isotropic subspace of V (i.e. the quadratic form vanishes identically on $f(V)$). Then if (V, q) is anisotropic, the only 0-similarity is the zero map.

We will restrict attention to those $S \subseteq \mathrm{Sim}(V)$ whose induced quadratic form is regular.

If S is a subspace of similarities with induced (regular) quadratic form σ, we write $(S, \sigma) \subseteq \mathrm{Sim}(V, q)$. If σ is a quadratic form over F we write

$$\sigma < \mathrm{Sim}(V, q)$$

if there exists a subspace $S \subseteq \mathrm{Sim}(V, q)$ whose induced quadratic form is isometric to σ.

Given q what is the largest possible σ?

Given σ what is the smallest possible q?

Various aspects of these questions comprise Part I of this book.

1.4 Proposition. *Suppose $(S, \sigma) \subseteq \mathrm{Sim}(q)$ is a regular subspace of similarities.*

(1) *If $a \in D_F(q)$ then $\langle a\rangle\sigma \subset q$. In particular,* $\dim \sigma \leq \dim q$.

(2) *If σ is isotropic then q is hyperbolic.*

Proof. (1) For any $v \in V$, the evaluation map $S \to V$ sending $f \mapsto f(v)$ is a $q(v)$-similarity. If $q(v) \neq 0$ then this map must be injective, since (S, σ) is regular.

(2) If $\dim V = n$ then any totally isotropic subspace of V has dimension $\leq n/2$, and (V, q) is hyperbolic if and only if there exist totally isotropic subspaces of dimension equal to $n/2$. We are given a 0-similarity $f \in S$ with $f \neq 0$. Since $\tilde{f}f = 0$, image $f = f(V)$ is totally isotropic. Also $\ker f$ is totally isotropic, for if $v \in V$ has $q(v) \neq 0$ then the proof of part (1) shows that $f(v) \neq 0$. Then dim image f and $\dim \ker f$ are both $\leq n/2$, but their sum equals n. Therefore these dimensions equal $n/2$ and (V, q) is hyperbolic. □

We can also generalize the Hurwitz Matrix Equations mentioned in Chapter 0.

1.5 Lemma. *Let (V, q) be a quadratic space and $(S, \sigma) \subseteq \mathrm{Sim}(V)$. If the form σ on S has Gram matrix $M_\sigma = (c_{ij})$, then there exist $f_i \in S$ satisfying*

$$\tilde{f_i} f_j + \tilde{f_j} f_i = 2c_{ij}1_V, \quad \text{for all } 1 \leq i, j \leq s.$$

Conversely if maps $f_i \in \mathrm{End}(V)$ satisfy these equations then they span a subspace $S \subseteq \mathrm{Sim}(V)$ where the induced form has Gram matrix $M = (c_{ij})$.

Proof. By definition of the Gram matrix, there is a basis $\{f_1, \dots, f_s\}$ of S such that $c_{ij} = B_\mu(f_i, f_j)$. The required equations are immediate from the definition of B_μ. Conversely, given f_i satisfying those equations, the remarks after Lemma 1.3 imply that $S = \mathrm{span}\{f_1, \dots, f_n\}$ is a subspace of $\mathrm{Sim}(V, q)$. □

Suppose that we choose an *orthogonal* basis $\{f_1, \dots, f_n\}$ of S. Then $\sigma \simeq \langle a_1, \dots, a_s\rangle$ and

$$\begin{aligned} \tilde{f_i} f_i &= a_i 1_V && \text{for } 1 \leq i \leq s, \\ \tilde{f_i} f_j + \tilde{f_j} f_i &= 0 && \text{for } 1 \leq i, j \leq s. \end{aligned}$$

When $\sigma = s\langle 1\rangle = \langle 1, 1, \dots, 1\rangle$ and $q = n\langle 1\rangle = \langle 1, 1, \dots, 1\rangle$ and we use matrices, these equations become the Hurwitz Matrix Equations mentioned in Chapter 0.

Here are some ways to manipulate subspaces of $\mathrm{Sim}(V)$.

1.6 Lemma. *Let $(S, \sigma) \subseteq \mathrm{Sim}(V, q)$ over the field F.*

(1) *If (W, δ) is another quadratic space then $\sigma < \mathrm{Sim}(q \otimes \delta)$.*

(2) *If K is an extension field of F then $\sigma \otimes K < \mathrm{Sim}(q \otimes K)$.*

(3) *If $f, g \in \mathrm{Sim}^\bullet(V, q)$ then $fSg \subseteq \mathrm{Sim}(V, q)$.*

Proof. (1) Use $S \otimes 1_w$ acting on $V \otimes W$.

(2) If $f_i \in S$ and $c_i \in K$ then Lemma 1.3 implies that $\sum f_i c_i$ lies in $\text{Sim}(V \otimes K)$.

(3) This is clear since $\text{Sim}(V, q)$ is closed under composition. □

The ideas developed so far can be generalized to subspaces of $\text{Sim}(V, W)$ for two quadratic spaces (V, q) and (W, q'). (See Exercise 2.) In this case the matrices are rectangular and the involution is replaced by the map $J = J_W^V : \text{Hom}(V, W) \to \text{Hom}(W, V)$. Since our main concern is the case $V = W$, we will not pursue this generality here. The main advantage of this restriction is that we can arrange the identity map 1_V to be in S.

1.7 Proposition. *Let $\sigma \simeq \langle 1, a_2, \dots, a_s\rangle$ be a quadratic form representing* 1.

Then $\sigma < \text{Sim}(V, q)$ if and only if there exist maps $f_2, \dots, f_s$ in $\text{End}(V)$ satisfying:

$$\left.\begin{aligned} \tilde{f}_i &= -f_i \\ f_i^2 &= -a_i 1_V \end{aligned}\right\} \quad \text{for } 2 \le i \le s,$$

$$f_i f_j = -f_j f_i \qquad \text{whenever } i \neq j.$$

Proof. Given $S \subseteq \text{Sim}(V)$ with induced form σ we can choose maps f_i as above. Replacing S by the isometric space $f_1^{-1}S$ we may assume that $f_1 = 1_V$. Then the equations above reduce to those given here. The converse follows similarly. □

The conditions above correspond to the second form of the Hurwitz Matrix Equations. With this formulation an experienced reader will notice that the algebra generated by the f_i is related to the Clifford algebra $C(\langle -a_2, \dots, -a_s\rangle)$. This connection is explored in Chapter 4.

1.8 Example. Let $q = \langle 1, a\rangle$ with corresponding basis $\{e_1, e_2\}$ of V. The adjoint involution I_q on $\text{End}(V)$ is translated into matrices as follows:

$$\text{if} \quad f = \begin{pmatrix} x & y \\ z & w \end{pmatrix} \quad \text{then} \quad \tilde{f} = \begin{pmatrix} x & az \\ a^{-1}y & w \end{pmatrix}.$$

Let $f_2 = \begin{pmatrix} 0 & -a \\ 1 & 0 \end{pmatrix}$, $g_1 = \begin{pmatrix} 1 & 0 \\ 0 & -1 \end{pmatrix}$ and $g_2 = \begin{pmatrix} 0 & a \\ 1 & 0 \end{pmatrix}$. One can quickly check that $\tilde{f}_2 = -f_2$ and $f_2^2 = -a1_V$. Then $S = \text{span}\{1_V, f_2\} = \left\{\begin{pmatrix} x & -ay \\ y & x \end{pmatrix} : x, y \in F\right\}$ is a subspace of $\text{Sim}(\langle 1, a\rangle)$ and the induced form is $(S, \sigma) \simeq \langle 1, a\rangle$. Similarly g_1 and g_2 are comparable (in fact, orthogonal) similarities and $T = \text{span}\{g_1, g_2\}$ is also a subspace of $\text{Sim}(\langle 1, a\rangle)$ with induced form $\langle 1, a\rangle$.

This 2-dimensional example contains the germs of several ideas exploited below. One way to understand this example is to interpret $\langle 1, a\rangle$ as the usual norm form on the quadratic extension $K = F(\theta)$ where $\theta^2 = -a$. (When $-a \notin F^{\bullet 2}$, K is a field.) Let $\mathcal{L} : K \to \mathrm{End}(K)$ be the regular representation: $\mathcal{L}(x)(y) = xy$. With $\{1, \theta\}$ as the F-basis of K, the matrix of $\mathcal{L}(\theta)$ is f_2, and the subspace S above is just $\mathcal{L}(K) \subseteq \mathrm{Sim}(K)$. Similarly to get T define the twisted representation $\mathcal{L}' : K \to \mathrm{End}(K)$ by $\mathcal{L}'(x)(y) = x\bar{y}$. Then $T = \mathcal{L}'(K) \subseteq \mathrm{Sim}(K)$.

More generally suppose there is an F-algebra A furnished with a "norm" quadratic form q which is multiplicative: $q(xy) = q(x)q(y)$ for all $x, y \in A$. Using the left regular representation in a similar way, we get $q < \mathrm{Sim}(q)$. For instance the quaternion algebra $A = \left(\frac{-a,-b}{F}\right)$ with basis $\{1, i, j, k\}$ has the norm form $q \simeq \langle 1, a, b, ab\rangle$. Conversely if $q < \mathrm{Sim}(q)$ then q does arise as the norm form of some "composition algebra" A. Any subspace $\sigma < \mathrm{Sim}(q)$ can be viewed as coming from a "partial multiplication" $S \times V \to V$.

1.9 Proposition. *Suppose (V, q) and (S, σ) are quadratic spaces over F. The following conditions are equivalent:*

(1) $\sigma < \mathrm{Sim}(q)$.

(2) *There is a bilinear pairing $* : S \times V \to V$ satisfying $q(f * v) = \sigma(f)q(v)$ for every $f \in S$ and $v \in V$.*

(3) *There is a formula $\sigma(X)q(Y) = q(Z)$ where each z_k is a bilinear form in the systems of indeterminates X, Y with coefficients in F.*

Proof. (1) $\iff$ (2). Given $S \subseteq \mathrm{Sim}(V, q)$ where the induced form on S is isometric to σ. For $f \in S$ and $v \in V$, define $f * v = f(v)$. Since f is a $\sigma(f)$-similarity we get required equation. Conversely, the map $*$ induces a linear map $\lambda : S \to \mathrm{End}(V)$ by $\lambda(f)(v) = f * v$. For each $f \in S$, $\lambda(f)$ is a $\sigma(f)$-similarity and therefore λ is an isometry from (S, σ) to the subspace $\lambda(S) \subseteq \mathrm{Sim}(V, q)$. Since (S, σ) is regular, this λ is injective and $\sigma < \mathrm{Sim}(q)$.

(1) $\Longrightarrow$ (3). Given $(S, \sigma) \subseteq \mathrm{Sim}(V, q)$ as before, choose bases $\{f_1, \dots, f_s\}$ of S and $\{v_1, \dots, v_n\}$ of V. Let $M = M_q$ be the $n \times n$ Gram matrix of q, $M_\sigma = (c_{ij})$ the $s \times s$ Gram matrix of σ, and A_i the $n \times n$ matrix of the map f_i. Then the matrix of $\tilde{f_i}$ is $M^{-1}A_i^\top M$ and the equations given in (1.5) become:

$$A_i^\top M A_j + A_j^\top M A_i = 2c_{ij}M, \quad \text{for } 1 = i, j = s.$$

Let $A = x_1A_1 + \cdots + x_sA_s$. Since the x_i are indeterminates the system of equations above is equivalent to the single equation:

$$A^\top M A = \sigma(X)M.$$

Since $q(Y) = Y^\top MY$ we see that $Z = AY$ satisfies the condition: $q(Z) = \sigma(X)q(Y)$.

(3) $\Longrightarrow$ (1). Reverse the steps above. □

If $\sigma < \mathrm{Sim}(q)$ and $\dim \sigma \geq 2$, then $\dim q$ must be even. One way to see this is to scale σ to represent 1 and get a map $f = f_2$ as in (1.7). Since f is nonsingular and skew-symmetric, the dimension must be even. (In matrix notation in the proof of (1.9) we see that MA is a skew-symmetric matrix.) For a different proof, let K be an extension field of F where σ becomes isotropic. Since $\sigma \otimes K < \mathrm{Sim}(q \otimes K)$, (1.4) implies $q \otimes K$ hyperbolic and hence of even dimension. The next proposition shows that more can be said about the structure of q.

1.10 Proposition. (1) *If* $\langle 1, a\rangle < \mathrm{Sim}(q)$ *then* $q \simeq \langle\langle a\rangle\rangle \otimes \varphi$ *for some form* φ.
(2) *If* $\langle 1, a, b\rangle < \mathrm{Sim}(q)$ *then* $q \simeq \langle\langle a, b\rangle\rangle \otimes \psi$ *for some form* ψ.

Proof. (1) By hypothesis there is $f \in \mathrm{Sim}(V)$ with $\mu(f) = a$ and $\tilde{f} = -f$. This skew-symmetry implies $B(v, f(v)) = 0$ for every $v \in V$. Choose v with $q(v) \neq 0$. Then the line $U = Fv$ is a regular subspace such that U and $f(U)$ are orthogonal. Let U_0 be maximal among such regular subspaces. Then $U_0 \perp f(U_0)$ is a regular subspace of V. If this subspace is proper, choose $w \in (U_0 \perp f(U_0))^\perp$ with $q(w) \neq 0$ and note that $U_0 + Fw$ contradicts the maximality. Therefore $V = U_0 \perp f(U_0) \simeq \langle 1, a\rangle \otimes \varphi$, where φ is the quadratic form on U_0.

(2) Given a basis $1_V, f, g$ corresponding to $\langle 1, a, b\rangle$, choose W_0 maximal among the regular subspaces W for which W, $f(W)$, $g(W)$ and $fg(W)$ are mutually orthogonal. The argument above generalizes to show that $V = W_0 \perp f(W_0) \perp g(W_0) \perp fg(W_0) \simeq \langle 1, a, b, ab\rangle \otimes \psi$, where ψ is the quadratic form on W_0. □

This elementary proof in part (1) can perhaps be better understood by considering the given f as inducing an action of $K = F(\sqrt{-a})$ on V. The formation of U_0 corresponds to choosing a K-basis of V. Similarly part (2) corresponds to the action of the quaternion algebra $\left(\frac{-a,-b}{F}\right)$ on V. These ideas are explored in Chapter 4 when we view V as a module over a certain Clifford algebra.

We conclude this chapter with an independence argument, due to Hurwitz, which suffices to prove the "1, 2, 4, 8 Theorem". Let us first set up a multi-index notation. Let $\mathbb{F}_2 = \{0, 1\}$ be the field of 2 elements. If $\delta \in \mathbb{F}_2$ define

$$a^\delta = \begin{cases} 1 & \text{if } \delta = 0 \\ a & \text{if } \delta = 1. \end{cases}$$

Let $\Delta = (\delta_1, \dots, \delta_n)$ be a vector in $\mathbb{F}_2^n$. If $f_1, f_2, \dots, f_n$ are elements of some ring, define

$$f^\Delta = f_1^{\delta_1} \dots f_n^{\delta_n}.$$

By convention $f^0 = 1$ here. Define $|\Delta|$ to be the number of indices i such that $\delta_i = 1$. Then f^Δ is a product of $|\Delta|$ elements f_i.

1.11 Proposition. *Suppose A is an associative F-algebra with 1 and $\{f_1, \dots, f_n\}$ is a set of pairwise anti-commuting invertible elements of A. If n is even then $\{f^\Delta : \Delta \in \mathbb{F}_2^n\}$ is a linearly independent set of 2^n elements of A.*

Proof. Suppose there exists a non-trivial dependence relation and let $\sum c_\Delta f^\Delta = 0$ be such a relation having the fewest non-zero coefficients $c_\Delta \in F$. We may assume $c_0 \neq 0$ by multiplying the relation by $(f^\Delta)^{-1}$ for some Δ where $c_\Delta \neq 0$. For fixed i, conjugate the given relation by f_i and subtract, noting that $f_i f^\Delta f_i^{-1} = \pm f^\Delta$. The result is a shorter relation among the f^Δ (since the f^0 terms cancel). By the minimality of the given relation, this shorter one must be trivial. Therefore if $c_\Delta \neq 0$ then f^Δ must commute with f_i. Since this holds for every index i it follows that either $\Delta = 0$ or else $\Delta = (1, 1, \dots, 1)$ and n is odd. Since n is even the dependence relation must have just one term: $c_0 f^0 = 0$, which is absurd. □

1.12 Corollary. *If* $\sigma < \mathrm{Sim}(V, q)$ *where* $\dim \sigma = s$ *and* $\dim q = n$, *then* $2^{s-2} \leq n^2$. *Consequently,* $s = n$ *is possible only when* $n = 1, 2, 4$ *or* 8.

Proof. By (1.7) there are $s - 1$ anti-commuting invertible elements in $\mathrm{End}(V)$. Since $\dim \mathrm{End}(V) = n^2$, Proposition 1.11 provides the inequality. If $s = n$ the inequality $2^{n-2} \leq n^2$ implies that $n \leq 8$. The restrictions given in (1.10) show that n must equal $1, 2, 4$ or 8. □

Appendix to Chapter 1. Composition algebras

This appendix contains another proof of the 1, 2, 4, 8 Theorem. This approach uses the classical "doubling process" to construct the composition algebras, so it provides information on the underlying algebras and not just their dimensions. This appendix is self-contained, using somewhat different notation for the quadratic and bilinear forms. The theorem here is well- known, first proved by Albert (1942a), who also handled the case when the characteristic is 2. The organization of the ideas here follows a lecture by J. H. Conway (1980).

Suppose F is a field with characteristic $\neq 2$ and let A be an F-algebra. This algebra is not assumed to be associative or finite dimensional: A is simply an F-vector space and the multiplication is an F-bilinear map $A \times A \to A$. We do assume that A has an identity element 1.

A.1 Definition. An F-algebra A with 1 is called a *composition algebra* if there is a regular quadratic form $A \to F$, denoted $a \mapsto [a]$, such that

$$[a] \cdot [b] = [ab] \quad \text{for every } a, b \in A. \qquad (*)$$

Let $[a, b]$ denote the associated symmetric bilinear form: $2[a, b] = [a+b] - [a] - [b]$. This differs from our previous notation: $[a] = q(a)$ and $[a, b] = B(a, b)$. This notation will not be used elsewhere in this book because the square brackets stand for so many other things (like quaternion symbols and cohomology classes).

We will determine all the possible composition algebras over F. The classical examples over the field $\mathbb{R}$ include the real numbers (dim $= 1$), the complex numbers (dim $= 2$), the quaternions (dim $= 4$) and the octonions (dim $= 8$). We assume no knowledge of these examples and consider an arbitrary composition algebra A over F. This appendix is organized as a sequence of numbered statements, with hints for their proofs.

A.2. $[ac, ad] = [a] \cdot [c, d]$. (Set $b = c + d$ in $(*)$).
Symmetrically, $[ac, bc] = [a, b] \cdot [c]$.

A.3. *The "Flip Law":* $[ac, bd] = 2[a, b] \cdot [c, d] - [ad, bc]$.
(Replace a by $a + b$ in (A.2)).

A.4. *Define "bar" by:* $\bar{c} = 2[c, 1] - c$.
Then $[ac, b] = [a, b\bar{c}]$. (Apply (A.3) with $d = 1$.)
Symmetrically, $[ca, b] = [a, \bar{c}b]$.

Repeating property (A.4) yields a "braiding sequence" of six equal quantities:

$$\cdots = [a, bc] = [a\bar{c}, b] = [\bar{c}, \bar{a}b] = [\bar{c}\bar{b}, \bar{a}] = [\bar{b}, c\bar{a}] = [\bar{b}a, c] = [a, bc] = \cdots$$

Basic Principle: to prove $X = Y$ show that $[X, t] = [Y, t]$ for every $t \in A$.

A.5. *Properties of "bar":*

$\overline{a + b} = \bar{a} + \bar{b}$.

c is scalar if and only if $\bar{c} = c$.

$[a, b] = [\bar{a}, \bar{b}]$. (Apply the braiding sequence when $c = 1$.)

$\bar{\bar{c}} = c$. (Use the Basic Principle.)

$\overline{bc} = \bar{c}\bar{b}$. (Use $[a, bc] = [\bar{c}\bar{b}, \bar{a}]$ from braiding.)

$\bar{b} \cdot ac = 2[a, b]c - \bar{a} \cdot bc$. (Use (A.3), isolate d, apply the Basic Principle: $[\bar{b} \cdot ac, d] = [2[a, b]c, d] - [\bar{a} \cdot bc, d]$.)

$\bar{a} \cdot ab = ba \cdot \bar{a} = [a]b$. (Set $a = b$ in previous line.)

$\bar{a}a = a\bar{a} = [a]$. (Set $b = 1$ in previous line.)

Since $\bar{a} = 2[a, 1] - a$ we have $a \cdot ab = a^2b$ and $ba \cdot a = ba^2$. These are the "Alternative Laws", a weak version of associativity.

Now suppose that $H \subseteq A$ is a composition subalgebra (that is, H is an F-subalgebra on which the quadratic form is regular, i.e. $H \cap H^\perp = 0$.) Suppose $H \neq A$. Then we may choose $i \in H^\perp$ with $[i] = \alpha \neq 0$.

A.6. *H and Hi are orthogonal, $H' = H + Hi$ is a subspace on which the form is regular and* $\dim H' = 2 \cdot \dim H$. *Moreover, if $a, b, c, d \in H$ then:*

$$(a + bi) \cdot (c + di) = (ac - \alpha \bar{d} b) + (da + b\bar{c})i.$$

Consequently H' is also a composition subalgebra of A.

Proof. H is invariant under "bar" since $1 \in H$. If $a, b \in H$ then $[a, bi] = [\bar{b}a, i] = 0$. Then H and Hi are orthogonal and $H \cap Hi = \{0\}$ since the form is regular.

To verify the formula for products it suffices to analyze three cases.

$$\text{(1)} \qquad [bi \cdot c, t] = [bi, t\bar{c}] = -[b\bar{c}, ti] \quad \text{using the Flip Law (A.3),}$$
$$= [b\bar{c} \cdot i, t].$$

Hence $bi \cdot c = b\bar{c} \cdot i$.

(2) If $x \in H$ then $\bar{x} \cdot iy = i \cdot xy$, since $[iy, xt] = -[it, xy]$ by the Flip Law. In particular, $\bar{x}i = ix$. Hence if $a, d \in H$ then $a \cdot di = a \cdot i\bar{d} = i \cdot \bar{a}\bar{d} = i \cdot \overline{da} = da \cdot i$.

$$\text{(3)} \qquad [bi \cdot di, t] = [di, \overline{bi} \cdot t] = [d \cdot i, -bi \cdot t] \quad \text{since } [bi, 1] = 0,$$
$$= [d \cdot t, bi \cdot i] \quad \text{by the Flip Law,}$$
$$= -[dt \cdot i, bi] = -\alpha[dt, b] = [t, -\alpha\bar{d}b].$$

Hence $bi \cdot di = -\alpha\bar{d}b$. □

This observation imposes severe restrictions on the structure of a composition algebra A. The smallest composition subalgebra is $A_0 = F$. If $A \neq F$ there must be a 2-dimensional subalgebra $A_1 \subseteq A$ built as $A_1 = F + Fi$ for some $i \in F^{\perp}$. If $A \neq A_1$ there must be a 4-dimensional subalgebra $A_2 \subseteq A$ built as $A_2 = A_1 + A_1 j$ for some $j \in A_2^{\perp}$. If $A \neq A_2$ there must be an 8-dimensional subalgebra $A_3 \subseteq A$, etc. This doubling process cannot continue very long.

A.7. *Suppose H is a composition subalgebra of A, and H' is formed as in* (A.6). *Then:*

H is associative.

H' is associative if and only if H is a commutative and associative.

H' is commutative and associative if and only if $H = F$.

Proof. We know that $H' = H \oplus Hi$ is a composition algebra. Then $[(a + bi) \cdot (c + di)] = [a + bi] \cdot [c + di]$, for every a, $b, c, d \in H$. The left side equals $[(ac - \alpha\bar{d}b) + (da + b\bar{c})i] = [ac - \alpha\bar{d}b] + \alpha[da + b\bar{c}]$ and the right side equals $([a] + \alpha[b]) \cdot ([c] + \alpha[d])$. Expanding the two sides and canceling like terms yields: $[ac, \bar{d}b] = [da, b\bar{c}]$. Then $[d \cdot ac, b] = [da \cdot c, b]$ and we conclude that $d \cdot ac = da \cdot c$ and H is associative. The converse follows since these steps are reversible.

The proofs of the other statements are similar direct calculations. □

A.8 Theorem. *If A is a composition algebra over F then A is obtained from the algebra F by "doubling" 0, 1, 2 or 3 times. In particular,* $\dim A = 1, 2, 4$ *or* 8.

Proof. As remarked before (A.7), if $\dim A \notin \{1, 2, 4, 8\}$ then there is a chain of subalgebras $F = A_0 \subset A_1 \subset A_2 \subset A_3 \subset A_4 \subseteq A$, where $\dim A_k = 2^k$. Applying the statements in (A.7) repeatedly we deduce that A_1 is commutative and associative but not equal to F; A_2 is associative but not commutative; A_3 is not associative and therefore A_4 cannot exist inside the composition algebra A. This contradiction proves the assertion. □

This argument does more than compute the dimensions. It characterizes all the composition algebras. These are

(0) F,

(1) $F\langle\alpha\rangle \simeq F[x]/(x^2 - \alpha)$, a quadratic extension of F,

(2) $\left(\frac{\alpha,\beta}{F}\right)$, a quaternion algebra over F,

(3) $\left(\frac{\alpha,\beta,\gamma}{F}\right)$, an octonion algebra over F.

These algebras are defined recursively by the formulas in (A.6). Further properties of quaternion algebras are mentioned in Chapter 3. Further properties of octonions algebras appear in Chapters 8 and 12.

This relationship between H' and H in (A.6) is clarified by formalizing the idea of doubling an F-algebra. A map $a \mapsto \bar{a}$ on an F-algebra is called an *involution* if it is F-linear, $\bar{\bar{a}} = a$ and $\overline{ab} = \bar{b}\bar{a}$. Suppose H is an F-algebra with 1, $a \mapsto [a]$ is a regular quadratic form on H, and $a \mapsto \bar{a}$ is an involution on H. Let $\alpha \in F^\bullet$.

Definition. The *α-double*, $\mathcal{D}_\alpha(H)$, is the F-algebra with underlying vector space $H \times H$, and with multiplication given by the formula in (A.6), viewing $1 = (1, 0)$ and $i = (0, 1)$. For $(a, b) = a + bi \in \mathcal{D}_\alpha(H)$, define $\overline{a + bi} = \bar{a} - bi$, and $[a + bi] = [a] + \alpha[b]$.

A.9. *$\mathcal{D}_\alpha(H)$ is an F-algebra containing H as a subalgebra;* $\dim \mathcal{D}_\alpha(H) = 2 \cdot \dim H$*; $[a]$ is a regular quadratic form on $\mathcal{D}_\alpha(H)$; and $a \mapsto \bar{a}$ is an involution on $\mathcal{D}_\alpha(H)$. If the elements of H satisfy the following properties then so do the elements of $\mathcal{D}_\alpha(H)$:*

$a = \bar{a}$ if and only if $a \in F$;

$[a, b] = [\bar{a}, \bar{b}]$;

$[ax, b] = [a, b\bar{x}]$ and $[xa, b] = [a, \bar{x}b]$;

$a\bar{a} = [a] = \bar{a}a$;

$a\bar{b} + b\bar{a} = 2[a, b] = \bar{a}b + \bar{b}a$;

$T(x) = x + \bar{x}$ satisfies $T(ab) = T(ba)$ and $T(a \cdot bc) = T(ab \cdot c)$.

Proof. These are straightforward calculations. □

The algebras built from F by repeated application of this doubling process are called Cayley–Dickson algebras. Given a sequence of scalars $\alpha_1, \dots, \alpha_n \in F^\bullet$ define

$$A(\alpha_1, \dots, \alpha_n) = \mathcal{D}_{\alpha_n} \dots \mathcal{D}_{\alpha_1}(F).$$

When $n \le 3$ we obtain the composition algebras mentioned above. For example, $A(\alpha, \beta) \simeq \left(\frac{-\alpha, -\beta}{F}\right)$ is the quaternion algebra with norm form $\langle\langle \alpha, \beta \rangle\rangle = \langle 1, \alpha, \beta, \alpha\beta \rangle$.

Every Cayley–Dickson algebra $A(\alpha_1, \dots, \alpha_n)$ satisfies the properties listed in (A.9), even though it is not a composition algebra when $n > 3$. Further properties are given in Exercise 25. Here are a few properties of composition algebras, not all valid for larger Cayley–Dickson algebras.

A.10. *If A is a composition algebra and $a, b, x, y, z \in A$ then:*

$a \cdot ba = ab \cdot a$. *(The "Flexible Law". Hence, it is unambiguous to write aba.)*

$aba \cdot x = a \cdot (b \cdot ax)$,
$a(xy)a = ax \cdot ya$. *(These are the Moufang identities.)*

Proof. From (A.5) $\bar{b} \cdot ac = 2[a, b]c - \bar{a} \cdot bc$. Substitute a for $\bar{b}$, b for a and ax for c, to deduce: $a \cdot (b \cdot ax) = (2[\bar{a}, b]a - [a]\bar{b}) \cdot x$. When $x = 1$ this is $a \cdot ba = 2[\bar{a}, b]a - [a]\bar{b}$, so that: $a \cdot (b \cdot ax) = (a \cdot ba) \cdot x$. Apply "bar" to the formula for $a \cdot ba$ and replace a, b by $\bar{a}, \bar{b}$ to find: $ab \cdot a = 2[a, \bar{b}]a - [\bar{a}]\bar{b} = a \cdot ba$.

The second Moufang identity also follows directly from the Flip Law.

$$\begin{aligned} [ax \cdot ya, t] &= [a \cdot x, t \cdot \overline{ya}] \\ &= 2[a, t] \cdot [xy, \bar{a}] - [a] \cdot [\bar{y}, tx] \quad \text{by the Flip Law and (A.5),} \\ &= 2[a, t] \cdot [\overline{xy}, a] - [a] \cdot [\overline{xy}, t]. \end{aligned}$$

Then for fixed a the value $ax \cdot ya$ depends only on the quantity xy. The stated formula is clear from this independence. □

These identities hold in any alternative ring. See Exercise 26.

Exercises for Chapter 1

1. **Composition algebras.** Suppose $S \subseteq \mathrm{Sim}(V, q)$ with $\dim S = \dim V$. Explain how this yields a composition algebra, as defined in (A.1) of the appendix.

(Hint. Scale q to assume it represents 1 and choose $e \in V$ with $q(e) = 1$. Identify S with V by $f \mapsto f(e)$. (1.9) provides a pairing. Apply Exercise 0.8 to obtain an algebra with 1.)

2. Let (V, q) and (W, q') be two quadratic spaces. Define the "adjoint" map $J = J_W^V : \mathrm{Hom}(V, W) \to \mathrm{Hom}(W, V)$.

(1) The inverse of J_W^V is J_V^W. How does J behave on matrices?

(2) Define subspaces of $\mathrm{Sim}(V, W)$ and generalize (1.3), (1.5), (1.6) and (1.9).

(3) $\sigma < \mathrm{Sim}(q, q')$ if and only if q$< \mathrm{Sim}(\sigma, q')$.

3. Let (V, q) be a quadratic space and $f, g \in \mathrm{End}(V)$.

(1) f is a c-similarity if and only if $q(f(v)) = cq(v)$ for every $v \in V$.

(2) $\tilde{f}g + \tilde{g}f = 0$ if and only if $f(v)$ and $g(v)$ are orthogonal for every $v \in V$.

(3) If f, g are comparable similarities, then $B(f(v), g(v)) = B_\mu(f, g)q(v)$. In euclidean space over $\mathbb{R}$, the notion of the angle between two vectors is meaningful. Similarities f and g are comparable iff the angle between $f(v)$ and $g(v)$ is constant, independent of v.

4. (1) Suppose $q \simeq \langle 1, a\rangle$. If $f \in \mathrm{Sim}^\bullet(q)$ then either $f = \begin{pmatrix} x & -ay \\ y & x \end{pmatrix}$ or $f = \begin{pmatrix} x & ay \\ y & -x \end{pmatrix}$ for some $x, y \in F$ with $\mu(f) = x^2 + ay^2$. Consequently, every (regular) subspace of $\mathrm{Sim}(q)$ is contained in one of the spaces S, T in Example 1.8. If $q \simeq \langle 1, -1\rangle$, list all the 0-similarities.

(2) Let $D = \left(\frac{-1,-1}{F}\right)$ be the quaternion algebra with norm form $q = \langle 1, 1, 1, 1\rangle$. Let $L_0 = \mathcal{L}(D) \subseteq \mathrm{Sim}(D, q)$ arising from the left regular representation of D. Then L_0 is the set of all $\begin{pmatrix} a & b & c & d \\ -b & a & d & -c \\ -c & -d & a & b \\ -d & c & -b & a \end{pmatrix}$. Similarly $R_0 = \mathcal{R}(D)$ is the set of all $\begin{pmatrix} a & b & c & d \\ -b & a & -d & c \\ -c & d & a & -b \\ -d & -c & b & a \end{pmatrix}$. Then L_0 and R_0 are subalgebras of $\mathrm{End}(D)$ which commute with each other and they are 4-dimensional subspaces of $\mathrm{Sim}(q)$. Also, $R_0 \cdot J = J \cdot L_0$ where J = "bar". These subspaces of similarities are unique in a strong sense:

Lemma. *If $U \subseteq \mathrm{Sim}(D, q)$ is a (regular) subspace and $f \in U^\bullet$, either $U \subseteq f \cdot L_0$ or $U \subseteq f \cdot R_0$.*

(3) What are the possibilities for the intersections $f \cdot L_0 \cap L_0$ and $f \cdot L_0 \cap R_0$?

(Hint. (2) Reduce to the case $1_V \in U$. If $A^2 \in F^\bullet$ for a 4×4 skew-symmetric matrix A show that $A \in L_0$ or $A \in R_0$. Deduce that $U \subseteq L_0$ or $U \subseteq R_0$.

5. **Characterizing similarities.** Let (V, q) be a quadratic space and $f \in \mathrm{End}(V)$.

(1) If f preserves orthogonality (i.e. $B(v, w) = 0$ implies $B(f(v), f(w)) = 0$), then f is a similarity.

(2) If q is isotropic and f preserves isotropy (i.e. $q(v) = 0$ implies $q(f(v)) = 0$), then f is a similarity. (This also follows from Exercise 14, at least if F is infinite.)

(3) If q represents 1 and f preserves the unit sphere (i.e. $q(v) = 1$ implies $q(f(v)) = 1$), then f is an isometry.

6. **Multiplication formulas.** (1) Suppose $q = 2^m\langle 1\rangle$. If q represents $c \in F^\bullet$ then $\langle c\rangle q \simeq q$. Equivalently, $D_F(q) = G_F(q)$.

(2) Follow the notations of Proposition 1.9. There is an equivalence:

(i) There is a formula $\sigma(X)q(Y) = q(Z)$ such that each $z_k \in F(X)[Y]$ is a linear form in Y.

(ii) $\sigma(X) \in G_{F(X)}(q \otimes F(X))$.

(Hint. (1) The matrix in Exercise 0.5 provides a c-similarity.)

7. Let (V, B) be an *alternating space*, that is, B is a nonsingular skew-symmetric bilinear form on the F-vector space V.

(1) The adjoint involution I_B is well defined. The set $\mathrm{Sim}(V, B)$ is defined as usual and if $S \subseteq \mathrm{Sim}(V, B)$ is a linear subspace then the norm map induces a quadratic form σ on S.

(2) When $\dim V = 2$ then $\mathrm{Sim}(V, B) = \mathrm{End}(V)$ and the map $\mu : \mathrm{End}(V) \to F$ is the determinant. That is, the 2-dimensional alternating space admits a 4-dimensional subspace of similarities.

(3) $\dim V = 2m$ is even and $(V, B) \simeq m\boldsymbol{H} \simeq \boldsymbol{H} \perp \cdots \perp \boldsymbol{H}$, where H is the 2-dimensional alternating space where the Gram matrix of the form is $\begin{pmatrix} 0 & -1 \\ 1 & 0 \end{pmatrix}$. If B is allowed to be singular then $(V, B) \simeq r\langle 0\rangle \perp m\boldsymbol{H}$.

(4) Let $J_{r,m}$ be the $n \times n$ matrix $\begin{pmatrix} 0_r & 0 & 0 \\ 0 & 0_m & -I_m \\ 0 & I_m & 0_m \end{pmatrix}$ where $n = r + 2m$. Any skew-symmetric $n \times n$ matrix is of the type $P^\top \cdot J_{r,m} \cdot P$ for some r, m and some $P \in \mathrm{GL}_n(F)$. In particular rank $M = 2m$ is even and det M is a square.

8. **Jordan forms.** Let (V, q) be a quadratic space over an algebraically closed field F. Let $f \in \mathrm{O}(V, q)$, that is, $\tilde{f}f = 1_V$. For each $a \in F$ define the general eigenspace $V((a)) = \{x \in V : (f - a1_V)^k(x) = 0 \text{ for some } k\}$.

(1) If $ab \neq 1$ then $V((a))$ and $V((b))$ are orthogonal. In particular if $a \neq \pm 1$ then $V((a))$ is totally isotropic.

(2) Therefore $V = V((1)) \perp V((-1)) \perp \perp(V((a)) \oplus V((a^{-1})))$, summed over some scalars $a \neq \pm 1$. Then $\det f = (-1)^m$ where $m = \dim V((-1))$.

(3) If $f \in \mathrm{End}(V)$, what condition on the Jordan form of f ensures the existence of a quadratic form q on V having $f \in \mathrm{O}(V, q)$? Certainly f must be similar to f^{-1}.

Proposition. *If $f \in \mathrm{GL}(V)$ then $f \in \mathrm{O}(V, q)$ for some regular quadratic form q if and only if $\tilde{f} \sim f^{-1}$ and every elementary divisor $(x \pm 1)^m$ for m even occurs with even multiplicity.*

(Hint. (3) ($\Longleftarrow$) If f has matrix $\begin{pmatrix} B & 0 \\ 0 & B^{-\top} \end{pmatrix}$ in block form, we can use $M = \begin{pmatrix} 0 & 1 \\ 1 & 0 \end{pmatrix}$ as the Gram matrix of q. By canonical form theory we need only find M in the case f has the single elementary divisor $(x-1)^m$ for odd m.

($\Longrightarrow$) Trickier to prove. See Gantmacher (1959), §11.5, Milnor (1969), §3, or Shapiro (1992).)

9. **More Jordan forms.** Let $f \in \text{End}(V)$. What conditions on the Jordan form of f are needed to ensure the existence of a quadratic form q on V for which f is symmetric: $I_q(f) = f$? (Answer. Such q always exists. In matrix terms this says that for any square matrix A there is a nonsingular symmetric matrix S such that $SAS^{-1} = A^\top$. This result was proved by Frobenius (1910) and has appeared in the literature in various forms. Similar questions arise: When does there exist a quadratic form q such that $I_q(f) = -f$? When does there exist an alternating form B on V such that $I_B(f) = \pm f$? These are questions are addressed in Chapter 10.)

10. **Determinants of skew matrices.** If (V, q) is a quadratic space then all the invertible skew-symmetric matrices have the same determinant, modulo squares. In fact:

$$\text{if } f \in \text{GL}(V) \text{ and } I_q(f) = -f \text{ then } \langle \det f \rangle = \det q \text{ in } F^\bullet/F^{\bullet 2}.$$

(Hint. The determinant of a skew-symmetric matrix is always a square.)

11. **Multi-indices.** In the notation of Proposition 1.11:

(1) For any Δ, Γ note that $f^\Delta \cdot f^\Gamma = \pm f^\Gamma \cdot f^\Delta$. Determine this sign explicitly.

(2) Suppose further that $f_i^2 = a_i \in F^\bullet$. Define $a_\Delta = a_1^{\delta_1} \dots a_n^{\delta_n}$. Then $(f^\Delta)^2 = \pm a_\Delta$. Determine exactly when that sign is "+". Further investigation of such signs appears in Exercise 3.18 below.

(3) The "multi-index" Δ can be viewed as the subset of $\{1, 2, \dots, n\}$ consisting of all indices i such that $\delta_i = 1$. Listing this subset as $\Delta = \{i_1, \dots, i_k\}$ such that $i_1 < i_2 < \dots < i_k$, we have $f^\Delta = f_{i_1} f_{i_2} \dots f_{i_k}$. Then $|\Delta|$ is the cardinality of Δ, $\Delta\Gamma = \Delta \cap \Gamma$, and $\Delta + \Gamma = (\Delta \cup \Gamma) - (\Delta \cap \Gamma)$, the symmetric difference.

(Hint. (1) The answer involves only $|\Delta|$, $|\Gamma|$ and $|\Delta\Gamma|$.)

12. **Anticommuting matrices.** (1) **Proposition.** *Suppose $n = 2^m n_0$ where n_0 is odd. There exist k elements of* $\text{GL}_n(F)$ *which anticommute pairwise, if and only if $k \le 2m+1$.*

(2) Suppose $f_1, \dots, f_{2m+1}$ is a maximal anticommuting system in $\text{GL}_n(F)$ as above. If $n = 2^m$ then f_i^2 must be a scalar. For other values of n this claim may fail.

(Hint. (1) ($\Longleftarrow$) Inductively construct such $f_1, \dots, f_{2m+1}$ in $\text{GL}(n)$ with $f_i^2 = \pm 1$. To get the system in $\text{GL}_{2n}(F)$ use block matrices: $\begin{pmatrix} 1 & 0 \\ 0 & -1 \end{pmatrix}, \begin{pmatrix} 0 & 1 \\ 1 & 0 \end{pmatrix}, \begin{pmatrix} 0 & f_i \\ -f_i & 0 \end{pmatrix}$.

($\Longrightarrow$) We may assume F is algebraically closed and $k = 2k_0$ is even. *To show*: $2^{k_0} \mid n$. Use notations of Exercise 8 with $V = F^n$. Then V is a direct sum of the $V((a))$'s and $f_j : V((a)) \to V((-a))$ is a bijection, for every $j > 1$. Let $n_a = \dim V((a))$, so that $n = \sum 2n_a$, summed over certain eigenvalues a. The maps $f_2 f_j$ induce $k-2$ anticommuting elements of $\mathrm{GL}\big(V((a))\big)$, and hence $2^{k_0-1} \mid n_a$, by induction hypothesis.

(2) By (1.11) the matrices f^Δ span all of $\mathbb{M}_n(F)$ and f_i^2 commutes with every f_j.)

13. **Trace forms.** The map $\mu : \mathrm{Sim}(V) \to F$ induces a quadratic form on every linear subspace.

(1) In fact μ extends to a quadratic form on $\mathrm{End}(V)$, at least if $\dim V$ is not divisible by the characteristic of F.

(2) If (V, q) is a quadratic space define $\tau : \mathrm{End}(V) \to F$ by $\tau(f) = \mathrm{trace}(\tilde{f} f)$. Then $(\mathrm{End}(V), \tau) \simeq q \otimes q$.

(3) Let $A = \{f \in \mathrm{End}(V) : \tilde{f} = -f\}$, the space of skew-symmetric maps. Compute the restriction τ_A of the trace form τ to $A \times A$.

(4) If (V, B) is a alternating space what can be said about the trace form τ?

(5) Suppose (V, α) and (W, β) are quadratic spaces and use the isomorphism $\theta_\alpha : V \to \hat{V}$ to find an isomorphism $V \otimes W \to \mathrm{Hom}(V, W)$. Does the quadratic form $\alpha \otimes \beta$ get carried over to the trace form τ on $\mathrm{Hom}(V, W)$ defined by $\tau(f) = \mathrm{trace}(J_W^V(f) \circ f)$? (See Exercise 2 for J_W^V.)

(Hint. (1) Consider $\mathrm{trace}(\tilde{f} f)$.

(2) The bilinear form $b = B_q$ induces an isomorphism $\theta : V \to \hat{V}$ where $\hat{V}$ is the dual vector space. Then $\varphi : V \otimes V \to \hat{V} \otimes V \to \mathrm{End}(V)$ is an isomorphism. Verify that

(i) $\varphi(v_1 \otimes w_1) \circ \varphi(v_2 \otimes w_2) = b(w_1, v_2) \cdot \varphi(v_1 \otimes w_2)$.

(ii) $\mathrm{trace}(\varphi(v \otimes w)) = b(v, w)$.

(iii) $I_q(\varphi(v \otimes w)) = \varphi(w \otimes v)$. Show that φ carries $q \otimes q$ to τ.

(3) If $q \simeq \langle a_1, \dots, a_n \rangle$ define $P_2(q) = \langle a_1 a_2, a_1 a_3, \dots, a_{n-1} a_n \rangle$. Then $\tau_A \simeq \langle 2 \rangle P_2(q)$. To see this let $\{v_1, \dots, v_n\}$ be the given orthogonal basis and note that $\varphi^{-1}(A)$ is spanned by $v_i \otimes v_j - v_j \otimes v_i$ for $i < j$.) Compare Exercise 3.13.)

14. **Geometry lemma.** Let $X = (x_1, \dots, x_n)$ be indeterminates. If $f \in F[X]$ define the zero-set $\mathcal{Z}(f) = \{a \in F^n : f(a) = 0\}$. If $f \mid p$ then p vanishes on $\mathcal{Z}(f)$, or equivalently, $\mathcal{Z}(f) \subseteq \mathcal{Z}(p)$. The converse is false in general even if f is irreducible. (Hilbert's Nullstellensatz applies when F is algebraically closed.) However, the converse does hold in some cases:

(1) **Lemma.** *Let F be an infinite field and let $x, Y = (y_1, \dots, y_n)$ be indeterminates. Suppose $f(x, Y) = x \cdot g(Y) + h(Y) \in F[x, Y]$ where $g(Y), h(Y)$ are non-zero and relatively prime in $F[Y]$. Then f is irreducible and if $p \in F[x, Y]$ vanishes on $\mathcal{Z}(f)$ then $f \mid p$.*

Proof outline. Express $p = c_0(Y)x^d + \cdots + c_d(Y)$. Since $xg \equiv -h \pmod f$, we have $g^d p \equiv Q \pmod f$ where $Q = c_0(Y)(-h(Y))^d + \cdots + c_d(Y)g(Y)^d \in F[Y]$. Since p vanishes on $\mathcal{Z}(f)$ it follows that $Q(B) = 0$ for every $B \in F^n$ with $g(B) \neq 0$. Therefore $g \cdot Q$ vanishes identically on F^n. Hence $Q = 0$ as a polynomial and $f \mid g^d p$ and we conclude that $f \mid p$.

(2) Suppose $\deg p = d$ and $\deg f = m$ above. If F is finite and $|F| \geq (m+1)\cdot(d+1)$, the conclusion still holds.

(3) Suppose q is an isotropic quadratic form over an infinite field F, and $\dim q > 2$. If $p \in F[X]$ vanishes on $\mathcal{Z}(q)$, then $q \mid p$. In particular if q, q' are quadratic forms with dimension > 2 and if $\mathcal{Z}(q) \subseteq \mathcal{Z}(q')$ then $q' = c \cdot q$ for some $c \in F$. For what finite fields F do these statements hold?

(Hint. (2) If $k(Y) \in F[Y]$ vanishes on F^n and if $|F| > \deg k$ then $k = 0$.

(3) Change variables to assume $q(X) = xy + h(Z)$ where h is a quadratic form of dim ≥ 1. Analyze (1) to show that $\deg Q \leq 2d$. The argument works if $|F| \geq 2 \cdot \deg p + 2$.)

15. **Transversality.** Suppose F is a field with more than 5 elements.

(1) If $a \in F^\bullet$ there exist non-zero $r, s \in F$ such that $r^2 + as^2 = 1$.

(2) Suppose $q \simeq \langle a_1, \ldots, a_n\rangle$ represents c. That is, there exists $0 \neq v \in F^n$ such that $q(v) = \sum a_i v_i^2 = c$. Then there exists $w \in F^n$ such that $q(w) = c$ and $w_i \neq 0$ for *every* i.

(3) **Transversality Lemma.** *Suppose (V, α) and (W, β) are quadratic spaces over F and $\alpha \perp \beta$ represents c. Then there exist $v \in V$ and $w \in W$ such that $c = \alpha(v) + \beta(w)$ and $\alpha(v) \neq 0$ and $\beta(w) \neq 0$.*

(4) Generalize (2) to non-diagonal forms. The answer reduces to the case $c = 0$:

Proposition. *Let (V, q) be an isotropic quadratic space with $\dim q = n \geq 3$ and let $H_1, \ldots, H_n$ be linearly independent hyperplanes in V. Then there exists an isotropic vector $v \in V$ such that $v \notin H_1 \cup \cdots \cup H_n$.*

For example, relative to a given basis the coordinate hyperplanes are linearly independent. If F is infinite we can avoid any finite number of hyperplanes in V. (Use Exercise 14 with p a product of linear forms.) If F is finite the result follows from Exercise 14 provided that $|F| \geq 2n + 2$. The stated result (valid if $|F| > 5$) is due to Leep (unpublished) and seems quite hard to prove.

(Hint. (1) Use the formulas $r = \frac{t^2-a}{t^2+a}$ and $s = \frac{2t}{t^2+a}$.

(2) Suppose $n = 2$. Re-index to assume $v_1 \neq 0$ and scale to assume $a_1 = 1$. If $v_2 = 0$ scale again to assume $c = 1$. Apply (1). What if $n > 2$?

(3) If α, β are anisotropic diagonalize them and apply (2). If α is isotropic choose w so that $\beta(w) \neq 0, c$ and note that α represents $c - \beta(w)$.)

16. **Conjugate subspaces. Lemma.** *Suppose $S, S' \subseteq \mathrm{Sim}(V, q)$ are subspaces containing 1_V and such that $S \simeq S'$ as quadratic spaces (relative to the induced quadratic form). If $\dim S \leq 3$ then $S' = gSg^{-1}$ for some $g \in \mathrm{O}(V, q)$.*

(Hint. Compare (1.10). Induct on $\dim V$. Restate the hypotheses in the case $\dim S = 2$: For $i = 1, 2$ we have isometric spaces (V_i, q_i) and $f_i \in \mathrm{End}(V_i)$ such that $\tilde{f}_i = -f_i$ and $f_i^2 = -a \cdot 1_V$. Choose $v_i \in V_i$ such that $q_1(v_1) = q_2(v_2) \neq 0$ and let $W_i = \mathrm{span}\{v_i, f_i(v_i)\}$. Define $g : W_1 \to W_2$ with $g(v_1) = v_2$ and $g(f(v_1)) = f(v_2)$. Then g is an isometry and $f_2 = gf_1g^{-1}$ on W_2. Apply induction to $W_i^{\perp}$.)

17. **Proper similarities.** If $f \in \mathrm{Sim}^\bullet(V, q)$ where $\dim V = n$ then $(\det f)^2 = \mu(f)^n$.

(1) If n is odd then $\mu(f) \in F^{\bullet 2}$ and $f \in F^\bullet \mathrm{O}(V)$.

(2) Suppose $n = 2m$. Define f to be *proper* if $\det f = \mu(f)^m$. The proper similarities form a subgroup $\mathrm{Sim}^+(V)$ of index 2 in $\mathrm{Sim}^\bullet(V)$. This is the analog of the special orthogonal group $\mathrm{O}^+(n) = \mathrm{SO}(n)$.

(3) Suppose $\tilde{f} = -f$. If $g = a1_V + bf$ for $a, b \in F$ then g is proper.

(4) **Wonenburger's Theorem.** *Suppose $f, g \in \mathrm{Sim}^\bullet(V)$ and $\tilde{f} = -f$. If g commutes with f, then g is proper. If g anticommutes with f and $4 \mid n$ then g is proper.*

(5) Let $L_0, R_0 \subseteq \mathrm{Sim}(\langle 1, 1, 1, 1\rangle)$ be the subspaces described in Exercise 4(2). Let G be the group generated by $L_0^\bullet$ and $R_0^\bullet$. Then G is the set of all maps $g(x) = axb$ for $a, b \in D^\bullet$.

Lemma. $G = \mathrm{Sim}^+(q)$.

(Hint. (1) Show $\mu(f) \in F^{\bullet 2}$.

(4) Assume F algebraically closed and $f^2 = 1_V$. The eigenspaces U^+ and U^- are totally isotropic of dimension m. Examine the matrix of g relative to $V = U^+ \oplus U^-$ using the Gram matrix $\begin{pmatrix} 0 & 1 \\ 1 & 0 \end{pmatrix}$.

(5) $G \subseteq \mathrm{Sim}^+(V)$ by Wonenburger. Conversely it suffices to show that $\mathrm{SO}(q) \subseteq G$. The maps τ_a generate $\mathrm{O}(q)$, where τ_a is the reflection fixing the hyperplane $(a)^\perp$. Therefore the maps $\tau_a\tau_1$ generate $\mathrm{SO}(q)$. Writing $[a]$ for $q(a)$ as in the appendix, we have $\tau_a(x) = x - \frac{2[x,a]}{[a]} \cdot a = x - [a]^{-1}(x\bar{a} + a\bar{x})a = -[a]a\bar{x}a$. Then $\tau_a\tau_1(x) = [a]^{-1}axa$ so that $\tau_a\tau_1$ lies in G.)

18. **Zero-similarities.** Let $h \in \mathrm{End}(V)$ where (V, q) is a quadratic space.

(1) $(\mathrm{image}\, h)^\perp = \ker \tilde{h}$ and $(\ker h)^\perp = \mathrm{image}\, \tilde{h}$.

(2) If $h \in \mathrm{Sim}(V)$, it is possible that $\tilde{h} \notin \mathrm{Sim}(V)$. However if h is comparable to some $f \in \mathrm{Sim}^\bullet(V)$ then $\tilde{h} \in \mathrm{Sim}(V)$.

(3) Suppose $h \in S \subseteq \mathrm{Sim}(V)$ where $\mu(h) = 0$ and S is a regular subspace. Then $h\tilde{h} = \tilde{h}h = 0$ and $\dim \ker h = \dim \mathrm{image}\, h$, as in the proof of (1.4). Conversely if $h \in \mathrm{End}(V)$ satisfies these conditions then h is in some regular $S \subseteq \mathrm{Sim}(V)$.

(Hint: (3) $U = \ker h = \mathrm{image}\, \tilde{h}$ and $U' = \mathrm{image}\, h = \ker \tilde{h}$ are totally isotropic of dimension $n/2$. Replace h by gh for a suitable $g \in \mathrm{O}(V, q)$, to assume $U = U'$. Choose a totally isotropic complement W and use matrices relative to $V = U \oplus W$ to construct a 0-similarity k with $\tilde{h}k + \tilde{k}h = 1_V$.)

19. **Singular subspaces.** (1) Find an example of a regular quadratic space (V, q) admitting a (singular) subspace $S \subseteq \mathrm{Sim}(V, q)$ where $\dim S > \dim V$.

(2) Find such an example where $1_V \in S$.

(Hint. Let $(V, q) = m\mathbb{H}$ be hyperbolic and choose a basis so that the matrix of q is $M = \begin{pmatrix} 0 & 1 \\ 1 & 0 \end{pmatrix}$ in $m \times m$ blocks. If $f = \begin{pmatrix} a & b \\ c & d \end{pmatrix}$ then $\tilde{f} = \begin{pmatrix} d^\top & b^\top \\ c^\top & a^\top \end{pmatrix}$. Consider the cases $f = \begin{pmatrix} a & b \\ 0 & 0 \end{pmatrix}$.)

20. **Extension Theorem.** (1) **Lemma.** *Suppose (V, q) is a quadratic space, $W \subseteq V$ is a regular subspace and $f : W \to V$ is a similarity. If $\mu(f) \in G_F(q)$. Then there exists $\hat{f} \in \mathrm{Sim}(V, q)$ extending f.*

(2) **Corollary.** *If D is an $n \times r$ matrix over F satisfying $D^\top \cdot D = aI_r$ and $a \in G_F(n\langle 1\rangle)$ then D can be enlarged to an an $n \times n$ matrix $\hat{D} = (D|D')$ which satisfies $\hat{D}^\top \cdot \hat{D} = aI_n$.*

(Hint. (1) There exists $g \in \mathrm{Sim}(V, q)$ with $\mu(g) = \mu(f)$. Then $g^{-1} \circ f : W \to V$ is an isometry and Witt's Extension Theorem for isometries applies. See Scharlau (1985) Theorem 1.5.3.)

21. **Bilinear terms in Pfister Form Formulas.** According to Pfister's Theory, if $n = 2^m$ and X, Y are systems of n indeterminates over F, there exists a formula

$$(x_1^2 + x_2^2 + \cdots + x_n^2)(y_1^2 + y_2^2 + \cdots + y_n^2) = z_1^2 + z_2^2 + \cdots + z_n^2, \qquad (*)$$

where each z_k is a linear form in Y with coefficients in the rational function field $F(X)$. In Exercise 0.5 we found formulas where several of the z_k's were also linear in X. Question. How many of the terms z_k can be taken to be bilinear in X, Y?

Proposition. *Suppose F is a formally real field, $n = 2^m$ and X, Y are systems of n indeterminates. There is an n-square identity $(*)$ as above with each z_k linear in Y and with $z_1, \dots, z_r$ also linear in X if and only if $r \le \rho(n)$.*

Proof outline. Suppose $z_1, \dots, z_r$ are also linear in X. Then $Z = AY$ for an $n \times n$ matrix A over $F(X)$ such that the entries in the first r rows of A are linear forms in X. Then $(*)$ becomes: $A^\top \cdot A = \left(\sum x_i^2\right) \cdot I_n$. Consequently $A \cdot A^\top = \left(\sum x_i^2\right) \cdot I_n$. Express A in block form as $A = \begin{pmatrix} B \\ C \end{pmatrix}$ where B is an $r \times n$ matrix whose entries are linear forms in X and C is an $(n-r) \times n$ matrix over $F(X)$. Then $B \cdot B^\top = \left(\sum x_i^2\right) \cdot I_r$, so that $B^\top$ satisfies the Hurwitz conditions for a formula of size $[n, r, n]$. The Hurwitz–Radon Theorem implies $r \le \rho(n)$.

Conversely if $r \le \rho(n)$ there exists an $r \times n$ matrix B, linear in X, with $B \cdot B^\top = \left(\sum x_i^2\right) \cdot I_r$. Note that $\left(\sum x_i^2\right)$ is in $G_{F(X)}(n\langle 1\rangle)$ by Pfister's theorem. (Use Exercise 6 or invoke (5.2) (1) below.) Apply Exercise 20(2) to $B^\top$ over $F(X)$ to find the matrix A.

22. **Isoclinic planes.** If $U \subseteq \mathbb{R}^n$ is a subspace and ℓ is a line in $\mathbb{R}^n$ let $\angle(\ell, U)$ be the angle between them. If $\ell \not\subseteq U^\perp$ then $\angle(\ell, U)$ is the angle between the lines ℓ and $\pi_U(\ell)$, where $\pi_U : \mathbb{R}^n \to U$ is the orthogonal projection. Subspaces $U, W \subseteq \mathbb{R}^n$ are *isoclinic* if $\angle(\ell, U)$ is the same for every line $\ell \subseteq W$.

(1) U, W are isoclinic if and only if π_U is a similarity when restricted to W.

(2) If $\dim U = \dim W$ this relation "isoclinic" is symmetric.

(3) Suppose $\mathbb{R}^{2n} = U \perp U'$ where $\dim U = \dim U' = n$. If $f \in \mathrm{Hom}(U, U')$ define its graph to be $U[f] = \{u + f(u) : u \in U\} \subseteq \mathbb{R}^{2n}$. Let $W \subseteq \mathbb{R}^{2n}$ be a subspace of dimension n.

Lemma. *U, W are isoclinic if and only if either $W = U'$ or $W = U[f]$ for some similarity f.*

Proposition. *Suppose $f, g \in \mathrm{Sim}(U, U')$. Then $U[f], U[g]$ are isoclinic if and only if f, g are comparable similarities.*

(4) If $T \subseteq \mathrm{Sim}(U, U')$ is a subspace define $\mathcal{S}(T) = \{U[f] : f \in T\} \cup \{U'\}$. Then $\mathcal{S}(T)$ is a set of mutually isoclinic n-planes in $\mathbb{R}^{2n}$. It is called an isoclinic sphere since it is a sphere when viewed as a subset of the Grassmann manifold of n-planes in $2n$-space.

(5) If $n \in \{1, 2, 4, 8\}$ there exists such T with $\dim T = n$. In these cases $\mathcal{S}(T)$ is "space filling": whenever $0 \neq x \in \mathbb{R}^{2n}$ there is a unique n-plane in $\mathcal{S}(T)$ containing x.

23. **Normal sets of planes.** Define two n-planes U, V in $\mathbb{R}^{2n}$ to be *normally related* if $U \cap V = \{0\} = U \cap V^\perp$. A set of n-planes is *normal* if every two distinct elements are either normally related or orthogonal.

(1) Suppose $\mathcal{S}$ is a maximal normal set of n-planes in $\mathbb{R}^{2n}$. If $U \in \mathcal{S}$ then $U^\perp \in \mathcal{S}$. A linear map $f : U \to U^\perp$ has a corresponding graph $U[f] \subseteq U \perp U^\perp = \mathbb{R}^{2n}$, as in Exercise 22. If $W \in \mathcal{S}$ then either $W = U^\perp$ or $W = U[f]$ for some bijective f. Note that $U[f]^\perp = U[-f^{-\top}]$, and that $U[f]$ and $U[g]$ are normally related iff $f - g$ and $f + g^{-\top}$ are bijective.

(2) Let $\mathbb{O} = \mathbb{R}^n \times \{0\}$ be the basic n-plane. If $T \subseteq \mathrm{M}_n(\mathbb{R})$ define

$$\mathcal{S}(T) = \{\mathbb{O}[A] : A \in T\} \cup \{\mathbb{O}^\perp\}.$$

Any maximal normal set of n-planes in $\mathbb{R}^{2n}$ containing $\mathbb{O}$ equals $\mathcal{S}(T)$ for some subset T such that: T is *nonsingular* (i.e. every non-zero element is in $\mathrm{GL}_n(\mathbb{R}^n)$), T is an additive subgroup and T is closed under the operation $A \mapsto A^{-\top}$.

(3) Consider the case $T \subseteq \mathrm{M}_n(\mathbb{R})$ is a linear subspace. (If $\mathcal{S}(T)$ is maximal normal must T be a subspace?)

Proposition. *If $T \subseteq \mathrm{M}_n(\mathbb{R})$ is a linear subspace such that $\mathcal{S}(T)$ is a maximal normal set of n-planes, then $T \subseteq \mathrm{Sim}(\mathbb{R}^n)$ and $\mathcal{S}(T)$ is a maximal isoclinic sphere.*

(Hint. (3) If $0 \neq A \in T$ express $A = PDQ^\top$ where $P, Q \in \mathrm{O}(n)$ and D is diagonal with positive entries. (This is a singular value decomposition.) If $a, b \in \mathbb{R}$ then $aA + bA^{-\top} \in T$. Then if $aD + bD^{-\top}$ is non-zero it must be nonsingular. Deduce $D =$ scalar so that $A \in \mathrm{Sim}(\mathbb{R}^n)$.

24. **Automorphisms.** Suppose C is a composition algebra and $a \in C$.

(1) If a is invertible, must $x \mapsto axa^{-1}$ be an automorphism of C?

Define $L_a(x) = ax$, $R_a(x) = xa$ and $B_a(x) = axa$.

(2) $B_a(xy) = L_a(x) \cdot R_a(y)$. Consequently $B_a B_b(xy) = L_a L_b(x) \cdot R_a R_b(y)$, etc. These provide examples of maps $\alpha, \beta, \gamma : C \to C$ satisfying: $\gamma(xy) = \alpha(x) \cdot \beta(y)$.

(3) For such α, β, γ if $\alpha(1) = \beta(1) = 1$ then $\alpha = \beta = \gamma$ is an automorphism of C. If C is associative this automorphism is the identity. Suppose C is octonion and $\gamma = B_{a_1} B_{a_2} \dots B_{a_n}$, etc. There exist non-trivial automorphisms built this way (but they require $n \geq 4$).

(4) Suppose C_1 and C_2 are composition algebras. If $\varphi : C_1 \to C_2$ is an isomorphism then φ commutes with the involutions: $\varphi(\bar{x}) = \overline{\varphi(x)}$, and φ is an isometry of the norm forms $[x]$. Conversely if the norm forms of C_1 and C_2 are isometric then the algebras must be isomorphic. (However not every isometry is an isomorphism.)

(Hint. (2) Moufang identities.

(3) $\alpha(1) = 1 \implies \gamma = \beta$, and $\beta(1) = 1 \implies \gamma = \alpha$. Case $n = 2$: $ab = 1$ implies $a \cdot bx = x$ since alternative. Case $n = 3$: Given $a \cdot bc = 1 = cb \cdot a$, show a, b, c lie in a commutative subalgebra. Then a, b, c, x lie in an associative subalgebra and $L_a L_b L_c(x) = x$.

(4) $a \in C$ is pure (i.e. $\bar{a} = -a$) iff $a \notin F$ and $a^2 \in F$. Any isomorphism preserves "purity". Given an isometry $\eta : C_1 \to C_2$ assume inductively that C_i is the double of a subalgebra B_i, that $\eta(B_1) = B_2$ and that the restriction of η to B_1 is an isomorphism. By Witt Cancellation the complements $B_i^\perp$ are isometric.)

25. Suppose $A = A(\alpha_1, \dots, \alpha_n)$ is a Cayley–Dickson algebra. Then elements of A satisfy the identities in (A.9).

(1) The elements of A also satisfy:

$yx \cdot x = x \cdot xy$

$a \cdot ba = ab \cdot a$ (the flexible law)

$[ab] = [\bar{a}b] = [ba]$

$a^n \cdot a^m = a^{n+m}$ (A is power-associative).

(2) If $ab = 1$ in A does it follow that $ba = 1$? If $a, b, c \in A$ does it follow that $[ab \cdot c] = [a \cdot bc]$?

(3) $\dim A = 2^n$ and A is a composition algebra if and only if $n \leq 3$. Define $A^\circ = \ker(T)$, the subspace of "pure" elements, so that $A = F \perp A^\circ$. If a, b are anisotropic, orthogonal elements of A° (using the norm form) then $a^2, b^2 \in F^\bullet$ and $ab = -ba$. These elements might fail to generate a quaternion subalgebra.

(4) There exists a basis $e_0, e_1, \dots, e_{2^n-1}$ of A such that:

$e_0 = 1$

$e_1, \dots, e_{2^n-1} \in A^\circ$ pairwise anti-commute and $e_i^2 \in F^\bullet$.

For each i, j: $e_i e_j = \lambda e_k$ for some index $k = k(i, j)$ and some $\lambda \in F^\bullet$.

The basis elements are "alternative": $e_ie_i \cdot x = e_i \cdot e_ix$ and $xe_i \cdot e_i = x \cdot e_ie_i$.

(5) The norm form $x \mapsto [x]$ on A is the Pfister form $\langle\langle -\alpha_1, \dots, -\alpha_n \rangle\rangle$.

(6) Let $A_n = A(\alpha_1, a_2, \dots \alpha_n)$. We proved that the identity $a \cdot bc = ab \cdot c$ holds in every A_2 and fails in every A_3. Similarly the identity $a \cdot ab = aa \cdot b$ holds in A_3 and fails in A_4.

Open question. Is there some other simple identity which holds in A_4 and fails in A_5?

(Hints. (1) If $\bar{a} = -a$ we know $[x, ax] = 0 = [x, xa]$. The flexible law for $\mathcal{D}_\alpha(H)$ follows from the properties of H, after some calculation. It suffices to prove power-associativity for pure elements, that is, when $\bar{a} = -a$. But then $a^2 \in F$.

(2) The $ab = 1$ property holds, at least when the form $[x]$ is anisotropic: Express $a = \alpha + e$ and $b = \beta + \gamma e + f$ where $1, e, f$ are orthogonal and $[e], [f] \neq 0$. Then $\{1, e, f, ef\}$ is linearly independent.

(4) Construct the basis inductively.)

26. **Alternative algebras.** Suppose A is a ring. If $x, y, z \in A$ define the associator $(x, y, z) = xy \cdot z - x \cdot yz$. Suppose A is alternative: $(a, a, b) = (a, b, b) = 0$.

(1) (x, y, z) is an alternating function of the three variables. In particular, $a \cdot ba = ab \cdot a$, which is the "Flexible Law".

(2) $xax \cdot y = x(a \cdot xy)$

$y \cdot xax = (yx \cdot a)x$ (the Moufang identities)

$bx \cdot yb = b \cdot xy \cdot b$

(3) $(y, xa, z) + (y, za, x) = -(y, x, a)z - (y, z, a)x$.

(4) **Proposition.** *Any two elements of A generate an associative subalgebra.*

(Hint. (2) $xax \cdot y - x(a \cdot xy) = (xa, x, y) + (x, a, xy) = -(x, xa, y) - (x, xy, a) = -x^2a \cdot y - x^2y \cdot a + x(xa \cdot y + xy \cdot a) = -(x^2, a, y) - (x^2, y, a) + x \cdot [(x, a, y) + (x, y, a)] = 0$. For the second, use the opposite algebra. Finally, $bx \cdot yb - b \cdot xy \cdot b = (b, x, yb) - b(x, y, b) = -(b, yb, x) - b(x, y, b) = b \cdot [(y, b, x) - (x, y, b)] = 0$, using the first identity.

(3) $(y, xa, x) = -(y, x, a)x$ by (2). Replace x by $x + z$.

(4) If $u, v \in A$, examine "words" formed from products of u's and v's. It suffices to show that $(p, q, r) = 0$ for any such words p, q, r. Induct on the sum of the lengths of p, q, r. Rename things to assume that words q and r begin with u. Apply (3) when $x = u$.)

27. **The nucleus.** The *nucleus* $\mathcal{N}(A)$ of an algebra A is the set of elements $g \in A$ which associate with every pair of elements in A. That is, $xy \cdot z = x \cdot yz$ whenever one of the factors x, y, z is equal to g. Then $\mathcal{N}(A)$ is an associative subalgebra of A.

(1) If A is alternative it is enough to require $g \cdot xy = gx \cdot y$ for every x, y.

(2) If A is an octonion algebra over F then $\mathcal{N}(A) = F$.

(3) Does this hold true for all the Cayley–Dickson algebras A_n when $n > 3$?

28. Suppose A is an octonion algebra and $a, b \in A$. Then: $(a, b, x) = 0$ for every $x \in A$ iff $1, a, b$ are linearly dependent.

(Hint. ($\Longrightarrow$) $a, b \in H$ for some quaternion subalgebra H. Use (A.6) to deduce that $ab = ba$.)

Notes on Chapter 1

The independence result in (1.11) was proved by Hurwitz (1898) for skew symmetric matrices. The general result (for matrices) is given in Robert's thesis (1912) and in Dickson (1919). Similar results are mentioned in the Notes for Exercise 12.

The notation $\langle\langle a_1, \dots, a_n\rangle\rangle$ for Pfister forms was introduced by T. Y. Lam. Other authors reverse the signs of the generators, writing $\langle\langle a\rangle\rangle$ for $\langle 1, -a\rangle$. In particular this is done in the monumental work on involutions by Knus, Merkurjev, Rost and Tignol (1998). We continue to follow Lam's notation, hoping that readers will not be unduly confused.

The topics in the appendix have been described in several articles and textbooks. The idea of the "doubling process", building the octonions from pairs of quaternions, is implicit in Cayley's works, but it was first formally introduced in Dickson's 1914 monograph. Dickson was also the first to note that the real octonions form a division ring. E. Artin conjectured that the octonions satisfy the alternative law. This was first proved by Artin's student M. Zorn (1930).

Perhaps the first study of the Cayley–Dickson algebras of dimension > 8 was given by Albert (1942a). He analyzed the algebras A_n of dimension 2^n over an arbitrary field F (allowing F to have characteristic 2), and proved a general version of Theorem A.8. Properties of A_n appear in Schafer (1954), Khalil (1993), Khalil and Yiu (1997), and Moreno (1998). Further information on composition algebras appears in Jacobson (1958), Kaplansky (1953), Curtis (1963).

Are there infinite dimensional composition algebras? Kaplansky (1953) proved that every composition algebra with (2-sided) identity element must be finite dimensional. However there do exist infinite dimensional composition algebras having only a left identity element. Further information is given in Elduque and Pérez (1997).

Define an algebra A over the real field $\mathbb{R}$ to be an *absolute-valued algebra* if A is a normed space (in the sense of real analysis) and $|xy| = |x| \cdot |y|$. Urbanik and Wright (1960) proved that an absolute-valued algebra with identity must be isomorphic to one of the classical composition algebras. Further information and references appear in Palacios (1992).

Exercise 3. (3) Let $V^\bullet = \{v \in V : q(v) \neq 0\}$ be the set of anisotropic vectors in (V, q). If $x, y \in V^\bullet$ define the angle-measure $\angle(x, y) = \frac{B(x,y)}{q(x)q(y)}$. If $f \in \operatorname{End}(V)$ preserves all such angle-measures, must f be a similarity? Alpers and Schröder (1991) investigate this question (without assuming f is linear).

Exercise 5. (1) In fact if $A, B : V \times W \to F$ are bilinear forms such that $A(x, y) = 0$ implies $B(x, y) = 0$, then B is a scalar multiple of A. This is proved in Rothaus (1978). A version of this result for p-linear maps appears in Shaw and Yeadon (1989). In a different direction, Alpers and Schröder (1991) study maps $f : V \to V$ (not assumed linear) which preserve the orthogonality of the vectors in $V^\bullet$ (the set of anisotropic vectors).

(2) See Samuel (1968), de Géry (1970), Lester (1977).

Exercise 6. This is part of the theory of Pfister forms. See Chapter 5.

Exercise 9. Compare Exercise 10.13. Also see Gantmacher (1959), §11.4, Taussky and Zassenhaus (1959), Kaplansky (1969), Theorem 66, Shapiro (1992).

Exercise 12. This generalizes (1.11). Variations include systems where the matrices are skew-symmetric, or have squares which are scalars. Results of these types were obtained by Eddington (1932), Newman (1932), Littlewood (1934), Dieudonné (1953), Kestelman (1961), Gerstenhaber (1964), and Putter (1967). The skew-symmetric case appears in Exercise 2.13. A system of anticommuting matrices whose squares are scalars becomes a representation of a Clifford algebra, and the bounds are determined by the dimension of an irreducible module. Another variation on the problem is considered by Eichhorn (1969), (1970).

Exercises 14. This Geometry Lemma was pointed out to me by A. Wadsworth with further comments by D. Leep.

Exercise 15 (1), (2) appears in Witt (1937). There is a related Transversality Theorem for quadratic forms over semi-local rings, due to Baeza and Knebusch.

Exercise 16. Compare (7.16).

Exercise 17. (4) Wonenburger (1962b).

(5) This lemma appears in Coxeter (1946). It is also valid in more general quaternion algebras.

Exercise 21. **Remark.** There is very little control on the denominators that arise in the process of extending the similarity B to the full matrix A. Even writing out an explicit 16-square identity having 9 bilinear terms seems difficult. There are so many choices for extending the given 9×16 matrix to a 16×16 matrix that nothing interesting seems to arise.

One can generalize these results to formulas $\varphi(X)\varphi(Y) = \varphi(Z)$ for any Pfister form φ. There are similar results on multiplication formulas for hyperbolic forms, but difficulties arise in cases of singular spaces of similarities.

Exercise 22. Details and further results about isoclinic planes are given in: Wong (1961), Wolf (1963), Tyrrell and Semple (1971), Shapiro (1978b), and Wong and Mok (1990). Isoclinic spaces are briefly mentioned after (15.23) below.

Exercise 23. These ideas follow Wong and Mok (1990), and Yiu (1993). Yiu proves:

Theorem. *Every maximal subspace in* $\mathrm{Sim}(\mathbb{R}^n)$ *is maximal as a subspace of nonsingular matrices.*

The proof uses homotopy theory. Related results appear in Adams, Lax and Philips (1965).

Exercise 24. Information on automorphisms appears in Jacobson (1958). Jacobson defines the inner automorphisms of an octonion algebra to be the ones defined in (3), and he proves that every automorphism is inner. That construction of inner automorphisms works for any alternative algebra. Also see Exercise 8.16 below. Automorphisms may also be considered geometrically. For the real octonion division algebra $\mathbb{K}$ the group $\mathrm{Aut}(\mathbb{K})$ is a compact Lie group of dimension 14, usually denoted G_2. The map $\sigma_a(x) = axa^{-1}$ is not often an automorphism of an octonion algebra. In fact, H. Brandt proved that if $[a] = 1$ then σ_a is an automorphism $\iff a^6 = 1$. Proofs appear in Zorn (1935) and Khalil (1993).

Exercise 25. See Schafer (1954), Adem (1978b) and Moreno (1998). The Cayley–Dickson algebras are also mentioned in Chapter 13. The 2^n basis elements e_i of Exercise 25 (4) can instead be indexed by $V = \mathbb{F}_2^n$, using notation as in (1.11). If $e_i^2 = -1$ then: $e^\Delta \cdot e^\Gamma = (-1)^{\beta(\Delta,\Gamma)} e^{\Delta+\Gamma}$ for some map $\beta : V \times V \to \mathbb{F}_2$. Gordan et al. (1993) discuss the following question: If $n = 3$, which maps β yield an octonion algebra? These results can also be cast in terms of intercalate matrices defined in Chapter 13. A related situation for Clifford algebras is mentioned in Exercise 3.19.

Exercise 26. The proposition (due to Artin) follows Schafer (1966), pp. 27–30. See also Zhevlakov et al. (1982), p. 36.

Chapter 2

Amicable Similarities

Analysis of compositions of quadratic forms leads us quickly to the Hurwitz–Radon function $\rho(n)$ as defined in Chapter 0. This function enjoys a property sometimes called "periodicity 8". That is: $\rho(16n) = 8 + \rho(n)$. This and similar properties of the function $\rho(n)$ can be better understood in a more general context. Instead of considering $s - 1$ skew symmetric maps as in the original Hurwitz Matrix Equations (1.7), we allow some of the maps to be symmetric and some skew symmetric. This formulation exposes some of the symmetries of the situation which were not evident at the start. For example we can use these ideas to produce explicit 8-square identities without requiring previous knowledge of octonion algebras.

2.1 Definition. Two (regular) subspaces $S, T \subseteq \text{Sim}(V, q)$ are *amicable* if

$$\tilde{f}g = \tilde{g}f \quad \text{for every } f \in S \text{ and } g \in T.$$

In this case we write $(S, T) \subseteq \text{Sim}(V, q)$. If σ and τ are quadratic forms, the notation

$$(\sigma, \tau) < \text{Sim}(V, q)$$

means that there is a pair $(S, T) \subseteq \text{Sim}(V, q)$ where the induced quadratic forms on S and T are isometric to σ and τ, respectively.

It follows from the definition that if $(S, T) \subseteq \text{Sim}(V)$ and $h, k \in \text{Sim}^\bullet(V)$, then $(hSk, hTk) \subseteq \text{Sim}(V)$. On the level of quadratic forms this says:

$$\text{If } (\sigma, \tau) < \text{Sim}(q) \text{ and } d \in G_F(q) \text{ then } (\langle d\rangle\sigma, \langle d\rangle\tau) < \text{Sim}(q).$$

If $S \neq 0$ we may translate by some f to assume $1_V \in S$. Since this normalization is so useful we give it a special name.

2.2 Definition. An *(s, t)-family* on (V, q) is a pair $(S, T) \subseteq \text{Sim}(V, q)$ where $\dim S = s$, $\dim T = t$ and $1_V \in S$. If $(\sigma, \tau) < \text{Sim}(V, q)$ and σ represents 1, we abuse the notation and say that (σ, τ) is an (s, t)-family on (V, q).

2.3 Lemma. *Suppose* $\sigma \simeq \langle 1, a_2, \dots, a_s\rangle$ *and* $\tau \simeq \langle b_1, \dots, b_t\rangle$. *Then* $(\sigma, \tau) <$ $\text{Sim}(V, q)$ *if and only if there exist* $f_2, \dots, f_s; g_1, \dots, g_t$ *in* $\text{End}(V)$ *satisfying the* *following conditions:*

$$\tilde{f}_i = -f_i, \quad f_i^2 = -a_i 1_V \quad \text{for } 2 \le i \le s.$$
$$\tilde{g}_j = g_j, \quad g_j^2 = b_j 1_V \quad \text{for } 1 \le j \le t.$$

The $s+t-1$ *maps* $f_2, \dots, f_s; g_1, \dots, g_t$ *pairwise anti-commute.*

Proof. If $(\sigma, \tau) < \mathrm{Sim}(V, q)$ let $(S, T) \subseteq \mathrm{Sim}(V, q)$ be the corresponding amicable pair. Since σ represents 1, we may translate by an isometry to assume $1_V \in S$. Let $\{1_V, f_2, \dots, f_s\}$ and $\{g_1, \dots, g_t\}$ be orthogonal bases of S and T, respectively, corresponding to the given diagonalizations. The conditions listed above quickly follow. The converse is also clear. □

An (s, t)-family corresponds to a system of $s + t - 1$ anti-commuting matrices where $s - 1$ of them are skew-symmetric and t of them are symmetric. Stated that way our notation seems unbalanced. The advantage of the terminology of (s, t)-families is that s and t behave symmetrically:

$$(\sigma, \tau) < \mathrm{Sim}(V, q) \quad \text{if and only if} \quad (\tau, \sigma) < \mathrm{Sim}(V, q).$$

Example 1.8 provides an explicit $(2, 2)$-family on a 2-dimensional space: $(\langle 1, d\rangle, \langle 1, d\rangle) < \mathrm{Sim}(\langle 1, d\rangle)$. More trivially, $(\langle 1\rangle, \langle 1\rangle) < \mathrm{Sim}(V)$ for every space V.

2.4 Basic sign calculation. Continuing the notation of (2.3), let $z = f_2 \dots f_s g_1 \dots g_t$. Let $\det(\sigma)\det(\tau) = \langle d\rangle$. Then $\tilde{z} = \pm z$ and $\mu(z) = \tilde{z}z = d$, up to a square factor. In fact, $\tilde{z} = z$ if and only if $s \equiv t$ or $t + 1 \pmod 4$.

Proof. The formula for $\tilde{z}z$ is clear since $\tilde{}$ reverses the order of products. If $e_1, e_2, \dots, e_n$ is a set of n anti-commuting elements, then $e_n \dots e_2 e_1 = (-1)^k e_1 e_2 \dots e_n$ where $k = (n-1) + (n-2) + \dots + 2 + 1 = n(n-1)/2$. Since z is a product of $s + t - 1$ anti-commuting elements and the tilde $\tilde{}$ produces another minus sign for $s - 1$ of those elements, we find that $\tilde{z} = (-1)^{s-1}(-1)^k z$ where $k = (s+t-1)(s+t-2)/2$. The stated calculation of the sign is now a routine matter. □

2.5 Expansion Lemma. *Suppose* $(\sigma, \tau) < \mathrm{Sim}(V)$ *with* $\dim\sigma = s$ *and* $\dim\tau = t$. *Let* $\det(\sigma)\det(\tau) = \langle d\rangle$.

If $s \equiv t - 1 \pmod 4$, *then* $(\sigma \perp \langle d\rangle, \tau) < \mathrm{Sim}(V)$.

If $s \equiv t + 1 \pmod 4$, *then* $(\sigma, \tau \perp \langle d\rangle) < \mathrm{Sim}(V)$.

If (σ', τ') *is the amicable pair obtained from* (σ, τ) *then* $\dim\sigma' \equiv \dim\tau' \pmod 4$ *and* $d\sigma' = d\tau'$.

Proof. First suppose σ represents 1, arrange $1_V \in S$ and choose bases as in (2.3). Let $z = f_2 \dots f_s g_1 \dots g_t$ as before. Then $z \in \mathrm{Sim}^\bullet(V)$ and $\mu(z) = d$, up to a square factor. If $s + t$ is odd, then z anti-commutes with each f_i and g_j. If $\tilde{z} = -z$ then z can be adjoined to S while if $\tilde{z} = z$ then it can be adjoined to T. The congruence conditions follow from (2.4) and the properties of (σ', τ') follow easily. □

2.6 Shift Lemma. *Let σ, τ, φ, ψ be quadratic forms and suppose* $\dim \varphi \equiv \dim \psi \pmod 4$. *Let* $\langle d \rangle = (\det \varphi)(\det \psi)$. *Then:*

$$(\sigma \perp \varphi, \tau \perp \psi) < \mathrm{Sim}(V, q) \text{ if and only if } (\sigma \perp \langle d\rangle\psi, \tau \perp \langle d\rangle\varphi) < \mathrm{Sim}(V, q).$$

This remains valid when φ or ψ is zero. That is, if α is a quadratic form with $\dim \alpha \equiv 0 \pmod 4$ and $\langle d \rangle = \det \alpha$ then:

$$(\sigma \perp \alpha, \tau) < \mathrm{Sim}(q) \text{ if and only if } (\sigma, \tau \perp \langle d\rangle\alpha) < \mathrm{Sim}(q).$$

This shifting result exhibits some of the flexibility of these families: an $(s+4, t)$-family is equivalent to an $(s, t+4)$-family.

Proof of 2.6. Suppose $(S \perp H, T \perp K) \subseteq \mathrm{Sim}(V, q)$ and $a \equiv b \pmod 4$, where $a = \dim H$ and $b = \dim K$. We may assume $S \neq 0$. (For if $T \neq 0$ interchange the eigenspaces. If $S = T = 0$ the lemma is clear.) Scale by suitable f to assume $1_V \in S$. Choose orthogonal bases $\{1_V, f_2, \dots, f_s, h_1, \dots, h_a\}$ and $\{g_1, \dots, g_t, k_1, \dots, k_b\}$ and define $y = h_1h_2 \dots h_ak_1 \dots k_b$. Then $\tilde{y} = y$ and y commutes with elements of S and T, and anticommutes with elements of H and K. Therefore $(S + yK, T + yH) \subseteq \mathrm{Sim}(V, q)$. The converse follows since the same operation applied again leads back to the original subspaces. □

2.7 Construction Lemma. *Suppose* $(\sigma, \tau) < \mathrm{Sim}(q)$ *where σ represents* 1. *If $a \in F^\bullet$ then*

$$(\sigma \perp \langle a\rangle, \tau \perp \langle a\rangle) < \mathrm{Sim}(q \otimes \langle\langle a\rangle\rangle).$$

Proof. Recall that $\langle\langle a\rangle\rangle = \langle 1, a\rangle$ is a binary form. Let (S, T) be an (s, t)-family on (V, q) corresponding to (σ, τ), and express $S = F1_V \perp S_1$. Recall the $(2, 2)$-family constructed in Example 1.8 given by certain 2×2 matrices f_2, g_1 and g_2 in $\mathrm{Sim}(\langle\langle a\rangle\rangle)$. We may verify that $S' = F(1_V \otimes 1) + S_1 \otimes g_1 + F(1_V \otimes f_2)$ and $T' = T \otimes g_1 + F(1_V \otimes g_2)$ does form an $(s+1, t+1)$-family on $q \otimes \langle\langle a\rangle\rangle$. Compare Exercise 1. □

2.8 Corollary. *If $\varphi \simeq \langle\langle a_1, \dots, a_m\rangle\rangle$ is a Pfister form of dimension 2^m, then there is an $(m+1, m+1)$-family $(\sigma, \sigma) < \mathrm{Sim}(\varphi)$, where $\sigma \simeq \langle 1, a_1, \dots, a_m\rangle$.*

Proof. Starting from the $(1, 1)$-family on $\langle 1\rangle$, repeat the Construction Lemma m times. □

The construction here is explicit. The $2m+2$ basis elements of this family can be written out as $2^m \times 2^m$ matrices. Each entry of one of these matrices is either 0 or $\pm a_{i_1} \dots a_{i_r}$ for some $1 \le i_1 < \dots < i_r \le m$. In particular if every $a_i = 1$ then the matrix entries lie in $\{0, 1, -1\}$.

These lemmas can be used to construct large spaces of similarities. Suppose φ is an m-fold Pfister form as above. Then there is an $(m+1, m+1)$-family $(\sigma, \sigma) < \mathrm{Sim}(\varphi)$. Now use the Shift Lemma to shift as much as possible to the left. The resulting sizes are:

$$\left.\begin{array}{rl} (2m+1, 1) & \text{if } m \equiv 0 \\ (2m, 2) & \text{if } m \equiv 1 \\ (2m-1, 3) & \text{if } m \equiv 2 \\ (2m+2, 0) & \text{if } m \equiv 3 \end{array}\right\} \pmod 4$$

Ignoring the "t" parts of these families we get some large subspaces of $\mathrm{Sim}(\varphi)$. In the case $m \equiv 2$ we have $2m-1 \equiv 3 \pmod 4$, and the Expansion Lemma furnishes a subspace of dimension $2m$. We have found subspaces of dimension $\rho(2^m)$.

2.9 Proposition. *Suppose $n = 2^m n_0$ where n_0 is odd. Suppose q is a quadratic form of dimension n expressible as $q \simeq \varphi \otimes \gamma$ where φ is an m-fold Pfister form and* $\dim \gamma = n_0$. *Then there exists $\sigma < \mathrm{Sim}(q)$ with* $\dim \sigma = \rho(n)$.

Proof. From the definition of $\rho(n)$ given in Chapter 0 and the remarks above we see that there exists $\sigma < \mathrm{Sim}(\varphi)$ with $\dim \sigma = \rho(2^m) = \rho(n)$. Then also $\sigma < \mathrm{Sim}(q)$ as mentioned in (1.5). □

Using a little linear algebra we prove the following converse to the Construction Lemma.

2.10 Eigenspace Lemma. *Suppose $(\sigma \perp \langle a \rangle, \tau \perp \langle a \rangle) < \mathrm{Sim}(q)$ is an $(s+1, t+1)$-family, where $s \ge 1$. Then $q \simeq \varphi \otimes \langle\langle a \rangle\rangle$ for some quadratic form φ such that $(\sigma, \tau) < \mathrm{Sim}(\varphi)$.*

Proof. Translating the given family if necessary we may assume it is given by $(S, T) \subseteq \mathrm{Sim}(V, q)$ where $\{1_V, f_2, \dots, f_s, f\}$ and $\{g_1, \dots, g_t, g\}$ are orthogonal bases and $\mu(f) = \mu(g) = a$. Then $\tilde{f} = -f$ and $f^2 = -a1_V$, $\tilde{g} = g$ and $g^2 = a1_V$. Then $h = f^{-1}g = -a^{-1}fg$ satisfies $\tilde{h} = h$ and $h^2 = 1_V$. Let U and U' be the ± 1-eigenspaces for h. Since $\tilde{h} = h$ these spaces are orthogonal and $V = U \perp U'$. Let φ, φ' be the quadratic forms on U and U' induced by q, so that $q \simeq \varphi \perp \varphi'$. Since f anti-commutes with h we have $f(U) = U'$ and $f(U') = U$. Consequently $\dim U = \dim U' = \frac{1}{2}\dim V$, $\varphi \simeq \langle a \rangle \varphi$ and $q \simeq \varphi \otimes \langle\langle a \rangle\rangle$. Furthermore $1_V, f_2, \dots, f_s, g_1, \dots, g_t$ all commute with h so they preserve U. Their restrictions to U provide the family $(\sigma, \tau) < \mathrm{Sim}(U, \varphi)$. □

We now have enough information to find all the possible sizes of (s, t)-families on quadratic spaces of dimension n.

2.11 Theorem. *Suppose $n = 2^m n_0$ where n_0 is odd. There exists an (s, t)-family on some quadratic space of dimension n if and only if $s \geq 1$ and one of the following holds:*

(1) $s + t \leq 2m$,

(2) $s + t = 2m + 1$ *and* $s \equiv t - 1$ *or* $t + 1 \pmod 8$,

(3) $s + t = 2m + 2$ *and* $s \equiv t \pmod 8$.

Proof. ("if") Suppose that there exist numbers s', t' such that $s \leq s', t \leq t', s' + t' = 2m + 2$ and $s' \equiv t' \pmod 8$. Then there is an (s', t')-family and hence an (s, t)-family on some quadratic space of dimension n. To see this first use the Construction Lemma to get an $(m + 1, m + 1)$-family on a space of dimension 2^m and tensor it up (as in (1.6)) to get such a family on an n-dimensional space (V, q). A suitable application of the Shift Lemma then provides an (s', t')-family in $\mathrm{Sim}(q)$.

If s, t satisfy one of the given conditions then such s', t' do exist, except in the case $s + t = 2m$ and $s \equiv t + 4 \pmod 8$. In this case, suppose $s \geq 2$. Then $s - 1 \equiv t + 3 \pmod 8$ and the work above implies that there is an $(s - 1, t + 3)$-family in $\mathrm{Sim}(q)$ for some n-dimensional form q. Restrict to an $(s - 1, t)$-family and apply the Expansion Lemma 2.5 to obtain an (s, t)-family. Similarly if $t \geq 1$ there is an $(s + 3, t - 1)$-family which restricts and expands to an (s, t)-family.

("only if") Suppose there is an (s, t)-family on some n-dimensional space and proceed by induction on m. If $m = 0$ Proposition 1.10 implies $s, t \leq 1$ and therefore $s + t \leq 2$. If $s + t = 2$ then certainly $s = t$. Similarly, if $m = 1$ Proposition 1.10 implies $s, t \leq 2$ so that $s + t \leq 4$. If $s + t = 4$ then $s = t$.

Suppose $m \geq 2$. If $s + t \leq 4$ the conditions are satisfied. Suppose $s + t > 4$ and apply the Shift Lemma and the symmetry of (s, t) and (t, s) to arrange $s \geq 2$ and $t \geq 1$. If $(\sigma, \tau) < \mathrm{Sim}(q)$ is the given (s, t)-family where $\dim q = n$, pass to an extension field of F to assume σ and τ represent a common value. The Eigenspace Lemma then implies that there is an $(s - 1, t - 1)$-family on some space of dimension $n/2$. The induction hypothesis implies the required conditions. □

2.12 Corollary. *If* $\sigma < \mathrm{Sim}(q)$ *where* $\dim q = n$ *then* $\dim \sigma \leq \rho(n)$.

Proof. Suppose $s = \dim \sigma > \rho(n)$ where $n = 2^m n_0$ and n_0 is odd. If $m \equiv 0 \pmod 4$ then $s > \rho(n) = 2m + 1$ so there is a $(2m + 2, 0)$-family in $\mathrm{Sim}(q)$. But $2m + 2 \equiv 2 \pmod 8$ contrary to the requirement in the Theorem 2.11. The other cases follow similarly. □

This corollary is the generalization of the Hurwitz–Radon Theorem to quadratic forms over any field (of characteristic not 2). A refinement of Theorem 2.11 appears in (7.8) below.

Theorem 2.11 contains all the information about possible sizes of families. This information can be presented in a number of ways. For example, given n and t we can determine the largest s for which an (s, t)-family exists on some n-dimensional quadratic space.

2.13 Definition. Given n and t, define $\rho_t(n)$ to be the maximum of 0 and the value indicated in the following table. Here $n = 2^m n_0$ where n_0 is odd.

$m \pmod 4$	$\rho_t(n)$
$m \equiv t$	$2m + 1 - t$
$m \equiv t + 1$	$2m - t$
$m \equiv t + 2$	$2m - t$
$m \equiv t + 3$	$2m + 2 - t$

2.14 Corollary. *Suppose (V, q) is an n-dimensional quadratic space.*

(1) *If there is an (s, t)-family in* $\mathrm{Sim}(V, q)$ *then $s \le \rho_t(n)$.*

(2) *Suppose $s = \rho_t(n)$. Then there is some (s, t)-family in* $\mathrm{Sim}(V, q)$, *provided that q can be expressed as a product $\varphi \otimes \gamma$ where φ is a Pfister form and* $\dim \gamma$ *is odd.*

The proof is left as an exercise for the reader.

Here is another way to codify this information. Given s, t we determine the minimal dimension of a quadratic space admitting an (s, t)-family.

2.15 Corollary. *Let $s \ge 1$ and $t \ge 0$ be given. The smallest n such that there is an (s, t)-family on some n-dimensional quadratic space is $n = 2^{\delta(s,t)}$ where the value $m = \delta(s, t)$ is defined as follows.*

Case $s + t$ is even:

if $s \equiv t \pmod 8$ then $s + t = 2m + 2$ and $\delta(s, t) = \frac{s+t-2}{2}$,

if $s \not\equiv t \pmod 8$ then $s + t = 2m$ and $\delta(s, t) = \frac{s+t}{2}$.

Case $s + t$ is odd:

if $s \equiv t \pm 1 \pmod 8$ then $s + t = 2m + 1$ and $\delta(s, t) = \frac{s+t-1}{2}$,

if $s \equiv t \pm 3 \pmod 8$ then $s + t = 2m - 1$ and $\delta(s, t) = \frac{s+t+1}{2}$.

Proof. This is a restatement of Theorem 2.11. The value $\delta(r) = \delta(r, 0)$ was calculated in Exercise 0.6. Further properties of $\delta(s, t)$ appear in Exercise 3. □

If $q = \varphi \otimes \gamma$ where φ is a large Pfister form, then (2.14) implies that $\mathrm{Sim}(V, q)$ admits a large (s, t)-family. We investigate the converse. By Proposition 1.10 subspaces of $\mathrm{Sim}(q)$ provide certain Pfister forms which are tensor factors of q. The following consequence of the Eigenspace Lemma is a similar sort of result: certain (s, t)- families in $\mathrm{Sim}(q)$ provide Pfister factors of q.

2.16 Corollary. *Suppose* $(\sigma \perp \alpha, \tau \perp \alpha) < \mathrm{Sim}(q)$ *where* $\dim \sigma \geq 1$ *and* $\alpha \simeq \langle a_1, \dots, a_k \rangle$. *Then* $q \simeq \langle\langle a_1, \dots, a_k \rangle\rangle \otimes \gamma$ *for some quadratic form* γ *such that* $(\sigma, \tau) < \mathrm{Sim}(\gamma)$.

Proof. Repeated application of (2.10). □

Suppose $\dim q = 2^m n_0$ where n_0 is odd. If there is an $(m+1, m+1)$- family $(\sigma, \tau) < \mathrm{Sim}(q)$ with $\sigma \simeq \tau \simeq \langle 1, a_1, \dots, a_m \rangle$ then $q \simeq \langle\langle a_1, \dots, a_m \rangle\rangle \otimes \gamma$ for some form γ of odd dimension n_0. The Pfister Factor Conjecture is a stronger version of this idea.

2.17 Pfister Factor Conjecture. *Suppose that* q *is a quadratic form of dimension* $n = 2^m n_0$ *where* n_0 *is odd. If there exists an* $(m+1, m+1)$*-family in* $\mathrm{Sim}(q)$ *then* $q \simeq \langle\langle a_1, \dots, a_m \rangle\rangle \otimes \gamma$ *for some quadratic form* γ *of dimension* n_0.

This conjecture is true for $m \leq 2$ by Proposition 1.10. We cannot get much further without more tools. We take up the analysis of this conjecture again in Chapter 9. The Decomposition Theorem and properties of unsplittable modules (Chapters 4–7) reduce the Conjecture to the case $\dim q = 2^m$. Properties of discriminants and Witt invariants of quadratic forms (Chapter 3) can then be used to prove the conjecture when $m \leq 5$. The answer is not known when $m > 5$ over arbitrary fields, but over certain nice fields (e.g. global fields) the conjecture can be proved for all values of m. After learning some properties of the invariants of quadratic forms (stated in (3.21)), the reader is encouraged to skip directly to Chapter 9 to see how they relate to the Pfister Factor Conjecture.

The Conjecture can be stated in terms of compositions of quadratic forms in the earlier sense: If $\dim q = n = 2^m n_0$ as usual and if there is a subspace $\sigma < \mathrm{Sim}(q)$ where $\dim \sigma = \rho(n)$, then there is an $(m+1, m+1)$-family in $\mathrm{Sim}(q)$. In fact in Proposition 7.6 we prove that $\dim \sigma \geq 2m - 1$ suffices to imply that there is an $(m+1, m+1)$-family in $\mathrm{Sim}(q)$. This "expansion" result seems to require some knowledge of algebras and involutions.

Exercises for Chapter 2

1. **Construction Lemma revisited.** (1) Write out the elements of the $(s+1, t+1)$-family in (2.7) using block matrices: if S has basis $\{1, h_2, \dots, h_s\}$ then for example $h_i \otimes g_1 = \begin{pmatrix} h_i & 0 \\ 0 & -h_i \end{pmatrix}$ while $1 \otimes f_2 = \begin{pmatrix} 0 & -a \\ 1 & 0 \end{pmatrix}$.

(2) Let (S, T) and (G, H) be commuting pairs of amicable subspaces of $\mathrm{Sim}(V)$. That is, every element of $S \cup T$ commutes with every element of $G \cup H$. Choose anisotropic $f \in S$, $g \in G$ and define $S_1 = (f)^{\perp}$ and $G_1 = (g)^{\perp}$. Then $(\tilde{f}S_1 + \tilde{g}H, \tilde{g}G_1 + \tilde{f}T)$ is an amicable pair in $\mathrm{Sim}(V)$.

(3) An (s,t)-family on V and an (a,b)-family on W where $s, a \geq 1$ yield an $(s+b-1, t+a-1)$-family on $V \otimes W$.

2. **When σ does not represent 1.** Define the *norm group* $G_F(q) = \{a \in F^{\bullet} : \langle a\rangle q \simeq q\}$, the group of norms of all elements of $\mathrm{Sim}^{\bullet}(q)$. For any form q, $G_F(q) \cdot D_F(q) \subseteq D_F(q)$. If $\sigma < \mathrm{Sim}(q)$ then $D_F(\sigma) \subseteq G_F(q)$. Let σ, τ be quadratic forms over F.

(1) Suppose $c \in D_F(\sigma)$ and let $\sigma_0 = \langle c\rangle\sigma$ and $\tau_0 = \langle c\rangle\tau$. Then $(\sigma, \tau) < \mathrm{Sim}(q)$ if and only if $c \in G_F(q)$ and $(\sigma_0, \tau_0) < \mathrm{Sim}(q)$.

(2) The Expansion Lemma 2.5 remains true without assuming σ represents 1.

(3) If $(S \perp A, T) \subseteq \mathrm{Sim}(V, q)$ where $\dim A = 4$, there is a shifted amicable family $(S, T \perp A') \subseteq \mathrm{Sim}(V, q)$. If $\{h_1, \dots, h_4\}$ is an orthogonal basis of A, what is an explicit basis of A'?

(4) The Construction Lemma remains true without assuming σ represents 1.

(Hint. (4) If $c \in G_F(q)$ then $q \otimes \langle\langle ca\rangle\rangle \simeq q \otimes \langle\langle a\rangle\rangle$.)

3. (1) Deduce the following formulas directly from the early lemmas about (s,t)-families.

$$\delta(s,t) = \delta(t,s) \quad \delta(s+1, t+1) = 1 + \delta(s,t) \quad \delta(s+4, t) = \delta(s, t+4).$$

(2) Recall the definition of $\delta(r)$ from Exercise 0.6: $r \leq \rho(n)$ iff $2^{\delta(r)} \mid n$. Note that $\delta(r+8) = \delta(r) + 4$ and use this to extend the definition to $\delta(-r)$.

Lemma. $\delta(s,t) = t + \delta(s-t)$.

$$(3)\ \delta(s) = \begin{cases} \frac{s-2}{2} & \text{if } s \equiv 0 \\ \frac{s-1}{2} & \text{if } s \equiv \pm 1 \\ \frac{s}{2} & \text{if } s \equiv \pm 2,\ 4 \\ \frac{s+1}{2} & \text{if } s \equiv \pm 3 \end{cases} \pmod 8.$$

4. **More Shift Lemmas.** (1) If $\dim \varphi \equiv 2 + \dim \psi \pmod 4$ in the Shift Lemma then $(\sigma \perp \varphi, \tau \perp \psi)$ is equivalent to $(\sigma \perp \langle d\rangle\varphi, \tau \perp \langle d\rangle\psi)$.

(2) Let $(\sigma, \tau) = (\langle 1, a, b, c\rangle, \langle x, y\rangle)$ and $d = abcxy$. Then we can shift it to $(\sigma', \tau') = (\langle 1, a, b, c, dx, dy\rangle, 0)$. Similarly if $(\sigma, \tau) = (\langle 1, a, b\rangle, \langle x, y, z\rangle)$, and $d = abxyz$ we can shift it to $(\sigma', \tau') = (\langle 1, a, bd\rangle, \langle x, y, zd\rangle)$.

(3) Generalize this idea to other (s, t)-pairs where $s + t$ is even. If $s \equiv t \pmod 4$ explain why $\delta(s, t+2) = \delta(s+2, t)$.

5. (1) If $(\langle 1, a, b\rangle, \langle x\rangle) < \mathrm{Sim}(q)$ then $\langle 1, abx\rangle < \mathrm{Sim}(q)$.

(2) Generalize this observation to $(\sigma, \tau) < \mathrm{Sim}(q)$ where $s \equiv t+2$ or $t+3 \pmod 4$.

(Hint. (1) Examine the element z as in (2.4).)

6. **Alternating spaces.** (1) **Lemma.** *Let σ, τ and α be quadratic forms and V a vector space. Suppose* $\dim \alpha \equiv 2 \pmod 4$. *Then $(\sigma \perp \alpha, \tau) < \mathrm{Sim}(V, q)$ for some quadratic form q on V if and only if $(\sigma, \tau \perp \alpha) < \mathrm{Sim}(V, B)$ for some alternating form B on V.*

(2) **Theorem.** *Suppose $n = 2^m n_0$ where n_0 is odd. There exists an (s, t)-family on some alternating space of dimension n if and only if one of the following holds:*

(i) $s + t \leq 2m$,

(ii) $s + t = 2m + 1$ *and* $s \equiv t - 3$ *or* $t + 3 \pmod 8$,

(iii) $s + t = 2m + 2$ *and* $s \equiv t + 4 \pmod 8$.

(3) Let $\delta'(s, t)$ be the corresponding function for alternating spaces. Then $\delta'(s+2, t) = \delta(s, t+2)$. Note that $\delta(s, t) = \delta'(s, t)$ iff $s \equiv t \pm 2 \pmod 8$. How does Exercise 4 help "explain" this? Does $\delta'(s, t) = t + \delta'(s - t)$?

(4) Let $\rho'(n)$ be the Hurwitz–Radon function for alternating spaces. Note that $\rho'(1) = 0$ and $\rho'(2) = 4$. (See Exercise 1.7.) The formula for $\rho'(n)$ in terms of $m \pmod 4$ is a "cycled" version of the formula for $\rho(n)$. In other words, $\rho'(4n) = 4 + \rho(n)$ whenever $n \geq 1$.

(Hint:. (1) Let $(S \perp A, T) \subseteq \mathrm{Sim}(V, q)$ and z be given as in the Shift Lemma 2.6. Define the form B' on V by: $B'(u, v) = B(u, z(v))$. Then $(S, T \perp A) \subseteq \mathrm{Sim}(V, B')$.)

7. **Hurwitz–Radon functions.** (1) Write out the formulas for $\rho_t'(n)$, the alternating version of the functions $\rho_t(n)$ given in (2.13) and prove the analog of (2.14).

(2) We write $\rho^\lambda(n)$ to denote $\rho(n)$ if $\lambda = 1$ and $\rho'(n)$ if $\lambda = -1$. The following properties of the Hurwitz–Radon functions are consequences of the formulas, assuming in each case that the function values are large enough.

$$\rho_{t+1}^\lambda(2n) = 1 + \rho_t^\lambda(n), \qquad \rho_t^\lambda(n) = 4 + \rho_{t+4}^\lambda(n), \qquad \rho_t^{-\lambda}(4n) = 4 + \rho_t^\lambda(n),$$

$$\rho_t^\lambda(n) = 2 + \rho_{t+2}^{-\lambda}(n), \qquad \max\{\rho(n), \rho'(n)\} = \begin{cases} 2m+1 & \text{if } m \text{ is even,} \\ 2m+2 & \text{if } m \text{ is odd.} \end{cases}$$

(3) Explain each of these formulas more theoretically.

8. **That element "z".** The element $z = z(S) \cdot z(T)$ was used in (2.4) and (2.5). What if a different orthogonal basis is chosen for S? Is there a suitable definition for z when S does not contain 1_V? We use the following result originally due to Witt:

Chain-Equivalence Theorem. *Let (V, q) be a quadratic space with two orthogonal bases $\mathcal{X}$ and $\mathcal{X}'$. Then there exists a chain of orthogonal bases $\mathcal{X} = \mathcal{X}_0, \mathcal{X}_1, \dots, \mathcal{X}_m = \mathcal{X}'$ such that $\mathcal{X}_i$ and $\mathcal{X}_{i-1}$ differ in at most 2 vectors.*

Proofs appear in Scharlau (1985), pp. 64–65 and O'Meara (1963), p. 150. Compare Satz 7 of Witt (1937).

(1) Let $S \subseteq \mathrm{Sim}(V, q)$ be a subspace of dimension s. If $\mathcal{B} = \{f_1, f_2, \dots, f_s\}$ is an orthogonal basis of S, define $z(\mathcal{B}) = f_1 \cdot \tilde{f_2} \cdot f_3 \cdot \tilde{f_4} \dots$ and $w(\mathcal{B}) = \tilde{f_1} \cdot f_2 \cdot \tilde{f_3} \cdot f_4 \dots$.

Lemma. *If $\mathcal{B}'$ is another orthogonal basis, then $z(\mathcal{B}') = \lambda \cdot z(\mathcal{B})$ and $w(\mathcal{B}') = \lambda \cdot w(\mathcal{B})$ for some $\lambda \in F^\bullet$.*

Define $z(S) = z(\mathcal{B})$ and $w(S) = w(\mathcal{B})$. These values are uniquely determined by the subspace S, up to non-zero scalar multiple. Note that $z(S) \cdot z(S)^\sim = w(S) \cdot w(S)^\sim = \det(\sigma) \cdot 1_V$.

Let $z = z(\mathcal{B})$ and $w = w(\mathcal{B})$.

(2) If s is odd: $w = (-1)^{(s-1)/2} \cdot \tilde{z}$. For every $f \in S^\bullet$, $z \cdot \tilde{f} = f \cdot \tilde{z}$.

If s is even: $\tilde{z} = (-1)^{s/2} \cdot z$ and $\tilde{w} = (-1)^{s/2} \cdot w$. For every $f \in S^\bullet$, $zf = fw$.

Consequently if s is even then $z^2 = w^2 = d\sigma \cdot 1_V$.

(3) Suppose $\varphi : S \to S$ is a similarity. Then $\varphi(\mathcal{B})$ is another orthogonal basis and

$$z(\varphi(\mathcal{B})) = (\det \varphi) \cdot z(\mathcal{B}).$$

(4) Let $g, h \in \mathrm{Sim}^\bullet(V, q)$ with $\alpha = \mu(g)\mu(h)$.

If s is odd: $z(g\mathcal{B}h) = \alpha^{(s-1)/2} \cdot g \cdot z(\mathcal{B}) \cdot h$.

If s is even: $z(g\mathcal{B}h) = \alpha^{s/2} \cdot g \cdot z(\mathcal{B}) \cdot g^{-1}$.

(5) Analyze the Expansion and Shift Lemmas using these ideas. (Compare Exercise 2.)

9. **Symmetric similarities.** (1) **Lemma** (Dieudonné). *Suppose $f \in \mathrm{Sim}^\bullet(V, q)$ where $\dim V = 2m$ is even. Then there exists $g \in \mathrm{O}(V, q)$ and a decomposition $V = V_1 \perp \dots \perp V_m$ such that $gf(V_i) = V_i$, $\dim V_i = 2$ and $\widetilde{gf} = gf$. Furthermore, given anisotropic $v \in V$, there is such a decomposition with $v \in V_1$.*

(2) $\langle a \rangle < \mathrm{Sim}(q)$ if and only if $(\langle 1 \rangle, \langle a \rangle) < \mathrm{Sim}(q)$.

(3) If $\dim q$ is even then $G_F(q) \subseteq D_F(\langle\langle -dq \rangle\rangle)$.

(Hint. (1) Assume $\mu(f) \notin F^{\bullet 2}$ and let $V_1 = \mathrm{span}\{v, f(v)\}$. Then V_1 is a regular 2-plane and there exists $g_1 \in \mathrm{O}(V)$ with $g_1 f(v) = f(v)$ and $g_1 f^2(v) = \mu(f)v$. Then $g_1 f$ preserves V_1. Apply induction to construct g. Note that $(gf)^2 = \mu(f)1_V$.

(3) If $a \in G_F(q)$ with $a \notin F^{\bullet 2}$ then $q \simeq \langle x_1 \rangle \langle\langle -d_1 \rangle\rangle \perp \dots \perp \langle x_m \rangle \langle\langle -d_m \rangle\rangle$ where $\langle\langle -d_j \rangle\rangle$ represents a. Then $\langle\langle -a \rangle\rangle$ represents $\langle d_1 d_2 \dots d_m \rangle = dq$.)

10. An *orthogonal design* of order n and type $(s_1, \dots, s_k)$ is an $n \times n$ matrix A with entries from $\{0, \pm x_1, \dots, \pm x_k\}$ such that the rows of A are orthogonal and each row

of A has s_i entries of the type $\pm x_i$. Here the x_i are commuting indeterminates and s_i is a positive integer. Then $A \cdot A^{\top} = \sigma \cdot I_n$ where $\sigma = \sum_{i=1}^{k} s_i x_i^2$.

Proposition. (1) *If there is such a design then* $\langle s_1, \ldots, s_k \rangle < \mathrm{Sim}(n\langle 1 \rangle)$. *In particular* $k \leq \rho(n)$.

(2) *If the* s_i *are positive integers and* $s_1 + \cdots + s_k \leq \rho(n)$ *then there is an orthogonal design of order* n *and type* $(s_1, \ldots, s_k)$.

(Hint. The Construction, Shift and Expansion Lemmas provide $A_i \in \mathbb{M}_n(\mathbb{Z})$, for $1 \leq i \leq \rho(n)$, which satisfy the Hurwitz Matrix Equations. This yields an integer composition formula as in Exercise 0.4, hence an orthogonal design of order n and type $(1, 1, \ldots, 1)$. Set some of the variables equal.)

11. **Constructing composition algebras.** (1) From the Construction and Expansion Lemmas there is a 4-dimensional subspace $\langle 1, a, b, ab \rangle < \mathrm{Sim}(\langle\langle a, b \rangle\rangle)$. This induces a 4-dimensional composition algebra (see Exercise 1.1). This turns out to be the usual quaternion algebra.

(2) The Construction and Shift Lemmas provide an explicit $\sigma < \mathrm{Sim}(\langle\langle a, b, c \rangle\rangle)$ with $\dim \sigma = 8$, and we get an induced 8-dimensional composition algebra. This turns out to be the standard octonion algebra.

12. **Amicable spaces of rectangular matrices.** Let V, W be two regular quadratic spaces and consider subspaces $S, T \subseteq \mathrm{Sim}(V, W)$ as in Exercise 1.2. Suppose S and T are "amicable" in the sense generalizing (2.1). If $\dim S = s$, $\dim T = t$, $\dim V = r$ and $\dim W = n$ we could call this an (s, t)-family of $n \times r$ matrices.

(1) If there is an (s, t)-family of $n \times r$ matrices then there is an $(s+1, t+1)$-family of $2n \times 2r$ matrices. This is the analog of the Construction Lemma.

(2) Why do the analogs of the Expansion and Shift Lemmas fail in this context?

13. **Systems of skew-symmetric matrices. Definition.** $\alpha_F(n) = \max\{t :$ there exist $A_1, \ldots, A_t \in \mathrm{GL}_n(F)$ such that $A_i^{\top} A_j + A_j^{\top} A_i = 0$ whenever $i \neq j\}$. Certainly $\rho(n) \leq \alpha_F(n)$.

Open question. Is this always an equality?

(1) $\alpha_F(n) - 1 = \max\{k :$ there exist k skew-symmetric elements of $\mathrm{GL}_n(F)$ which pairwise anticommute$\}$.

(2) If $n = 2^m n_0$ where n_0 is odd, then $\alpha_F(n) \leq 2m + 2$.

(3) If n_0 is odd then $\alpha_F(n_0) = 1$ and $\alpha_F(2n_0) = 2$.

(4) **Proposition.** $\alpha_F(2^m) = \rho(2^m)$.

(5) Let $\{I, f_2, \ldots, f_s\}$ be a system over F as above. Let $V = F^n$ with the standard inner product, and view $f_i \in \mathrm{End}(V)$. Define $W \subseteq V$ to be invariant if $f_i(W) \subseteq W$ for every i. The system is "decomposable" if V is an orthogonal sum of non-zero invariant subspaces.

Lemma. *If* $\{f_i\}$ *is an indecomposable system over* $\mathbb{R}$, *then* $f_i^2 =$ *scalar.*

(6) **Proposition.** *If* F *is formally real then* $\alpha_F(n) = \rho(n)$.

(Hint. (2) See Exercise 1.12.

(3) For $2n_0$ apply the "Pfaffians" defined in Chapter 10. If f, g are nonsingular, skew-symmetric, anti-commuting, then $pf(fg) = pf(gf) = pf(-fg)$.

(4) If $\alpha_F(2^m) > \rho(2^m)$ there exist $2m$ skew-symmetric, anti-commuting elements $f_i \in \mathrm{GL}_{2^m}(F)$. Then $f_i^2 =$ scalar as in Exercise 1.12(2), and Hurwitz–Radon applies.

(5) The Spectral Theorem implies V is the orthogonal sum of the eigenspaces of the symmetric matrix f_j^2. Since every f_i commutes with f_j^2 these eigenspaces are invariant.

(6) Generalize (5) to real closed fields and note that F embeds in a real closed field. Apply Hurwitz–Radon. Over $\mathbb{R}$ we could apply Adams' theorem (see Exercise 0.7) to the orthogonal vector fields $f_2(v), \dots, f_s(v)$ on S^{n-1}.)

14. A *Hadamard design* of order n on k letters $\{x_1, \dots, x_k\}$ is an $n \times n$ matrix H such that each entry is some $\pm x_j$ and the inner product of any two distinct rows is zero. If there is such a design then there exist $n \times n$ matrices $H_1, \dots, H_k$ with entries in $\{0, 1, -1\}$ such that $H_j \cdot H_i^\top + H_i \cdot H_j^\top = 0$ if $i \neq j$ and $H_i \cdot H_i^\top =$ diagonal.

Proposition. *If there exists a Hadamard design of order n on k letters, each of which occurs at least once in every column of H, then $k \leq \rho(n)$.*

(Hint. Note that each H_i is nonsingular and apply Exercise 13.)

15. **Hermitian compositions.** A *hermitian (r, s, n)-formula* is:

$$(|x_1|^2 + \cdots + |x_r|^2) \cdot (|y_1|^2 + \cdots + |y_s|^2) = |z_1|^2 + \cdots + |z_n|^2$$

where $X = (x_1, \dots, x_r)$ and $Y = (y_1, \dots, y_s)$ are systems of complex indeterminants over $\mathbb{C}$, and each z_k is bilinear in X, Y. Such a formula can be viewed as a bilinear map $f : \mathbb{C}^r \times \mathbb{C}^s \to \mathbb{C}^n$ satisfying $|f(x, y)| = |x| \cdot |y|$. We consider three versions of bilinearity:

Type 1: each z_k is bilinear in $(X, \bar{X})$ and $(Y, \bar{Y})$.

Type 2: each z_k is bilinear in $(X, \bar{X})$ and Y.

Type 3: each z_k is bilinear in X and Y.

Note that if $X = X_1 + iX_2$ then z is linear (over $\mathbb{C}$) in $(X, \bar{X})$ if and only if z is $\mathbb{C}$-linear in the system of real variables (X_1, X_2). For example $z_1 = x_1y_1 + x_2y_2$ and $z_2 = \bar{x}_1y_2 - \bar{x}_2y_1$ provides a $(2, 2, 2)$-formula of types 1 and 2.

Proposition. (1) *A hermitian (r, s, n)-formula of type* 1 *exists if and only if there exists an ordinary $(2r, 2s, 2n)$-formula over $\mathbb{R}$.*

(2) *A hermitian (r, s, n)-formula of type* 2 *exists if and only if there exist two amicable subspaces of dimension r in* $\mathrm{Sim}^{\mathrm{herm}}(\mathbb{C}^s, \mathbb{C}^n)$, *the set of hermitian similarities.*

(3) *A hermitian (r, s, n)-formula of type* 3 *exists if and only if $n \geq rs$.*

(Hints. (2) The formula exists iff there is an $n \times s$ matrix A whose entries are linear forms in $(X, \bar{X})$ and satisfying $A^* \cdot A = \left(\sum_1^r \bar{x}_j x_j\right) I_s$, where $*$ denotes the conjugate-transpose. Express $x_j = u_j + v_j\sqrt{-1}$ where u_j, v_j are real variables. Then

$A = \sum_1^r u_j B_j + v_j\sqrt{-1}C_j$, where B_j and C_j are $n \times s$ matrices over $\mathbb{C}$. Then $S = \mathrm{span}\{B_1, \dots, B_r\}$ and $T = \mathrm{span}\{C_1, \dots, C_r\}$ are the desired subspaces.

(3) Here we get the same equation for A, where $A = x_1A_1 + \dots + x_rA_r$ and each A_j is a complex $n \times s$ matrix. Then $A_j^*A_j = I_s$ and $A_j^*A_k = 0$ if $j \neq k$. Choose $A_1 = \begin{pmatrix} I_s \\ 0 \end{pmatrix}$.)

16. Consider the analogous composition formulas of the type

$$(x_1^2 + \dots + x_r^2) \cdot (|y_1|^2 + \dots + |y_s|^2) = |z_1|^2 + \dots + |z_n|^2$$

where X is a system over $\mathbb{R}$, Y is a system over $\mathbb{C}$ and each z_k is bilinear in X, Y. Is the existence of such a formula equivalent to the existence of $A_2, \dots, A_r \in \mathrm{GL}_n(\mathbb{C})$ which are anti-hermitian ($A_j^* = -A_j$), unitary ($A_j^* \cdot A_j = 1$) and pairwise anticommute?

Notes on Chapter 2

The notion of amicable similarities was pointed out to me by W. Wolfe in 1974 (see Wolfe (1976)). He introduced the term "amicable" in analogy with a related idea in combinatorics.

The idea of allowing some symmetric elements in the Hurwitz Matrix Equations has occurred independently in Ławrynowicz and Rembieliński (1990). They do this to obtain some further symmetries in their approach to the theory.

Exercise 9 follows Dieudonné (1954). Compare the appendix of Elman and Lam (1974). The decomposition of V is closely related to the "β-decomposition" in Corollary 2.3 of Elman and Lam (1973b). See Exercise 5.7 below.

Exercise 10. Orthogonal designs are investigated extensively in Geramita and Seberry (1979).

Exercise 13. The lemma in (5) follows Putter (1967). Further information on anticommuting matrices is given in the Notes on Chapter 1.

Exercise 14 generalizes ideas of Storer (1971).

Exercise 15–16. The observation on $\mathbb{C}$-bilinear hermitian compositions in 15 (3) is due to Alarcon and Yiu (1993). Hermitian compositions are discussed further in Exercise 4.7. Compositions as in Exercise 16 are also considered in Furuoya et al. (1994). They work with a slightly more general situation, allowing forms of arbitrary signature over $\mathbb{R}$. Compare Exercise 4.7.

Chapter 3

Clifford Algebras

This is essentially a reference chapter, containing the definitions and basic properties of Clifford algebras and Witt invariants, along with some related technical results that will be used later. The reader should have some acquaintance with central simple algebras, the Brauer group $\mathrm{Br}(F)$ and the Witt ring $W(F)$. This background is presented in a number of texts, including Lam (1973), Scharlau (1985).

Clifford algebras have importance in algebra, geometry and analysis. We need the basic algebraic properties of Clifford algebras over an arbitrary field F. We include the proofs of some of the basic results, assuming familiarity with the classical theory of central simple algebras. The exposition is simplified since we assume that the characteristic of F is not 2.

Every F-algebra considered here is a finite dimensional, associative F-algebra with an identity element denoted by 1. The field F is viewed as a subset of the algebra. An unadorned tensor product $\otimes$ always denotes $\otimes_F$, the tensor product over F.

The first non-commutative algebra was the real quaternion algebra discovered by Hamilton in 1843. That motivates the general definition of a quaternion algebra over F.

Definition. If $a, b \in F^\bullet$, the quaternion algebra $A = \left(\frac{a,b}{F}\right)$ is the associative F-algebra with generators i, j satisfying the multiplication rules:

$$i^2 = a, \quad j^2 = b \quad \text{and} \quad ij = -ji.$$

The associativity implies that A is spanned by $\{1, i, j, ij\}$ and it follows that $\dim A = 4$. An element of A is called "pure" if its scalar part is 0. Then the set $A^\circ = A_0$ of pure quaternions is the span of $\{i, j, ij\}$. Direct calculation shows that

$$A_0 = \{u \in A : u \notin F^\bullet \text{ and } u^2 \in F\}.$$

Consequently A_0 is independent of the choice of generators i, j. Define the "bar" map on A to be the linear map which acts as 1 on F and as -1 on A_0. Then $\bar{\bar{a}} = a$ and: $\bar{a} = a$ if and only if $a \in F$. Another calculation shows: $\overline{uv} = \bar{v} \cdot \bar{u}$. Then "bar" is the unique anti-automorphism of A with $\bar{i} = -i$ and $\bar{j} = -j$.

The norm $N : A \to F$, defined by $N(a) = \bar{a} \cdot a$, is a quadratic form on A and calculation on the given basis shows that $(A, N) \simeq \langle\langle -a, -b \rangle\rangle$. Moreover, $N(ab) = N(a) \cdot N(b)$ so that A is a composition algebra as described in Chapter 1, Appendix.

3.1 Lemma. (1) $\left(\frac{a,b}{F}\right)$ *is split if and only if its norm form* $\langle\langle -a, -b \rangle\rangle$ *is hyperbolic.*

(2) *The isomorphism class of* $\left(\frac{a,b}{F}\right)$ *is determined by the isometry class of* $\langle\langle -a, -b \rangle\rangle$.

Proof. An explicit isomorphism $\left(\frac{1,-1}{F}\right) \to \mathbb{M}_2(F)$ is provided by sending $i \mapsto \begin{pmatrix} 0 & 1 \\ 1 & 0 \end{pmatrix}$ and $j \mapsto \begin{pmatrix} 0 & 1 \\ -1 & 0 \end{pmatrix}$. It suffices to prove (2). An isomorphism of quaternion algebras preserves the pure parts, so it commutes with "bar" and is an isometry for the norms. If A is a quaternion algebra and $(A, N) \simeq \langle\langle -a, -b \rangle\rangle$, Witt Cancellation implies $(A_0, N) \simeq \langle -a, -b, ab \rangle$, so there exist orthogonal elements i, j in A_0 with $N(i) = -a$ and $N(j) = -b$. Therefore $i^2 = a$, $j^2 = b$ and $ij + ji = 0$ and these generators provide an isomorphism $A \cong \left(\frac{a,b}{F}\right)$. □

These results on quaternion algebras, together with the systems of equations in (1.6) and (2.3), help to motivate the investigation of algebras having generators $\{e_1, e_2, \dots, e_n\}$ which anticommute and satisfy $e_i^2 \in F^\bullet$. An efficient method for defining these algebras is to use their universal property.

Suppose (V, q) is a quadratic space over F and A is an F-algebra. A linear map $\iota : V \to A$ is *compatible with* q if it satisfies:

$$\iota(v)^2 = q(v) \quad \text{for every } v \in V.$$

For such a map ι, the quadratic structure of (V, q) is related to the algebra structure of A. For example, if $v, w \in V$ are orthogonal then $\iota(v)$ and $\iota(w)$ anticommute in A. (For $2B_q(v, w) = q(v + w) - q(v) - q(w) = \iota(v + w)^2 - \iota(v)^2 - \iota(w)^2 = \iota(v)\iota(w) + \iota(w)\iota(v)$.)

The Clifford algebra $C(V, q)$ is the F-algebra universal with respect to being compatible with q. More formally, define a *Clifford algebra* for (V, q) to be an F-algebra C together with an F-linear map $\iota : V \to C$ compatible with q and such that for any F-algebra A and any F-linear map $\varphi : V \to A$ which is compatible with q, there exists a unique F-algebra homomorphism $\hat{\varphi} : C \to A$ such that $\varphi = \hat{\varphi} \circ \iota$.

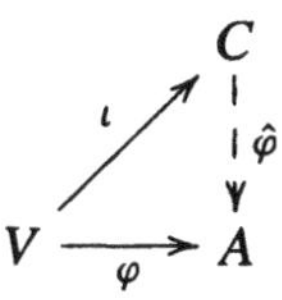

3.2 Lemma. *For any quadratic space (V, q) over F there is a Clifford algebra $(C(V, q), \iota)$, which is unique up to a unique isomorphism.*

Proof. The uniqueness of the Clifford algebra follows by the standard argument for universal objects. (See Exercise 1.) To prove existence we use the tensor algebra

$$T(V) = \oplus_k T^k(V) \quad \text{where } T^k(V) \text{ is the } k\text{-fold tensor product of } V.$$

Then $T(V) = F \oplus V \oplus (V \otimes V) \oplus (V \otimes V \otimes V) \oplus \cdots$. Let $C = T(V)/\mathcal{I}$ where $\mathcal{I}$ is the two-sided ideal of $T(V)$ generated by all elements $v \otimes v - q(v) \cdot 1$ for $v \in V$. Let $\iota : V \to C$ be the canonical map $V \to T(V) \to T(V)/\mathcal{I} = C$.

Claim. (C, ι) is a Clifford algebra for (V, q). For if $\varphi : V \to A$ is an F-linear map compatible with q then the universal property of tensor algebras implies that φ extends to a unique F-algebra homomorphism $\tilde{\varphi} : T(V) \to A$. Since φ is compatible with q we find that $\tilde{\varphi}(\mathcal{I}) = 0$ and therefore $\tilde{\varphi}$ induces a unique homomorphism $\hat{\varphi} : C \to A$ such that $\varphi = \hat{\varphi} \circ \iota$. □

Since the Clifford algebra is unique we are often sloppy about the notations, writing $C(q)$ rather than $C(V, q)$. A major advantage of using the universal property to define Clifford algebras is that the "functorial" properties follow immediately:

3.3 Lemma. (1) *An isometry $f : (V, q) \to (V', q')$ induces a unique algebra homomorphism $C(f) : C(V, q) \to C(V', q')$. Consequently if $q \simeq q'$ then $C(q) \cong C(q')$.*

(2) *If K is an extension field of F then there is a canonical isomorphism $C(K \otimes_F (V, q)) \cong K \otimes_F C(V, q)$.*

Proof. Exercise 1. □

The isomorphism class of $C(q)$ depends only on the isometry class of the quadratic form q. It is natural to ask the converse question: If $C(q) \cong C(q')$ does it follow that the quadratic forms q and q' are isometric? The answer is "no" in general, but the study of those quadratic forms which have isomorphic Clifford algebras is one of the major themes of this theory.

With the universal definition of Clifford algebras given above it is not immediately clear what the dimensions are. We spend some time presenting a proof that if $\dim q = n$ then $\dim C(q) = 2^n$.

3.4 Lemma. (1) *$C(V, q)$ is an F-algebra generated by $\iota(V)$.*

(2) *If $\dim q = n$ then $\dim C(q) \leq 2^n$.*

Proof. (1) Exercise 1.

(2) If $\{v_1, \dots, v_n\}$ is an orthogonal basis of V let $e_i = \iota(v_i)$. As mentioned earlier, $e_1, \dots, e_n$ anticommute and $e_j^2 = q(v_j) \in F$. By part (1), $C(q)$ is spanned by the products $e_1^{\delta_1} \dots e_n^{\delta_n}$ where each $\delta_i = 0$ or 1. There are 2^n of these products. □

3.5 Lemma. *Suppose q is a quadratic form on V, A is an F-algebra and $\varphi : V \to A$ is a linear map compatible with q such that A is generated by $\varphi(V)$. If* $\dim q = n$ *and* $\dim A = 2^n$ *then $A \cong C(q)$. If such an algebra A exists then* $\dim C(q) = 2^n$ *and $\iota : V \to C(q)$ is injective.*

Proof. There is an algebra homomorphism $\hat{\varphi} : C(q) \to A$ such that $\varphi = \hat{\varphi} \circ \iota$. Since $\iota(V)$ generates $C(q)$, $\varphi(V) = \hat{\varphi}(\iota(V))$ generates $\hat{\varphi}(C(q)) \subseteq A$. By hypothesis $\varphi(V)$ generates A and we see that $\hat{\varphi}$ is surjective. Then $\dim C(q) = 2^n + \dim(\ker \hat{\varphi})$ and (3.4) implies that $\dim C(q) = 2^n$ and $\ker \hat{\varphi} = \{0\}$. Therefore $\hat{\varphi}$ is an isomorphism. If ι is not injective then $\dim \iota(V) < n$ and the argument in (3.4) would imply that $\dim C(q) < 2^n$. □

This criterion can be used to provide some explicit examples of Clifford algebras. If $a \in F^\bullet$ define the quadratic extension to be $F\langle\sqrt{a}\rangle = F[x]/(x^2 - a)$. If $\langle a \rangle \not\simeq \langle 1 \rangle$ (that is, a is not a square in F) this algebra is just the quadratic field extension $F(\sqrt{a})$. However if $\langle a \rangle \simeq \langle 1 \rangle$ then $F\langle\sqrt{a}\rangle \cong F \times F$, the direct product of two copies of F.

3.6 Examples. (1) $C(\langle a \rangle) \cong F\langle\sqrt{a}\rangle$, the quadratic extension.

(2) $C(\langle a, b \rangle) \cong \left(\frac{a,b}{F}\right)$, the quaternion algebra.

(3) If q is the zero form on V then $C(V, q) = \Lambda(V)$ is the exterior algebra.

Proof. (1) The space $\langle a \rangle$ is given by $V = Fe$ with $q(xe) = ax^2$. If $A = F\langle\sqrt{a}\rangle$ define $\varphi : Fe \to A$ by $\varphi(xe) = x\sqrt{a}$. Then φ is compatible with q and (3.5) applies.

(2) The space $\langle a, b \rangle$ is given by $V = Fe_1 + Fe_2$ with $q(xe_1 + ye_2) = ax^2 + by^2$. Define $\varphi : V \to \left(\frac{a,b}{F}\right)$ by $\varphi(xe_1 + ye_2) = xi + yj$. Then φ is compatible with q and (3.5) applies.

(3) If $q = 0$ the definition of the Clifford algebra coincides with the definition of the exterior algebra. □

Perhaps the best way to prove the general result that $\dim C(q) = 2^n$ is to develop the theory of graded tensor products, prove that $C(\alpha \perp \beta) \cong C(\alpha) \hat{\otimes} C(\beta)$ and use induction. That approach has the advantage that it gives a unified treatment of the theory, combining the "even" and "odd" cases. Furthermore it is valid for quadratic forms over a commutative ring, provided that the forms can be diagonalized. Graded tensor products are discussed in the books by Lam and Scharlau, but see the booklets by Chevalley (1955) and Knus (1988) for further generality.

Rather than repeating that treatment, we provide an elementary, direct argument using the independence result (1.11). This method works only over fields of characteristic not 2, but that is the case of interest here anyway. The quadratic form q may be singular here.

3.7 Proposition. *If q is a quadratic form on V with* $\dim q = n$ *then* $\dim C(q) = 2^n$ *and the map $\iota : V \to C(q)$ is injective.*

Proof. First suppose q is a regular form of even dimension n. Choose an orthogonal basis $\{v_1, \dots, v_n\}$ yielding the diagonalization $q \simeq \langle a_1, \dots, a_n \rangle$. Then $C(q)$ contains elements $e_i = \iota(v_i)$ which anticommute and with $e_i^2 = a_i \in F^\bullet$. Then (1.11) implies that the 2^n elements e^Δ are linearly independent. Therefore $2^n \le \dim C(q)$ and (3.4) and (3.5) complete the argument.

If n is odd let $q' = q \perp \langle 1 \rangle$ and choose an isometry $\psi : (V, q) \to (V', q')$. Then (V', q') has an orthogonal basis $\{w_1, \dots, w_n, w_{n+1}\}$ where $w_i = \psi(v_i)$ for $i = 1, \dots, n$. Let $e_i = \iota(w_i)$ in $C(q')$. As before (1.11) implies that the elements e^Δ are linearly independent. Let A be the subalgebra generated by $e_1, \dots, e_n$, so that $\dim A = 2^n$. Then $\iota \circ \psi : V \to A$ is compatible with q and (3.5) completes the argument.

If (V, q) is singular the same idea works. Choose an isometry $\varphi : (V, q) \to (V', q')$ where q' is regular. (Why does such φ exist? See Exercise 20.) First step: if $\{w_1, \dots, w_m\}$ is any basis of V' (not necessarily orthogonal) and $f_i = \iota(w_i)$ then the 2^n elements f^Δ are linearly independent. Second step: choose $\{w_1, \dots, w_m\}$ so that $w_i = \varphi(v_i)$ for $1 \le i \le n$ and let A be the subalgebra generated by $f_1, \dots, f_n$. Then $\dim A = 2^n$ and $\iota \circ \psi : V \to A$ is compatible with q. □

Since $\iota : V \to C(V, q)$ is always injective we simplify the notation by considering V as a subset of $C(V, q)$. If $U \subseteq V$ is a subspace and q induces the quadratic form φ on U, then $C(U, \varphi)$ is viewed as a subalgebra of $C(V, q)$. If $\{e_1, \dots, e_n\}$ is a basis of V then $\{e^\Delta : \Delta \in \mathbb{F}_2^n\}$ is sometimes called the *derived basis* of $C(V, q)$.

If $f : (V, q) \to (V', q')$ is an isometry the universal property implies that there is a unique algebra homomorphism $C(f) : C(V, q) \to C(V', q')$ extending f. Consequently, if $g \in \mathrm{O}(V, q)$ there is an automorphism $C(g)$ of $C(V, q)$ extending g. When $g = -1_V$ we get an automorphism $\alpha = C(-1_V)$ of particular importance.

3.8 Definition. The *canonical automorphism* α of $C(V, q)$ is the automorphism with $\alpha(x) = -x$ for every $x \in V$. Define $C_0(V, q)$ to be the 1-eigenspace of α and $C_1(V, q)$ to be the (-1)-eigenspace of α. This subalgebra $C_0(V, q)$ is called the *even Clifford algebra* (or the second Clifford algebra) of q.

This notation α for an automorphism should not cause confusion even though we sometimes use α to denote a quadratic form. The meaning of the 'α' ought to be clear from the context.

Note that $\alpha^2 = \alpha \circ \alpha$ is the identity map on $C(q)$ and therefore $C(q) = C_0(q) \oplus C_1(q)$. Note that $\alpha(v_1 v_2 \dots v_m) = (-1)^m v_1 v_2 \dots v_m$ for any $v_1, v_2, \dots, v_m \in V$.

Therefore $C_0(V, q)$ is the span of all such products where m is even. Suppose $\{e_1, e_2, \dots, e_n\}$ is an orthogonal basis of V corresponding to $q \simeq \langle a_1, a_2, \dots, a_n\rangle$. Then

$$\alpha(e^{\Delta}) = (-1)^{|\Delta|}e^{\Delta}.$$

Therefore $\{e^{\Delta} : |\Delta| \text{ is even}\}$ is a basis of $C_0(q)$ while $\{e^{\Delta} : |\Delta| \text{ is odd}\}$ is a basis of $C_1(q)$. In particular, $\dim C_0(q) = \dim C_1(q) = 2^{n-1}$.

Furthermore, $C_0 \cdot C_1 \subseteq C_1$ and $C_1 \cdot C_1 \subseteq C_0$. If $u \in C_1(q)$ is an invertible element then $C_1(q) = C_0(q) \cdot u$. The next lemma shows that this subalgebra $C_0(q)$ can itself be viewed as a Clifford algebra, at least if q is regular.

3.9 Lemma. *If $q \simeq \langle a\rangle \perp \beta$ and $a \neq 0$ then $C_0(q) \simeq C(\langle -a\rangle\beta)$.*

Proof. Let $\{e_1, e_2, \dots, e_n\}$ be an orthogonal basis of V where e_1 corresponds to $\langle a\rangle$. Then the elements $e_1e_i \in C_0(q)$ anticommute and $(e_1e_i)^2 = -a \cdot q(e_i)$ for $i = 2, \dots, n$. Then the inclusion map from $W = \text{span}\{e_1e_2, \dots, e_1e_n\}$ to $C_0(q)$ is compatible with the form $\langle -a\rangle\beta$ on W and the universal property provides an algebra homomorphism $\hat{\varphi} : C(\langle -a\rangle\beta) \to C_0(q)$. Since $a \neq 0$ the elements e_1e_i generate $C_0(q)$ so that $\hat{\varphi}$ is surjective. Counting dimensions we conclude that $\hat{\varphi}$ is an isomorphism. □

Let us now restrict attention again to regular quadratic forms. The next goal is to define the "Witt invariant" $c(q)$ of a regular quadratic form and to derive some of its properties. The first step is to prove that if q has even dimension then $C(q)$ is a central simple algebra. Then $c(q)$ will be defined to be the class of $C(q)$ in the Brauer group $\text{Br}(F)$.

To begin this sequence of ideas we determine the centralizer of $C_0(q)$ in $C(q)$. The argument here is reminiscent of the proof of (1.11).

3.10 Definition. If $\{e_1, \dots, e_n\}$ is an orthogonal basis of (V, q) define the element $z(V, q) = e_1e_2\dots e_n \in C(V, q)$.

Define the subalgebra $Z(V, q) = \text{span}\{1, z(V, q)\} \subseteq C(V, q)$.

We sometimes write $z(q)$ or $z(V)$ for the element $z(V, q)$. If $e_i^2 = a_i \in F^{\bullet}$ then $z(q)^2 = (e_1 \dots e_n) \cdot (e_1 \dots e_n) = (-1)^{n(n-1)/2}a_1 \dots a_n$. Abusing notation slightly we have

$$z(q)^2 = dq \quad \text{and} \quad Z(q) \cong F\langle\sqrt{dq}\rangle.$$

Then if $dq \not\simeq \langle 1\rangle$ the subalgebra $Z(q) \cong F(\sqrt{dq})$ is a field. If $dq \simeq \langle 1\rangle$ then $Z(q) \cong F \times F$. From the next proposition it follows that $Z(q)$ is independent of the choice of basis. Furthermore, the element $z(q)$ is unique up to a non-zero scalar multiple.

3.11 Proposition. *Suppose q is a regular form on V and* $\dim q = n$.

(1) *$Z(V,q)$ is the centralizer of $C_0(V,q)$ in $C(V,q)$.*

(2) *The center of $C(V,q)$ is* $\begin{cases} F & \text{if } n \text{ is even,} \\ Z(V,q) & \text{if } n \text{ is odd.} \end{cases}$

(3) *The center of $C_0(V,q)$ is* $\begin{cases} Z(V,q) & \text{if } n \text{ is even,} \\ F & \text{if } n \text{ is odd.} \end{cases}$

Proof. (1) Suppose c is an element of that centralizer. Let $\{e_1, e_2, \dots, e_n\}$ be the orthogonal basis used in (3.10) and express $c = \sum c_\Delta e^\Delta$ for coefficients $c_\Delta \in F$. Since c commutes with every $e_i e_j$ and $(e_i e_j)^{-1} e^\Delta (e_i e_j) = \pm e^\Delta$ it follows that if $c_\Delta \neq 0$ then e^Δ commutes with every $e_i e_j$. If $\Delta = (\delta_1, \dots, \delta_n)$ and $\delta_r \neq \delta_s$ for some r, s then $e_r e_s$ anticommutes with e^Δ. Therefore either $\Delta = (0, \dots, 0)$ and $e^\Delta = 1$ or $\Delta = (1, \dots, 1)$ and $e^\Delta = z(q)$. Hence c is a combination of 1 and $z(q)$ so that $c \in Z(q)$.

(2) If c is in the center then $c \in Z(q)$ by part (1). If n is even every e_i commutes with $z(q)$ and the center is $Z(q)$. If n is odd every e_i anticommutes with $z(q)$ and the center is F.

(3) $z(q)$ is in $C_0(q)$ if and only if n is even. □

3.12 Lemma. *If n is odd and q is regular then $C(V,q) \cong C_0(V,q) \otimes Z(V,q)$.*

Proof. Since $C_0(q)$ and $Z(q)$ are subalgebras of $C(q)$ which centralize each other there is an induced algebra homomorphism $\psi : C_0(q) \otimes Z(q) \to C(q)$. Since n is odd these two subalgebras generate $C(q)$ so that ψ is surjective. Counting dimensions we see that ψ is an isomorphism. □

3.13 Structure Theorem. *Let $C = C(V,q)$ be the Clifford algebra of a regular quadratic space over the field F. Let $C_0 = C_0(V,q)$ and $Z = Z(V,q)$.*

(1) *If $\dim V$ is even then C is a central simple algebra over F, and C_0 has center Z.*

(2) *If $\dim V$ is odd then C_0 is a central simple algebra over F and $C \cong C_0 \otimes Z$.*

If $dq \not\simeq \langle 1 \rangle$ then C is a central simple algebra over Z. If $dq = \langle 1 \rangle$ then $C \cong C_0 \times C_0$.

Proof. Suppose $n = \dim V$ so that $\dim C = 2^n$. The centers of these algebras are given in (3.11).

(1) Suppose n is even and $\mathcal{I}$ is a proper ideal of C so that $\bar{C} = C/\mathcal{I}$ is an F-algebra with $\dim \bar{C} = 2^n - \dim \mathcal{I}$. If $\{e_1, \dots, e_n\}$ is an orthogonal basis of (V,q), the images $\bar{e}_1, \dots, \bar{e}_n$ are anticommuting invertible elements of $\bar{C}$. Since n is even (1.11) implies that $\dim \bar{C} \geq 2^n$ and therefore $\mathcal{I} = 0$. Hence C is simple.

(2) Apply (3.9), part (1) and (3.12). The final assertions follow since Z is a field if $dq \not\simeq \langle 1 \rangle$ and $Z \cong F \times F$ if $dq \simeq \langle 1 \rangle$. □

If q is a regular quadratic form of even dimension then $C(q)$ is a central simple algebra. In fact it is isomorphic to a tensor product of quaternion algebras. Before beginning the analysis of the Witt invariant we describe an explicit decomposition of $C(q)$ which will be useful later on. This decomposition provides another proof that $C(q)$ is central simple, since the class of central simple F-algebras is closed under tensor products.

3.14 Proposition. *If $q \simeq \langle a_1, \ldots, a_{2m}\rangle$ define $u_k = e_1 e_2 \ldots e_{2k-1}$ and $v_k = e_{2k-1}e_{2k}$ for $k = 1, 2, \ldots, m$. The subalgebra Q_k generated by u_k and v_k is a quaternion algebra and $C(q) \cong Q_1 \otimes \cdots \otimes Q_m$.*

Proof. Check that u_k anticommutes with v_k but commutes with every u_i and with every v_j for $j \neq k$. Since $u_k^2 = (-1)^{k-1}a_1 a_2 \ldots a_{2k-1}$ and $v_k^2 = -a_{2k-1}a_{2k}$ are scalars, each Q_k is a quaternion subalgebra. The induced map $Q_1 \otimes \cdots \otimes Q_m \to C(q)$ is injective since the domain is simple. By counting dimensions we see it is an isomorphism. □

If A is an F-algebra, the "opposite algebra" A^{op} is the algebra defined as the vector space A with the multiplication $*$ given by: $a * b = ba$. An algebra isomorphism $\varphi : A \to A^{\text{op}}$ can be interpreted as an anti-automorphism $\varphi : A \to A$. That is, φ is a vector space isomorphism and $\varphi(ab) = \varphi(b)\varphi(a)$ for every $a, b \in A$. If (V, q) is a quadratic space and $g \in \mathrm{O}(V, q)$ then the map $\iota \circ g : V \to C(V, q)^{\text{op}}$ is compatible with q. The universal property provides a homomorphism $\hat{\varphi} : C(V, q) \to C(V, q)^{\text{op}}$. This map is surjective and therefore it is an isomorphism. As above we may interpret this map as an anti-automorphism

$$C'(g) : C(V, q) \to C(V, q).$$

This is the unique anti-automorphism of $C(V, q)$ which extends $g : V \to V$.

The involutions of $C(V, q)$ will be particularly important for our work. By definition, an *involution* of an F-algebra is an F-linear anti-automorphism whose square is the identity. (These are sometimes called "involutions of the first kind".) For example, the transpose map on the matrix algebra $\mathbb{M}_n(F)$ and the usual "bar" on a quaternion algebra are involutions. If $g \in \mathrm{O}(V, q)$ satisfies $g^2 = 1_V$ then $C'(g)$ is an involution on $C(V, q)$.

3.15 Definition. If $V = R \perp T$, define $J_{R,T}$ to be the involution on $C(V, q)$ which is -1_R on R and 1_T on T. The *canonical involution* (denoted J_0 or $^\top$) is $J_{0,V}$ and the *bar involution* (denoted J_1 or $^-$) is $J_{V,0}$.

This $J_{R,T}$ is the anti-automorphism of $C = C(V, q)$ extending the reflection $(-1_R) \perp (1_T)$ on $V = R \perp T$. These involutions will be important when we analyze (s, t)- families using Clifford algebras.

The canonical involution acts as the identity on V. If $v_1, v_2, \ldots, v_m \in V$ then $(v_1v_2\ldots v_m)^\top = v_m \ldots v_2v_1$, reversing the order of the product. A simple sign calculation shows that $(e^\Delta)^\top = (-1)^{k(k-1)/2}e^\Delta$ where $k = |\Delta|$. The bar involution is the composition of α and $^\top$, and similar remarks hold. For the quaternion algebras, the bar involution is the usual "bar".

For the discussion of the Witt invariant below we assume some familiarity with the theory of central simple algebras. If A is a central simple algebra over F, then Wedderburn's theory implies that A is a matrix ring over a division algebra. That is, $A \cong \mathbb{M}_k(D)$ where D is a central division algebra over F, which is uniquely determined (up to isomorphism) by A. Two central simple algebras A, B are equivalent ($A \sim B$) if their corresponding division algebras are isomorphic. That is, $A \sim B$ if and only if $\mathbb{M}_m(A) \cong \mathbb{M}_n(B)$ for some m, n. Let $[A]$ denote the equivalence class of a central simple algebra A. Each such class contains a unique division algebra. The Brauer group $\mathrm{Br}(F)$ is the set of these equivalence classes, with a multiplication induced by $\otimes$.

The inverse of $[A]$ in $\mathrm{Br}(F)$ is $[A^{\mathrm{op}}]$, the class of the opposite algebra. (For there is a natural isomorphism $A \otimes A^{\mathrm{op}} \cong \mathrm{End}_F(A)$.) Since quaternion algebras possess involutions, $[A]^2 = 1$ whenever A is a tensor product of quaternion algebras. Let $\mathrm{Br}_2(F)$ be the subgroup of $\mathrm{Br}(F)$ consisting of all $[A]$ with $[A]^2 = 1$.

3.16 Definition. Let (V, q) be a (regular) quadratic space over F. The *Witt invariant* $c(q)$ is the element of $\mathrm{Br}_2(F)$ defined as:

$$c(q) = \begin{cases} [C(V,q)] & \text{if } \dim V \text{ is even,} \\ [C_0(V,q)] & \text{if } \dim V \text{ is odd.} \end{cases}$$

By the Structure Theorem the indicated algebras are central simple, so $c(q)$ is well-defined. To derive the basic properties of $c(q)$ we must recall some of the properties of quaternion algebras. The class of the quaternion algebra is written $[a, b]$.

3.17 Lemma. (1) $[a, b] = [a', b']$ *if and only if* $\langle\langle -a, -b\rangle\rangle \simeq \langle\langle -a', -b'\rangle\rangle$. *In particular,* $[a, b] = 1$ *if and only if* $\langle\langle -a, -b\rangle\rangle$ *is hyperbolic if and only if* $\langle a, b\rangle$ *represents* 1.

(2) $[a, b] = [b, a]$, $[a, a] = [-1, a]$, $[a, b]^2 = 1$ *and* $[a, b] \cdot [a, c] = [a, bc]$.

Proof. (1) This is a standard result about quaternion algebras: the isomorphism class of the algebra $\left(\frac{a,b}{F}\right)$ is determined by the isometry class of its norm form $\langle\langle -a, -b\rangle\rangle$. In particular the symbol $[a, b]$ depends only on the square classes $\langle a\rangle$ and $\langle b\rangle$.

(2) The first and second statements follow from (1), and the third one is clear. The last statement is equivalent to the isomorphism:

$$\left(\frac{a,b}{F}\right) \otimes \left(\frac{a,c}{F}\right) \cong \left(\frac{a,bc}{F}\right) \otimes \mathbb{M}_2(F).$$

To prove this let i, j be the generators of the algebra $B = \left(\frac{a,b}{F}\right)$ so that $i^2 = a$ and $j^2 = b$. Similarly let i', j' be the generators of the algebra $C = \left(\frac{a,c}{F}\right)$ so that $i'^2 = a$ and $j'^2 = c$. Then the subalgebra of $B \otimes C$ generated by $i \otimes 1$ and $j \otimes j'$ is isomorphic to $\left(\frac{a,bc}{F}\right)$ and the subalgebra generated by $i \otimes i'$ and $1 \otimes j'$ is isomorphic to $\left(\frac{a^2,c}{F}\right) \cong \mathbb{M}_2(F)$. Since those subalgebras centralize each other and together span the algebra $B \otimes C$ the claim follows. □

The same sorts of arguments are used to prove the various isomorphisms of Clifford algebras stated below.

3.18 Lemma. (1) *If* $\dim \alpha$ *and* $\dim \beta$ *are even then* $c(\alpha \perp \beta) = c(\alpha) \cdot c((d\alpha) \cdot \beta)$.
(2) *If* $\dim \alpha$ *is odd and* $\dim \beta$ *is even then* $c(\alpha \perp \beta) = c(\alpha) \cdot c((-d\alpha) \cdot \beta)$.

Proof. (1) We must prove that $C(\alpha \perp \beta) \cong C(\alpha) \otimes C((d\alpha) \cdot \beta)$. In fact this holds whenever $\dim \alpha$ is even. Let $v_1, \dots, v_r, w_1, \dots, w_s$ be an orthogonal basis corresponding to $\alpha \perp \beta$. The subalgebra of $C(\alpha \perp \beta)$ generated by $\{v_1, \dots, v_r\}$ is isomorphic to $C(\alpha)$. Since r is even the element $u = z(\alpha) = v_1 \dots v_r$ anticommutes with each v_i, commutes with each w_j and $\langle u^2 \rangle = d\alpha$. Then the subalgebra generated by $\{uw_1, \dots, uw_s\}$ is isomorphic to $C((d\alpha)\beta)$ and centralizes $C(\alpha)$. Since these subalgebras together generate the whole algebra $C(\alpha \perp \beta)$ the isomorphism follows.

(2) We must show that $C_0(\alpha \perp \beta) \cong C_0(\alpha) \otimes C((d\alpha) \cdot \beta)$. In fact this holds whenever $\dim \alpha$ is odd. Either use an argument similar to (1) above or apply (1) and Lemma 3.9. We omit the details. □

By applying (3.18) (1) successively to $q \simeq \langle a_1, \dots, a_{2m} \rangle$ we find that $C(q)$ is isomorphic to a tensor product of quaternion subalgebras. These are the same subalgebras found in (3.14) above.

3.19 Proposition. *Let* α, β *be (regular) quadratic forms over* F *and let* $x, y, z \in F^\bullet$.

(1) $c(\alpha \perp \beta) = \begin{cases} c(\alpha) \cdot c(\beta) \cdot [d\alpha, d\beta] & \text{if } \dim \alpha \text{ and } \dim \beta \text{ are both even} \\ & \text{or both odd,} \\ c(\alpha) \cdot c(\beta) \cdot [-d\alpha, d\beta] & \text{if } \dim \alpha \text{ is odd and } \dim \beta \text{ is even.} \end{cases}$

(2) $c(\langle x \rangle \alpha) = \begin{cases} c(\alpha)[x, d\alpha] & \text{if } \dim \alpha \text{ is even,} \\ c(\alpha) & \text{if } \dim \alpha \text{ is odd.} \end{cases}$

(3) $c(\alpha \perp \mathbb{H}) = c(\alpha)$ *where* $\mathbb{H} = \langle 1, -1 \rangle$ *is the hyperbolic plane.*

(4) $c(\langle\langle x \rangle\rangle \otimes \alpha) = [-x, d\alpha]$. *Hence* $c(\langle\langle x, y \rangle\rangle) = [-x, -y]$ *and* $c(\langle\langle x, y, z \rangle\rangle) = 1$.

Proof. (3) follows immediately from (3.18) when $\beta = \mathbb{H}$. To prove (2) first note that $C_0(\langle x \rangle q) \cong C_0(q)$ for any form q. This settles the case $\dim \alpha$ is odd. For the even case of (2) we apply (3.18) in two ways: $c(\alpha \perp \langle -1, x \rangle) = c(\alpha) \cdot c((d\alpha) \cdot \langle -1, x \rangle) =$

$c(\langle\langle -1, x\rangle\rangle) \cdot c(\langle x\rangle \cdot \alpha)$. Therefore $c(\langle x\rangle \cdot \alpha) = c(\alpha)[-d\alpha, (d\alpha) \cdot x] \cdot [-1, x] = c(\alpha) \cdot [d\alpha, x]$, using the properties in (3.17).

(1) If $\dim \alpha$ and $\dim \beta$ are even then $c(\alpha \perp \beta) = c(\alpha) \cdot c((d\alpha) \cdot \beta) = c(\alpha) \cdot c(\beta) \cdot [d\alpha, d\beta]$ by (3.18) and part (2). A similar argument works when $\dim \alpha$ is odd and $\dim \beta$ is even. Suppose $\dim \alpha$ and $\dim \beta$ are odd. One way to proceed is to express $\alpha = \langle a\rangle \perp \alpha'$ and $\beta = \langle b\rangle \perp \beta'$ so that $c(\alpha \perp \beta) = c(\langle a, b\rangle \perp \alpha' \perp \beta')$ and $c(\alpha) = c(\langle -a\rangle\alpha')$, $c(\beta) = c(\langle -b\rangle\beta')$ by (3.9). The desired equality follows after expanding both sides using the properties for forms of even dimension. Alternatively we can prove the isomorphism $C(\alpha \perp \beta) \cong C_0(\alpha) \otimes C_0(\beta) \otimes \left(\frac{d\alpha, d\beta}{F}\right)$ by directly examining basis elements. Further details are left to the reader.

The formula for $c(\langle\langle x\rangle\rangle \otimes \alpha) = c(\alpha \perp \langle x\rangle\alpha)$ in (4) follows from (1) and (2). □

Recall that if a (regular) quadratic form q is isotropic, then $q \simeq \mathbb{H} \perp \alpha$ for some form α. A form φ is *hyperbolic* if $\varphi \simeq m\mathbb{H} \simeq \mathbb{H} \perp \mathbb{H} \perp \cdots \perp \mathbb{H}$. Then every form q can be expressed as $q \simeq q_0 \perp H$ where q_0 is anisotropic and H is hyperbolic. (This is the "Witt decomposition" of q.) Witt's Cancellation Theorem implies that this form q_0 is uniquely determined by q (up to isometry). Two forms α and β are *Witt equivalent* (written $\alpha \sim \beta$) if $\alpha \perp -\beta$ is hyperbolic. If $\dim \alpha > \dim \beta$ and $\alpha \sim \beta$ then $\alpha \simeq \beta \perp H$ for some hyperbolic space H. Consequently every Witt class contains a unique anisotropic form. These classes form the elements of the Witt ring $W(F)$, where the addition is induced by $\perp$ and the multiplication by $\otimes$.

We often abuse the notations, using symbols like q, φ, α to stand for regular quadratic forms, and writing $q \in W(F)$ rather than stating that the Witt class of q lies in $W(F)$.

Define IF to be the ideal in the Witt ring $W(F)$ consisting of all quadratic forms of even dimension. (This is well-defined since $\dim \mathbb{H}$ is even.) Then IF is additively generated by all the forms $\langle\langle a\rangle\rangle = \langle 1, a\rangle$ for $a \in F^\bullet$. The square I^2F is additively generated by the 2-fold Pfister forms $\langle\langle a, b\rangle\rangle$, and similarly for higher powers.

The determinant $\det \alpha \in F^\bullet/F^{\bullet 2}$ does not generally induce a map on $W(F)$. The "correct" invariant is the signed discriminant $d\alpha = (-1)^{n(n-1)/2} \det \alpha$, because $d(\alpha \perp \mathbb{H}) = d\alpha$. This discriminant induces a map $d : W(F) \to F^\bullet/F^{\bullet 2}$. The ideal I^2F is characterized by this discriminant: $\alpha \in I^2F$ if and only if $\dim \alpha$ is even and $d\alpha = \langle 1\rangle$.

The Witt invariant $c(q)$ induces a map $c : W(F) \to \mathrm{Br}(F)$. The formulas in (3.19) imply that the restriction $c : I^2F \to \mathrm{Br}(F)$ is a homomorphism. One natural question is: Which quadratic forms q have all three invariants trivial? That is:

$$\dim q = \text{even}, \quad dq = \langle 1\rangle, \quad c(q) = 1.$$

It is easy to check that any 3-fold Pfister form $\langle\langle a, b, c\rangle\rangle$ has trivial invariants. Consequently so does everything in the ideal I^3F. For convenience we let $J_3(F)$ be the ideal of elements in $W(F)$ which have trivial invariants. That is:

$$J_3(F) = \ker(c : I^2F \to \mathrm{Br}(F)).$$

Does $J_3(F) = I^3F$ for every field F? This was a major open question until Merkurjev proved it true in 1981 using techniques from K-theory (which are well beyond the scope of this book). Before 1981 the only result in this direction valid over arbitrary fields was the one of Pfister (1966): if $q \in J_3(F)$ and $\dim q \leq 12$ then $q \in I^3F$. An important tool used in the proof is the following easy lemma about the behavior of forms under quadratic extensions.

3.20 Lemma. *Let q be an anisotropic quadratic form over F.*

(1) *$q \otimes F(\sqrt{d})$ is isotropic iff $q \simeq \langle x\rangle\langle 1, -d\rangle \perp \alpha$ for some $x \in F^\bullet$ and some form α over F.*

(2) *$q \otimes F(\sqrt{d})$ is hyperbolic iff $q \simeq \langle\langle -d\rangle\rangle \otimes \beta$ for some form β over F.*

Proof. Exercise 8. □

With this lemma, and some clever arguments with quaternion algebras, Pfister (1966) characterized the small forms in I^3F up to isometry.

3.21 Pfister's Theorem. *Let φ be a regular quadratic form over F with $\dim \varphi$ even, $d\varphi = \langle 1\rangle$ and $c(\varphi) = 1$.*

(1) *If $\dim \varphi < 8$ then $\varphi \sim 0$.*

(2) *If $\dim \varphi = 8$ then $\varphi \simeq \langle a\rangle\langle\langle x, y, z\rangle\rangle$ for some $a, x, y, z \in F^\bullet$.*

(3) *If $\dim \varphi = 10$ then φ is isotropic.*

(4) *If $\dim \varphi = 12$ then $\varphi \simeq \langle\langle x\rangle\rangle \otimes \delta$ for some $x \in F^\bullet$ and some quadratic form δ where $\dim \delta = 6$ and $d\delta = \langle 1\rangle$. Furthermore if $\langle a\rangle\langle\langle -b\rangle\rangle \subset \varphi$ then $\varphi \simeq \varphi_1 \perp \varphi_2 \perp \varphi_3$ where $\dim \varphi_i = 4$, $d\varphi_i = \langle 1\rangle$ and $\langle a\rangle\langle\langle -b\rangle\rangle \subset \varphi_1$.*

The proof is given in Exercises 9 and 10.

The three basic invariants described above induce group homomorphisms

$$\overline{\dim} : W(F)/IF \to \mathbb{Z}/2\mathbb{Z}, \quad \bar{d} : IF/I^2F \to F^\bullet/F^{\bullet 2}, \quad \bar{c} : I^2F/I^3F \to \mathrm{Br}_2(F).$$

The first two maps above are easily seen to be isomorphisms. Milnor (1970) conjectured that these maps are the first three of a sequence of well defined isomorphisms $e_n : I^nF/I^{n+1}F \to H^n(F)$, where H^n is a suitable Galois cohomology group. Many mathematicians have worked on various aspects of these conjectures. Merkurjev (1981) proved that $e_2 = \bar{c}$ is always an isomorphism. Recently Voevodsky proved that Milnor's conjecture is always true. For an outline of these ideas and further references see Morel (1998).

Exercises for Chapter 3

1. **Universal property.** (1) If $f : (V, q) \to (V', q')$ is an isometry then there is a unique algebra homomorphism $\psi : C(V, q) \to C(V', q')$ such that $\psi \circ \iota = \iota' \circ f$. If f is bijective then ψ is an isomorphism. The uniqueness of the Clifford algebra follows.

(2) $1 \otimes \iota : K \otimes V \to K \otimes C(V, q)$ is compatible with $K \otimes q$ and satisfies the universal property. This proves (3.3) (2).

(3) $C(V, q)$ is generated as an F-algebra by $\iota(V)$. This proves (3.4) (1):

(Hint. (3) Apply the proof of (3.2). Or directly from the definitions: Let A be the subalgebra generated by $\iota(V)$ so that ι induces a map $\iota' : V \to A$ which is compatible with q. Apply the definition of $C(q)$ to get an induced algebra homomorphism $\psi : C(q) \to C(q)$ with $\psi \circ \iota = \iota$. The uniqueness implies $\psi = 1_C$.)

2. **Homogeneous components.** (1) Let $\{e_1, e_2, \ldots, e_n\}$ be an orthogonal basis of (V, q), and define the subspace $V_{(k)} = \operatorname{span}\{e^\Delta : |\Delta| = k\}$. Then $\dim V_{(k)} = \binom{n}{k}$. For instance, $V_{(0)} = F$ and $V_{(1)} = V$. Each $V_{(k)}$ is independent of the choice of orthogonal basis.

(2) Since $V_{(n)} = F \cdot z(q)$, the element $z(q)$ in Definition 3.10 is uniquely determined up to a non-zero scalar multiple. For any subspace $U \subseteq V$, the line spanned by $z(U)$ is uniquely determined.

(Hint. (1) Use the Chain-Equivalence Theorem stated in Exercise 2.8.)

3. Prove (3.19) (1), (2) by exhibiting explicit algebra isomorphisms, similar to the proof of (3.18). For instance if $\dim \alpha$ and $\dim \beta$ are odd,

$$C(\alpha \perp \beta \perp \mathbb{H}) \cong C(\alpha) \otimes C(\beta) \otimes \left(\frac{d\alpha, d\beta}{F}\right).$$

4. **Discriminants.** Suppose $\dim \alpha = m$ and $\dim \beta = n$.

(1) $d(\alpha \perp \beta) = (-1)^{mn} d\alpha \cdot d\beta$. In particular $d(\alpha \perp \mathbb{H}) = d\alpha$.

(2) $d(\langle c \rangle \alpha) = \langle c \rangle^m \cdot d\alpha$.

(3) $d(\alpha \otimes \beta) = (d\alpha)^n \cdot (d\beta)^m$.

(4) If $q \simeq p\langle 1 \rangle \perp r\langle -1 \rangle$ define the signature $\operatorname{sgn}(q) = p - r$. Then

$$dq = \begin{cases} \langle 1 \rangle & \text{if } \operatorname{sgn}(q) \equiv 0, 1 \\ \langle -1 \rangle & \text{if } \operatorname{sgn}(q) \equiv 2, 3 \end{cases} \pmod 4.$$

5. **Witt invariant calculations.** (1) $c(\langle a, b \rangle) = [a, b]$, and $c(\langle a, b, c \rangle) = [-ab, -ac]$.

(2) If $d\alpha = d\beta$, $c(\alpha) = c(\beta)$ and $\dim \alpha = \dim \beta \leq 3$ then $\alpha \simeq \beta$.

(3) Let $\beta = \langle 1 \rangle \perp \beta_1$. Then $d(\beta) = d(-\beta_1)$ and $c(\beta) = c(-\beta_1)$.

(4) $c(\alpha \otimes \beta) = \begin{cases} [d\alpha, d\beta] & \text{if } \dim \alpha \text{ and } \dim \beta \text{ are even,} \\ c(\alpha)c(\beta) & \text{if } \dim \alpha \text{ and } \dim \beta \text{ are odd,} \\ c(\beta)[d\alpha, d\beta] & \text{if } \dim \alpha \text{ is odd and } \dim \beta \text{ is even.} \end{cases}$

(5) If $q \simeq p\langle 1 \rangle \perp r\langle -1 \rangle$ define the signature $\operatorname{sgn}(q) = p - r$. Then

$$c(q) = \begin{cases} 1 & \text{if } \operatorname{sgn}(q) \equiv -1, 0, 1, 2 \\ [-1, -1] & \text{if } \operatorname{sgn}(q) \equiv 3, 4, 5, 6 \end{cases} \pmod 8.$$

6. **Hasse invariant.** If $q \simeq \langle a_1, \ldots, a_n \rangle$ define the Hasse invariant

$$h(q) = \prod_{i \leq j} [a_i, a_j].$$

(1) If $\dim q = n$ then $h(q) = c(q \perp n\langle -1 \rangle)$. Consequently $h(q)$ depends only on the isometry class of q.

(2) This independence of $h(q)$ can be proved without Clifford algebras. Use the Chain-Equivalence Theorem stated in Exercise 2.8.

(3) $h(\alpha \perp \beta) = h(\alpha) \cdot h(\beta) \cdot [\det \alpha, \det \beta]$.

(4) Another version of the Hasse invariant is given by $s(q) = \prod_{i<j} [a_i, a_j]$. How are $h(q)$ and $s(q)$ related?

(Hint. (1) Apply (3.14) to $q \perp n\langle -1 \rangle \simeq \langle a_1, -1, a_2, -1, \ldots, a_n, -1 \rangle$.)

7. (1) Prove from the definition of I^2F that:
If $\dim q$ is even then $q \equiv \langle 1, -dq \rangle \pmod{I^2F}$.
If $\dim q$ is odd then $q \equiv \langle dq \rangle \pmod{I^2F}$.
(2) $q \in I^2F$ if and only if $\dim q$ is even and $dq = \langle 1 \rangle$.
(3) If $\dim q$ is even and $\langle a \rangle q \simeq q$ then $a \in D_F(\langle\langle -dq \rangle\rangle)$.
(4) If α is a form of even dimension then $\langle\langle d\alpha \rangle\rangle \otimes \alpha \in I^3F$.
(5) Suppose $q \equiv \langle\langle x \rangle\rangle \otimes \alpha \pmod{I^3F}$, where $dq = \langle 1 \rangle$ and $c(q) = 1$. Then $q \in I^3F$.

(Hint. (3) Use Witt invariants. Compare Exercise 2.9(3).)

8. **Quadratic extensions.** (1) Prove Lemma 3.20.

(2) If A is a central simple F-algebra and $K = F(\sqrt{d})$, then:

$$A \otimes K \sim 1 \iff A \sim \left(\frac{d, x}{F}\right) \quad \text{for some } x \in F^\bullet.$$

(3) $c(q) =$ quaternion $\iff c(q_K) = 1$ in $\operatorname{Br}(K)$ for some quadratic extension K/F.

(Hint. (1) If $q(v) = 0$ for $v \in V \otimes K$, express $v = v_0 + v_1\sqrt{d}$ and conclude: $q(v_0) + dq(v_1) = 0$ and $B_q(v_0, v_1) = 0$. The second statement follows by repeated application of the first.

(2) The theory of division algebras implies: If D is a central F-division algebra of degree n and K is a splitting field, then $[K : F] \geq n$. A central simple algebras of degree 2 must be quaternion.)

9. **Trivial invariants.** Suppose φ is a quadratic form over F.

(1) If $\dim \varphi = 2, 4$ or 6, $d\varphi = \langle 1 \rangle$ and $c(\varphi) = 1$, then $\varphi \sim 0$.

(2) Suppose $\dim \varphi = 6$, $d\varphi = \langle 1 \rangle$ and $c(\varphi) =$ quaternion, then φ is isotropic.

(3) Prove the cases where $\dim \varphi \leq 10$ in (3.21).

(Hint. (1) If $\dim \varphi = 6$, express $\varphi \simeq \langle a \rangle \langle 1, -b \rangle \perp \psi$. If $K = F(\sqrt{b})$ then $\varphi_K \sim 0$ because ψ_K has trivial invariants over K. If φ is anisotropic, (3.20) implies $\varphi \simeq \langle\langle -b \rangle\rangle \otimes \beta$. Discriminants show that $\langle b \rangle = \langle 1 \rangle$ so that $\varphi \sim 0$.

(2) $c(\varphi)$ is split over some $K = F(\sqrt{b})$. Finish as in part (1).

(3) Let $\dim \varphi = 8$, anisotropic with trivial invariants. Express $\varphi \simeq \langle a \rangle \langle\langle -b \rangle\rangle \perp \psi$, so that $d\psi = \langle b \rangle$ and $c(\psi) = [-a, b]$. Part (1) over $F(\sqrt{b})$ and (3.20) imply that $\psi \simeq \langle\langle -b \rangle\rangle \otimes \langle u, v, w \rangle$. Witt invariants show that $[b, auvw] = 1$ and $\langle\langle -b \rangle\rangle$ represents $auvw$. Therefore $\psi \simeq \langle\langle -b \rangle\rangle \otimes \langle u, v, auv \rangle$ and $\varphi \simeq \langle a \rangle \langle\langle -b, au, av \rangle\rangle$.

Let $\dim \varphi = 10$, anisotropic with trivial invariants. Express $\varphi \simeq \langle w \rangle \langle 1, a, b, c \rangle \perp \psi$. Then $\dim \psi = 6$, $d\psi = \langle abc \rangle$. Let $K = F(\sqrt{abc})$ so that $c(\psi_K) =$ quaternion. Part (2) implies ψ_K is isotropic, so that $\psi \simeq \langle u \rangle \langle\langle -abc \rangle\rangle \perp \delta$ where $\dim \delta = 4$ and $d\delta = \langle 1 \rangle$. Then $\varphi \cong \omega \perp \delta$ where $\omega = \langle w \rangle \langle 1, a, b, c \rangle \perp \langle u \rangle \langle\langle -abc \rangle\rangle$. Since $d\omega = \langle 1 \rangle$ and $c(\omega) = c(-\delta)$ is quaternion, (2) leads to a contradiction.)

10. **Linked Pfister forms.** If a Pfister form $\varphi \simeq \langle 1 \rangle \perp \varphi'$ then φ' is called the *pure part* of φ. For instance, if $\varphi = \langle\langle a, b \rangle\rangle$ then $\varphi' = \langle a, b, ab \rangle$. In this case, if φ' represents d then we may use d as one of the "slots": $\varphi \simeq \langle\langle d, x \rangle\rangle$ for some x. The 2-fold Pfister forms φ and ψ are *linked* if there is a common slot: $\varphi \simeq \langle\langle a, x \rangle\rangle$ and $\psi \simeq \langle\langle a, y \rangle\rangle$ for some $a, x, y \in F^\bullet$. This occurs if and only if φ' and ψ' represent a common value, if and only if $\varphi' \perp -\psi'$ is isotropic.

(1) φ and ψ are linked $\iff c(\varphi)c(\psi) =$ quaternion.

(2) Suppose ψ_1, ψ_2, ψ_3 are 2-fold Pfister forms and $c(\psi_1)c(\psi_2)c(\psi_3) = 1$. Then $\psi_1 \simeq \langle\langle a, x \rangle\rangle$, $\psi_2 \simeq \langle\langle a, y \rangle\rangle$ and $\psi_3 \simeq \langle\langle a, -xy \rangle\rangle$ for some $a, x, y \in F^\bullet$.

(3) Suppose β is anisotropic, $\dim \beta = 8$, $d\beta = \langle 1 \rangle$ and $c(\beta) =$ quaternion. Then $\beta \simeq \langle\langle a \rangle\rangle \otimes \gamma$ for some $a \in F^\bullet$ and some form γ.

(4) Complete the proof of (3.21).

(5) Let Q_1 and Q_2 be quaternion algebras with norm forms φ_1 and φ_2. Let $\alpha = \varphi_1' \perp -\varphi_2'$. Then $Q_1 \otimes Q_2$ is a division algebra if and only if α is anisotropic.

(Hint. (1) $q = \varphi' \perp -\psi'$ has $\dim q = 6$, $dq = \langle 1 \rangle$ and $c(q) = c(\varphi)c(\psi)$. Apply Exercise 9(2).

(2) $c(\psi_1)c(\psi_2) = c(\psi_3)$ and (1) imply $\psi_1 \simeq \langle\langle a, x \rangle\rangle$, $\psi_2 \simeq \langle\langle a, y \rangle\rangle$ for some a, x, y. Witt invariants then imply $\psi_3 \simeq \langle\langle a, -xy \rangle\rangle$.

(3) Let $\beta = \langle x \rangle \langle\langle -b \rangle\rangle \perp \delta$. Exercise 9(2) over $K = F(\sqrt{b})$ and (3.20) imply $\langle y \rangle \langle\langle -b \rangle\rangle \subset \delta$ for some y. Then $\beta \simeq \varphi_1 \perp \varphi_2$ where $\dim \varphi_i = 4$ and $d\varphi_i = \langle 1 \rangle$.

(Here $\varphi_1 \simeq \langle\langle -b\rangle\rangle \otimes \langle x, y\rangle$.) Express $\varphi_i \simeq \langle x_i\rangle\psi_i$ for some x_i and some 2-fold Pfister forms ψ_i. Apply part (1).

(4) Suppose $\dim\varphi = 12$, and φ is anisotropic with trivial invariants. Express $\varphi = \langle a\rangle\langle\langle -b\rangle\rangle \perp \alpha$. By (3.21) (3), α is isotropic over $K = F(\sqrt{b})$. Then $\varphi \simeq \varphi_1 \perp \beta$ where $\varphi_1 = \langle\langle -b\rangle\rangle \otimes \langle a, u\rangle$, $\dim\beta = 8$, $d\beta = \langle 1\rangle$ and $c(\beta) = [b, -au]$. Part (3) implies $\beta = \varphi_2 \perp \varphi_3$ where $\dim\varphi_i = 4$ and $d\varphi_i = \langle 1\rangle$. Express $\varphi_i = \langle x_i\rangle\psi_i$ and apply part (2).)

11. **Graded algebras.** Let A be an associative F-algebra with 1. Then A is graded (or more precisely, "$\mathbb{Z}/2\mathbb{Z}$-graded") if $A = A_0 \oplus A_1$, a direct sum as F-vector spaces, such that $A_i \cdot A_j \subseteq A_{i+j}$, where the subscripts are taken mod 2. It follows that $1 \in A_0$ and A_0 is a subalgebra of A. Every Clifford algebra $C(q)$ is a graded algebra using $C(q) = C_0(q) \oplus C_1(q)$. An element of A is *homogeneous* if it lies in $A_0 \cup A_1$. If a is homogeneous define the *degree* $\partial(a) = i$ if $a \in A_i$. The graded F-algebras A and B are *graded-isomorphic*, if there exists $f : A \to B$ which is an F-algebra isomorphism satisfying $f(A_i) = B_i$ for $i = 0, 1$.

Define the *graded tensor product* $A\hat{\otimes}B$ by taking the vector space $A \otimes B$ with the new multiplication induced by: $(a \otimes b) \cdot (a' \otimes b') = (-1)^{\partial(a)\partial(b)}aa' \otimes bb'$. Then $A \hat{\otimes} B$ is a graded algebra with $(A \hat{\otimes} B)_0 = A_0 \otimes B_0 + A_1 \otimes B_1$ and $(A \hat{\otimes} B)_1 = A_1 \otimes B_0 + A_0 \otimes B_1$. Then $A \hat{\otimes} 1$ and $1 \hat{\otimes} B$ are graded subalgebras of $A \hat{\otimes} B$. In the category of graded F-algebras the graded tensor product is commutative and associative, and it distributes through direct sums.

(1) **Lemma.** *If α, β are quadratic forms then $C(\alpha \perp \beta) \cong C(\alpha)\hat{\otimes}C(\beta)$ as graded algebras.*

(2) **Lemma.** *The Clifford algebras $C(\alpha)$ and $C(\beta)$ are graded-isomorphic if and only if*

$$\dim\alpha = \dim\beta, \quad d\alpha = d\beta \quad \textit{and} \quad c(\alpha) = c(\beta).$$

(3) Let A be a graded F-algebra. An A-module V is a *graded A-module* if V has a decomposition $V = V_0 \oplus V_1$ such that $A_i \cdot V_j \subseteq V_{i+j}$ whenever $i, j \in \mathbb{Z}/2\mathbb{Z}$. Let C be the Clifford algebra of some regular quadratic form. If V is a graded C-module then $V \cong C \otimes_{C_0} V_0$. Thus there is a one-to-one correspondence: {graded C-modules} $\leftrightarrow$ {C_0-modules}.

12. **4-dimensional forms.** Suppose $\dim\alpha = 4$, $d\alpha = \langle d\rangle$, and $L = F(\sqrt{d})$.

(1) $c(\alpha \otimes L) = 1$ iff α is isotropic.

(2) $c(\alpha) =$ quaternion iff $\alpha \perp \langle -1, d\rangle$ is isotropic.

(3) Suppose $\dim\alpha = \dim\beta = 4$, $d\alpha = d\beta = \langle d\rangle$, and $L = F(\sqrt{d})$.
Then α and β are similar over F if and only if $\alpha \otimes L$ and $\beta \otimes L$ are similar over L.

(4) If $\dim\alpha = \dim\beta = 4$, $d\alpha = d\beta$ and $c(\alpha) = c(\beta)$ then α and β are similar.

(5) If α and β are 4-dimensional forms then α, β are similar if and only if $C_0(\alpha) \cong C_0(\beta)$.

(Hint. (3) Scale α, β to assume $\alpha \simeq \langle 1\rangle \perp \alpha'$ and $\beta \simeq \langle 1\rangle \perp \beta'$. Since α_L and β_L are similar Pfister forms they are isometric and Exercises 8(3) and 9(2) imply that $\alpha' \perp -\beta'$ is isotropic (see Exercise 10). If $\alpha \simeq \langle\langle u\rangle\rangle \perp \langle a\rangle\langle\langle ud\rangle\rangle$ and $\beta \simeq \langle\langle u\rangle\rangle \perp \langle b\rangle\langle\langle ud\rangle\rangle$ then by (3.20), $\langle 1, -ab, u, -abud\rangle$ is isotropic. Choose $x \in D_F(\langle\langle u\rangle\rangle) \cap D_F(\langle ab\rangle\langle\langle ud\rangle\rangle)$, to obtain $\langle x\rangle\alpha \simeq \beta$.

(4) If $\langle d\rangle \neq \langle 1\rangle$ apply (3).)

13. (1) Suppose $\dim\varphi = 5$. Then $\varphi \subset \langle a\rangle\rho$ for some 3-fold Pfister form ρ if and only if φ represents $d\varphi$, if and only if $c(\varphi) =$ quaternion.

(2) Suppose $\dim\varphi = 6$. Then $\varphi \subset \langle a\rangle\rho$ for some 3-fold Pfister form ρ if and only if $\varphi \simeq \langle\langle x\rangle\rangle \otimes \delta$ for some $x \in F^\bullet$ and some form δ if and only if $c(\varphi)$ is split by $F(\sqrt{d\varphi})$.

14. **Trace forms.** Let $C = C(V, q)$ be a Clifford algebra of dimension 2^m. Define the "trace" map $\ell : C \to F$ to be the scalar multiple of the regular trace having $\ell(1) = 1$. Then for a derived basis $\{e^\Delta\}$ we have $\ell(e^\Delta) = 0$ whenever $\Delta \neq 0$. Recall that "bar" $= J_0$ is the involution extending -1_V. Define the associated trace form $B_0 : C \times C \to F$ by $B_0(x, y) = \ell(x\bar{y})$.

(1) B_0 is a regular symmetric bilinear form on C. If $q \simeq \langle a_1, \dots, a_m\rangle$ then $(C, B_0) \simeq \langle\langle -a_1, \dots, -a_m\rangle\rangle$. In particular the isometry class of this Pfister form is independent of the basis chosen for q.

(2) Suppose $\beta \simeq \langle b_1, \dots, b_m\rangle$ and define $P(\beta) = \langle\langle b_1, \dots, b_m\rangle\rangle$, the associated Pfister form.

Lemma. *If $\beta \simeq \gamma$ then $P(\beta) \simeq P(\gamma)$.*

This follows from part (1). Also prove it using Witt's Chain-Equivalence Theorem (Exercise 2.8), without mentioning Clifford algebras.

(3) Define $P_i(\beta)$ to be the "degree i" part of the Pfister form $P(\beta)$, so that $\dim P_i(\beta) = \binom{m}{i}$. For example, $P_0(\beta) = \langle 1\rangle$, $P_1(\beta) = \beta$ and $P_2(\beta) = \langle b_1b_2, b_1b_3, \dots, b_{m-1}b_m\rangle$. The lemma generalizes: If $\beta \simeq \gamma$ then $P_i(\beta) \simeq P_i(\gamma)$ for each i.

(4) If $C = C(-\alpha \perp \beta)$ and $J = J_{A,B}$ is the involution extending $(-1) \perp (1)$ on $-\alpha \perp \beta$, define the trace form B as before. Then $(C, B) \simeq P(\alpha \perp \beta)$.

15. **More trace forms.** Let $C = C(V, q)$ with the associated trace form $B_0 : C \times C \to F$ defined in Exercise 14.

(1) Let $\mathcal{L}$ and $\mathcal{R} : C \to \mathrm{End}_F(C)$ be the left and right regular representations. If $c \in C$ then $\mathcal{L}(c)$ is a similarity of (C, B_0) if and only if $\bar{c}c \in F$. Consequently $\mathcal{L}(F + V) \subseteq \mathrm{Sim}(C, B_0)$ is a subspace of dimension $m + 1$. Similarly for $\mathcal{R}(F+V)$. These two subspaces can be combined to provide an $(m+1, m+1)$-family $(\mathcal{L}(F + V), \mathcal{R}(F + V) \circ \alpha)$ where α is the canonical automorphism of C. How does this compare to the Construction Lemma 2.7?

(2) **Clifford and Cayley.** Let $C = C(\langle a_1, a_2, a_3\rangle)$ be an 8-dimensional Clifford algebra. Let $U = \mathrm{span}\{1, e_1, e_2, e_3\}$ so that $C = U + Uz$. Shift the $(4, 4)$-family constructed in (1) to an $(8, 0)$-family and identify it with C as in (1.9). This provides

a new multiplication $\diamond$ on C given by: $(x+y)\diamond c = xc + \alpha(c)y$, for $x \in U$, $y \in Uz$ and $c \in C$. Using $N_0(c) = B_0(c,c)$ we have $N_0(ab) = N_0(a)N_0(b)$. This $\diamond$ defines an octonion algebra. Compare the multiplication tables of the two algebras to see that they differ only by a few $\pm$ signs.

(3) Let $W \subseteq C$ be a linear subspace spanned by elements of degree $\equiv 2, 3 \pmod 4$. If $\bar{w}w = w\bar{w} \in F$ for every $w \in W$ then $\mathcal{L}(W) + \mathcal{R}(W) \circ \alpha \subseteq \mathrm{Sim}(C, B_0)$. For example we find $\langle\langle a_1\rangle\rangle \otimes \langle 1, a_2, \dots, a_m\rangle < \mathrm{Sim}(\langle\langle a_1, \dots, a_m\rangle\rangle)$.

16. **Clifford division algebras.** Let $\alpha = \langle 1\rangle \perp \alpha_1$ where $\dim\alpha = m+1$, and let $C = C(-\alpha_1)$. Then $\dim C = 2^m$. Note that $d\alpha = d(-\alpha_1)$.

(1) Here are necessary and sufficient conditions for C to be a division algebra:

$m = 1$: α is anisotropic.

$m = 2$: α is anisotropic.

$m = 3$: α and $\langle 1, -d\alpha\rangle$ are anisotropic.

$m = 4$: $\alpha \perp \langle -d\alpha\rangle$ is anisotropic.

$m = 5$: $\langle 1, -d\alpha\rangle$ is anisotropic and $\alpha \otimes F(\sqrt{da})$ is anisotropic.

No similar result is known for $m = 6$.

(2) Let q be a form over F and t an indeterminate. Then $C(q \perp \langle t\rangle)$ is a division algebra over $F(t)$ if and only if $C(q)$ is a division algebra over F.

(Hint. (1) For $m = 4$ use Exercise 10(5). (2) If $A = C(q)$ over F, let $A(t) = A \otimes F(t)$. Then $C = C(q \perp \langle t\rangle) = A(t) \oplus A(t)e$ where $e^2 = t$ and $e^{-1}xe = \alpha(x)$ for $x \in A(t)$. Suppose A is a division algebra and C is not, and choose $c = x(t) + y(t)e$ with $c^2 = 0$. Relative to a fixed basis of A assume that $x(t)$ and $y(t)$ have polynomial coefficients of minimal degree. From $x^2 + y\alpha(y)t = 0$ argue that t divides everything, contrary to the minimality.)

17. **Albert forms.** (1) Suppose q is a form with $\dim q = 6$ and $dq = \langle 1\rangle$. Then $q \perp \mathbb{H} \simeq \varphi_1 \perp -\varphi_2$, where φ_1 and φ_2 are 2-fold Pfister forms, and $c(q)$ is a tensor product of two quaternion algebras.

(2) **Lemma.** *Suppose α and β are forms with* $\dim\alpha = \dim\beta = 6$, $d\alpha = d\beta$ *and* $c(\alpha) = c(\beta)$. *Then α and β are similar.*

(3) Suppose $A = Q_1 \otimes Q_2$ is a tensor product of two quaternion algebras whose norm forms are φ_1 and φ_2. Define the *Albert form* $\alpha = \alpha_A = \varphi_1' \perp -\varphi_2'$. Then $\dim\alpha = 6$, $d\alpha = \langle 1\rangle$, $c(\alpha) = [A]$. Also: A is a division algebra if and only if α_A is anisotropic (by Exercise 10(5)).

Proposition. *The similarity class of α_A depends only on the algebra A and not on Q_1 and Q_2.*

(4) If $A = C(q)$ where $\dim q = 4$ then $\alpha_A = q \perp \langle -1, dq\rangle$.

(Hint. (2) Given $\alpha \equiv \beta \pmod{J_3(F)}$, we may assume $\alpha \simeq \langle 1\rangle \perp \alpha_1$ and $\beta \simeq \langle 1\rangle \perp \beta_1$. By (3.12), $\alpha_1 \perp -\beta_1$ is isotropic, so there exists d such that $\alpha_1 \simeq \langle d\rangle \perp \alpha_2$ and $\beta_1 \simeq \langle d\rangle \perp \beta_2$, where α_2 and β_2 are 4-dimensional. Apply Exercise 12(4).)

18. **Generalizing Albert forms.** (1) If $\dim q = 2m$ then $C(q)$ is isomorphic to a tensor product of m quaternion algebras. Conversely, if A is a tensor product of m quaternion algebras then $A \cong C(q)$ for some form q with $\dim q = 2m$. Possibly $C(q)$ is similar to a product of fewer than m quaternion algebras (e.g. if q is isotropic).

(2) An F-algebra A is similar to a tensor product of $m-1$ quaternion algebras if and only if $[A] = c(\varphi)$ for some form φ with $\dim \varphi = 2m$ and $d\varphi = \langle 1 \rangle$. Such a form φ is called an *Albert form* for A. Compare Exercise 17.

(3) Suppose A is a tensor product of quaternion algebras. If A is a division algebra then every Albert form for A is anisotropic.

(4) There exists some A having two Albert forms which are not similar. There exists some A which is not a division algebra but every Albert form of A is anisotropic.

(Hint. (2) ($\Longleftarrow$) Express $\varphi \simeq \langle a \rangle \perp \alpha$, note that $d\alpha = \langle -a \rangle$ and compute $c(\varphi) = c(\alpha) = [C_0(\alpha)]$. ($\Longrightarrow$) Express $A \cong C(q)$ where $\dim q = 2m-2$. Let $\varphi = q \perp \langle -1, dq \rangle$.

(4) For the second statement let D be the example of Amitsur, Rowen and Tignol (1979) mentioned in Theorem 6.15(2) below. Then $A = \mathbb{M}_2(D)$ is a product of 4 quaternion algebras but D contains no quaternion subalgebras. If φ is an isotropic Albert form for A then $[D] = [A] = c(\varphi)$ is a product of 3 quaternion algebras in $\mathrm{Br}(F)$.)

19. **The signs in $C(V, q)$.** Let $\{e_1, \dots, e_n\}$ be a basis of (V, q) corresponding to $q \simeq \langle a_1, \dots, a_n \rangle$, and with $\{e^\Delta\}$ the derived basis of $C = C(V, q)$. Then $e^\Delta e^\Gamma = \pm a_{\Delta\Gamma} e^{\Delta+\Gamma}$ where $a_\Delta = a_1^{\delta_1} \dots a_n^{\delta_n} \in F^\bullet$, as in Exercise 1.11.

(1) This $\pm$ sign is $(-1)^{\beta(\Delta,\Gamma)}$ where $\beta : \mathbb{F}_2^n \times \mathbb{F}_2^n \to \mathbb{F}_2$ is a bilinear form. In fact, if $\{\varepsilon_1, \dots, \varepsilon_n\}$ is the standard basis for $\mathbb{F}_2^n$ then $\beta(\varepsilon_i, \varepsilon_j) = 1$ if $i > j$ and 0 otherwise.

(2) Conversely, given a bilinear form β on $\mathbb{F}_2^n$ define an F-algebra $A(\beta)$ of dimension 2^n with basis $\{e^\Delta\}$ using the formula above. Then $A(\beta)$ is an associative algebra with 1. (For the β specified above, this observation leads to a proof of the existence of the Clifford algebra.)

(3) Let Q_β be the quadratic form on $\mathbb{F}_n^2$ defined by β, that is, $Q_\beta(x) = \beta(x, x)$. If β' is another form and $Q_\beta = Q_{\beta'}$ then $A(\beta) \cong A(\beta')$. Our algebra $A(\beta)$ could be called $A(Q)$.

(4) If Q is a regular form (i.e. the bilinear form $Q(x+y) - Q(x) - Q(y)$ is regular) then the algebra $A(Q)$ is central simple over F. (For the simplicity, suppose I is an ideal and choose $0 \neq c \in I$ with $c = \sum c_\Delta e^\Delta$ of minimal length. Compare Proposition 1.11.)

(5) If $Q \simeq Q_1 \perp Q_2$ then $A(Q) \cong A(Q_1) \otimes A(Q_2)$. Therefore if Q is regular, $A(Q)$ is a tensor product of quaternion algebras.

20. **Singular forms.** (1) Suppose q is a quadratic form on V and define the *radical* of V to be $V^\perp = \{v \in V : B(v, x) = 0 \text{ for every } x \in V\}$. Then q induces a quadratic form $\bar{q}$ on the quotient space $\bar{V} = V/V^\perp$ and $\bar{q}$ is regular. If W is any complement

of $V^\perp$ in V (i.e. $V = V^\perp \oplus W$ as vector spaces) the restriction $(W, q|W)$ is isometric to $(\bar{V}, \bar{q})$. Then $q \simeq r\langle 0\rangle \perp q_1$ for a regular form q_1, unique up to isometry.

(2) For q as above there exists an isometry $(V, q) \to (V', q')$ where q' is a regular form. In fact we can use $q' = r\mathbb{H} \perp q_1$. (Note: isometries are injective by definition.) What is the minimal value for dim V' here? Is that minimal form q' unique?

(3) Complete the proof of (3.7).

(4) If $q = r\langle 0\rangle \perp q_1$ for a regular form q_1, what is the center of $C(q)$? In particular what is the center of the exterior algebra $\Lambda(V)$?

Notes on Chapter 3

In 1878 W. K. Clifford defined his algebras using generators and relations. He expressed such an algebra as a tensor product of quaternion algebras and the center (but using different terminology). Clifford's algebras were used by R. Lipschitz in his study of orthogonal groups in 1884. These algebras were rediscovered in the case $n = 4$ by the physicist P. A. M. Dirac, who used them in his theory of electron spin. The algebraic theory of Clifford algebras was presented in a more general form by Brauer and Weyl (1935), by E. Cartan (1938) and by C. Chevalley (1954). Also compare Lee (1945) and Kawada and Iwahori (1950). Most of the algebraic information about Clifford algebras that we need appears in various texts, including: Bourbaki (1959), Lam (1973), Jacobson (1980), Scharlau (1985) and Knus (1988).

Terminologies and notations often vary with the author, and sometimes the notations are inconsistent within one book. For example we used A° for the "pure" part of the Cayley–Dickson algebra in Exercise 1.25 (3) and we used C_0 for the even Clifford algebra in (3.8). (Some authors use A^+ and C^+ for these objects.) Moreover, a quaternion algebra A is both a Cayley–Dickson algebra and a Clifford algebra, but we sometimes write A_0 for the set of pure quaternions.

Here are some further examples of confusing notations in this subject. As mentioned earlier, the Pfister form $\langle\langle a_1, \dots, a_n\rangle\rangle$ in this book (and in Lam (1973) and Scharlau (1985)) is written as $\langle\langle -a_1, \dots, -a_n\rangle\rangle$ in Knus et al. (1998). For a quadratic form q, Lam uses dq for our determinant $\det(q)$ and $d_\pm q$ for the discriminant. In Clifford algebra theory, the names for the canonical automorphism α, and for the involutions $x^\top = J_0(x)$ and $\bar{x} = J_1(x)$ are even less standard. Their names and notations vary widely in the literature.

The quaternion algebra decompositions in (3.14) follow Dubisch (1940). Similar explicit formulas were given by Clifford (1878).

Proofs of (3.18) also appear in Lam (1973), p. 121 or implicitly in Scharlau (1985), p. 81, Theorem 12.9.

The ideal $J_3(F)$ coincides with the ideal defined by Arason and Knebusch using the "degree" of a quadratic form. See Scharlau (1985), p. 164.

The proof of Theorem 3.21 first appeared in Satz 14 of Pfister (1966). It is also given in Scharlau (1985), pp. 90–91. Some work has been done recently on 14-dimensional forms with trivial invariants. See the remarks before (9.12) below.

Exercise 2. There is a more abstract treatment of the subspaces $V_{(k)}$. It uses the canonical bijection $\Lambda(V) \to C(V, q)$, and the exterior power $\Lambda^k(V)$ corresponds to $V_{(k)}$. In the usual product on $C(V, q)$, if $x \in V_{(r)}$ and $y \in V_{(s)}$ then $xy \in V_{(r+s)} + V_{(r+s-2)} + \cdots$. Using that bijection to transfer the exterior product "$\wedge$" to $C(V, q)$, it turns out that $x \wedge y$ is exactly the $V_{(r+s)}$-component of xy. See Bourbaki (1959), Wonenburger (1962a) or Marcus (1975).

Exercise 6. This first appeared in Satz 9 of Witt (1937).

Exercise 8. These are old results going back at least to Albert (1939). The simple proof of part (1) appears in Lam (1973), p. 200 and Scharlau (1985), p. 45.

Exercise 10. (5) Albert (1931) first discovered when a tensor product of two quaternion algebras is a division algebra. If $Q_1 \otimes Q_2$ is not a division algebra then α is isotropic and Q_1 and Q_2 have a common maximal subfield (as in 10(1)). A more direct proof of this was given by Albert (1972). A different approach appears in Knus et al. (1998) in Corollary 16.29.

Exercise 11. Further information on such graded algebras appears in Lam (1973), Chapter 4 and in Knus (1988), Chapter 4.

Exercise 12 follows Wadsworth (1975). See Exercise 5.23 below for another proof. A different method and related results appear in Knus (1988), pp. 76–78. Tignol has found a proof involving the corestriction of algebras.

Exercise 14. Some trace forms in the split case are considered in Exercise 1.13. See also Exercise 7.14.

Exercise 16. (1) Compare Edwards (1978).
(2) A more general result is stated in Mammone and Tignol (1986).

Exercise 17. Albert forms were introduced by Albert (1931) in order to characterize when a tensor product of two quaternion algebras is a division algebra. The proposition about the similarity class of α_A was first proved by Jacobson (1983) using Jordan structures. The quadratic form proof here was extended to characteristic 2 in Mammone and Shapiro (1989). The Albert form arises more naturally as the form on the alternating elements of A induced by a "Pfaffian" as mentioned in Chapter 9 below. This application of Pfaffians originated with Knus, Parimala and Sridharan (1989). This whole theory is also presented in Knus et al. (1998), §16.

Exercise 18 is part of the preliminary results for Merkurjev's construction of a non-formally real field F having $u(F) = 2n$. See Lam (1989) and Merkurjev (1991).

Exercise 19 follows ideas told to me by F. Rodriguez-Villegas.

Chapter 4

C-Modules and the Decomposition Theorem

An (s, t)-family on (V, q) provides V with the structure of a C-module, for a certain Clifford algebra C. Moreover, the adjoint involution I_q on $\mathrm{End}(V)$ is compatible with an involution J on C. We examine a more general sort of (C, J)-module, discuss hyperbolic modules and derive the basic Decomposition Theorem for (s, t)-families.

Before pursuing these general ideas let us state the main result. If (σ, τ) is a pair of quadratic forms over F we say that a quadratic space (V, q) is a (σ, τ)-*module* if $(\sigma, \tau) < \mathrm{Sim}(V, q)$. A (σ, τ)-module (V, q) is *unsplittable* if there is no decomposition $(V, q) \simeq (V_1, q_1) \perp (V_2, q_2)$ where each (V_i, q_i) is a non-zero (σ, τ)-module. Clearly every (σ, τ)-module is isometric to an orthogonal sum of unsplittables.

4.1 Decomposition Theorem. *Let (σ, τ) be a pair of quadratic forms where σ represents* 1. *All unsplittable (σ, τ)-modules have the same dimension 2^k, for some k.*

Without the condition that σ represents 1 the result fails. Examples appear in Exercise 5.9.

The proof of this theorem involves the development of the theory of quadratic (C, J)-modules, where C is a Clifford algebra with an involution J. In order to pinpoint the properties of the Clifford algebras used in the proof, we will develop the theory of quadratic modules over a semisimple F-algebra with involution. But before introducing those ideas we point out some simple consequences of (4.1).

First of all, this theorem explains why the Hurwitz–Radon function $\rho(n)$ depends only on the 2-power part: if $n = 2^m n_0$ where n_0 is odd, then $\rho(n) = \rho(2^m)$. In fact, suppose $(\sigma, \tau) < \mathrm{Sim}(q)$ where $\dim q = n$. By the theorem, an unsplittable (σ, τ)-module (W, φ) must have dimension 2^k dividing n. Then $k \le m$ and $(\sigma, \tau) < \mathrm{Sim}(q')$ for some form q' of dimension 2^m. Here q' can be taken to be $2^{m-k}\varphi$.

The theorem also provides a more conceptual proof of Proposition 1.10. Suppose (W, α) is unsplittable for $\langle 1, a\rangle$. (I.e., it is an unsplittable $(\langle 1, a\rangle, 0)$-module.) By (1.9) we know that $\langle x\rangle\langle\langle a\rangle\rangle \subset \alpha$ for some $x \in F^\bullet$, and in particular $\dim \alpha \ge 2$. By explicit construction, $\langle 1, a\rangle < \mathrm{Sim}(\langle\langle a\rangle\rangle)$, so $\langle\langle a\rangle\rangle$ is an unsplittable module for $\langle 1, a\rangle$. The Decomposition Theorem then implies that $\dim \alpha = 2$, so that $\alpha \simeq \langle x\rangle\langle\langle a\rangle\rangle$. It quickly follows that if $\langle 1, a\rangle < \mathrm{Sim}(q)$ then q is a sum of unsplittables, or equivalently $q \simeq \langle\langle a\rangle\rangle \otimes \varphi$ for some form φ.

Several small (s,t)-families can be analyzed in this way. These arguments are presented in Chapter 5, and the interested reader can skip there directly. For the rest of this Chapter we discuss quadratic modules.

Suppose $(S,T) \subseteq \mathrm{Sim}(V,q)$ is an (s,t)-family and σ and τ are the quadratic forms induced on S and T. Then $1_V \in S$ and we define $S_1 = (1_V)\perp$. If $f \in S_1$ and $g \in T$ then:

$$I_q(f+g) = -f+g$$
$$(f+g)^2 = f^2+g^2 = (-\sigma(f)+\tau(g))1_V.$$

Therefore the inclusion map $S_1 \perp T \to \mathrm{End}(V)$ is compatible with the quadratic form on $-\sigma_1 \perp \tau$ on $S_1 \perp T$. By the definition of Clifford algebras, there is an induced F-algebra homomorphism

$$\pi : C \to \mathrm{End}(V),$$

where $C = C(-\sigma_1 \perp \tau)$ is the Clifford algebra of dimension 2^{s+t-1}. This π provides an action of C on V, making V into a C-module.

To be a little more careful let us define the space $\check{S}_1 \perp \check{T}$ to be the generating space of C, so that $S_1 = \pi(\check{S}_1)$ and $T = \pi(\check{T})$. Setting $\check{S} = F\cdot 1 + \check{S}_1 \subseteq C$, we have $S = \pi(\check{S})$. For example, suppose $\sigma \simeq \langle 1, a_2, \dots, a_s\rangle$ and $\tau \simeq \langle b_1,\dots,b_t\rangle$, and let $f_2,\dots,f_s,g_1,\dots,g_t$ be given as in (2.1). The algebra $C = C(-\sigma_1 \perp \tau)$ has generators $e_2,\dots,e_s,d_1,\dots,d_t$ corresponding to the given diagonalization of $-\sigma_1 \perp \tau$. Then $\check{S} = \mathrm{span}\{1,e_2,\dots,e_s\}$, $\check{T} = \mathrm{span}\{d_1,\dots,d_t\}$ and $\pi(e_i) = f_i$, $\pi(d_j) = g_j$. We often identify S with $\check{S}$ and T with $\check{T}$ to simplify notations. We do this even though the identification of S and T with subspaces of C is sometimes misleading, since this embedding depends on the representation π.

4.2 Proposition. *Let (V,q) be a quadratic space of dimension $n = 2^m n_0$, where n_0 is odd. Then any (s,t)-family on (V,q) must have $s+t \le 2m+2$.*

Proof. Suppose $(\sigma,\tau) < \mathrm{Sim}(V,q)$ and let $C = C(-\sigma_1 \perp \tau)$ be the associated Clifford algebra with representation $\pi : C \to \mathrm{End}(V)$. By the Structure Theorem 3.6 either $\pi(C)$ or $\pi(C_0)$ is a central simple subalgebra of $\mathrm{End}(V)$. The double centralizer theorem implies that the dimension of this subalgebra must divide $\dim(\mathrm{End}(V)) = n^2$. Then 2^{s+t-2} divides $n^2 = 2^{2m}n_0^2$ so that $s+t-2 \le 2m$. □

The original 1, 2, 4, 8 Theorem is an immediate consequence of this inequality. (Compare Exercise 1.1.) The proof of this proposition uses only the information that V is a C-module. To get sharper information we will take the involutions into account and analyze the "unsplittable components" of (V,q).

4.3 Definition. Define $J_S = J_{\check{S}_1,\check{T}}$ to be the involution on C as in Definition 3.15. Then J_S acts as -1 on $\check{S}_1$ and as 1 on $\check{T}$.

This involution is chosen to match the behavior of the involution I_q on $\mathrm{End}(V)$. That is, the action of I_q on the maps f_i, g_j in $\mathrm{End}(V)$ matches the action of J_S on the generators e_i, d_j in C. Therefore, for every $c \in C$, $\pi(J_S(c)) = I_q(\pi(c))$. Hence, π is a homomorphism of algebras-with-involution:

$$\pi : (C, J_S) \to (\mathrm{End}(V), I_q).$$

Such a map π is sometimes called a *similarity representation* or a *spin representation*.

The map π is usually not written explicitly. We write the action of an element $c \in C$ as a multiplication: $cv = \pi(c)(v)$, for $v \in V$. Then the compatibility of the involutions says exactly that:

$$B_q(cv, w) = B_q(v, J_S(c)w), \tag{$*$}$$

for every $c \in C$ and $v, w \in V$. Conversely, if V is a C-module and q is a quadratic form on V satisfying the compatibility condition $(*)$, then $(\sigma, \tau) < \mathrm{Sim}(V, q)$. Therefore (V, q) is a (σ, τ)-module if and only if V is a C-module and the form q satisfies the compatibility condition $(*)$ above.

Different C-module structures on V can arise from the same family (S, T) in $\mathrm{Sim}(V, q)$. To see this let δ be an automorphism of C which preserves the subspaces $\check{S}_1$ and $\check{T}$. If π is a similarity representation coming from (S, T), define $\pi' = \pi \circ \delta$. Then π' is another similarity representation associated to (S, T). This ambiguity should cause little trouble, since we usually fix one representation.

Let us now set up the theory of quadratic modules. To see where the special properties of Clifford algebras are used, we describe the theory for a wider class of algebras.

Notation. Let C be a finite dimensional semisimple F-algebra with involution J. The involution is often written simply as "bar". Unless explicitly stated otherwise, every module is a left C-module which is finite dimensional over F.

It is useful to consider alternating spaces in parallel with quadratic spaces. To handle these cases together, let $\lambda = \pm 1$ and define a *λ-form* B on a vector space V to be a bilinear form $B : V \times V \to F$ which is λ-symmetric, that is: $B(y, x) = \lambda B(x, y)$ for every $x, y \in V$. The λ-form B is *regular* if $V^\perp = 0$, (or equivalently if the induced map θ_B from V to its dual is a bijection). Then a *λ-space* (V, B) is a vector space V with a regular λ-form B.

Since $2 \neq 0$ in F, a quadratic space is the same as a 1-space. That is, a quadratic form q is determined by its associated symmetric bilinear form B_q. An *alternating* space is another name for a (-1)-space. For any λ-space (V, B) there is an associated adjoint involution I_B as in (1.2).

It is well known that alternating spaces over the field F must have even dimension, and any two alternating spaces of the same dimension are isometric. However in the category of alternating spaces admitting C, such an easy characterization no longer applies.

4.4 Definition. Let C be an algebra with involution as above. Suppose B is a regular λ-form on a C-module V. Then B *admits* C if

$$B(cu, v) = B(u, \bar{c}v)$$

for every $u, v \in V$ and $c \in C$. In this case (V, B) is called a *λ-space admitting C*, or a *λ-symmetric (C, J)-module*. If V, V' are λ-spaces admitting C, then they are *C-isometric* (written $V \approx V'$) if there is a C-module isomorphism $V \to V'$ which is also an isometry.

We will extend the standard definitions and techniques of the theory of quadratic forms over F to this wider context. Much of this theory can be done more generally, allowing F to be a commutative ring having $2 \in F^{\bullet}$ and considering λ-hermitian modules. For example see Fröhlich and McEvett (1969), Shapiro (1976), or for more generality, Quebbemann et al. (1979). The category of λ-spaces admitting C is equivalent to the category of λ-hermitian C-modules. This equivalence is proved in the appendix to this chapter.

If $U \subseteq V$ then $U^{\perp}$ is the "orthogonal complement" in the usual sense: $U^{\perp} = \{x \in V : B(x, U) = 0\}$. Then a subspace $U \subseteq V$ is regular if and only if $U \cap U^{\perp} = 0$, and is totally isotropic iff $U \subseteq U^{\perp}$.

4.5 Lemma. *Let (V, B) be a λ-space admitting C, and let $T \subseteq V$ be a C-submodule.*

(1) *Then $T^{\perp}$ is also a C-submodule.*

(2) *If T is a regular subspace of V then $V \approx T \perp T^{\perp}$.*

(3) *If T is an irreducible submodule then either T is regular or T is totally isotropic.*

Proof. The same argument as in the classical cases. For (3) consider $T \cap T^{\perp}$. □

We can consider these bilinear forms in terms of dual spaces. Let $\hat{V} = \mathrm{Hom}_F(V, F)$ be the dual vector space. The dual pairing is written

$$\langle \mid \rangle : V \times \hat{V} \to F.$$

The space $\hat{V}$ has a natural structure as a right C-module, defined by: $\langle v|\hat{v}c\rangle = \langle cv|\hat{v}\rangle$. We will use the involution to change hands and make $\hat{V}$ into a left C-module.

4.6 Definition. Let V be a left C-module. For $\hat{v} \in \hat{V}$ and $c \in C$ define $c\hat{v}$ by:

$$\langle v|c\hat{v}\rangle = \langle \bar{c}v|\hat{v}\rangle \quad \text{for all } v \in V.$$

A λ-form B on V induces a linear map $\theta_B : V \to \hat{V}$ by defining $\theta_B(v) = B(-, v)$. That is, $\langle u|\theta_B(v)\rangle = B(u, v)$. By definition, B is regular if and only if θ_B is bijective. Furthermore the definitions imply that: B admits C if and only if θ_B is a C-module isomorphism.

Consequently if (V, B) is a λ-space admitting C then $\hat{V} \cong V$ as C-modules. Conversely, if $\hat{V} \cong V$ must there exist such a λ-form? Here is one special case when a form always exists.

4.7 Definition. Let T be a C-module and $\lambda = \pm 1$. Define $H_\lambda(T)$ to be the C-module $T \oplus \hat{T}$ together with the λ-form B_H defined by: $B_H(s + \hat{s}, t + \hat{t}) = \langle s|\hat{t}\rangle + \lambda\langle t|\hat{s}\rangle$, for $s, t \in T$ and $\hat{s}, \hat{t} \in \hat{T}$. A λ-space admitting C is *C-hyperbolic* if it is C-isometric to some $H_\lambda(T)$.

One easily checks that B_H is regular, λ-symmetric and admits C. If $T \cong T'$ as C-modules then $H_\lambda(T) \approx H_\lambda(T')$. Also $H_\lambda(S \oplus T) \approx H_\lambda(S) \perp H_\lambda(T)$.

4.8 Lemma. *Let (V, B) be a λ-space admitting C.*

(1) *V is C-hyperbolic iff $V = T_1 + T_2$ where T_1 and T_2 are totally isotropic submodules.*

(2) *$V \perp \langle -1\rangle V \approx H_\lambda(V)$ is C-hyperbolic.*

Proof. (1) If $V = H_\lambda(T)$ let $T_1 = T \oplus 0$ and $T_2 = 0 \oplus \hat{T}$. Conversely, suppose $V = T_1 + T_2$. Then $T_i \subseteq T_i^\perp$, and since $V^\perp = 0$ we have $T_1^\perp \cap T_2^\perp = 0$. The map $\psi : T_2 \to \hat{T}_1$, defined by $\psi(x) = B(-, x)$, is a C-homomorphism and $\ker \psi = 0$. The surjectivity of ψ follows from the surjectivity of θ_B (or by comparing dimensions). Then ψ is a C-isomorphism and $1 \oplus \psi$ is a C-isometry $V \to H_\lambda(T_1)$.

(2) We are given a space $U = V \oplus V'$ and a C-isomorphism $f : V \to V'$ which is also a (-1)-similarity. Let $T_+ = \{x + f(x) : x \in V\} \subseteq U$ be the graph of f, and similarly let T_- be the graph of $-f$. It is easy to check that T_+ and T_- are totally isotropic submodules and $U = T_+ + T_-$ so part (1) applies. □

4.9 Proposition. *Let (V, B) be a λ-space admitting C and $T \subseteq V$ a totally isotropic submodule. Then there is another totally isotropic submodule $T' \subseteq V$ with $T + T' \approx H_\lambda(T)$, a regular submodule of V.*

Proof. Since C is semisimple there exists a submodule W of V complementary to the submodule $T^\perp$. Then the induced pairing $B_0 : W \times T \to F$ induces a C-isomorphism $W \cong \hat{T}$. If W is totally isotropic then $T + W \approx H_\lambda(T)$ by Lemma 4.8. Therefore the proposition is proved once we find a totally isotropic complement to $T^\perp$.

Given W, all other complements to $T^\perp$ appear as graphs of C-homomorphisms $f : W \to T^\perp$. Define the homomorphism $f : W \to T \subseteq T^\perp$ by the equation:

$$B_0(w_1, f(w_2)) = -\tfrac{1}{2}B(w_1, w_2),$$

for $w_1, w_2 \in W$. The nonsingularity of the induced pairing B_0 above shows that f is well-defined and C-linear. The graph of this f provides the complement we need. □

This proposition gives information about the structure of unsplittables. We define a λ-space V admitting C to be *C-unsplittable* if there is no expression $V \approx V_1 \perp V_2$ where V_1 and V_2 are non-zero submodules. Equivalently, V is C-unsplittable if and only if V has no regular proper submodules. The following result contains most of the information needed to prove the Decomposition Theorem 4.1.

4.10 Theorem. *Let (V, B) be an unsplittable λ-space admitting C. Then either V is irreducible or $V \approx H_\lambda(T)$ for an irreducible module T. Moreover, $H_\lambda(T)$ is unsplittable if and only if T is irreducible and possesses no regular λ-form admitting C.*

Proof. If V is reducible let $T \subseteq V$ be an irreducible submodule. Since V is unsplittable T must be singular, and hence totally isotropic by (4.5) (3). Then by (4.9) we have $T \subseteq H \subseteq V$, where $H \approx H_\lambda(T)$. Since V is unsplittable, $V = H$.

Now suppose T is an irreducible C-module. If T possesses a regular λ-form admitting C, then by (4.8) (2) we have $H_\lambda(T) \approx T \perp \langle -1 \rangle T$ is splittable. □

Let (V, B) be any λ-space admitting C. Then there is a decomposition

$$V = H_u \perp H_s \perp V_a,$$

where H_u is an orthogonal sum of unsplittable C-hyperbolic subspaces, and H_s is an orthogonal sum of splittable hyperbolic subspaces, and V_a is C-anisotropic. Here we define a λ-space to be C-anisotropic if it has no totally isotropic irreducible submodules. From (4.10) we conclude that no irreducible submodule of H_u can be isomorphic to a submodule of $H_s \perp V_a$. Therefore H_u and $H_s \perp V_a$ are uniquely determined submodules of V.

Moreover, as in the classical case, the submodules H_s and V_a are unique up to isometry because there is a Cancellation Theorem:

If U, V and W are λ-spaces admitting C and $U \perp V \approx U \perp W$ then $V \approx W$.

We omit the proof of this theorem because it does not seem to have a direct application to the study of (s, t)-families. Proofs of more general results appear in McEvett (1969), Shapiro (1976), and Quebbemann et al. (1979). Knowing the Cancellation Theorem it is natural to investigate the Witt ring $W^\lambda(C, J)$ of λ-spaces admitting (C, J). We will not pursue this investigation here.

When does an irreducible C-module W possess a regular λ-form admitting C? If such a form exists then certainly $\hat{W} \cong W$ as C-modules.

4.11 Lemma. *If W is an irreducible C-module with $\hat{W} \cong W$, then W has a regular λ-form admitting C, for some sign λ.*

Proof. If $g : U \to V$ is a C-module homomorphism, the transpose $g^\top : \hat{V} \to \hat{U}$ is defined by the equation $\langle u | g^\top(\hat{v}) \rangle = \langle g(u) | \hat{v} \rangle$. Any C-isomorphism $\theta : W \to \hat{W}$

induces a regular bilinear form B on W by setting $B(x, y) = \langle x|\theta(y)\rangle$. After identifying W with its double dual, we see that B is a regular λ-form if and only if $\theta^\top = \lambda\theta$. Now from the given C-isomorphism θ, define $\theta_\lambda = \frac{1}{2}(\theta + \lambda\theta^\top)$. These two maps are C-homomorphisms and $\theta_\lambda^\top = \lambda \cdot \theta_\lambda$. They are not both zero since $\theta = \theta_+ + \theta_-$, so at least one of them must be an isomorphism, by the irreducibility. The corresponding form B is then a regular λ-form admitting C. □

We now return to the original situation of Clifford algebras. Let (σ, τ) be a pair of forms as usual and $C = C(-\sigma_1 \perp \tau)$ with the involution J_S. For a λ-space (V, B) we see that $(\sigma, \tau) < \mathrm{Sim}(V, B)$ if and only if V can be expressed as a C-module where B admits (C, J_S). If C is simple then up to isomorphism there is only one irreducible module, and the ideas above are easy to apply. The non-simple case requires an additional remark.

Suppose C is not simple. Then $s + t$ is even, $d(-\sigma_1 \perp \tau) = \langle 1\rangle$ and C_0 is simple. We can choose $z = z(S_1 \perp T)$ with $z^2 = 1$. Then $Z = F + Fz$ is the center of C and the non-trivial central idempotents are $e_\varepsilon = \frac{1}{2}(1 + \varepsilon z)$ for $\varepsilon = \pm 1$. Then $C = Ce_+ \times Ce_- \cong C_0 \times C_0$ and there are two natural projection maps $p_+, p_- : C \to C_0$. Let V be an irreducible C_0-module with associated representation $\pi : C_0 \to V$. Define V_ε to be the C-module associated to the representation $\pi_\varepsilon = \pi \circ p_\varepsilon$. It follows that every irreducible C-module is isomorphic to either V_+ or V_-. These two module structures differ only by an automorphism of C.

4.12 Lemma. *Let C, J_S be as above. Let $\lambda = \pm 1$ be a fixed sign.*

(1) *Suppose C is not simple. If $s \equiv t \pmod 4$ then $\hat{V}_+ \cong V_+$ and $\hat{V}_- \cong V_-$. If $s \equiv t + 2 \pmod 4$ then $\hat{V}_+ \cong V_-$.*

(2) *If one irreducible C-module possesses a regular λ-form admitting C, then they all do.*

Proof. (1) By the definition of $\hat{W}$ as a left C-module, z acts on $\hat{W}$ as $J_S(z)$ acts on W. Therefore if $J_S(z) = z$ then $\hat{V}_\varepsilon \cong V_\varepsilon$. If $J_S(z) = -z$ then $\hat{V}_\varepsilon \cong V_\varepsilon$. A sign calculation completes the proof.

(2) If C is simple the claim is trivial. Suppose C is not simple and V_ε possesses a regular λ-form B admitting C. The module $V_{-\varepsilon}$ can be viewed as the same vector space V with a twisted representation: $\pi_{-\varepsilon}(c) = \pi_\varepsilon(\alpha(c))$ where α is the canonical automorphism of C. It follows that the form B admits this twisted representation (since α and J_S commute), so $V_{-\varepsilon}$ also has a regular λ-form admitting C. □

4.13 Corollary. *Let (C, J_S) be the Clifford algebra with involution as above. Let $\lambda = \pm 1$ be a fixed sign. Then all the C-unsplittable λ-spaces have the same dimension 2^k, for some k.*

Proof. By the remarks above, all irreducible C-modules have the same dimension 2^m, for some m. It is a power of 2 since C is a direct sum of irreducibles and $\dim C = 2^{r+s-1}$. If one irreducible module possesses a regular λ-form admitting C, then they all do and by (4.10) every unsplittable is irreducible of dimension 2^m. Otherwise (4.10) implies that every unsplittable is isometric to $H_\lambda(T)$ for some irreducible T, so that the unsplittables all have dimension 2^{m+1}. □

Finally we can complete the proof of the original Decomposition Theorem 4.1. The only remaining step is to show that the two notions of "unsplittable" coincide. The problem is that the definition of unsplittable (σ, τ)-module involves isometry of quadratic spaces over F (written $\simeq$) while the definition of unsplittable quadratic space admitting C involves C-isometry (written $\approx$). If a space is (σ, τ)-unsplittable it certainly must be C-unsplittable. Conversely suppose (V, q) is C-unsplittable, but $V \simeq V_1 \perp V_2$ for some non-zero (σ, τ)-modules V_i. Then V_i is a quadratic space admitting C, so it is C-isometric to a sum of C-unsplittables. Comparing dimensions we get a contradiction to (4.13). This completes the proof. □

4.14 Definition. An (s, t)-pair (σ, τ) is of *regular type* if an unsplittable quadratic (σ, τ)-module is irreducible. Otherwise it is of *hyperbolic type*. In working with alternating forms we say that (σ, τ) is of *(-1)-hyperbolic type* or of *(-1)-regular type*.

By (4.12) this condition does not depend on the choice of unsplittable module. If (σ, τ) is of hyperbolic type then (4.10) implies that every unsplittable (σ, τ)-module is $H_\lambda(T)$ for some irreducible C-module T. Consequently, every (σ, τ)-module is C-hyperbolic and (σ, τ)-modules are easy to classify. For example, if $s \equiv t + 2 \pmod 4$ then (4.12) implies that (σ, τ) is of hyperbolic type (and of (-1)-hyperbolic type). The other cases are not as easy to classify. Further information about these types is obtained in Chapter 7.

When C is a division algebra with involution the irreducible module is just C itself. In this case it is sometimes useful to analyze all the λ-forms on C which admit C. This can be done by comparing everything to a given trace form.

4.15 Definition. Let C be an F-algebra with involution. An F-linear map $\ell : C \to F$ is an *involution trace* if

(1) $\ell(\bar{c}) = \ell(c)$ for every $c \in C$;

(2) $\ell(c_1c_2) = \ell(c_2c_1)$ for every $c_1, c_2 \in C$;

(3) The F-bilinear form $L : C \times C \to F$, defined $L(c_1, c_2) = \ell(c_1c_2)$, is regular.

For example if $C = \mathrm{End}(V)$ and J is any involution on C, then the ordinary trace is an involution trace. Suppose $C = C(-\sigma_1 \perp \tau)$ is a Clifford algebra as above and

$J = J_S$. Then the usual trace is an involution trace. Generally every semisimple F-algebra with involution does possess an involution trace. (See Exercise 14 below.)

4.16 Proposition. *Let C be an F-algebra with involution J and with an involution trace ℓ. If B is a regular λ-form on C such that (C, B) admits C then there exists $d \in C^\bullet$ such that $J(d) = \lambda d$ and $B(x, y) = \ell(xdJ(y))$ for all $x, y \in C$.*

Proof. Let $B_1(x, y) = \ell(xJ(y))$. Since ℓ is an involution trace, B_1 is a regular 1-form on C. It is easy to check that (C, B_1) admits C. Since B_1 is regular, the given bilinear form B can be described in terms of B_1: there exists $f \in \mathrm{End}_F(C)$ such that $B(x, y) = B_1(f(x), y)$ for all $x, y \in C$. Since B is regular this f is invertible. The condition that B admits C becomes $B_1(f(cx), y) = B(cx, y) = B(x, J(c)y) = B_1(f(x), J(c)y) = B_1(cf(x), y)$. Therefore $f(cx) = cf(x)$, and f is determined by its value $d = f(1)$. Therefore $B(x, y) = B_1(xd, y) = \ell(xdJ(y))$. Since B is a λ-form we find $J(d) = \lambda d$. □

For example, one such trace form was considered in Exercise 3.14. In the case that $C = \left(\frac{-a,-b}{F}\right)$ is a quaternion division algebra with the usual involution J = "bar" there is a *unique* unsplittable quadratic (C, J)-module. This follows since the only J-symmetric elements of C are the scalars. The "uniqueness" here is up to a C-similarity. In terms of the quadratic forms over F this says that if $\langle 1, a, b\rangle < \mathrm{Sim}(q)$ is unsplittable then q is similar to $\langle\langle a, b\rangle\rangle$. This re-proves Proposition 1.10.

Appendix to Chapter 4. λ-Hermitian forms over C

In this appendix we return to the more general set-up where C is a semisimple F-algebra with involution J and describe the one-to-one correspondence between the λ-spaces admitting C and the λ-hermitian forms over C. This equivalence of categories provides a different viewpoint for this whole theory. Since these ideas are not heavily used later we just sketch them in this appendix.

We could allow the involutions here to be non-trivial on the base ring F. In that case there can be λ-hermitian forms for any $\lambda \in F$ with $\bar{\lambda}\lambda = 1$. The details were worked out by Fröhlich and McEvett (1969), and the reader is referred there for further information.

A.1 Definition. Let V be a C-module and $\lambda = \pm 1$. A *λ-hermitian* form on V (over C) is a mapping $h : V \times V \to C$ satisfying

(1) h is additive in each slot,

(2) $h(cx, y) = ch(x, y)$ for every $x, y \in V$ and $c \in C$,

(3) $h(y, x) = \lambda\overline{h(x, y)}$ for every $x, y \in V$.

It follows that $h(x, cy) = h(x, y)\bar{c}$. For a simple example let $V = C$ and for $a \in C$ define $h_a : C \times C \to C$ by $h_a(x, y) = xa\bar{y}$. If $\bar{a} = \lambda a$ then h_a is a λ-hermitian form.

Define the C-dual module $\tilde{V} = \mathrm{Hom}_C(V, C)$. For fixed $x \in V$ the map $\theta_h(v) = h(-, v)$ is in $\tilde{V}$. This map $\theta_h : V \to \tilde{V}$ is F-linear. The form h is said to be *regular* if θ_h is bijective. Define a *λ-hermitian space* over C to be a C-module V equipped with a regular λ-hermitian form h. One can now define isometries, similarities, orthogonal sums and tensor products in the category of λ-hermitian spaces.

In analogy with our treatment of the F-dual, we write the dual pairing as

$$[\ |\ ] : V \times \tilde{V} \to C.$$

Then $[\ |\ ]$ is F-bilinear, and by definition $[cx|\tilde{x}] = c[x|\tilde{x}]$. With this notation we have: $[u|\theta_h(v)] = h(u, v)$.

As before we use the involution on C to change hands and make $\tilde{V}$ a left C-module: for $\tilde{x} \in \tilde{V}$ and $c \in C$ define $c\tilde{x}$ by

$$[x|c\tilde{x}] = [x|\tilde{x}]\bar{c}$$

for all $x \in V$. Therefore if h is a λ-hermitian form then $\theta_h : V \to \tilde{V}$ is a homomorphism of left C-modules.

The hyperbolic functor H_λ can be introduced in this new context, in analogy with the discussion for λ-spaces admitting C. All the results proved above for λ-spaces admitting C can be done for λ-hermitian modules. In fact, when there is an involution trace map on C, these two contexts are equivalent. This equivalence arises because the two notions of "dual" module, $\tilde{V}$ and $\hat{V}$, actually coincide.

A.2 Proposition. *Let C be a semisimple F-algebra with involution J and possessing an involution trace ℓ.*

(1) *Let (V, h) be a λ-hermitian space over C and define $B = \ell \circ h$. Then (V, B) is a λ-space admitting C, denoted by $\ell_*(V, h)$.*

(2) *Let (V, B) be a λ-space admitting C. Then there is a unique regular λ-hermitian form h on V having $\ell_*(V, h) = (V, B)$.*

(3) *This correspondence ℓ_* preserves isometries, orthogonal sums and the H_λ construction.*

Proof sketch. (This is Theorem 7.11 of Fröhlich and McEvett.) (1) It is easy to see that B is F-bilinear, λ-symmetric and admits C. Composition with ℓ induces a map on the dual spaces $\ell_0 : \tilde{V} \to \hat{V}$, that is: $\langle x|\ell_0(\tilde{x})\rangle = \ell([x|\tilde{x}])$. The properties of ℓ imply that this ℓ_0 is an isomorphism of left C-modules. Furthermore $\ell_0 \circ \theta_h = \theta_B$, and we conclude that B is regular.

(2) Given B we can construct θ_h as $\ell_0^{-1} \circ \theta_B$. It then follows that h is λ-hermitian and $\ell \circ h = B$.

(3) Checking these properties is routine. □

By using hermitian forms over C we sometimes obtain a better insight into a problem. For instance the simplest hermitian spaces over C are the "1-dimensional" forms obtained from the left C-module C itself. If $a \in C^\bullet$ satisfies $\bar{a} = \lambda a$ then the form

$$h_a : C \times C \to C \quad \text{defined by} \quad h_a(x, y) = xa\bar{y},$$

is a regular λ-hermitian form on C. Let $\langle a \rangle_C$ denote this λ-hermitian space (C, h_a). In this notation, the trace form considered in Proposition 4.16 is just $\ell_*(\langle d \rangle_C)$. This transfer result (A.2) quickly yields another proof of (4.16).

A.3 Lemma. *Let C be an F-algebra with involution and $a, b \in C^\bullet$ with $\bar{a} = \lambda a$ and $\bar{b} = \lambda b$. Then $\langle a \rangle_C \simeq \langle b \rangle_C$ if and only if $b = ca\bar{c}$ for some $c \in C^\bullet$.*

Proof. Suppose $\varphi : (C, h_b) \to (C, h_a)$ is an isometry. Then φ is an isomorphism of left C-modules, so that $\varphi(x) = xc$ where $c = \varphi(1) \in C^\bullet$. Since $h_a(\varphi(x), \varphi(y)) = h_b(x, y)$ the claim follows. The converse is similar. □

A.4 Corollary. *Suppose D is an F-division algebra with involution.*

(1) *Every hermitian space over D has an orthogonal basis.*

(2) *If the involution is non-trivial then every (-1)-hermitian space over D has an orthogonal basis.*

Proof. (1) The irreducible left D-module D admits a regular hermitian form (e.g. $\langle 1 \rangle_D$). Then by (A.2) and (4.10) we conclude that every hermitian space over D is an orthogonal sum of unsplittable submodules, each of which is 1-dimensional over D. These provide an orthogonal basis.

(2) There exists $a \in D^\bullet$ such that $\bar{a} = -a$. Then $\langle a \rangle_D$ is a regular (-1)-hermitian form on D. The conclusion follows as before. □

Of course this corollary can be proved more directly, without transferring to the theory of λ-spaces admitting D.

Exercises for Chapter 4

1. **Group rings.** Suppose G be a finite group with order n and the characteristic of F does not divide n. Then the group algebra $C = F[G]$ is semisimple (Maschke's theorem). Define the involution J on C by sending $g \mapsto g^{-1}$ for every $g \in G$. There is a one-to-one correspondence between orthogonal representations $G \to \mathrm{O}(V, q)$ and quadratic (C, J)-modules. These algebras provide examples where unsplittable (C, J)-modules may have different dimensions.

2. **Proposition.** *Suppose V, W are C-anisotropic λ-spaces admitting C. If $V \perp W$ is C-hyperbolic then $V \approx \langle -1 \rangle W$. This is a converse of* (4.8) (2).

(Hint. Let $V \perp W = H$ and suppose $T \subseteq H$ is a totally isotropic submodule with $2 \cdot \dim T = \dim H$. Examine the projections to V and W to see that T is the graph of some $f : V \to W$. This f must be a (-1)-similarity.)

3. **Averaging Process.** (1) Let F be an ordered field and suppose σ, τ are positive definite forms over F with $\sigma \simeq \langle 1 \rangle \perp \sigma_1$ as usual. Let $C = C(-\sigma_1 \perp \tau)$ and let V be a C-module. Then there exists a positive definite quadratic form q on V making $(\sigma, \tau) < \mathrm{Sim}(q)$.

(2) Let $\mathbb{R}$ be the real field, and $C = C((r-1)\langle -1 \rangle)$. Then $r\langle 1 \rangle < \mathrm{Sim}(n\langle 1 \rangle)$ over $\mathbb{R}$ if and only if there is a C-module of dimension n. This explains why the Hurwitz–Radon Theorem can be done over $\mathbb{R}$ without considering the involutions.

(Hint. (1) Let φ be a positive definite form on V and define q by averaging. For instance if $\sigma \simeq \langle 1, a_2, \dots, a_n \rangle$ and $\tau = 0$, define $q(x) = \sum a_\Delta^{-1} \varphi(e^\Delta x)$.)

4. **Commuting similarities.** What are the possible dimensions of two subspaces of $\mathrm{Sim}(V, q)$ which commute elementwise?

Definition. $\kappa_r(n) = \max\{s$: there exist commuting subspaces of dimensions r and s in $\mathrm{Sim}(V, q)$ for some n-dimensional quadratic space $(V, q)\}$.

Let $R, S \subseteq \mathrm{Sim}(V, q)$ be commuting subspaces which we may assume contain 1_V. Let D be the Clifford algebra corresponding to R so that V is a left D-module. Define $\mathrm{Sim}_D(V, q)$ and note that $S \subseteq \mathrm{Sim}_D(V, q)$. If C is the Clifford algebra for S then this occurs if and only if there is a homomorphism $C \otimes D \to \mathrm{End}(V)$ which preserves the involutions.

(1) If $n = 2^m \cdot$ (odd) then $\kappa_r(n) = \kappa_r(2^m)$.

(2) Define (s, t)-families in $\mathrm{Sim}_D(V, q)$ and prove the Expansion, Shift and Construction Lemmas.

(3) Define $\kappa'_r(n)$ analogously for alternating forms (V, B).

(i) $\kappa'_r(4n) \geq 4 + \kappa_r(n)$ and $\kappa_r(4n) \geq 4 + \kappa'_r(n)$.

(ii) If $r \equiv 2 \pmod 4$ then $\kappa'_r(n) = \kappa_r(n)$.

(4) $\kappa_2(2^m) = 2m$.

(5) **Proposition.** *Suppose $\kappa_r(n) > 0$ where $n = 2^m \cdot$ (odd). Then*

$$\kappa_r(n) = \begin{cases} \rho(n) + 1 - r & \text{if } r \equiv 1 \pmod 4 \\ 2m + 2 - r & \text{if } r \equiv 2 \pmod 4 \\ 1 + \rho_r(n) & \text{if } r \equiv 3 \pmod 4. \end{cases}$$

(Hint. (3) (i) If $\sigma < \mathrm{Sim}_D(q)$ then $(\sigma \perp \langle 1, 1 \rangle, \langle 1, 1 \rangle) < \mathrm{Sim}(\langle\langle 1, 1 \rangle\rangle \otimes q)$. Shift by 2 as in Exercise 2.6(1). (ii) Let $R, S \subseteq \mathrm{Sim}(V, q)$. Then $z = z(R_1)$ commutes with R and S, $\tilde{z} = -z$ and $R, S \subseteq \mathrm{Sim}(V, B')$ where $B'(u, v) = B(u, zv)$. In fact an (s, t)-family in this case is equivalent to an $(s + t, 0)$-family.

(4) $\kappa_2(2^m) \le 2m$ for otherwise the representation $C \to \operatorname{End}_F(V)$ is surjective and there is no room for D. Check $\kappa_2(1) = 0$ and $\kappa_2(2) = 2$, then apply (3), (4) and induction.

(5) When r is odd, consider commuting $R, S \subseteq \operatorname{Sim}(V)$ and examine $z(R_1) \circ S$. For the even case use (3) to see $\kappa_r(n) \le \min\{\kappa_{r-1}(n), \kappa'_{r-1}(n)\} = 2m + 2 - r$. Equality follows as in (4).)

5. **More commuting similarities.** Let $D = C(\langle -1\rangle\alpha_1)$ where $\alpha = \langle 1\rangle \perp \alpha_1$ is a given form with $\dim \alpha = r$. We examine subspaces $S \subseteq \operatorname{Sim}_D(V, q)$.

Definition. $\kappa(\alpha; n) = \max\{s$: there exists a subspace of dimension s in $\operatorname{Sim}_D(V, q)$ for some n-dimensional quadratic space for which $\alpha < \operatorname{Sim}(V, q)\}$.

Then certainly $\kappa(\alpha; n) \le \kappa_r(n)$ with equality if F is algebraically closed.

(1) If $n = 2^m\cdot$ (odd) then $\kappa(\alpha; n) = \kappa(\alpha; 2^m)$. Also $\kappa^{-\lambda}(\alpha; 4n) \ge 4 + \kappa^{\lambda}(\alpha; n)$.

(2) If $\alpha \simeq \langle 1, a\rangle$ then $\kappa(\alpha; n) = \kappa_2(n)$. If $\alpha \simeq \langle 1, a, b\rangle$ or $\alpha \simeq \langle 1, a, b, ab\rangle$ then $\kappa(\alpha; n) = \kappa_3(n)$.

(3) Suppose K is a field with a non-trivial involution and F is the fixed field of that involution, so that $K = F(\sqrt{-a})$. If (V, h) is a hermitian space over K define $\operatorname{Sim}_K(V, h)$ and note that comparable similarities span F-quadratic spaces. Define the corresponding Hurwitz–Radon function $\rho^{\text{herm}}(n)$, where $n = \dim_K(V)$.

Let $B = \ell \circ h$ be the underlying F-bilinear form and let g be the action of $\sqrt{-a}$ on V. Then $\{1_V, g\}$ spans a subspace $R \subseteq \operatorname{Sim}(V, B)$ with $R \simeq \langle 1, a\rangle$. Then $S \subseteq \operatorname{Sim}_K(V, h)$ if and only if $S \subseteq \operatorname{Sim}_F(V, B)$ and S commutes with R. Consequently, if $n = 2^m\cdot$ (odd) then

$$\rho^{\text{herm}}(n) = \kappa(\langle 1, a\rangle; 2n) = \kappa_2(2n) = 2m + 2.$$

(4) How does this analysis generalize to hermitian forms over a quaternion algebra?

(Hint. (2) The equalities for small values follow by considering the quadratic and quaternion algebras with prescribed norm forms.)

6. **More Hurwitz–Radon functions.** (1) Define $\rho^+(n) = \max\{k : C((k-1)\langle 1\rangle)$ has an n-dimensional module over $\mathbb{R}\}$. Let $n = 2^{4a+b}n_0$ where n_0 is odd and $0 \le b \le 3$. According to Lam (1973), p. 132, Theorem 4.8′, $\rho^+(n) = 8a+b+[b/3]+2$. (Here $[x]$ denotes the greatest integer function.) In our notation, $\rho^+(n) = 1 + \rho_1(n)$. "Explain" this result as in Exercise 3.

(2) Let D be the quaternion division algebra over $\mathbb{R}$. Let $\rho^D(n) = \max\{r$: $C((r-1)\langle -1\rangle)$ has an n-dimensional module over $D\}$. Then $\rho^D(n) = 8a + 2b + \frac{1}{2}(b+2)(3-b)$, according to Wolf (1963a), p. 437. Modify Exercise 3 to see that $\rho^D(n) = \max\{r : r\langle 1\rangle < \operatorname{Sim}_D(n\langle 1\rangle)\}$. Use Exercises 3, 4, 5 to show $\rho^D(n) = \kappa(3\langle 1\rangle; n) = \kappa_3(4n) = 1 + \rho_3(4n)$, which coincides with Wolf's formula.

7. **Hermitian compositions again.** Suppose $n = 2^m\cdot$ (odd).

(1) Recall the compositions (over $\mathbb{C}$) considered in Exercise 2.15. The formula has type 2 if each z_k if bilinear in $(X, \bar{X})$ and Y.

Proposition. *A type 2 composition of size (r, n, n) exists if and only if $r \le m+1$.*

(2) $s \le \rho^{\text{herm}}(n) = 2m+2$ if and only if there is a formula

$$(x_1^2 + x_2^2 + \cdots + x_s^2) \cdot (|y_1|^2 + \cdots + |y_n|^2) = |z_1|^2 + \cdots + |z_n|^2$$

where $X = (x_1, \dots, x_s)$ is a system of real variables, $Y = (y_1, \dots, y_n)$ is a system of complex variables and each z_k is $\mathbb{C}$-bilinear in X, Y. Write out some examples of such formulas.

(Hint. (1) From Exercise 2.15 such a composition exists iff there is an (r, r)-family in $\text{Sim}^{\text{herm}}(V, h)$ where $V = \mathbb{C}^n$ and h is the standard hermitian form. Clifford algebra representations imply $2r - 2 \le 2m$. Conversely constructions over $\mathbb{R}$ provide an $(m+1, m+1)$-family in $\text{Sim}^{\text{herm}}(V, h)$.)

8. **Matrix factorizations.** Suppose $\sigma(X) = \sigma(x_1, \dots, x_r)$ is a quadratic form. Recall that: $\sigma < \text{Sim}(n\langle 1\rangle)$ if and only if there exists an $n \times n$ matrix A whose entries are linear forms in X satisfying: $A^{\top} \cdot A = \sigma(X) \cdot I_n$. (Compare (1.9).) Define a somewhat weaker property: σ admits a *matrix factorization in order* n if there exist $n \times n$ matrices A, B whose entries are linear forms in X satisfying: $A \cdot B = \sigma(X) \cdot I_n$. If σ has such a factorization over F then so does every quadratic form similar to σ.

(1) **Proposition.** *Let σ be a quadratic form over F. Then σ admits a matrix factorization in order n if and only if there is a $C_0(\sigma)$-module of dimension n.*

In fact, any regular quadratic form σ possesses "essentially" just one matrix factorization.

(2) If $\sigma = \langle 1\rangle \perp \sigma_1$ represents 1 then $C_0(\sigma) \cong C(-\sigma_1)$ as ungraded F-algebras. Suppose F is an ordered field and σ is a positive definite form over F. Then σ admits a matrix factorization in order n over F if and only if $\sigma < \text{Sim}(q)$ for some n-dimensional positive definite form q over F.

(Hint. (1) ($\Longrightarrow$): Let (S, σ) be the given space. View A, B as linear maps $\alpha, \beta : S \to \text{End}(V)$, where $\dim V = n$. Define $\lambda : S \to \text{End}(V \oplus V)$ by: $\lambda = \begin{pmatrix} 0 & \alpha \\ \beta & 0 \end{pmatrix}$. Then $\lambda(f)^2 = \sigma(f) \cdot 1_{V \oplus V}$ for every $f \in S$ so that $V \oplus V$ becomes $C(S, \sigma)$-module. It is a graded module as in Exercise 3.10(3), determined by the $C_0(\sigma)$-module V.

(2) See Exercise 3.)

9. **Conjugate families.** Let C and J_S be as usual. Suppose (V, q) and (V', q') are quadratic spaces admitting (C, J_S), with associated (s, t)-families $(S, T) \subseteq \text{Sim}(V, q)$ and $(S', T') \subseteq \text{Sim}(V', q')$. Then V and V' are C-similar if and only if $(S', T') = (fSf^{-1}, fTf^{-1})$ for some invertible $f \in \text{Sim}(V, V')$.

10. **Quaternion algebras.** Let A be a quaternion algebra over F, with the usual bar-involution. Recall that the norm and trace on A are defined by: $N(a) = a \cdot \bar{a}$ and $T(a) = a + \bar{a}$. Let φ be the norm form of A, so that $D_F(\varphi) = \{N(d) : d \in A^\bullet\}$ is the group of all norms. Let A_0 be the subspace of "pure" quaternions.

(1) If $a, b \in F^\bullet$ then $\langle a\rangle_A \simeq \langle b\rangle_A$ if and only if the classes of a, b coincide in $F^\bullet/D_F(\varphi)$.

(2) Two λ-hermitian spaces (V_i, h_i) are *similar* if $(V_2, h_2) \simeq (V_1, r \cdot h_1)$ for some $r \in F^\bullet$.

Lemma. *Let $a, b \in A_0^\bullet$. The following statements are equivalent.*

(i) *$\langle a\rangle_A$ and $\langle b\rangle_A$ are similar as skew-hermitian spaces.*

(ii) *$b = t \cdot da\bar{d}$ for some $t \in F^\bullet$ and some $d \in A^\bullet$.*

(iii) *$\langle N(a)\rangle = \langle N(b)\rangle$ in $F^\bullet/F^{\bullet 2}$.*

(3) It is harder to characterize isometry. The lemma above reduces the question to determining the "similarity factors" of the space $\langle a\rangle_A$. Suppose $a \in A_0^\bullet$ is given and let $x = N(a)$, so that the norm form is $\varphi \simeq \langle\langle x, y\rangle\rangle$ for some $y \in F^\bullet$.

If $t \in F^\bullet$ then: $\langle t \cdot a\rangle_A \simeq \langle a\rangle_A$ if and only if $t \in D_F(\langle\langle x\rangle\rangle) \cup -y \cdot D_F(\langle\langle x\rangle\rangle)$.

In particular: $\langle t \cdot a\rangle_A \simeq \langle a\rangle_A$ for every $t \in F^\bullet$ if and only if $D_F(\langle\langle x\rangle\rangle)$ is a subgroup of index 1 or 2 in $F^\bullet$.

(4) Let D be the quaternion division algebra over $\mathbb{R}$ with the "bar" involution. Isometry of hermitian spaces over D is determined by the dimension and the signature. Isometry of skew-hermitian spaces over D is determined by the dimension.

(Hint. (2) (iii) $\Longrightarrow$ (ii). Given $N(b) = s^2 \cdot N(a)$ for $s \in F^\bullet$, alter a to assume $N(b) = N(a)$.

Claim. There exists $u \in A^\bullet$ such that $b = uau^{-1}$. (If A is split this is standard linear algebra. Suppose A is a division algebra. If $b = -a$ choose $u \in F(a)^\perp$, otherwise let $u = a + b$.)

(3) Choose $b \in A_0$ with $ab = -ba$ and $N(b) = y$. From the isometry we have $t \cdot a = da\bar{d}$ for some $d \in A^\bullet$. Then $t = \lambda N(d)$ where $\lambda = \pm 1$, so that $\lambda ad = da$. If $\lambda = 1$ then $d \in F(a)$ while if $\lambda = -1$ then $d \in b \cdot F(a)$.)

11. **Associated Hermitian Forms.** Suppose $\sigma \simeq \langle 1, a_2, \ldots, a_s\rangle$ and let $C = C(-\sigma_1)$ and $J = J_S$ as usual. Let ℓ be the trace map with $\ell(1) = 1$ (as in Exercise 3.14). Then $J(e^\Delta)e^\Delta = a_\Delta$ where $\{e^\Delta\}$ is the derived basis. If (V, h) is a λ-hermitian space over C, the associated bilinear form $B = \ell \circ h$ is defined on V as in Proposition A.2 above. Given B we can reconstruct the hermitian form h explicitly as: $h(x, y) = \sum B((e^\Delta)^{-1}x, y)e^\Delta = \sum B(x, e^\Delta y)J(e^\Delta)^{-1}$.

12. Let (V, B) be a regular λ-symmetric bilinear space over F. Then the ring $E = \mathrm{End}(V)$ acts on V as well, and the form B admits $(\mathrm{End}(V), I_B)$. Use Proposition A.2 (with the usual trace map on $\mathrm{End}(V)$) to lift the form $B : V \times V \to F$ to a unique λ-hermitian form $h : V \times V \to \mathrm{End}(V)$. Exactly what is this form h?

(*Answer*: $h(u, v)(x) = B(x, v)u$ for all $x, u, v \in V$. Compare Exercise 1.13.)

13. Let C be semisimple F-algebra with involution. Then $C \cong A_1 \times \cdots \times A_k$ where the A_j are the simple ideals of C. Let e_j be the identity element of A_j and let V_j be an irreducible A_j-module, viewed as a C-module by setting $A_i V_j = 0$ if $i \neq j$.

(1) Every C-module is isomorphic to a direct sum of some of these V_j.

(2) Any involution J on C permutes $\{e_1, \ldots, e_k\}$. If $J(e_i) = e_j$ then $\hat{V}_i \cong V_j$.

(3) Under what conditions do all the unsplittable λ-spaces admitting (C, J) have the same dimension?

14. **Existence of an involution trace.** Generalize Definition 4.15, allowing K to be a field with involution ("bar"), requiring the involution on the algebra C to be compatible ($\overline{rc} = \bar{r} \cdot \bar{c}$), and replacing condition (4.15) (1) by: $\ell(\bar{c}) = \overline{\ell(c)}$

Proposition. *Every semisimple K-algebra C with involution has an involution trace $C \to K$.*

The proof is done in several steps:

(1) Every central simple K-algebra C with involution has an involution trace $C \to K$.

(2) If E/K is a finite extension of fields with involution, there is an involution trace $E \to K$. Consequently, if C is an E-algebra with involution having an involution trace $C \to E$ then there is an involution trace $C \to K$.

(3) The proposition follows by considering the simple components.

(Hint. (1) The reduced trace Trd always works. Every K-linear map $\ell : C \to K$ with $\ell(xy) = \ell(yx)$ must be a scalar multiple of Trd. (For $[C, C] = \mathrm{span}\{xy - yx\}$ has codimension 1.)

15. **Homometric elements.** Let A be a ring with involution J = "bar". Elements $a, b \in A$ are called *homometric* if $\bar{a}a = \bar{b}b$. If $u \in A$ is a "spectral unit", that is if $\bar{u}u = 1$, then a and ua are homometric. We say that (A, J) has the *homometric property* if the converse holds: if $\bar{a}a = \bar{b}b$ then a and b differ by a spectral unit.

(1) If A is a division ring then (A, J) has the homometric property.

(2) Suppose (A, J) has the homometric property and $a \in A$. If $\bar{a}a$ is nilpotent, then $a = 0$. Consequently A has no non-trivial nil ideals.

(3) Let $A = \mathrm{End}(V)$ and $J = I_h$, where (V, h) is an anisotropic hermitian space over a field F with involution. (For example, $A \cong \mathbb{M}_n(\mathbb{C})$ with J = conjugate-transpose.) Then (A, J) has the homometric property.

(4) What other semisimple rings with involution have the homometric property?

(Hint. (3) Given $f, g \in \mathrm{End}(V)$. Then $\ker(\tilde{f}f) = \ker(f)$ and hence $\tilde{f}f(V) = \tilde{f}(V)$. If $\tilde{f}f = \tilde{g}g$ then $\tilde{f}(V) = \tilde{g}(V)$. Construct an isometry $\sigma : f(V) \to g(V)$. By Witt Cancellation, σ extends to an isometry $\varphi : V \to V$. Then $\tilde{\varphi}\varphi = 1$ and $\varphi f = g$.)

Notes on Chapter 4

The proof of Proposition 4.9 follows Knebusch (1970), especially (3.2.1) and (3.3.2).

Exercises 4 and 5 are derived from §6 of Shapiro (1977a).

Exercise 8. Matrix factorizations of quadratic forms are investigated in Buchweitz, Eisenbud and Herzog (1987), where they are related to graded Cohen–Macauley modules over certain rings. A similar situation was studied by Eichhorn (1969), (1970). Matrix factorizations of forms of higher degrees are analyzed using generalized Clifford algebras by Backelin, Herzog and Sanders (1988).

Exercise 14. This follows Shapiro (1976), Proposition 4.4.

Exercise 15. See Rosenblatt and Shapiro (1989).

Chapter 5

Small (s, t)-Families

As a break from the general theory of algebras with involution we present some explicit examples of (σ, τ)-modules for small (s, t)-pairs. Since we are concerned with these (σ, τ)-modules up to F-similarity, we will work only with quadratic forms over F.

Good information is obtained for (σ, τ)-modules when the unsplittables have dimension at most 4. In these cases we can classify the (σ, τ)-modules in terms of certain Pfister factors. The smallest case where non-Pfister behavior can occur is for $(2, 2)$-families. The unsplittable $(\langle 1, a\rangle, \langle x, y\rangle)$-modules are analyzed using a new "trace" method. We obtain concrete examples where the unsplittable module is not similar to a Pfister form.

As a convenience to the reader we provide the proofs of the basic properties of Pfister forms, even though this theory appears in a number of texts. If q is a quadratic form recall that the *value set* and the *norm group* are:

$$D_F(q) = \{c \in F^\bullet : q \text{ represents } c\} \quad \text{and} \quad G_F(q) = \{c \in F^\bullet : \langle c\rangle q \simeq q\}.$$

One easily checks that $G_F(q) \cdot D_F(q) \subseteq D_F(q)$. A (regular) quadratic form φ is defined to be *round* if $G_F(\varphi) = D_F(\varphi)$. In particular this implies that the value set $D_F(n)$ is a multiplicative group. We will prove below that every Pfister form is round.

We need the notion of "divisibility" of quadratic forms:

$$\alpha \mid \beta \text{ means that } \beta \simeq \alpha \otimes \delta, \text{ for some quadratic form } \delta.$$

For anisotropic forms, we have seen in (3.20) that divisibility by a binary form $\langle 1, b\rangle = \langle\langle b\rangle\rangle$ is determined by behavior under a quadratic extension. We restate that result here since it is so important for motivating some of the later work.

5.1 Lemma. *Let q be an anisotropic quadratic form over F. If $q \otimes F(\sqrt{-b})$ is isotropic then $q \simeq \langle x\rangle\langle 1, b\rangle \perp q_1$ for some $x \in F^\bullet$ and some form q_1 over F. Consequently, $q \otimes F(\sqrt{-b})$ is hyperbolic iff $\langle\langle b\rangle\rangle \mid q$.*

Proof. See Lemma 3.20. □

5.2 Proposition. *Let ψ be a round form over F, and $\varphi = \langle\langle a\rangle\rangle \otimes \psi$ for some $a \in F^\bullet$.*

(1) *Then φ is also round. Consequently, every Pfister form is round.*

(2) *If φ is isotropic then it is hyperbolic.*

(3) *Suppose φ is a Pfister form and define the pure subform φ' by $\varphi \simeq \langle 1\rangle \perp \varphi'$. If $b \in D_F(\varphi')$ then $\langle\langle b\rangle\rangle \mid \varphi$.*

Proof. (1) Suppose φ represents c, say $c = x + ay$ where $x, y \in D_F(\psi) \cup \{0\}$. Suppose $x, y \neq 0$. (The other cases are easier and are left to the reader.) Comparing determinants we see that $\langle x, ay\rangle \simeq \langle c, caxy\rangle$. Since ψ is round $D_F(\psi) = G_F(\psi)$ is a group so that x, y and xy lie in $G_F(\psi)$. Then $\varphi \simeq \langle\langle a\rangle\rangle \otimes \psi \simeq \langle x, ay\rangle \otimes \psi \simeq \langle c\rangle\langle 1, axy\rangle \otimes \psi \simeq \langle c\rangle\langle\langle a\rangle\rangle \otimes \psi \simeq \langle c\rangle\varphi$ and consequently φ is round. An induction proof now shows that a Pfister form φ is round, for if $\dim\varphi = 2^m > 1$ then $\varphi \cong \langle\langle a\rangle\rangle \otimes \psi$ where ψ is another Pfister form.

(2) Since φ is isotropic there exist $x, y \in D_F(\psi)$ such that $x + ay = 0$. Then $-a = xy^{-1} \in G_F(\psi)$ so that $\varphi \simeq \langle\langle -xy^{-1}\rangle\rangle \otimes \psi \simeq \langle\langle -1\rangle\rangle \otimes \psi$ is hyperbolic.

(3) We are given $\varphi \simeq \langle\langle a\rangle\rangle \otimes \psi$ for a Pfister form ψ. Note that ψ remains round under any field extension. By hypothesis, $\varphi \simeq \langle 1, b, \dots\rangle$. If φ is isotropic then by (2) it is hyperbolic and the conclusion is clear. Otherwise φ is anisotropic but $\varphi \otimes F(\sqrt{-b})$ is isotropic. But then $\varphi \otimes F(\sqrt{-b})$ is hyperbolic by (2) applied over this larger field, and (5.1) implies $\langle\langle b\rangle\rangle \mid \varphi$. □

The fundamental fact here is that Pfister forms are round. That is, a Pfister form φ has multiplicative behavior:

$$D_F(\varphi) \text{ is a subgroup of } F^{\bullet}.$$

Applying this to the form $2^m\langle 1\rangle$ we see that the set $D_F(2^m)$ of all non-zero sums of 2^m squares in F is closed under multiplication. (See Exercise 0.5 for another proof of this fact.)

If φ is any m-fold Pfister form over F, the element $\varphi(X)$ in $F(X) = F(x_1, \dots, x_{2^m})$ is represented by the form $\varphi \otimes F(X)$, and the proposition implies that $\varphi(X)$ lies in $G_{F(X)}(\varphi \otimes F(X))$. Writing V for the underlying space of $\varphi \otimes F(X)$, this says that there exists a linear mapping $f : V \to V$ with $\varphi(f(v)) = \varphi(X)\varphi(v)$ for every $v \in V$. Writing this out in terms of matrices as done in Chapter 0, we obtain a multiplication formula

$$\varphi(X) \cdot \varphi(Y) = \varphi(Z),$$

where each entry z_k is linear in Y with coefficients in $F(X)$. Further information appears in the texts by Lam and Scharlau.

When the unsplittable (σ, τ)-modules have dimension ≤ 4 we characterize the unsplittables in terms of certain Pfister forms. In the discussion below we use the quadratic forms over F rather than working directly with modules over the Clifford algebras. The module approach provides more information but the proofs tend to be longer.

5.3 Proposition. *In the following table, every unsplittable (σ,τ)-module is F-similar to one of the forms q listed.*

(σ,τ)	q
$(\langle 1\rangle, \langle x\rangle)$ *where* $\langle x\rangle \not\simeq \langle 1\rangle$	$q \simeq \langle\langle w\rangle\rangle$ *where* $x \in G_F(q)$
$(\langle 1, a\rangle, 0)$	$q \simeq \langle\langle a\rangle\rangle$
$(\langle 1, a, b\rangle, 0)$	$q \simeq \langle\langle a, b\rangle\rangle$
$(\langle 1, a\rangle, \langle x\rangle)$ *where* $\langle 1, a, -x\rangle$ *is anisotropic*	$q \simeq \langle\langle a, w\rangle\rangle$ *where* $x \in G_F(q)$
$(\langle 1, a\rangle, \langle x, y\rangle)$ *where* $\langle axy\rangle \not\simeq \langle 1\rangle$ *and* $\langle 1, a, -x, -y\rangle$ *is isotropic*	$q \simeq \langle\langle a, w\rangle\rangle$ *where* $\langle\langle xy\rangle\rangle \mid q$
$(\langle 1, a, b, c\rangle, 0)$ *where* $\langle abc\rangle \not\simeq \langle 1\rangle$	$q \simeq \langle\langle a, b, w\rangle\rangle$ *where* $abc \in G_F(\langle\langle w\rangle\rangle)$
$(\langle 1, a, b\rangle, \langle x\rangle)$ *where* $\langle 1, a, b, -x\rangle$ *is anisotropic*	$q \simeq \langle\langle a, b, w\rangle\rangle$ *where* $\langle\langle abx\rangle\rangle \mid q$

Proof. We will do a few of these cases in detail, leaving the rest to the reader. First suppose $(\sigma,\tau) \simeq (\langle 1, a, b\rangle, 0)$, and $\sigma < \mathrm{Sim}(q)$ is unsplittable where q represents 1. Then (1.9) implies $\sigma \subset q$ so that $\dim q \geq 3$. Since $\sigma < \mathrm{Sim}(\langle\langle a, b\rangle\rangle)$, the Decomposition Theorem shows that $\dim q = 4$. Therefore $q \simeq \langle 1, a, b, d\rangle$ for some $d \in F^\bullet$. Since $\langle 1, a\rangle < \mathrm{Sim}(q)$ we know from an earlier case (or from (1.10)) that $\langle\langle a\rangle\rangle \mid q$, so that $\det q = \langle 1\rangle$. Then $\langle d\rangle = \langle ab\rangle$ and $q \simeq \langle\langle a, b\rangle\rangle$.

Suppose $(\sigma,\tau) = (\langle 1, a\rangle, \langle x, y\rangle)$ where $\langle axy\rangle \not\simeq \langle 1\rangle$ and $\langle 1, a, -x, -y\rangle$ is isotropic. Then $\langle 1, a\rangle$ and $\langle x, y\rangle$ represent some common value e. Scaling by e we get an equivalent pair of forms $(\langle e\rangle\langle 1, a\rangle, \langle e\rangle\langle x, y\rangle) \simeq (\langle 1, a\rangle, \langle 1, xy\rangle)$. There exist 4-dimensional (σ,τ)-modules, for example $\langle\langle a, xy\rangle\rangle$. Since $\langle axy\rangle \not\simeq \langle 1\rangle$ the unsplittables cannot have dimension 2, and the Decomposition Theorem implies that every unsplittable q has $\dim q = 4$. Since $\langle 1, a\rangle < \mathrm{Sim}(q)$ we know $q \simeq \langle\langle a, w\rangle\rangle$ for some w. Also since $\langle 1, xy\rangle < \mathrm{Sim}(q)$ we have $\langle\langle xy\rangle\rangle \mid q$. Conversely suppose $q \simeq \langle\langle a, w\rangle\rangle$ and $\langle\langle xy\rangle\rangle \mid q$. Then the form $\langle a, w, aw\rangle$ represents xy, so we can express $xy = ar^2 + wd$, for some $r \in F$ and $d \in D_F(\langle\langle a\rangle\rangle) \cup \{0\}$. Then $d \neq 0$, since $\langle axy\rangle \not\simeq \langle 1\rangle$, so that $q \simeq \langle\langle a, wd\rangle\rangle$ and $\langle a, wd\rangle$ represents xy. Therefore $(\langle 1, a\rangle, \langle 1, xy\rangle) \subset (\langle 1, a, wd\rangle, \langle 1, a, wd\rangle) < \mathrm{Sim}(q)$, and the result follows.

Suppose $(\sigma,\tau) \simeq (\langle 1, a, b, c\rangle, 0)$ where $\langle abc\rangle \not\simeq \langle 1\rangle$. There exist (σ,τ)-modules of dimension 8, for instance $\langle\langle a, b, c\rangle\rangle$. If q is an unsplittable module which represents 1, then $\langle 1, a, b, c\rangle \subset q$ and since $\langle 1, a, b\rangle < \mathrm{Sim}(q)$ we also have $\langle\langle a, b\rangle\rangle \mid q$. If $\dim q = 4$ we contradict the hypothesis $\langle abc\rangle \not\simeq \langle 1\rangle$. Therefore $\dim q = 8$ and $q \simeq \langle\langle a, b, u\rangle\rangle$ for some u. Since $\langle 1, a, b, c\rangle \subset q$ we find $\langle ab\rangle \perp \langle u\rangle\langle\langle a, b\rangle\rangle$ represents c and therefore $\langle 1\rangle \perp \langle u\rangle\langle\langle a, b\rangle\rangle$ represents abc. Express $abc = r^2 + ue$ where $r \in F$ and $e \in D_F(\langle\langle a, b\rangle\rangle) \cup \{0\}$. Since $\langle abc\rangle \not\simeq \langle 1\rangle$ we find $e \neq 0$ and therefore $q \simeq \langle\langle a, b, ue\rangle\rangle$ and $\langle 1, ue\rangle$ represents abc. The desired shape for q follows when we set $w = ue$. The converse follows as before. □

In the small cases analyzed above we can go on to characterize *arbitrary* (σ, τ)-modules. For instance it immediately follows from (5.2) that $\langle 1, a\rangle < \mathrm{Sim}(q)$ if and only if $\langle\langle a\rangle\rangle \mid q$, and that $\langle 1, a, b\rangle < \mathrm{Sim}(q)$ if and only if $\langle\langle a, b\rangle\rangle \mid q$. For the other cases we need a decomposition theorem for Pfister factors analogous to the Decomposition Theorem 4.1.

5.4 Definition. Let $\mathcal{M}$ be a set of (isometry classes of) quadratic forms over F. A quadratic form $q \in \mathcal{M}$ is *$\mathcal{M}$-indecomposable* if there is no non-trivial decomposition $q \simeq q_1 \perp q_2$ where $q_i \in \mathcal{M}$.

Certainly any form q in $\mathcal{M}$ can be expressed as $q \simeq q_1 \perp \cdots \perp q_k$ where each q_j is $\mathcal{M}$-indecomposable. We will get some results about the $\mathcal{M}$-indecomposables for some special classes $\mathcal{M}$. Generally if the φ_i are round forms and $b_j \in F^\bullet$ we consider the classes of the type

$$\mathcal{M} = \mathcal{M}(\varphi_1, \dots, \varphi_k, b_1, \dots, b_n) = \{q : \varphi_i \mid q \text{ and } b_j \in G_F(q) \text{ for every } i, j\}.$$

The $\mathcal{M}$-indecomposables are easily determined in a few small cases. For instance, for a single round form φ we see that q is $\mathcal{M}(\varphi)$-indecomposable if and only if q is similar to φ. For a single scalar $b \in F^\bullet$ where $\langle b\rangle \not\simeq \langle 1\rangle$, every $\mathcal{M}(b)$-indecomposable has dimension 2. (This is Dieudonné's Lemma of Exercise 2.9; also see Exercise 7.) The Proposition 5.6 below generalizes these two cases.

We first prove a lemma about "division" by round forms which is of some interest in its own right. Recall that $\mathbb{H} = \langle 1, -1\rangle$ is the hyperbolic plane and that any quadratic form q has a unique "Witt decomposition" $q = q_a \perp q_h$ where q_a is anisotropic and $q_h \simeq m\mathbb{H}$ is hyperbolic.

5.5 Lemma. *Suppose φ is a round form.*

(1) *If $\varphi \mid q$ and $a \in D_F(q)$ then $q \simeq \varphi \otimes \alpha$ for some form α which represents a. If $\varphi \mid q$ and q is isotropic with $\dim q > \dim \varphi$ then $q \simeq \varphi \otimes \alpha$ for some isotropic form α.*

(2) *If $\varphi \mid \alpha \perp \beta$ and $\varphi \mid \alpha$ then $\varphi \mid \beta$.*

(3) *Suppose φ is anisotropic. Then: $\varphi \mid m\mathbb{H}$ if and only if $\dim \varphi \mid m$. If $\varphi \mid q$ then $\varphi \mid q_a$ and $\varphi \mid q_h$, where $q = q_a \perp q_h$ is the Witt decomposition.*

Proof. (1) If $q \simeq \varphi \otimes \langle b_1, \dots, b_n\rangle$ represents a then $a = b_1x_1 + \cdots + b_nx_n$ for some $x_j \in D_F(\varphi) \cup \{0\}$. Define $y_j = x_j$ if $x_j \neq 0$ and $y_j = 1$ if $x_j = 0$ and set $\alpha = \langle b_1y_1, \dots, b_ny_n\rangle$. Then α represents a and since $\varphi \otimes \langle y_j\rangle \simeq \varphi$ we have $q \simeq \varphi \otimes \alpha$. If q is isotropic we use a non-trivial representation of $a = 0$. If the terms x_i above are not all 0 the previous argument works. Otherwise the non-triviality of the representation implies that φ must be isotropic and hence universal. Since φ is round this implies that $\langle c\rangle\varphi \simeq \varphi$ for every $c \in F^\bullet$. In particular, $\varphi \otimes \langle b_1, b_2\rangle \simeq \varphi \otimes \langle 1, -1\rangle$ and the result follows.

(2) Apply induction on $\dim\alpha$. If $a \in D_F(\alpha)$ then part (1) implies that $\alpha \perp \beta \simeq \langle a\rangle\varphi \perp \delta$ and $\alpha \simeq \langle a\rangle\varphi \perp \alpha_0$ for some forms δ and α_0 such that $\varphi \mid \delta$ and $\varphi \mid \alpha_0$. Cancelling we find that $\delta \simeq \alpha_0 \perp \beta$ and the induction hypothesis applies.

(3) Let $k = \dim\varphi$. The "if" part is clear since $\varphi \otimes \mathbb{H} \simeq k\mathbb{H}$. For the "only if" part we use induction on m, assuming $\varphi \mid m\mathbb{H}$. Since φ is anisotropic we know $k \le m$. Part (1) implies that $m\mathbb{H} \simeq \varphi \otimes \alpha$ where α is isotropic. Expressing $\alpha \simeq \mathbb{H} \perp \alpha'$ we have $m\mathbb{H} \simeq k\mathbb{H} \perp (\varphi \otimes \alpha')$. If $k = m$ we are done. Otherwise, $\varphi \mid (m-k)\mathbb{H}$ and the induction hypothesis applies.

For the last statement let $q_h \simeq m\mathbb{H}$ and use induction on m. We may assume $m > 0$. Then q is isotropic and $\dim q > \dim\varphi$ (since φ is anisotropic). Part (1) implies that $q \simeq \varphi \otimes \alpha$ for some isotropic α. Expressing $\alpha \simeq \alpha' \perp \mathbb{H}$ we have $q_a \perp m\mathbb{H} \simeq q \simeq (\varphi \otimes \alpha') \perp k\mathbb{H}$ where $k = \dim\varphi$. Therefore $k \le m$ and cancellation implies $\varphi \mid (q_a \perp (m-k)\mathbb{H})$. The result follows using the induction hypothesis. □

5.6 Proposition. *Suppose φ is a Pfister form and $b \in F^\bullet$. Then all $\mathcal{M}(\varphi, b)$-indecomposables have the same dimension and all $\mathcal{M}(\varphi, \langle\langle b\rangle\rangle)$-indecomposables have the same dimension.*

Proof. We will consider the case $\mathcal{M} = \mathcal{M}(\varphi, \langle\langle b\rangle\rangle)$ here, leaving the other to the reader. Suppose q is an $\mathcal{M}$-indecomposable which represents 1. If $\langle\langle b\rangle\rangle \mid \varphi$ it is clear that $q \simeq \varphi$.

Suppose $\langle\langle b\rangle\rangle \nmid \varphi$. By (5.5), $q \simeq \varphi \perp q_1$ and $\langle\langle b\rangle\rangle \subset q$. Then $b \in D_F(\varphi' \perp q_1)$ so that $b = x + y$ where $x \in D_F(\varphi') \cup \{0\}$ and $y \in D_F(q_1) \cup \{0\}$. If $y = 0$ then $b = x \in D_F(\varphi')$ and (5.2)(3) implies that $\langle\langle b\rangle\rangle \mid \varphi$, contrary to hypothesis. Then $y \ne 0$ and by (5.5) again we have $q_1 \simeq \langle y\rangle\varphi \perp q_2$ where $\varphi \mid q_2$, and therefore $q \simeq \varphi \otimes \langle\langle y\rangle\rangle \perp q_2$. Since $\alpha = \varphi \otimes \langle\langle y\rangle\rangle$ is a Pfister form and $\alpha' = \varphi' \perp \langle y\rangle\varphi$ represents b we know that $\langle\langle b\rangle\rangle \mid \alpha$ so that $\alpha \in \mathcal{M}$. By (5.5)(2) we also have $q_2 \in \mathcal{M}$. Since q is $\mathcal{M}$-indecomposable, q_2 must be 0 and $\dim q = 2 \cdot \dim\varphi$. □

Since φ is a Pfister form here we see that every $\mathcal{M}$-indecomposable is also a Pfister form.

5.7 Proposition. (1) $(\langle 1\rangle, \langle x\rangle) < \mathrm{Sim}(q)$ *iff* $x \in G(q)$.

(2) $\langle 1, a\rangle < \mathrm{Sim}(q)$ *iff* $\langle\langle a\rangle\rangle \mid q$.

(3) $\langle 1, a, b\rangle < \mathrm{Sim}(q)$ *iff* $\langle\langle a, b\rangle\rangle \mid q$.

(4) $(\langle 1, a\rangle, \langle x\rangle) < \mathrm{Sim}(q)$ *iff* $\langle\langle a\rangle\rangle \mid q$ *and* $x \in G(q)$.

(5) *If* $\langle 1, a, -x, -y\rangle$ *is isotropic, then* $(\langle 1, a\rangle, \langle x, y\rangle) < \mathrm{Sim}(q)$ *iff* $\langle\langle a\rangle\rangle \mid q$ *and* $\langle\langle xy\rangle\rangle \mid q$.

(6) $\langle 1, a, b, c\rangle < \mathrm{Sim}(q)$ *iff* $q \simeq \langle\langle a, b\rangle\rangle \otimes \gamma$ *where* $abc \in G_F(\gamma)$.

(7) $(\langle 1, a, b\rangle, \langle x\rangle) < \mathrm{Sim}(q)$ *iff* $\langle\langle a, b\rangle\rangle \mid q$ *and* $\langle\langle abx\rangle\rangle \mid q$.

Proof. We will prove the last two, omitting the others. Suppose $\langle 1,a,b,c\rangle < \mathrm{Sim}(q)$. If $\langle abc\rangle \simeq \langle 1\rangle$ the result is easy, so suppose $\langle abc\rangle \not\simeq \langle 1\rangle$. If $\langle 1,a,b,c\rangle < \mathrm{Sim}(q)$ then q is a sum of unsplittables of the type listed in (5.3), and it follows that $q \simeq \langle\langle a,b\rangle\rangle \otimes \gamma$ where $abc \in G_F(\gamma)$. Conversely suppose q is given in this way. Since $\langle abc\rangle \not\simeq \langle 1\rangle$, Proposition 5.6 implies that the $\mathcal{M}(abc)$-indecomposables all have dimension 2. Therefore $\gamma \simeq \gamma_1 \perp \cdots \perp \gamma_r$ where $\dim\gamma_j = 2$ and $\gamma_j \in \mathcal{M}(abc)$. Then $\gamma_j \simeq \langle u_j\rangle\langle\langle w_j\rangle\rangle$ where $abc \in G_F(\langle\langle w_j\rangle\rangle)$ and we get $q \simeq q_1 \perp \cdots \perp q_r$ where $q_j \simeq \langle u_j\rangle\langle\langle a,b,w_j\rangle\rangle$. Again by Proposition 5.3 we conclude that $\langle 1,a,b,c\rangle < \mathrm{Sim}(q)$.

Now consider the case $(\langle 1,a,b\rangle, \langle x\rangle)$. If $\langle 1,a,b\rangle$ represents x then $\langle\langle abx\rangle\rangle \mid \langle\langle a,b\rangle\rangle$ and $\langle 1,a,b\rangle < \mathrm{Sim}(q)$ if and only if $(\langle 1,a,b\rangle, \langle x\rangle) < \mathrm{Sim}(q)$. Therefore we may assume $\langle 1,a,b,-x\rangle$ is anisotropic. If $(\langle 1,a,b\rangle\langle x\rangle) < \mathrm{Sim}(q)$ then $q \simeq q_1 \perp \cdots \perp q_r$ where each $(\langle 1,a,b\rangle, \langle x\rangle) < \mathrm{Sim}(q_j)$ is unsplittable. By Proposition 5.3 we have $\langle\langle a,b\rangle\rangle \mid q_j$ and $\langle\langle abx\rangle\rangle \mid q_j$, and the claim follows. Conversely suppose that $q \in \mathcal{M} = \mathcal{M}(\langle\langle a,b\rangle\rangle, \langle\langle abx\rangle\rangle)$. Then $q \simeq q_1 \perp \cdots \perp q_r$ where each q_j is $\mathcal{M}$-indecomposable. Since $\langle 1,a,b,-x\rangle$ is anisotropic we see that $\langle\langle abx\rangle\rangle \nmid \langle\langle a,b\rangle\rangle$ and Proposition 5.6 implies that $\dim q_j = 8$. Therefore $q_j \simeq \langle u_j\rangle\langle\langle a,b,w_j\rangle\rangle$ where $\langle\langle abx\rangle\rangle \mid q_j$. Apply (5.3) again to conclude that $(\langle 1,a,b\rangle, \langle x\rangle) < \mathrm{Sim}(q_j)$ for each j and therefore $(\langle 1,a,b\rangle, \langle x\rangle) < \mathrm{Sim}(q)$. □

The rest of this chapter is concerned with the more difficult case of (2, 2)-families. Let $(\sigma,\tau) = (\langle 1,a\rangle, \langle x,y\rangle)$. The case when $\langle 1,a,-x,-y\rangle$ is isotropic is included in Proposition 5.7. If $\langle axy\rangle \simeq \langle 1\rangle$ then $(\langle 1,a\rangle, \langle x,y\rangle) < \mathrm{Sim}(q)$ iff $(\langle 1,a\rangle, \langle x\rangle) < \mathrm{Sim}(q)$, by the Expansion Lemma. This case is also included in (5.7). Therefore let us assume that

$$\langle 1,a,-x,-y\rangle \text{ is anisotropic and } \langle axy\rangle \not\simeq \langle 1\rangle.$$

Let $C = C(\langle -a,x,y\rangle)$ be the associated Clifford algebra. Then the center of C is isomorphic to the field $E = F(\sqrt{axy})$ and C is a quaternion algebra over E. It follows that C is a division algebra (this is part of Exercise 3.16) and every unsplittable (σ,τ)-module has dimension 8. Let J_S denote the usual involution on C.

If $(\langle 1,a\rangle, \langle x,y\rangle) < \mathrm{Sim}(V,q)$ then we have $\langle\langle a\rangle\rangle \mid q$, $\langle\langle xy\rangle\rangle \mid q$ and $x \in G_F(q)$. It is not so clear whether the converse holds: do those "divisibility" conditions on q always imply the existence of the (2, 2)-family? Those conditions do provide some motivation for the following approach.

To say that $(\langle 1,a\rangle, \langle x,y\rangle) < \mathrm{Sim}(V,q)$ is equivalent to saying that (V,q) is a quadratic (C, J_S)-module. In this case we have $f_2, g_1, g_2 \in \mathrm{End}(V)$ which satisfy the familiar rules listed in Lemma 2.3. Then $f = f_2$ satisfies $\tilde{f} = -f$ and $f^2 = -a\cdot 1$ so it corresponds to the subspace $\langle 1,a\rangle < \mathrm{Sim}(q)$. Similarly $g = g_1g_2$ satisfies $\tilde{g} = -g$ and $g^2 = -xy\cdot 1$ so that g corresponds to $\langle 1,xy\rangle < \mathrm{Sim}(q)$. Since f and g commute, they induce an action of the field $K = F(\sqrt{-a}, \sqrt{-xy})$ on the vector space V. Let J be the involution of K sending $\sqrt{-a}$ and $\sqrt{-xy}$ to their negatives. Then (V,q) becomes a quadratic (K,J)-module. Naturally we may view K as a subfield of C

where J is the restriction of J_S to K. Note that $E = F(\sqrt{axy})$ is the subfield of K which is fixed by J. We often write "bar" for J when there is no ambiguity.

Conversely, if (V, q) is a quadratic (K, J)-module, what further information is needed to make it a (C, J_S)-module?

5.8 Lemma. *Suppose (K, J) is the field with involution described above and (V, q) is a quadratic (K, J)-module. This structure extends to make (V, q) into a (C, J_S)-module if and only if there exists $k \in \mathrm{End}_F(V)$ such that k is (K, J)-semilinear, $\tilde{k} = k$, and $k^2 = x \cdot 1$.*

Proof. The (K, J)-semilinearity of k means that $k(\alpha v) = \bar{\alpha}k(v)$ for every $\alpha \in K$ and $v \in V$. This is equivalent to saying that k anticommutes with f and g. If the (C, J_S)-module structure is given, just define $k = g_1$. Conversely, given the (K, J)-module structure and given k, define $f_2 = f$, $g_1 = k$, $g_2 = k^{-1}g$ and verify that they provide the desired (2, 2)-family. □

Suppose (V, q) is a (K, J)-module. Then the symmetric bilinear form $b_q : V \times V \to F$ is the "transfer" of a hermitian form over K. This is a special case of Proposition A.2 of Chapter 4, applied to the algebra $C = K$. In fact if $s : K \to F$ is an involution trace then there exists a unique hermitian form $h : V \times V \to K$ such that $b_q = s \circ h$. (See Exercise 16.) This hermitian space (V, h) is an orthogonal sum of some 1-dimensional spaces over K:

$$(V, h) \simeq \langle\theta_1, \theta_2, \ldots, \theta_m\rangle_K \quad \text{for some } \theta_i \in E^\bullet.$$

These diagonal entries θ_i lie in E since they must be symmetric: $\bar{\theta}_i = \theta_i$.

To do calculations we must choose an involution trace. First note that $K = E(\sqrt{-a})$ and define $\mathrm{tr} : K \to E$ by setting $\mathrm{tr}(1) = 1$ and $\mathrm{tr}(\sqrt{-a}) = 0$. (This is the unique involution trace from K to E, up to scalar multiple.) Since the involution on $E = F(\sqrt{axy})$ is trivial there are many involution traces from E to F. We will use the standard one employed in quadratic form theory, namely $\ell : E \to F$ defined by setting $\ell(1) = 0$ and $\ell(\sqrt{axy}) = 1$. If $\theta \in E^\bullet$ then the 1-dimensional hermitian space $\langle\theta\rangle_K$ over K transfers down to a 4-dimensional quadratic space $\ell \circ \mathrm{tr}(\langle\theta\rangle_K)$ over F.

5.9 Lemma. *Suppose $\theta \in E^\bullet$. Then $\ell \circ \mathrm{tr}(\langle\theta\rangle_K) \simeq \langle s\rangle\langle\langle a, -N\theta\rangle\rangle$ over F.*

Proof. Here N is the field norm $N_{E/F}$. If $\theta = r + s\sqrt{axy}$ then $N(\theta) = r^2 - axys^2$. The 1-dimensional hermitian space $\langle\theta\rangle_K$ can be viewed as the K-vector space K with the form $h : K \times K \to K$ given by $h(x, y) = \theta x\bar{y}$. If $b = \mathrm{tr} \circ h$ then $b(u, v) = \theta \cdot \mathrm{tr}(u\bar{v})$. If $u = u_1 + u_2\sqrt{-a}$ and $v = v_1 + v_2\sqrt{-a}$ we find that $\mathrm{tr}(u\bar{v}) = u_1v_1 + au_2v_2$, so that $\mathrm{tr}(\langle\theta\rangle_K) \simeq \langle\theta, a\theta\rangle_E$ as quadratic forms over E. To transfer the quadratic form $\langle\theta\rangle_E$ from $E = F(\sqrt{axy})$ down to F, we compute the inner products (relative to this form) of the basis elements $\{1, \sqrt{axy}\}$ to find the

Gram matrix $\begin{pmatrix} s & r \\ r & axys \end{pmatrix}$. Since it represents s and has determinant $-N(\theta)$, we have $\ell(\langle\theta\rangle_E) \simeq \langle s\rangle\langle\langle -N\theta\rangle\rangle$, provided $s \neq 0$. If $s = 0$ that form is isotropic, hence is $\mathbb{H}$. The result now follows since $\ell \circ \mathrm{tr}(\langle\theta\rangle_K) = \ell(\langle\theta\rangle_E) \perp \langle a\rangle\ell(\langle\theta\rangle_E)$. □

Remark. If $\theta, \theta' \in E^\bullet$ then $\langle\theta\rangle_K \simeq \langle\theta'\rangle_K$ as hermitian spaces over K if and only if $\theta' = \alpha\bar{\alpha}\theta$ for some $\alpha \in K^\bullet$. (For a K-linear map $K \to K$ must be multiplication by some $\alpha \in K$.) If such α exists then the transferred quadratic forms over F must also be isometric.

Our goal is to construct unsplittable modules (V, q) for the (2, 2)-pair $(\langle 1, a\rangle, \langle x, y\rangle)$. Then $\dim_F V = 8$ and $\dim_K V = 2$ using the induced (K, J)-action, and we view (V, q) as the transfer of a hermitian space $(V, h) = \langle\theta_1, \theta_2\rangle_K$. Given such a hermitian space over K, we will find conditions on θ_1 and θ_2 which imply that this (K, J)-action can be extended to an action of (C, J_S).

5.10 Lemma. *Suppose (V, q) is the transfer of the hermitian space $(V, h) = \langle\theta_1, \theta_2\rangle_K$. The following statements are equivalent, where $\theta = \theta_1\theta_2$.*

(1) *(V, q) is a (C, J_S)-module in a way compatible with the given (K, J)-action.*

(2) *$\langle 1, \theta\rangle_K$ represents x, that is, $\alpha\bar{\alpha} + \theta\beta\bar{\beta} = x$ for some $\alpha, \beta \in K^\bullet$.*

(3) *$\langle 1, a, \theta, a\theta\rangle$ represents x over E.*

(4) *$-\theta \in D_E(\langle\langle a, -x\rangle\rangle)$.*

Proof. We use a matrix formulation of (1) to show its equivalence with (2). By (5.8), condition (1) holds if and only if there exists $k \in \mathrm{End}_F(V)$ which is (K, J)-semilinear, $\tilde{k} = k$ and $k^2 = x \cdot 1$. We are given a K-basis $\{v_1, v_2\}$ of V such that $h(v_i, v_i) = \theta_i$ and $h(v_1, v_2) = 0$. Representing a vector $v = x_1v_1 + x_2v_2$ in V as a column vector $X = \begin{pmatrix} x_1 \\ x_2 \end{pmatrix}$, the (K, J)-semilinear map k is represented by a matrix $A = \begin{pmatrix} \alpha & \gamma \\ \beta & \delta \end{pmatrix}$ where $\alpha, \beta, \gamma, \delta \in K$. This is done so that $k(v) = x_1'v_1 + x_2'v_2$ is represented by the column vector $X' = A\bar{X}$.

The adjoint map $\tilde{k}$ is also (K, J)-semilinear and has matrix $\tilde{A} = M^{-\top}A^\top M$ where $M = \begin{pmatrix} \theta_1 & 0 \\ 0 & \theta_2 \end{pmatrix}$ is the matrix of the hermitian form. (See Exercise 15 for more details.) The symmetry condition $\tilde{k} = k$ is equivalent to the symmetry of the matrix $MA = \begin{pmatrix} \theta_1\alpha & \theta_1\gamma \\ \theta_2\beta & \theta_2\delta \end{pmatrix}$. This condition holds if and only if $\theta_2\beta = \theta_1\gamma$. Define $\theta = \theta_2/\theta_1$ (which has the same square class in $E^\bullet$ as $\theta_1\theta_2$). Then the symmetry condition becomes: $\gamma = \theta\beta$.

The condition $k^2 = x \cdot 1$ becomes the matrix equation $A\bar{A} = xI$ (see Exercise 15). On multiplying this out when $A = \begin{pmatrix} \alpha & \theta\beta \\ \beta & \delta \end{pmatrix}$ we find it to be equivalent to the following equations:

$$\alpha\bar{\beta} = -\beta\bar{\delta}, \quad \alpha\bar{\alpha} = \delta\bar{\delta}, \quad \alpha\bar{\alpha} + \theta\beta\bar{\beta} = x.$$

If $\beta = 0$ then $\alpha\bar{\alpha} = x$ so that $x \in D_E(\langle 1, a\rangle)$ and $\langle 1, a, -x\rangle_E$ is isotropic over E. We know that $C = \left(\frac{-a,x}{E}\right)$ is a quaternion division algebra, so its norm form $\langle\langle a, -x\rangle\rangle_E$ is anisotropic over E. This is a contradiction.

Therefore $\beta \neq 0$ and $\delta = -\bar{\alpha}\beta\bar{\beta}^{-1}$. With this formula for δ the first two equations above are automatic. Therefore statement (1) holds if and only if there exist $\alpha, \beta \in K$ such that $\alpha\bar{\alpha} + \theta\beta\bar{\beta} = x$. This is statement (2).

The equivalence of statements (2) and (3) is clear since $\{\alpha\bar{\alpha} : \alpha \in K\} = D_E(\langle 1, a\rangle)$. Finally note that (3) holds if and only if $\langle\langle a, -x, \theta\rangle\rangle$ is hyperbolic over E, if and only if $\langle\langle a, -x\rangle\rangle$ represents $-\theta$ over E. Therefore (3) and (4) are equivalent. □

In the statement of Lemma 5.10 the symmetry between x and y is not apparent. However, since $\langle axy\rangle_E \simeq \langle 1\rangle_E$ we may note that $\langle 1, a, -x, -y\rangle_E \simeq \langle\langle a, -x\rangle\rangle_E \simeq \langle\langle a, -y\rangle\rangle_E$.

The payoff of these calculations can now be summarized.

5.11 Proposition. *Suppose $a, x, y \in F^\bullet$ such that $\langle axy\rangle \not\simeq \langle 1\rangle$ and $\langle 1, a, -x, -y\rangle$ is anisotropic. Let $E = F(\sqrt{axy})$ and let $N = N_{E/F}$ be the norm. If q is a quadratic form over F with $\dim q = 8$, then the following statements are equivalent:*

(1) $(\langle 1, a\rangle, \langle x, y\rangle) < \mathrm{Sim}(q)$.

(2) *There exist $\theta_i = r_i + s_i\sqrt{axy}$ in $E^\bullet$ such that* $q \simeq \langle s_1\rangle\langle\langle a, -N\theta_1\rangle\rangle \perp \langle s_2\rangle\langle\langle a, -N\theta_2\rangle\rangle$ *and such that* $-\theta_1\theta_2 \in D_E(\langle\langle a, -x\rangle\rangle)$.

Proof. Here, as before, if $s_i = 0$ the corresponding term in the expression for q is interpreted as $2\mathbb{H}$. This equivalence is obtained by combining (5.9) and (5.10). □

As one immediate consequence we see that $(\langle 1, a\rangle, \langle x, y\rangle) < \mathrm{Sim}(4\mathbb{H})$ for every a, x, y. To prove the next corollary we use the following "Norm Principle".

5.12 Lemma. *Let $E = F(\sqrt{d})$ be a quadratic extension, let φ be a Pfister form over F and let $\theta \in E^\bullet$. Then $\theta \in F^\bullet \cdot D_E(\varphi_E)$ if and only if $N\theta \in D_F(\varphi)$.*

Proof. See Elman and Lam (1976). The proof is outlined in Exercise 17. □

5.13 Corollary. *Suppose $a, x, y \in F^\bullet$ as above and suppose φ is a 2-fold Pfister form. The following are equivalent.*

(1) $(\langle 1, a\rangle, \langle x, y\rangle) < \mathrm{Sim}(\varphi \perp 2\mathbb{H})$.

(2) $\varphi \cong \langle\langle a, -c\rangle\rangle$ *for some* $c \in D_F(\langle\langle -axy\rangle\rangle) \cap D_F(\langle\langle a, -x\rangle\rangle)$.

Proof. (1) $\Longrightarrow$ (2). By (5.5) $\langle\langle a\rangle\rangle \mid \varphi$ so that $\varphi \simeq \langle\langle a, w\rangle\rangle$ for some w. Furthermore $\varphi \perp 2\mathbb{H} \simeq \langle s_1\rangle\langle\langle a, -N\theta_1\rangle\rangle \perp \langle s_2\rangle\langle\langle a, -N\theta_2\rangle\rangle$ for θ_i as in (5.11). Computing Witt invariants we find that $\varphi \simeq \langle\langle a, -c\rangle\rangle$ where $c = N(\theta_1\theta_2) \in D_F(\langle\langle -axy\rangle\rangle)$. Since $-\theta_1\theta_2 \in D_E(\langle\langle a, -x\rangle\rangle)$ the lemma implies that $c \in D_F(\langle\langle a, -x\rangle\rangle)$.

(2) $\Longrightarrow$ (1). We may express $c = N\theta \in D_F(\langle\langle a, -x\rangle\rangle)$. If $\langle c\rangle \simeq \langle 1\rangle$ the claim is vacuous, so we may assume that $\theta \notin F$. By the lemma we find that $\theta = t \cdot \theta_1$ where $t \in F^\bullet$ and $-\theta_1 \in D_E(\langle\langle a, -x\rangle\rangle)$. Let $\theta_2 = 1$ and apply (5.11) to conclude that $(\langle 1, a\rangle, \langle x, y\rangle) < \mathrm{Sim}(q)$ where $q \simeq \langle s_1\rangle\langle\langle a, -N\theta_1\rangle\rangle \perp 2\mathbb{H}$. Then $\langle s_1\rangle q \simeq \langle\langle a, -c\rangle\rangle \perp 2\mathbb{H}$ and the result follows. □

Example 1. Let $a = 1$, $x = -1$ and $y = -2$ over the rational field $\mathbb{Q}$. Then $\langle 1, a, -x, -y\rangle \simeq \langle 1, 1, 1, 2\rangle$ is anisotropic and $\langle axy\rangle \simeq \langle 2\rangle \not\simeq \langle 1\rangle$. If φ is a 2-fold Pfister form then (5.13) says: $(\langle 1, 1\rangle, \langle -1, -2\rangle) < \mathrm{Sim}(\varphi \perp 2\mathbb{H})$ if and only if $\varphi \simeq \langle\langle 1, -c\rangle\rangle$ for some $c \in D_{\mathbb{Q}}(\langle\langle -2\rangle\rangle) \cap D_{\mathbb{Q}}(\langle\langle 1, 1\rangle\rangle)$. For example

$$(\langle 1, 1\rangle, \langle -1, -2\rangle) < \mathrm{Sim}(q) \quad \text{where } q = \langle\langle 1, -7\rangle\rangle \perp 2\mathbb{H}.$$

To get anisotropic examples we use the criterion in (5.11). To deduce that a form $\langle 1, a, -x, -ax, \theta\rangle_E$ is isotropic over an algebraic number field we need only check (by the Hasse–Minkowski Theorem) that it is indefinite relative to every ordering of E.

Example 2. Let $a = 1$, $x = 7$ and $y = 14$ over $\mathbb{Q}$. Then $\langle axy\rangle \simeq \langle 2\rangle \not\simeq \langle 1\rangle$ and $\langle 1, a, -x, -y\rangle \simeq \langle 1, 1, -7, -14\rangle$ is anisotropic. (In fact it is anisotropic over the field $\mathbb{Q}_7$.) Furthermore $E = \mathbb{Q}(\sqrt{2})$, and $K = \mathbb{Q}(\sqrt{-1}, \sqrt{-2})$. For any $\theta \in E^\bullet$ the form $\langle\langle a, -x\rangle\rangle \perp \langle -\theta\rangle = \langle 1, 1, -7, -7, \theta\rangle$ is isotropic over $E = \mathbb{Q}(\sqrt{2})$ since it is indefinite at both orderings. Using $\theta_1 = 1 + \sqrt{2}$ and $\theta_2 = 1 + 2\sqrt{2}$ we find:

$$(\langle 1, 1\rangle, \langle 7, 14\rangle) < \mathrm{Sim}(q) \quad \text{where } q = \langle\langle 1, 1\rangle\rangle \perp \langle\langle 1, 7\rangle\rangle \simeq \langle\langle 1\rangle\rangle \otimes \langle 1, 1, 1, 7\rangle.$$

This form q is anisotropic but is not similar to a Pfister form.

Many further examples can be constructed along these lines. Non-Pfister unsplittables for larger (s,t)-families can be found using the Construction and Shift Lemmas. The smaller families considered earlier in this chapter were all characterized by certain "division" properties of quadratic forms. If $(\langle 1, a\rangle, \langle x, y\rangle) < \mathrm{Sim}(q)$ then certainly $\langle\langle a\rangle\rangle \mid q$, $\langle\langle xy\rangle\rangle \mid q$ and $x \in G_F(q)$. Is it possible that these three independent conditions suffice? Certainly this converse fails for simple reasons of dimensions: the 4-dimensional form $2\mathbb{H}$ always satisfies these divisibility conditions, but the dimensions of the unsplittables may be 8. We modify the conjecture as follows:

5.14 Question. If $\dim q = 8$ and $\langle\langle a\rangle\rangle \mid q$, $\langle\langle xy\rangle\rangle \mid q$ and $x \in G_F(q)$ does it follow that $(\langle 1, a\rangle, \langle x, y\rangle) < \mathrm{Sim}(q)$?

We may assume that $\langle 1, a, -x, -y\rangle$ is anisotropic and $\langle axy\rangle \not\simeq \langle 1\rangle$ since the other cases are settled by Proposition 5.7. The answer is unknown in general. In Chapter 10 we succeed in proving the answer to be "yes" when F is a global field. As the ideas used to prove (5.7) indicate, the following question is relevant.

5.15 Question. If $\mathcal{M} = \mathcal{M}(\langle\langle a\rangle\rangle, \langle\langle xy\rangle\rangle, x)$ what are the dimensions of the $\mathcal{M}$-indecomposables?

We will see in Chapter 10 that over a global field the indecomposables must have dimension 2 or 4. It is unknown what dimensions are possible over arbitrary fields.

The following observation is interesting (and perhaps surprising) in light of the $F \subseteq E \subseteq K$ set-up used above. To simplify notations we use b in place of xy here.

5.16 Proposition. *Suppose $\langle a\rangle \not\simeq \langle b\rangle$ over F. Let $K = F(\sqrt{-a}, \sqrt{-b})$ with involution J as above. The following statements are equivalent for a quadratic space (V, q) over F.*

(1) *$\langle\langle a\rangle\rangle \mid q$ and $\langle\langle b\rangle\rangle \mid q$.*

(2) *(V, q) can be made into a (K, J)-module.*

Proof. (2) $\Longrightarrow$ (1). Given the (K, J)-module (V, q) let $f = \mathcal{L}(\sqrt{-a})$ be the multiplication map on V. Then $f \in \mathrm{End}(V)$ and $f^2 = -a \cdot 1$. Since q admits (K, J) and $J(\sqrt{-a}) = -\sqrt{-a}$ we know that $\tilde{f} = -f$. Therefore $\{1_V, f\}$ span a space of similarities $\langle 1, a\rangle < \mathrm{Sim}(V, q)$ and we conclude that $\langle\langle a\rangle\rangle \mid q$ by (1.10). Similarly using $g = \mathcal{L}(\sqrt{-b})$ we find $\langle\langle b\rangle\rangle \mid q$.

(1) $\Longrightarrow$ (2). It suffices to settle the case $\dim q = 4$. This follows from (5.6) since the $\mathcal{M}(\langle\langle a\rangle\rangle, \langle\langle b\rangle\rangle)$-indecomposables are 4-dimensional. Given $\dim q = 4$ and $\langle\langle a\rangle\rangle \mid q$ we know there exists $f \in \mathrm{End}(V)$ with $\tilde{f} = -f$ and $f^2 = -a \cdot 1$. Similarly since $\langle\langle b\rangle\rangle \mid q$ there exists $g \in \mathrm{End}(V)$ with $\tilde{g} = -g$ and $g^2 = -b \cdot 1$. The difficulty is to find such f, g which commute.

We may assume q represents 1. By hypothesis, $q \simeq \langle\langle a, c\rangle\rangle \simeq \langle\langle b, d\rangle\rangle$ for some $c, d \in F^\bullet$. In fact we may assume $c = d$, (see Exercise 19) so that

$$q \simeq \langle\langle c, a\rangle\rangle \simeq \langle\langle c, b\rangle\rangle \text{ and } \langle a, ac\rangle \text{ represents } b.$$

Let $\{v_1, w_1, v_2, w_2\}$ be the orthogonal basis corresponding to $q \simeq \langle 1, c, a, ac\rangle$ and define f by setting $f(v_1) = v_2$, $f(v_2) = -av_1$ and similarly for the w_i's. Then $\tilde{f} = -f$ and $f^2 = -a \cdot 1$. The matrix of f can be expressed as a Kronecker product: $f = \begin{pmatrix} 0 & -a \\ 1 & 0 \end{pmatrix} \otimes \begin{pmatrix} 1 & 0 \\ 0 & 1 \end{pmatrix}$.

Let $\{v_1, w_1, \hat{v}_2, \hat{w}_2\}$ be a basis corresponding to $q \simeq \langle 1, c, b, bc\rangle$ and define g analogously using this new basis, so that $\tilde{g} = -g$ and $g^2 = -b \cdot 1$. To compute g explicitly, express $b = ax^2 + acy^2$ for some $x, y \in F$, and use $\hat{v}_2 = xv_2 + yw_2$

and $\hat{w}_2 = ycv_2 - xw_2$. The matrix of g (relative to the original basis) becomes: $g = \begin{pmatrix} 0 & -a \\ 1 & 0 \end{pmatrix} \otimes \begin{pmatrix} x & cy \\ y & -x \end{pmatrix}$. It is now easy to see that $fg = gf$. □

Exercises for Chapter 5

1. Suppose $(\sigma, \tau) < \mathrm{Sim}(q)$ is an unsplittable (s, t)-family where σ represents 1 and $\dim q \leq 8$. If $(s, t) \neq (2, 2)$ then q must be similar to a Pfister form.

2. Complete Proposition 5.7 by listing all q such that $(\sigma, \tau) < \mathrm{Sim}(q)$, where (σ, τ) equals:

(i) $(\langle 1, a\rangle, \langle x, y\rangle)$ with $\langle axy\rangle \simeq \langle 1\rangle$.

(ii) $(\langle 1\rangle, \langle x, y\rangle)$.

(iii) $(\langle 1\rangle, \langle x, y, z\rangle)$.

3. (i) Give a simple direct proof that if $(\langle 1, a, b\rangle, \langle x\rangle) < \mathrm{Sim}(q)$ then $\langle 1, abx\rangle < \mathrm{Sim}(q)$.

(ii) Find some a, b, x, q such that $\langle\langle a, b\rangle\rangle \mid q$ and $x \in G_F(q)$ but $(\langle 1, a, b\rangle, \langle x\rangle)$ is not realizable in $\mathrm{Sim}(q)$.

4. **Round forms.** (1) **Lemma.** *A quadratic space (V, φ) is round iff the group $\mathrm{Sim}^\bullet(V, \varphi)$ acts transitively on the set $V^\bullet$ of anisotropic vectors.*

(2) Recall that any (regular) quadratic form q has a *Witt decomposition* $q \simeq q_a \perp q_h$ where q_a is anisotropic and q_h is hyperbolic. These components are unique up to isometry. An isotropic form φ is round iff φ_a is round and universal.

5. **Level of a field.** If $d \in F$ define $\mathrm{length}_F(d)$ to be the smallest n such that d is a sum of n squares in F. That is, $n = \mathrm{length}_F(d) \iff d \in D_F(n) - D_F(n-1)$. If d is not a sum of squares then $\mathrm{length}_F(d) = \infty$. The *level* (or Stufe) of F is: $s(F) = \mathrm{length}_F(-1)$.

(1) **Proposition.** *If $s(F)$ is finite then $s(F) = 2^m$ for some m.*

(2) Suppose $K = F(\sqrt{-d})$. Then $s(K)$ is finite $\iff \mathrm{length}_F(d)$ is finite. It is each to check that $s(K) \leq \mathrm{length}_F(d)$.

Proposition. *Suppose $K = F(\sqrt{-d})$ and define m by: $2^m \leq \mathrm{length}_F(d) < 2^{m+1}$. Then $s(K) = 2^m$.*

(Hint. (1) Suppose $-1 = a_1^2 + \cdots + a_s^2$ and suppose $2^m \leq s < 2^{m+1}$. To prove: $-1 \in D_F(2^m)$. If $n = 2^m$ then $-(1 + a_{n+1}^2 + \cdots + a_s^2) = (a_1^2 + \cdots + a_n^2)$. By (5.2) or Exercise 0.5, $D_F(2^m)$ is a group.

(2) $s(K) \leq \mathrm{length}_F(d)$ implies $s(K) \leq 2^m$ by (1). If $s(K) = n$ then $-1 = \sum_{i=1}^n (a_i + b_i\sqrt{-d})^2$ so that $d \cdot \sum_{i=1}^n b_i^2 = 1 + \sum_{i=1}^n a_i^2$ and $\sum_{i=1}^n a_i b_i = 0$. Then

$d = \left(\sum_{i=1}^n b_i^2\right)^{-1} + \left(\sum_{i=1}^n a_i^2\right) \cdot \left(\sum_{i=1}^n b_i^2\right)^{-1}$, and the first term of a sum of n squares. Since n is a 2-power the second term is a sum of $n-1$ squares, using Exercise 0.5(4). Therefore $n \le \mathrm{length}_F(d) < 2n$ implying $n = 2^m$.)

6. **$\mathcal{M}$-indecomposables.** Suppose $\mathcal{M} = \mathcal{M}(\varphi_1, \dots, \varphi_k, \langle b_1\rangle, \dots, \langle b_n\rangle)$ for some $b_i \in F^\bullet$ and some round forms φ_j, following the notations used before (5.5).

(1) Every $\mathcal{M}$-indecomposable which is isotropic must actually be hyperbolic.

(2) There is a unique hyperbolic $\mathcal{M}$-indecomposable form $m\mathbb{H}$.

(3) When can there exist an $\mathcal{M}$-indecomposable with dimension $< 2m$?

7. (1) **Lemma.** *If $\langle\langle x\rangle\rangle$ is anisotropic and $\langle\langle x\rangle\rangle \otimes q$ is isotropic then there exists $\beta \subset q$ such that* $\dim \beta = 2$ *and $\langle\langle x\rangle\rangle \otimes \beta$ is hyperbolic.*

(2) **Corollary.** *If $\langle a\rangle q \simeq q$ then $q \simeq q_1 \perp \cdots \perp q_n$ for subforms q_i with* $\dim q_i = 2$ *and $\langle a\rangle q_i \simeq q_i$.*

(3) If $\langle\langle x, y\rangle\rangle \otimes q$ is isotropic, does the analog of (1) hold?

(Hint. (1) Mimic the argument in (5.5).)

8. (1) If $(\langle 1, a, b\rangle, \tau) < \mathrm{Sim}(\langle\langle a, b\rangle\rangle)$, then $\tau \subset \langle 1, a, b\rangle$.

(2) List all pairs (σ, τ) having an unsplittable module of dimension ≤ 4.

(3) If $(\langle 1, a, b, c\rangle, \tau) < \mathrm{Sim}(\langle\langle a, b, c\rangle\rangle)$, then $\tau \subset \langle 1, a, b, c\rangle$. Characterize the forms τ such that $(\langle 1, a, b, c\rangle, \tau) < \mathrm{Sim}(\langle\langle a, b, w\rangle\rangle)$. Here $abc \in G_F(\langle\langle w\rangle\rangle)$ as in (5.3).

(Hint. (1) Show $\dim \tau \le 3$ and use (5.7)(7) if $\dim \tau = 1$. By Expansion we may assume $\dim \tau = 3$. Then $\det \tau = \langle ab\rangle$ since the Clifford algebra is not simple.)

9. **When σ does not represent 1.** Recall Exercise 2.2(1).

(1) Let $\mathcal{M} = \mathcal{M}(a, b) = \{q : a, b \in G_F(q)\}$. Then $q \in \mathcal{M}(a, b)$ iff $(\langle a\rangle, \langle b\rangle) <$ $\mathrm{Sim}(q)$. If $\langle a\rangle \not\simeq \langle 1\rangle$ then the hyperbolic plane is a 2-dimensional $\mathcal{M}$-indecomposable.

(2) Over the rational field $\mathbb{Q}$ the forms $\mathbb{H}$, $\langle\langle 1\rangle\rangle$ and $\langle\langle 2, 5\rangle\rangle$ are $\mathcal{M}(2, 5)$-indecomposables. Find an $\mathcal{M}(2, 5)$-indecomposable which is not similar to a Pfister form. (Note. These proofs involve the Hasse–Minkowski Theorem over $\mathbb{Q}$.)

(3) **Open question.** What are the possible dimensions of $\mathcal{M}(a, b)$-indecomposables?

10. The following are equivalent:

(i) $\langle x, y\rangle < \mathrm{Sim}(q)$.

(ii) $(\langle 1\rangle, \langle x, y\rangle) < \mathrm{Sim}(q)$.

(iii) $(\langle 1, xy\rangle, \langle x\rangle) < \mathrm{Sim}(q)$.

(iv) $\langle\langle xy\rangle\rangle \mid q$ and $x \in G_F(q)$.

11. (1) The following are equivalent:

(i) $(\langle 1, a\rangle, \langle 1, x\rangle) < \mathrm{Sim}(q)$.

(ii) $\langle\langle a\rangle\rangle \mid q$ and $\langle\langle x\rangle\rangle \mid q$.

(iii) $q \simeq \langle\langle a\rangle\rangle \otimes \beta$ for some form β such that $ax \in G_F(\beta)$.

(2) Find a direct proof of (ii) $\iff$ (iii), not using results on similarities.

(Hint. (1) To see (i) $\iff$ (iii) scale by a and use the Eigenspace Lemma 2.10.)

12. **Proposition.** $(\langle 1, a, b\rangle, \langle 1, x\rangle) < \mathrm{Sim}(q)$ *if and only if* $\langle\langle a, b\rangle\rangle \mid q$ *and* $\langle\langle ab, x\rangle\rangle \mid q$.

The proof is outlined below, following the same steps as (5.7).

(1) $(\langle 1, a, b\rangle, \langle 1, x\rangle) < \mathrm{Sim}(q)$ if and only if $(\langle 1, a, b\rangle, \langle 1, ab, abx\rangle) < \mathrm{Sim}(q)$. The "only if" part of the proposition follows.

(2) For the "if" we may assume $\langle a, b\rangle$ does not represent x, so that $\langle\langle a, b\rangle\rangle \not\simeq \langle\langle ab, x\rangle\rangle$.

(3) (8-dim case.) Suppose $q \simeq \langle\langle a, b, w\rangle\rangle$ and $\langle\langle ab, x\rangle\rangle \mid q$. Then $\langle a, b\rangle \perp \langle w\rangle\langle\langle a, b\rangle\rangle$ represents x, so that $x = ar^2 + bs^2 + u$ where $u \in D_F(\langle w\rangle\langle\langle a, b\rangle\rangle)$. Then $q \simeq \langle\langle a, b, u\rangle\rangle$ and $(\langle 1, a, b\rangle, \langle 1, x\rangle) \subset (\langle 1, a, b, u\rangle, \langle 1, a, b, u\rangle) < \mathrm{Sim}(q)$.

(4) If $\varphi = \langle\langle a, b\rangle\rangle$ and $\psi = \langle\langle ab, x\rangle\rangle$, the $\mathcal{M}(\varphi, \psi)$-indecomposables are all 8-dimensional. More generally suppose $\varphi = \alpha \otimes \langle\langle b\rangle\rangle$ and $\psi = \alpha \otimes \langle\langle c_1, \dots, c_k\rangle\rangle$ where α is an r-fold Pfister form and $\varphi \nmid \psi$. Then the $\mathcal{M}(\varphi, \psi)$-indecomposables all have dimension 2^{r+k+1}.

(5) If $\langle 1, a, b, -x, -y\rangle$ is isotropic, for what q is $(\langle 1, a, b\rangle, \langle x, y\rangle) < \mathrm{Sim}(q)$?

(Hint. (1) Use the generators f_2, f_3, g_1, g_2.)

13. The following are equivalent:

(i) $\langle\langle a, b\rangle\rangle \mid q$ and $\langle\langle ab, x\rangle\rangle \mid q$.

(ii) $q \simeq \langle\langle a\rangle\rangle \otimes \gamma$ for some form γ where $\langle\langle ab\rangle\rangle \mid \gamma$ and $ax \in G_F(\gamma)$.

(Hint. Use (5.7), Exercise 11 and the Eigenspace Lemma 2.10.)

Open question. Is there some generalization which includes the Pfister factor results of Exercises 11, 12 and 13 ?

14. Suppose that the trace map ℓ used in (5.9) is replaced by $\ell' : E \to F$ where $\ell'(1) = 1$ and $\ell'(\sqrt{axy}) = 0$. If $\theta = r + s\sqrt{axy}$ determine the form $\ell'(\langle\theta\rangle_E)$.

15. Suppose (K, J) is a field with non-trivial involution, where we write $\bar{\alpha}$ for $J(\alpha)$. Suppose V is a K-vector space and $f : V \to V$ is (K, J)-semilinear.

(1) Let $\{v_1, \dots, v_n\}$ be a K-basis of V and express $f(v_j) = \sum_{i=1}^n a_{ij} v_i$. Then $A = (a_{ij})$ is the matrix associated to f. A vector $v = \sum_{i=1}^n x_i v_i$ is represented by the column vector $X = (x_1, \dots, x_n)^\top$ so that $f(v) = \sum_{i=1}^n x_i' v_i$ is represented by the column vector $X' = A\bar{X}$.

(2) If f and g are (K, J)-semilinear maps on V represented by matrices A and B, then $f \circ g$ is K-linear and is represented by the matrix $A\bar{B}$.

(3) Suppose $h : V \times V \to K$ is a regular hermitian form. Let $M = (h(v_i, v_j))$ be the matrix of h, so that $M^\top = \bar{M}$. If $v, w \in V$ correspond to the column vectors X, Y then $h(v, w) = X^\top M \bar{Y}$. To define the adjoint involution $\sim$ applied to a (K, J)-semilinear map f the usual formula makes no sense: $h(f(v), w) = h(v, \tilde{f}(w))$. (Why?) It is replaced by the definition:

$$\overline{h(f(v), w)} = h(v, \tilde{f}(w)).$$

Then $\tilde{f}$ is also (K, J)-semilinear and $\sim$ is a K-linear involution on the space of all (K, J)-semilinear maps of V. (I.e. $\widetilde{(\alpha f)} = \alpha \tilde{f}$, $\widetilde{(f + g)} = \tilde{f} + \tilde{g}$ and $\tilde{\tilde{f}} = f$ when f is (K, J)-semilinear and $\alpha \in K$.)

(4) If $\tilde{A}$ is the matrix corresponding to $\tilde{f}$ then $\tilde{A} = M^{-\top} A^\top M$. Consequently, $\tilde{f} = f$ if and only if the matrix $M^\top A$ is symmetric.

(5) Does any of this become easier if we use the other definition of "hermitian", where $h(v, w)$ is (K, J)-semilinear in v and K-linear in w?

16. Suppose F, E, K are as described before (5.9) and the involution trace $\ell \circ \mathrm{tr} : K \to F$ is given. Suppose V is a K-vector space and $b_q : V \times V \to F$ is a symmetric bilinear form which admits (K, J). Then there exists a unique hermitian form $h : V \times V \to K$ such that $\ell \circ \mathrm{tr} \circ h = b_q$. Find an explicit formula for h.

(Hint. Say $b : V \times V \to E$ is the corresponding form over E. For $v, w \in V$ show that $b(v, w) = b_q(\sqrt{axy} \cdot v, w) + b_q(v, w) \cdot \sqrt{axy}$. Now build b up to h.)

17. **Norm principle.** Suppose $K = F(\sqrt{d})$ is a quadratic extension of F and define $s : K \to F$ by $s(x + y\sqrt{d}) = y$. If α is a quadratic form over K let $s_*(\alpha)$ denote the transfer to F. (See Lam (1973), p. 201 or Scharlau (1985), p. 50 for discussions of this s_*.)

Lemma. *$s_*(\alpha)$ is isotropic iff α represents some element of $F^\bullet$.*

We also need the following analog of "Frobenius reciprocity":

If φ is a form over F and α is a form over K then $s_*(\varphi_E \otimes \alpha) \simeq \varphi \otimes s_*(\alpha)$.

(1) **Norm Principle.** Let φ be a form over F and $x \in K$. Then $N(x) \in D_F(\varphi) \cdot D_F(\varphi)$ if and only if $x \in F^\bullet \cdot D_K(\varphi_K)$.

(2) Deduce Lemma 5.12.

(Hint. (1) $\varphi \perp \langle -Nx \rangle \varphi$ is F-isotropic iff $s_*(\langle x \rangle \varphi)$ is F-isotropic.)

18. **Examples.** (1) Give an example of an unsplittable $\sigma < \mathrm{Sim}(q)$ where q is anisotropic but is not similar to a Pfister form.

(2) Give an example of an unsplittable $\sigma < \mathrm{Sim}(8\mathbb{H} \perp 16\langle 1 \rangle)$ over $\mathbb{Q}$ where $\dim \sigma = 8$.

19. **Common slot.** Suppose $\alpha \simeq \langle\langle a, a' \rangle\rangle$ and $\beta \simeq \langle\langle b, b' \rangle\rangle$ are 2-fold Pfister forms. If $\alpha \simeq \beta$ then there exists $x \in F^\bullet$ such that $\alpha \simeq \langle\langle a, x \rangle\rangle$ and $\beta \simeq \langle\langle b, x \rangle\rangle$.

20. **Contradiction? Conjecture.** Suppose $(\sigma,\tau) < \mathrm{Sim}(q)$ is an (s,t)-family and q represents $a \in F^\bullet$. Then there is a decomposition $q = q_1 \perp \cdots \perp q_n$ such that for every i, $(\sigma,\tau) < \mathrm{Sim}(q_i)$ is unsplittable, and such that q_1 represents a.

(1) If q is a Pfister form the Conjecture is true. Suppose there is a Pfister form φ such that: $(\sigma,\tau) < \mathrm{Sim}(q)$ iff $\varphi \mid q$. Then the Conjecture is true.

(2) Consider the set-up of (C,J)-modules and suppose $V = U \perp U'$ where U is an irreducible submodule. If $W \subseteq V$ is irreducible with $W \not\subseteq U'$, then $W = U[f] = \{u + f(u) : u \in U\}$ is the graph of some C-homomorphism $f : U \to U'$. Now specialize to the case that $\mathrm{End}_C(U) = F$ and $U' \cong U$. Then any value represented by an irreducible submodule W must lie in $(1+\lambda^2)\cdot D_F(U)$ for some $\lambda \in F$. For a specific case let $(\sigma,\tau) = (\langle 1,1\rangle, \langle 1\rangle)$ and $V \simeq \langle\langle 1,1\rangle\rangle$. Then any irreducible submodule of V represents only values in $D_F(\langle\langle 1\rangle\rangle)$, and the Conjecture is false.

(3) Resolve the apparent contradiction between parts (1) and (2).

(Hint. (1) For the first statement, choose any unsplittable decomposition and let $b \in D_F(q_1)$. Then $q \simeq \langle ab\rangle q$.)

21. **Transfer ideals.** Suppose (K,J) is a field with involution, F is a subfield fixed by J and $t : K \to F$ is an involution trace (that is, t is F-linear and $t(\bar{a}) = t(a)$). If (V,h) is a (K,J)-hermitian space then the transfer $t_*(V,h) = (V, t\circ h)$ is a quadratic space over F. Let $\mathcal{I}((K,J)/F)$ be the set of (isometry classes of) all such transferred spaces. Then $\mathcal{I}((K,J)/F)$ does not depend on the choice of t and its image in the Witt ring $W(F)$ is an ideal.

Suppose $a, b \in F^\bullet$ and $K = F(\sqrt{-a}, \sqrt{-b})$ is an extension field of degree 4. Let J be the involution on K which induces non-trivial involutions J_a and J_b on the subfields $A = F(\sqrt{-a})$ and $B = F(\sqrt{-b}$ respectively. Let $t : K \to F$ be an involution trace which induces the (unique) involution traces $t_a : A \to F$ and $t_b : B \to F$.

Proposition. $\mathcal{I}((K,J)/F) = \mathcal{I}((A,J_a)/F) \cap \mathcal{I}((B,J_b)/F)$.

(Hint. This is a restatement of Proposition 5.16. First check that $\mathcal{I}((A,J_a)/F) = \mathcal{M}(\langle\langle a\rangle\rangle)$ and similarly for b.)

22. **Forms of odd dimension.** Assume the following result, due originally to Pfister (1966).

Proposition. *If* $\dim\delta$ *is odd then* δ *is not a zero-divisor in the Witt ring* $W(F)$.

(1) If α is not hyperbolic then $\alpha \mid m\mathbb{H}$ if and only if $\dim\alpha \mid m$. (Generalizing (5.5)(3).)

(2) If $a \in G_F(\alpha\otimes\delta)$ where $\dim\delta$ is odd, then $a \in G_F(\alpha)$.

(3) If φ is a Pfister form and $\varphi \mid \alpha\otimes\delta$ where $\dim\delta$ is odd, then $\varphi \mid \alpha$.

(4) If (σ,τ) has unsplittables of dimension ≤ 4, the answer to the following question is "yes".

Odd Factor Question. If $(\sigma,\tau) < \mathrm{Sim}(\alpha\otimes\delta)$ where $\dim\delta$ is odd, does it follow that $(\sigma,\tau) < \mathrm{Sim}(\alpha)$?

(Hint. (3) This seems to require the theory of function fields described in the appendix to Chapter 9. Express $\alpha = \alpha_0 \perp k\mathbb{H}$ where α_0 is anisotropic. Apply (9.A.6) and (5.5).)

23. **Pfister factors.** (1) If φ is a Pfister form and $\langle 1, b\rangle \subset \varphi$ then $\varphi \simeq \langle\langle b, c_2, \dots, c\rangle\rangle$ for some $c_j \in F^\bullet$. This was proved in (5.2) (1).

Lemma. *If φ is a 3-fold Pfister form and $\langle 1, a, b\rangle \subset \varphi$ then $\varphi \simeq \langle\langle a, b, w\rangle\rangle$ for some w.*

(2) If $\dim \alpha = \dim \beta = 4$, $d\alpha = d\beta$ and $c(\alpha) = c(\beta) = 1$ then α and β are similar.

(Hint. (1) Given $\varphi \simeq \langle\langle a, x, y\rangle\rangle$ such that $\langle x\rangle\langle\langle a\rangle\rangle \perp \langle y\rangle\langle\langle a, x\rangle\rangle$ represents b. We may assume $b = xu + yv$ for some $u \in D_F(\langle\langle a\rangle\rangle)$ and $v \in D_F(\langle\langle a, x\rangle\rangle)$. Then $\varphi \simeq \langle\langle a, xu, yv\rangle\rangle$.

(2) Let $d\alpha = \langle d\rangle$ and let $\varphi = \alpha \perp \langle d\rangle\beta$. Then $\dim \varphi = 8$, $d\varphi = \langle 1\rangle$ and $c(\varphi) = 1$ so that φ is similar to a Pfister form, by (3.20) (2). We may assume $\alpha = \langle 1, a, b, abd\rangle$ and find $\varphi \simeq \langle\langle a, b, w\rangle\rangle$ for some w. Then d is represented by $\langle 1\rangle \perp \langle w\rangle\langle\langle a, b\rangle\rangle$ so that $d = t^2 + u$ for some $t, u \in F^\bullet$ such that $\varphi \simeq \langle\langle a, b, u\rangle\rangle$. Then $\varphi \simeq \langle 1, a, b, ab\rangle \otimes \langle\langle u\rangle\rangle \simeq \alpha \otimes \langle\langle u\rangle\rangle$. Cancel α to finish the proof.)

Notes on Chapter 5

In the proof of Lemma 5.2 we assumed that $x, y \neq 0$, leaving the other cases to the reader. Actually that non-zero case is sufficient if we invoke the Transversality Lemma of Exercise 1.15

Lemma 5.5 and Proposition 5.6 follow Wadsworth and Shapiro (1977b). Lemma 5.5 is also treated in Szymiczek (1977). More recent results on round forms appear in Alpers (1991) and Hornix (1992).

Exercise 5. These results on the level $s(F)$, due to Pfister, helped to motivate the investigation of the multiplicative properties of quadratic forms. The second result leads to examples of fields which have prescribed level 2^m. See Exercise 9.11 below.

Exercise 7. See Elman and Lam (1973b), pp. 288–289. Compare Exercise 2.9.

Exercise 9. The different dimensions possible for unsplittable $(\langle a\rangle, \langle b\rangle)$-modules contrast with the Decomposition Theorem 4.1. The image of $\mathcal{M}(a, b)$ in $W(F)$ is the ideal $\mathcal{A} = \text{ann}(\langle\langle -a\rangle\rangle) \cup \text{ann}(\langle\langle -b\rangle\rangle)$. It is known that $\mathcal{A}$ is generated by 1-fold and 2-fold Pfister forms. See Elman, Lam and Wadsworth (1979). For the case of global fields see Exercise 11.6.

Exercise 12 (4) follows Wadsworth and Shapiro (1977b).

Exercise 17. The Norm Principle appears in Elman and Lam (1976), 2.13.

Exercise 19. Compare Exercise 3.10.

Exercise 21. If $E = F(\sqrt{ab})$ with trivial involution then $\mathcal{I}(E/F) = \mathcal{M}(ab)$ is contained in $\mathcal{M}(\langle\langle a\rangle\rangle, \langle\langle b\rangle\rangle)$. The analog of this proposition for biquadratic extensions with trivial involution is proved in Leep and Wadsworth (1990).

Exercise 22. Proofs of the proposition appear in Lam (1973) on pp. 250 and 310, in Scharlau (1985), p. 54, and in D. W. Lewis (1989).

Exercise 23. Compare Exercise 3.12(4) and the references given in Chapter 3. The lemma here is a special case of Exercise 9.15.

Chapter 6

Involutions

If (C, J) is an algebra with involution, when does a given C-module V possess a λ-form admitting C? A regular λ-form on V induces an adjoint involution on $\mathrm{End}(V)$, and every involution on $\mathrm{End}(V)$ arises from some λ-form. This sign λ is called the "type" of the involution. The question posed above is then equivalent to asking whether there is an involution on $\mathrm{End}(V)$ which is compatible with (C, J). If C is central simple it splits off as a tensor factor: $\mathrm{End}(V) \cong C \otimes A$, for some central simple algebra A. The involutions on $\mathrm{End}(V)$ compatible with (C, J) are then exactly the maps $J \otimes K$, where K is an involution on A. The focus of our work has then moved to an analysis of this algebra A and its involutions.

In this short chapter we describe the basic results about involutions on central simple algebras, postponing the applications to later chapters. Those results on involutions have appeared in various textbooks. In fact, most of the ideas we use go back at least to the 1930s and are summarized in Albert's book *Structure of Algebras* (1939). We assume the reader is familiar with the general theory of central simple algebras, including the Wedderburn Theorems, the Double Centralizer Theorem, the existence of splitting fields, and the Skolem–Noether Theorem. However it seems worthwhile to derive the tools we need concerning involutions. Further information about algebras and involutions is available in the books by Rowen (1980), Scharlau (1985), Knus (1988), and Knus et al. (1998).

If A is a ring we let $A^\bullet$ denote the group of units, and if $S \subseteq A$ we write $S^\bullet$ for the subset $S \cap A^\bullet$. However, following standard practice we write $\mathrm{GL}(V)$ rather than $\mathrm{End}^\bullet(V)$. If A is an F-algebra an *involution* J on A is defined to be an anti-automorphism such that J^2 is the identity map. When F is the center of A then J preserves F and the restriction is an involution on the field F. The involution is said to be of the first kind or second kind, depending on whether or not it fixes F.

Unless explicitly stated otherwise, involutions in this book are F-linear. That is, we assume they are of the "first kind", inducing the identity map on the ground field.

6.1 Definition. Let A be an F-algebra with involution J. If $a \in A^\bullet$ define the map $J^a : A \to A$ by

$$J^a(x) = a^{-1}J(x)a \quad \text{for } x \in A.$$

6.2 Lemma. *Let A, J and a be given as above and suppose A has center F. Then J^a is an involution if and only if $J(a) = \pm a$. The element a is uniquely determined, up to non-zero scalar multiple, by J and J^a.*

Proof. If J^a is an involution then $x = J^a J^a(x) = a^{-1}J(a)xJ(a^{-1})a$ for every $x \in A$. Then $a^{-1}J(a)$ is central so that $J(a) = \varepsilon a$ for some $\varepsilon \in F^{\bullet}$. Applying J again we find that $\varepsilon^2 = 1$. The converse follows from the same formula. If $J^b = J^a$ for some $b \in A^{\bullet}$ then $a^{-1}b$ is central and $b \in aF^{\bullet}$. □

We now make a key observation: every involution on the split algebra $\mathrm{End}(V)$ comes from a regular λ-form on V.

6.3 Lemma. *Let V be an F-vector space.*

(1) *If B is a regular λ-form on V and $f \in \mathrm{GL}(V)$, define the bilinear form $B^f : V \times V \to F$ by*
$$B^f(x, y) = B(f(x), y)$$
for $x, y \in V$. If $I_B(f) = \varepsilon f$ where $\varepsilon = \pm 1$, then B^f is a regular $\varepsilon\lambda$-form and $I_{B^f} = I_B^f$. Every regular $\varepsilon\lambda$-form on V arises from B in this way.

(2) *If J is an involution on $\mathrm{End}(V)$ then $J = I_B$ for some regular λ-form B on V. This form B is uniquely determined, up to non-zero scalar multiple.*

Proof. (1) It is easy to see that B^f is a regular $\varepsilon\lambda$-form. To prove the formula for the involutions note that $B^f(x, h(y)) = B(I_B(h)f(x), y) = B^f(f^{-1}I_B(h)f(x), y)$. Recall that the map $\theta_B : V \to \hat{V}$ is defined by $\langle x|\theta_B(y)\rangle = B(x, y)$. If B' is any regular $\varepsilon\lambda$-form on V, let $f = (\theta_{B'} \circ \theta_B^{-1})^{\top}$. Then $B' = B^f$.

(2) Let B_0 be a regular 1-form on V with adjoint involution I_0. By the Skolem–Noether Theorem and (6.2) we have $J = I_0^f$ for some $f \in \mathrm{GL}(V)$ with $I_0(f) = \lambda f$ for some $\lambda = \pm 1$. Then $B = B_0^f$ is a λ-form on V having $I_B = J$. If B' is another regular form having $I_{B'} = J$, then (1) implies that $B' = B^g$ for some $g \in \mathrm{GL}(V)$ and $J = I_B^g = J^g$. Then g is in the center of $\mathrm{End}(V)$, and B' is a scalar multiple of B. □

An involution J is the adjoint involution of some λ-form on V. We define the *type* of J to be this sign λ, and say that J is a *λ-involution*. Some authors say that J has *orthogonal type* if its type is 1 and J has *symplectic type* if its type is -1.

The notion of type can be generalized by considering the behavior of involutions under extension of scalars. If L/F is a field extension and J is an involution of the F-algebra A, then $J \otimes 1_L$ is an involution of the L-algebra $A \otimes L$. If A is a central simple F-algebra then there are "splitting fields" L such that $A \otimes L \cong \mathrm{End}_L(V)$, for some L-vector space V. One well-known consequence is that $\dim A$ is a square. The algebra A is said to have *degree n* if $\dim A = n^2$ (and $\dim_L V = n$).

6.4 Definition. Suppose (A, J) is a central simple F-algebra with involution and L is a splitting field for A. Then the involution $J \otimes 1_L$ on $A \otimes L \cong \text{End}_L(V)$ is the adjoint involution of some λ-form B on V. The *type* of J is this sign λ, and J is called a *λ-involution.*

For a given splitting field L Lemma 6.3 implies that this sign λ is uniquely determined. Since any two splitting fields can be embedded in a larger field extension, it follows that the type λ is independent of the choice of L. This independence is also clear from the next lemma.

6.5 Lemma. *Let A be a central simple F-algebra of degree n, so that* $\dim A = n^2$.

(1) *If J and J' are involutions on A then $J' = J^a$ for some $a \in A^\bullet$ with $J(a) = \pm a$. Furthermore, J and J' have the same type if and only if $J(a) = a$.*

(2) *If J is an involution on A define $\mathcal{S}^\varepsilon(A, J) = \{x \in A : J(x) = \varepsilon x\}$, the subspace of elements which are ε-symmetric for J. If J has type λ then* $\dim \mathcal{S}^\varepsilon(A, J) = \frac{n(n+\varepsilon\lambda)}{2}$.

Proof. (1) The existence and uniqueness (up to scalar multiple) of the element a follow as in (6.3) (2) and (6.2). We may extend scalars to assume $A \cong \text{End}(V)$ for some vector space V. If $J(a) = \varepsilon a$ then by (6.3) $J = I_B$ for some λ-form B on V and $J' = I_{B'}$ where $B' = B^a$ is an $\varepsilon\lambda$-form on V.

(2) We may assume that $A = \text{End}(V)$. The quadratic form $n\langle 1\rangle$ on V has adjoint involution I which is just the transpose map on matrices. The dimensions are easily found: $\dim \mathcal{S}^\varepsilon(A, J) = \frac{n(n+\varepsilon)}{2}$. By (1) $J = I^a$ for some $a \in A^\bullet$ with $I(a) = \lambda a$. The claim follows from the general observation that

$$\mathcal{S}^\varepsilon(A, I^a) = \mathcal{S}^{\lambda\varepsilon}(A, I) \cdot a. \qquad \square$$

We are working here in the category of "central simple F-algebras with involution." If (A_1, J_1) and (A_2, J_2) are in that category we write $\varphi : (A_1, J_1) \to (A_2, J_2)$ to indicate an F-algebra homomorphism $\varphi : A_1 \to A_2$ which preserves the involutions: $J_2 \circ \varphi = \varphi \circ J_1$. Similarity representations (as in Chapter 4) are examples of such homomorphisms. Let us analyze some special cases of isomorphisms in this category.

6.6 Proposition. *Suppose (V_i, B_i) is a regular λ_i-space for $i = 1, 2$. Let I_i denote the involution I_{B_i} on* $\text{End}(V_i)$. *Then* $(\text{End}(V_1), I_1) \cong (\text{End}(V_2), I_2)$ *if and only if* (V_1, B_1) *and* (V_2, B_2) *are similar spaces.*

Proof. Suppose $h : (V_1, B_1) \to (V_2, B_2)$ is a bijective similarity. Define the map $\varphi : \text{End}(V_1) \to \text{End}(V_2)$ by: $\varphi(f) = hfh^{-1}$. To show that $I_2 \circ \varphi = \varphi \circ I_1$ we check that for $x, y \in V$ the expressions $B_2(I_2(\varphi(f))(h(x)), h(y))$ and $B_2(\varphi(I_1(f))(h(x)), h(y))$ both reduce to the same value $\mu(h)B_1(x, f(y))$. Conversely suppose $\varphi : (\text{End}(V_1), I_1) \to (\text{End}(V_2), I_2)$ is an isomorphism. Since the dimensions are equal there is some linear bijection $g : V_1 \to V_2$. By Skolem–Noether, the map $f \mapsto g^{-1}\varphi(f)g$ is an inner automorphism of $\text{End}(V_1)$, so there is a linear bijection $h : V_1 \to V_2$ with $\varphi(f) = hfh^{-1}$. Define B' on V by setting $B'(x, y) = B_2(h(x), h(y))$. Then h is an isometry $(V_1, B') \to (V_2, B_2)$ and the calculation above shows that $I_{B'} = \varphi^{-1} \circ I_2 \circ \varphi = I_1$. Therefore $B' = aB_1$ for some $a \in F^\bullet$, and $(V_2, B_2) \simeq (V_1, aB_1)$. □

When considering isomorphisms of two algebras with involution we often identify the algebras and concentrate on the involutions.

6.7 Lemma. *Let (A, J) be a central simple F-algebra with involution, and let $a, b \in A^\bullet$. Then $(A, J^a) \cong (A, J^b)$ if and only if $b = rJ(u)au$ for some $r \in F^\bullet$ and $u \in A^\bullet$.*

Proof. If $\alpha : (A, J^a) \to (A, J^b)$ is the given isomorphism then α is an F-algebra isomorphism and $J^b = \alpha \circ J^a \circ \alpha^{-1}$. By Skolem–Noether there exists $u \in A^\bullet$ such that $\alpha(x) = u^{-1}xu$ and the claim follows. The converse is similar. □

For quaternion algebras we get a complete characterization of the involutions.

6.8 Lemma. *Let A be a quaternion algebra with bar involution J_0. Express $A = F + A_0$ where A_0 is the set of pure quaternions.*

(1) *J_0 is the only (-1)-involution on A.*

(2) *If J is a 1-involution then $J = J_0^e$ for some $e \in A_0^\bullet$. For any $e \in A_0^\bullet$, the only involutions sending $e \mapsto -e$ are J_0 and J_0^e.*

(3) *For J as above the value Ne is uniquely determined up to a square factor. Define $\det(J) = \langle Ne \rangle$ in $F^\bullet/F^{\bullet 2}$. Suppose J_1, J_2 are 1-involutions on A. Then $(A, J_1) \cong (A, J_2)$ if and only if $\det(J_1) = \det(J_2)$.*

Proof. (1) By (6.5) J_0 has type -1. Any involution J on A must equal J_0^e for some $e \in A^\bullet$ with $J_0(e) = \pm e$. If J has type -1 then $J_0(e) = e$ so that $e \in F^\bullet$ and $J = J_0$.

(2) If J has type 1 then $e \in A_0^\bullet$ and $J(e) = -e$. The uniqueness follows since $\dim \mathcal{S}^-(A, J) = 1$.

(3) If $J = J_0^e$, the element e is determined up to a factor in $F^\bullet$. Hence the norm Ne is determined up to a factor in $F^{\bullet 2}$, and $\det(J)$ is well defined. Suppose $J_1 = J_0^a$ and $J_2 = J_0^b$ for some $a, b \in A_0^\bullet$. If $J_1 \cong J_2$ use (6.7). Conversely suppose $\det(J_1) = \det(J_2)$. Altering b by a scalar we may assume that $Na = Nb$. Standard facts about quaternion algebras (see Exercise 2) imply that there exists $u \in A^\bullet$ such that $b = u^{-1}au = (Nu)^{-1}J(u)au$ and (6.7) applies. □

Our next task is to show that the type behaves well under tensor products.

6.9 Proposition. *Let A_i be a central simple F-algebra with λ_i-involution J_i, for $i = 1, 2$. Then $J_1 \otimes J_2$ is a $\lambda_1\lambda_2$-involution on $A_1 \otimes A_2$.*

Proof. We may replace the field F by a splitting field to assume that $A_i \cong \mathrm{End}(V_i)$ and that J_i is the adjoint involution of a λ_i-form B_i on V_i. Suppose ψ is the natural isomorphism

$$\psi : \mathrm{End}(V_1) \otimes \mathrm{End}(V_2) \to \mathrm{End}(V_1 \otimes V_2).$$

To complete the proof we must verify that ψ carries $I_{B_1} \otimes I_{B_2}$ to $I_{B_1 \otimes B_2}$. To see this recall that by definition, $\psi(f_1 \otimes f_2)(x_1 \otimes x_2) = f_1(x_1) \otimes f_2(x_2)$ whenever $f_i \in \mathrm{End}(V_i)$ and $x_i \in V_i$. One can then check directly that $\psi(I_{B_1}(f_1) \otimes I_{B_2}(f_2))$ does act as the adjoint of $\psi(f_1 \otimes f_2)$ relative to the form $B_1 \otimes B_2$. □

6.10 Corollary. *Suppose (V_i, B_i) is a regular λ_i-space for $i = 1, 2$. Let I_i denote the involution I_{B_i} on* $\mathrm{End}(V_i)$.

(1) *(V, B) is similar to $(V_1 \otimes V_2, B_1 \otimes B_2)$ if and only if*

$$(\mathrm{End}(V), I_B) \cong (\mathrm{End}(V_1), I_1) \otimes (\mathrm{End}(V_2), I_2).$$

(2) *There is a homomorphism $(\mathrm{End}(V_1), I_1) \to (\mathrm{End}(V_2), I_2)$ if and only if (V_1, B_1) "divides" (V_2, B_2) in the sense that $(V_2, B_2) \simeq (V_1, B_1) \otimes (W, B)$ for some $\lambda_1\lambda_2$-space (W, B).*

Proof. For (1) apply (6.6) and (6.9). We prove a sharper version of (2) in the next corollary. □

6.11 Corollary. *Suppose (C, J) is a central simple algebra with involution and $A \subseteq C$ is a central simple subalgebra preserved by J. Then $(C, J) \cong (A, J|_A) \otimes (C', J')$ for some central simple subalgebra C' with involution J'.*

Suppose further that A is split so that $(A, J|_A) \cong (\mathrm{End}(U), I_B)$ for some λ-form B on U. If (V, q) is a quadratic (C, J)-module, one then obtains:
$(V, q) \simeq (U, B) \otimes (U', B')$ where (U', B') is some λ-space admitting (C', J').

Proof. The algebra C' is the centralizer of A in C and the Double Centralizer Theorem implies that C' is central simple and $A \otimes C' \cong C$. Since J preserves A it also preserves C' and induces some involution J' there. Since C is simple the given homomorphism $(C, J) \to (\mathrm{End}(V), I_q)$ is injective and we view C as a subalgebra of $\mathrm{End}(V)$. Then as above there is a decomposition $(C, J) \otimes (C'', J'') \cong (\mathrm{End}(V), I_q)$. Therefore $A \otimes C' \otimes C'' \cong \mathrm{End}(V)$ and since A is split Wedderburn's Theorem implies that $C' \otimes C'' \cong \mathrm{End}(U')$ for some U'. The involution $J' \otimes J''$ then induces an involution $I_{B'}$ for some form B' on U'. Therefore $(\mathrm{End}(U), I_B) \otimes (\mathrm{End}(U'), I_{B'}) \cong (A, J|_A) \otimes (C' \otimes C'', J' \otimes J'') \cong (\mathrm{End}(V), I_q)$ and (6.10)(1) implies that (V, q) is similar to $(U, B) \otimes (U', B')$. We may alter B' by a scalar to assume this is an isometry.

Since q is quadratic and B is λ-symmetric, (6.9) implies that B' is λ-symmetric. By construction (U', B') admits (C', J'). □

This corollary gives another proof of the Eigenspace Lemma 2.10. See Exercise 4(3) below. It also provides an interpretation of "Pfister factors" entirely in terms of algebras, as follows.

6.12 Corollary. *Suppose (V, q) is a quadratic space and $a_1, \dots, a_m \in F^\bullet$. Then $\langle\langle a_1, \dots, a_m\rangle\rangle$ is a tensor factor of q if and only if there is a homomorphism $(Q_1, J_1)\otimes\cdots\otimes(Q_m, J_m) \to (\mathrm{End}(V), I_q)$ where each (Q_k, J_k) is a split quaternion algebra with involution of type 1 such that there exists $f_k \in Q_k$ such that $J_k(f_k) = -f_k$ and $f_k^2 = -a_k$.*

Proof. Note that $(Q_k, J_k) \cong (\mathrm{End}(F^2), I_{\varphi_k})$ where $\varphi_k \simeq \langle\langle a_k\rangle\rangle$. The equivalence follows from (6.11). □

Suppose C is a central simple F-algebra with an ε-involution J, and V is a C-module. The relevant question is:

When is there a regular λ-form B on V admitting C?

The C-module structure provides a homomorphism $\pi : C \to \mathrm{End}(V)$ which is injective since C is simple. We may view π as an inclusion $C \subseteq \mathrm{End}(V)$ and let A be the centralizer of C, that is, $A = \mathrm{End}_C(V)$. By the Double Centralizer Theorem, A is also a central simple F-algebra and

$$C \otimes A \cong \mathrm{End}(V).$$

In particular, the dimension of A can be found from $\dim C$ and $\dim V$.

If V possesses a regular λ-form B admitting C then there is an involution I_B on $\mathrm{End}(V)$ which is compatible with the involution J on C. That is, I_B extends J and in particular it preserves the subspace $C \subseteq \mathrm{End}(V)$. Therefore I_B preserves the centralizer A and induces an involution K on A. Then $J \otimes K = I_B$, and by (6.9) the involution K has type $\varepsilon\lambda$. Conversely if A possesses an $\varepsilon\lambda$-involution K then $J \otimes K$ on $C \otimes A \cong \mathrm{End}(V)$ provides an involution on $\mathrm{End}(V)$. Then by (6.3) and (6.9) this involution must be I_B for some regular λ-form B on V. This form B does admit C since I_B is compatible with J. Therefore, the existence of a λ-form B admitting C is equivalent to the existence of an $\varepsilon\lambda$-involution on A.

We can use these methods to prove that A must possess an involution.

6.13 Proposition. *Suppose A and C are central simple algebras which are equivalent in the Brauer group. If C has an involution then so does A.*

Proof. By Wedderburn, $C \cong D \otimes \mathrm{End}(U)$ and $A \cong D \otimes \mathrm{End}(W)$ where D is some F-central division algebra and U, W are F-vector spaces. Since $\mathrm{End}(W)$ always has a 1-involution it suffices to prove that D possesses an involution. Since J is an anti-automorphism, we know C is isomorphic to its opposite algebra C^{op}, so that $C \otimes C \cong C \otimes C^{\text{op}}$ is split. Therefore $C \otimes D$ is also split, say $C \otimes D \cong \text{End}(V)$. Since D is a division algebra, V is an irreducible C-module. The dual $\hat{V}$ is also a C-module (as defined in Chapter 4) and has the same dimension as V. Therefore $\hat{V} \cong V$ and Lemma 4.11 implies that V has some regular λ-form B admitting (C, J), for some $\lambda = \pm 1$. The adjoint involution I_B on $\text{End}(V)$ preserves the subalgebra C, so it must also preserve D, the centralizer of C. The restriction of I_B to D is an involution. □

Actually (6.13) is part of a famous theorem of Albert (1939). If A is a central simple algebra admitting an involution then it certainly has an anti-automorphism. If A has an anti-automorphism then there is an isomorphism $A \cong A^{\text{op}}$, and therefore $[A]^2 = 1$ in the Brauer group $\text{Br}(F)$. Albert proved the converse.

6.14 Theorem. *If A is a central simple algebra with $[A]^2 = 1$ then A has an involution.*

We refer the reader to the beautiful proof appearing as Theorem 8.8.4 in Scharlau (1985). Several proofs have appeared in the literature. For example see Knus et al. (1998), §3. The original version, given as Theorem 10.19 of Albert (1939), was proved using the theory of crossed products.

6.15 Corollary. *Let A be a central simple algebra with involution. There exist involutions of both types on A unless A is a split algebra of odd degree.*

Proof. Let D be the "division algebra part" of A. Then $A \cong D \otimes \text{End}(U)$ for some vector space U. By (6.14) the algebra D has an involution and there is always a 1-involution on $\text{End}(U)$. Therefore there is an involution J on A which preserves the subalgebras D and $\text{End}(U)$. If there exists $c \in A^{\bullet}$ with $J(c) = -c$ then J and J^c have opposite type. If $D \neq F$ there exists $d \in D$ with $J(d) \neq d$, and we use $c = J(d) - d$. If $\dim U$ is even then there exists a regular (-1)-form on U so there must exist $c \in \text{GL}(U)$ with $J(c) = -c$. The only exception is when $D = F$ and $\dim U$ is odd. □

We noted in Chapter 4 that unsplittable (C, J)-modules are usually irreducible. For a central simple algebra C the exceptions are now easy to describe.

6.16 Corollary. *Let C be a central simple algebra with an ε-involution J and let V be a C-module. The hyperbolic module $H_\lambda(V)$ is (C, J)-unsplittable if and only if $C \cong \text{End}(V)$ and $\lambda \neq \varepsilon$. In this case all λ-symmetric (C, J)-modules are hyperbolic.*

Proof. By Theorem 4.10 we know that $H_\lambda(V)$ is unsplittable if and only if V is irreducible and possesses no regular λ-form admitting C. The "if" part is clear. Conversely, we know that $C \otimes A \cong \text{End}(V)$ where $A = \text{End}_C(V)$. Then (6.9) implies that A has no $\varepsilon\lambda$-involution. Since V is irreducible Schur's Lemma implies A is a division algebra and (6.15) implies that $A = F$. □

The standard examples of central simple algebras with involution are quaternion algebras and matrix algebras. So if $A \cong \mathbb{M}_n(D)$ where D is a tensor product of quaternion algebras, then A has an involution. In the 1930s Albert considered the following converse question:

If D is an F-central division algebra with involution then must D be isomorphic to a tensor product of quaternions?

There has been considerable work on this question since then. The next theorem summarizes some major results in this area.

6.17 Theorem. *Suppose D is an F-central division algebra with involution.*

(1) *D has degree 2^m for some m. If $m = 1$ then D is a quaternion algebra. If $m = 2$ then D is a tensor product of two quaternion algebras.*

(2) *There exists a division algebra D of degree 8 over its center F such that D has an involution but has no quaternion subalgebras. For any such D the algebra $\mathbb{M}_2(D)$ is isomorphic to a tensor product of 4 quaternion algebras.*

(3) *$[D]$ is a product of quaternion algebras in the Brauer group.*

Here are references where the proofs of these statements can be found.

If $\deg(D) = n$, Albert showed that $[D]^n = 1$ in $\mathrm{Br}(F)$, and that $\deg(D)$ and the order of [D] involve the same prime factors. (See Albert (1939), Theorem 5.17, p. 76, or Draxl (1983), Theorem 11, p. 66.) Consequently if D has an involution then $[D]^2 = 1$ and $\deg(D)$ must be a 2-power. The stronger result when $m = 2$ is due to Albert (1932), with various different proofs given by Racine (1974), Jančevskiĭ (1974) and Rowen (1978). Several proof are presented by Knus et al. (1998), §16. We prove it in (10.21) below following Rowen's method.

(2) Such examples were found by Amitsur, Rowen and Tignol (1979), where the center is a purely transcendental extension of $\mathbb{Q}$ of degree 4. The criteria involved in constructing this counterexample were generalized by Elman, Lam, Tignol and Wadsworth (1982) and further counterexamples were found (all of characteristic 0). The second statement was proved by Tignol (1978).

(3) This is part of an important theorem of Merkurjev (1981) which states that the quaternion symbol map $k_2F \to \mathrm{Br}_2(F)$ is an isomorphism. This implies that some matrix algebra over D is isomorphic to a tensor product of quaternion algebras.

Exercises for Chapter 6

1. **The type of J_S.** Let $\sigma \cong \langle 1 \rangle \perp \sigma_1$ be a quadratic form of dimension $s = 2m + 1$. Then $C = C(-\sigma_1)$ is central simple of degree 2^m and has the involution J_S.

Lemma. *J_S has type* 1 *if and only if* $s \equiv \pm 1 \pmod 8$.

(1) Proof #1. Apply (6.5) directly by computing $\dim \mathcal{S}^+(C, J_S)$ to be the sum of all $\binom{n}{j}$ where $j \equiv 0, 3 \pmod 4$. Such sums can be evaluated using the binomial theorem with appropriate roots of unity. (See Knuth (1968), 1.2.6, Exercise 38.)

(2) Proof #2. An explicit decomposition of C as a product of quaternions is given in (3.14). Note that J_S preserves each quaternion algebra, compute the type and apply (6.9).

A third proof appears in (7.5) below.

2. **Quaternion conjugates.** Let A be a quaternion algebra over F and recall the usual definitions of the norm and trace of an element a: $Na = a\bar{a}$ and $Ta = a + \bar{a}$. If $a, b \in A$ we write $a \sim b$ to mean that a and b are conjugate, i.e. $b = cac^{-1}$ for some $c \in A^\bullet$.

Lemma. *If $a, b \in A$ then $a \sim b$ if and only if $Na = Nb$ and $Ta = Tb$.*

(Hint. See Exercise 4.10(2).)

3. **Two Quaternions.** Suppose (A, J) is a central simple F-algebra with involution and with $\dim A = 16$. Suppose J is "decomposable", in the sense that there exists a J-invariant quaternion subalgebra $Q_1 \subseteq A$. For every such subalgebra there is a decomposition

$$(A, J) \simeq (Q_1, J_1) \otimes (Q_2, J_2).$$

(1) If J_1 and J_2 both have type 1, then $(A, J) \cong (A_1, K_1) \otimes (A_2, K_2)$ where each A_j is a quaternion algebra and each K_j is the "bar" involution, of type -1.

(2) Suppose J_1 and J_2 both have type -1. Then those quaternion subalgebras Q_1, Q_2 are unique in a strong sense: If B is any J-invariant quaternion subalgebra on which the induced involution has type -1, then either $B = Q_1$ or $B = Q_2$.

(Hint. (1) Re-arrange the generators $i_1 \otimes i_2, i_1 \otimes j_2$, etc.

(2) Compare Exercise 1.4.)

4. **Explicit quaternions.** Suppose (σ, τ) is an (s, t)-pair where $s + t = 2m + 1$. Let (C, J) be the associated Clifford algebra with involution. Let $\{e_1, \dots, e_{2m}\}$ be an orthogonal basis of the generating subspace such that $J(e_j) = \pm e_j$. Then $\{e^\Delta : \Delta \in \mathbb{F}_2^{2m}\}$ forms the derived basis of C. If e^Γ and e^Δ anticommute then they generate a quaternion subalgebra Q preserved by J and $C \cong Q \otimes C'$ where C' is the centralizer of Q. Then J induces an involution J' on C'.

(1) (C', J') is the Clifford algebra with involution associated to some (s', t')-family (σ', τ') where $s' + t' = 2m - 1$.

(2) Suppose $(\sigma, \tau) < \mathrm{Sim}(q)$ and Q is split. If $J|_Q$ has type -1 then q is hyperbolic (but not necessarily (C, J)-hyperbolic). If $J|_Q$ has type 1 then $(Q, J|_Q)$ is the Clifford algebra associated to some (2, 2)-family $(\langle 1, a\rangle, \langle 1, a\rangle)$ and $q \simeq \langle\langle a\rangle\rangle \otimes q'$ for some q' such that $(\sigma', \tau') < \mathrm{Sim}(q')$. Moreover in this case we may assume $(s', t') = (s-1, t-1)$.

(3) The Eigenspace Lemma 2.10 follows by these methods.

(4) Suppose $\sigma < \mathrm{Sim}(q)$ where $\sigma = \langle 1, a_1, \dots, a_{2m}\rangle$. Decompose the associated (C, J) into quaternion subalgebras with involution:
$(C, J) \cong (Q_1, J_1) \otimes \cdots \otimes (Q_m, J_m)$ as in (3.14). Then $[Q_k] = [d\alpha_k, -a_{2k-1}a_{2k}]$ where $\alpha_k = \langle 1, a_1, \dots, a_{2k-1}\rangle$ and J_k has type $(-1)^k$. Deduce some consequences of (2). For instance: If $\alpha \subset \sigma < \mathrm{Sim}(q)$ where $\dim \alpha \equiv 2 \pmod 4$, $\alpha \neq \sigma$ and $d\alpha = \langle 1\rangle$, then q is hyperbolic. (Compare Yuzvinsky (1985).) Many results of this nature follow more easily from Exercise 2.5.

(5) Suppose C is split and J has type 1 so that $(C, J) \cong (\mathrm{End}(V), I_q)$ where (V, q) is a quadratic space of dimension 2^m. Further suppose $C \cong Q_1 \otimes \cdots \otimes Q_m$ where each Q_k is a split quaternion algebra preserved by the involution J. Then q is similar to a Pfister form.

5. **Trace forms once more.** (1) Let A be a central simple F-algebra with involution. There is an algebra isomorphism $\varphi : A \otimes A \xrightarrow{\cong} \mathrm{End}_F(A)$ defined as follows, using an anti-autormophism ι of A: $\varphi(a \otimes b)(x) = ax\iota(b)$ for every $a, b, x \in A$. Let J_1 and J_2 be involutions of the same type on A so that $J_1 \otimes J_2$ is a 1-involution on $A \otimes A$, inducing an involution I_B on $\mathrm{End}_F(A)$. The isometry class of this symmetric bilinear form B on A depends only on the isomorphism classes of the involutions J_1, J_2, and is independent of the choice of ι.

(2) The form $B : A \times A \to F$ can be chosen to satisfy:
$B(axb, y) = B(x, J_1(a)yJ_2(b))$ for every $a, b, x \in A$. Express B as a trace form.

(3) Suppose $A = C(-\sigma_1 \perp \tau)$ is the Clifford algebra for an (s, t)-pair (σ, τ) such that $s + t$ is odd. Let $J_1 = J_2$ be the corresponding (s, t)-involution. Then B is a Pfister form.

(4) Let $A = \left(\frac{-a, x}{F}\right) \cong \left(\frac{-b, y}{F}\right)$ be a quaternion algebra, so that $\langle\langle a, -x\rangle\rangle \simeq \langle\langle b, -y\rangle\rangle$. Let J_1 be the involution corresponding to $(\langle 1, a\rangle, \langle x\rangle)$, and J_2 the involution for $(\langle 1, b\rangle, \langle y\rangle)$. Then $J_1 \otimes J_2$ yields I_B on $\mathrm{End}_F(A)$. Then $(A, B) \simeq \langle\langle a, xb\rangle\rangle \simeq \langle\langle b, ya\rangle\rangle$.

(Hint. (2) Let $J_1 = J_2^w$ for $w \in A^\bullet$ with $J_1(w) = w$. Then $B(x, y) = \mathrm{tr}(wJ_1(x)y) = \mathrm{tr}(wyJ_2(x))$.

(3) Use Exercise 3.14.)

6. **$\otimes$ of irreducibles.** (1) Suppose A_1 and A_2 are central simple F-algebras with irreducible modules V_1, V_2, respectively. Then $V_1 \otimes V_2$ becomes an $A_1 \otimes A_2$-module where the action is defined "diagonally": $(a_1 \otimes a_2)(v_1 \otimes v_2) = (a_1v_1) \otimes (a_2v_2)$. Let D_i be the "division algebra part" of A_i. That is $A_i \cong \mathbb{M}_{n_i}(D_i)$.

Lemma. *$V_1 \otimes V_2$ is an irreducible $A_1 \otimes A_2$-module if and only if $D_1 \otimes D_2$ is a division algebra.*

(2) Here is an analog to Corollary 6.11: Suppose $(C, J) \cong (A_1, J_1) \otimes (A_2, J_2)$ in the category of central simple algebras with involution. Suppose V_k is an A_k-module so that $V = V_1 \otimes V_2$ is a C-module. If q is a quadratic form on V which admits (C, J) does it follow that $(V, q) \cong (U_1, q_1) \otimes (U_2, q_2)$ for some quadratic spaces (U_k, q_k) admitting A_k?

(Hint. (1) Count the dimensions. Suppose D_i has degree d_i over F. Then $\dim V_i = n_i d_i^2$. If $D_1 \otimes D_2 \cong \mathbb{M}_r(D)$ for a division algebra D of degree d over F then $d_1 d_2 = rd$. Compute that an irreducible $A_1 \otimes A_2$-module has dimension $n_1 n_2 r d^2$. Then $V_1 \otimes V_2$ is irreducible if and only if $\dim V_1 \otimes V_2 = n_1 n_2 r d^2$.)

7. **Uniqueness of the forms.** Suppose q and q' are regular quadratic forms on the vector space V where $\dim V = n$.

(1) Suppose $S \subseteq \mathrm{End}(V)$ is a linear subspace which is a (regular) subspace of similarities for both forms q and q'. Must the induced forms σ, σ' on S coincide?

(2) Suppose $S, T \subseteq \mathrm{End}(V)$ are linear subspaces and that (S, T) is an (s, t)-family relative to both q and q'. Then the induced forms (σ, τ) and (σ', τ') coincide. Express $n = 2^m n_0$ where n_0 is odd and suppose further that $s + t \geq 2m + 1$. Then $J = J'$ and $q' = c \cdot q$ for some $c \in F^\bullet$.

(Hint. (1) Let J, J' be the involutions and express $J' = J^g$. For each $f \in S$, $\sigma'(f) = \zeta \cdot \sigma(f)$ for some $\zeta \in F$ with $\zeta^n = 1$. This ζ is independent of f. Are there examples where $\zeta \neq 1$?

(2) Let C be the associated Clifford algebra and note that the similarity representation $C \to \mathrm{End}(V)$ is surjective. In fact this uniqueness holds true whenever the given family is "minimal" as defined in the next chapter.)

Notes on Chapter 6

The analysis of central simple algebras with involution was covered in some depth by Albert (1939), who used somewhat different terminology. Most of the results in this chapter have appeared in other books. See especially Knus et al. (1998), §3.

The invariant $\det(J)$ in (6.8) is generalized in (10.24) below.

The ideas for (6.11), Exercise 4 and Exercise 6 follow Yuzvinsky (1985).

Exercise 1. The computation of the type of the standard involution of a central simple Clifford algebra was done by Chevalley (1954) using a different technique. The dimension counting method is mentioned in Jacobson (1964).

Chapter 7

Unsplittable (σ, τ)-Modules

Given (σ, τ), what is the dimension of an unsplittable (σ, τ)-module? We present a complete answer when the associated Clifford algebra C is split or reduces to a quaternion algebra. We also characterize the (s, t)-pairs (σ, τ) which have unsplittables of minimal dimension.

Notations. Let (σ, τ) be a pair of quadratic forms where $\dim \sigma = s$, $\dim \tau = t$. Assume σ represents 1 and define σ_1 by $\sigma = \langle 1 \rangle \perp \sigma_1$. Define $\beta = \sigma \perp -\tau$ and $\beta_1 = \sigma_1 \perp -\tau$. Let $C = C(-\beta_1)$ be the associated Clifford algebra with involution $J = J_S$. Then $\dim C = 2^{s+t-1}$. Let z be an "element of highest degree" in C and $Z = F + Fz$.

The Basic Sign Calculation (2.4) says: $J(z) = z$ if and only if $s \equiv t$ or $t + 1 \pmod 4$. A direct calculation shows that $d\beta = d(-\beta_1)$ and $c(\beta) = c(-\beta_1)$. As noted in (4.2) an unsplittable (σ, τ)-module has dimension 2^k for some k where $s + t \leq 2k + 2$. When can equality hold?

7.1 Lemma. *Suppose $s + t = 2m + 2$. Then $(\sigma, \tau) < \mathrm{Sim}(V, B)$ for some 2^m-dimensional λ-space (V, B) (for some $\lambda = \pm 1$) if and only if $d\beta = \langle 1 \rangle$, $c(\beta) = 1$ and $s \equiv t \pmod 4$.*

Proof. If such (V, B) exists let $\pi : C \to \mathrm{End}(V)$ be the representation. By comparing dimensions we must have $C \cong C_0 \times C_0$ and $\pi(C_0) = \mathrm{End}(V)$. Therefore $d\beta = \langle 1 \rangle$ and $c(\beta) = [C_0] = 1$. Furthermore $\pi(z)$ must be a scalar, so that $J(z) = z$ since the involutions are compatible. The Basic Sign Calculation (2.4) then implies that $s \equiv t \pmod 4$.

Conversely since $s + t - 1$ is odd and $c(\beta) = 1$ we find $[C_0] = 1$ so that $C_0 \cong \mathrm{End}(V)$ for some V with $\dim V = 2^m$. Since $d\beta = \langle 1 \rangle$ the Structure Theorem implies that $C \cong C_0 \times C_0$ and the restriction of J to C_0 induces an involution I on $\mathrm{End}(V)$, corresponding to a λ-form B on V by (6.3). From $s \equiv t \pmod 4$ we find $J(z) = z$, so the composite map $C \to C_0 \cong \mathrm{End}(V)$ is compatible with the involutions and (V, B) becomes a (C, J)-module. □

Note. The conditions $\dim \beta =$ even, $d\beta = \langle 1 \rangle$ and $c(\beta) = 1$ are equivalent to: $\beta \in J_3(F)$. (Recall that $J_3(F)$ is the ideal of the Witt ring introduced at the end of

Chapter 3, and that $J_3(F) = I^3F$ by Merkurjev's Theorem.) Since $\beta = \sigma \perp -\tau$, those conditions say: $\sigma \equiv \tau \pmod{J_3(F)}$, or equivalently: $\dim\sigma \equiv \dim\tau \pmod 2$, $d\sigma = d\tau$ and $c(\sigma) = c(\tau)$.

7.2 Lemma. *Let (V, B) be a λ-symmetric (C, J)-module where* $\dim V = 2^m$ *and* $s + t = 2m + 1$. *Then I_B is the unique involution on* $\mathrm{End}(V)$ *compatible with (C, J). Consequently every (C, J)-module of dimension 2^m is C-similar to (V, B).*

Proof. The uniqueness of the involution is clear since the representation $C \to \mathrm{End}(V)$ is bijective. If (V', B') is another (C, J)-module of dimension 2^m then $V' \cong V$ as C-modules. Let $h : V \to V'$ be a C-isomorphism and define the form B_1 on V by: $B_1(x, y) = B'(h(x), h(y))$. Then h is a C-isometry $(V, B_1) \to (V', B')$ and the forms here admit (C, J). By the uniqueness of the involution, $I_{B_1} = I_B$ so that $B_1 = aB$ for some $a \in F^\bullet$. Then h is an a-similarity $(V, B) \to (V', B')$. (Compare the proof of (6.6).) □

This result is also true if $s + t = 2m + 2$, except that the C-module may have to be "twisted" by the main automorphism of C to ensure that $V' \cong V$. (There are two irreducible C-modules as described in (4.12).)

The next step is to separate the types of the involutions used above. This refinement of (7.1) is equivalent to computing the type of the involution J_S.

7.3 Proposition. *Suppose $s+t = 2m+2$. Then $(\sigma, \tau) < \mathrm{Sim}(V, q)$ where (V, q) is a quadratic space of dimension 2^m if and only if $d\beta = \langle 1\rangle$, $c(\beta) = 1$ and $s \equiv t \pmod 8$. For the case of alternating forms the congruence changes to $s \equiv t + 4 \pmod 8$.*

Proof. Suppose that $d\beta = \langle 1\rangle$, $c(\beta) = 1$ and $s \equiv t \pmod 4$. Then $(\sigma, \tau) < \mathrm{Sim}(V, B)$ for some 2^m-dimensional λ-space (V, B). If $s \equiv t \pmod 8$ we will show $\lambda = 1$. By (2.8) we have an example of an $(m+1, m+1)$-family $(\alpha, \alpha) < \mathrm{Sim}(W, \varphi)$ where $\dim W = 2^m$. Since $s \equiv t \pmod 8$ the Shift Lemma (2.6) produces $(\sigma', \tau') < \mathrm{Sim}(W, \varphi)$ where $\dim\sigma' = s$ and $\dim\tau' = t$. Extending scalars to an algebraic closure K of F we see that $\sigma \simeq \sigma'$ and $\tau \simeq \tau'$ over K, and Lemma 7.2 implies that (V, B) and (W, φ) are similar over K and we conclude that $\lambda = 1$. Analogously if $s \equiv t + 4 \pmod 8$ then $\lambda = -1$.

Conversely, suppose $(\sigma, \tau) < \mathrm{Sim}(V, q)$ where $\dim V = 2^m$. Then $d\beta = \langle 1\rangle$, $c(\beta) = 1$ and $s \equiv t \pmod 4$, by (7.1). If $s \equiv t+4 \pmod 8$ we obtain a contradiction from the proof above. Therefore $s \equiv t \pmod 8$. A similar argument works when $\lambda = -1$. □

7.4 Corollary. (1) *If $s + t$ is odd then C is central simple, and J_S has type* 1 *iff* $s \equiv t \pm 1 \pmod 8$.

(2) *If $s+t$ is even then C_0 is central simple, and the restriction J^+ of J_S has type* 1 *iff* $s \equiv t$ *or* $t + 2 \pmod 8$.

Proof. (1) Suppose $s + t = 2m + 1$. Extending to a splitting field we may assume $C \cong \text{End}(V)$ where $\dim V = 2^m$. If J_S has type λ there is an induced λ-form B on V so that $(\sigma, \tau) < \text{Sim}(V, B)$. By the Expansion Lemma 2.5, (σ, τ) expands to either an $(s + 1, t)$-family or an $(s, t + 1)$-family in $\text{Sim}(V, B)$. Apply (7.3).

(2) As in (3.9) C_0 becomes a Clifford algebra and J^+ is the involution corresponding to an $(s-1, t)$-family. Now apply part (1) to compute the type. A similar argument works in the case $t \geq 1$, viewing C_0 as the algebra for a $(t, s - 1)$-family. □

So far in this chapter we have analyzed cases where $c(\beta) = 1$. We push these ideas one step further by allowing $c(\beta) =$ quaternion. This means that $c(\beta)$ is represented by a (possibly split) quaternion algebra in the Brauer group.

7.5 Corollary. (1) *Suppose* $(\sigma, \tau) < \text{Sim}(V, B)$ *where* $\dim V = 2^m$. *If* $s+t \geq 2m-1$ *then* $c(\beta) =$ quaternion.

(2) *If* $c(\beta) =$ quaternion *and* $s + t \leq 2m - 1$ *then there are* λ*-symmetric* (σ, τ)*-modules of dimension* 2^m*, for both values of* λ.

Proof. (1) Generally $s + t \leq 2m + 2$. We have seen that if $s + 1 \geq 2m + 1$ then $c(\beta) = 1$. If $s + t = 2m$ then C_0 is central simple and we have $C_0 \otimes A \cong \text{End}(V)$ where A is the centralizer of C_0. Counting dimensions we find $\dim A = 4$ so that A is a quaternion algebra and $c(\beta) = [C_0] = [A] =$ quaternion. If $s + t = 2m - 1$ a similar argument works.

(2) Suppose $s+t$ is odd. It suffices to settle the case $s+t = 2m-1$. If $c(\beta) = [A]$ where A is a quaternion algebra, then $[C \otimes A] = 1$ so that $C \otimes A \cong \text{End}(V)$ where $\dim V = 2^m$. Since involutions of both types exist on A there are regular λ-forms on V which admit C, for both values of λ. Suppose $s + t$ is even. Then $s+t+1 \leq 2m - 1$ and we can apply the odd case to $(\sigma, \tau \perp \langle 1\rangle)$ after noticing that $c(\beta \perp \langle -1\rangle) = c(\beta) =$ quaternion. □

Next we consider expansions of a given (s, t)-family, generalizing the Expansion Lemma 2.5. Recall that when $s + t$ is odd we can "adjoin z" to $(S, T) \subseteq \text{Sim}(V, q)$ to form a family (S_0, T_0) which is one dimension larger. This larger family has $s_0 \equiv t_0 \pmod 4$ and $d\beta_0 = \langle 1\rangle$. Furthermore the module V is not a faithful $C_{(0)}$-module, for the larger Clifford algebra $C_{(0)}$. This means that $C_{(0)} \to \text{End}(V)$ is not injective, so that the element "z" for the larger family acts as a scalar. Conversely every non-faithful family arises this way from a smaller family.

7.6 Expansion Proposition. *Suppose* $(S, T) \subseteq \text{Sim}(V, q)$ *is an* (s, t)*-family where* $\dim V = 2^m$ *and* $s+t = 2m-1$. *Then* (S, T) *expands to an* (s', t')*-family* $(S', T') \subseteq \text{Sim}(V, q)$ *where* $s' + t' = 2m+2$. *Moreover, any expansion of* (S, T) *either is inside* (S', T') *or is obtained from* (S, T) *by adjoining* z.

Proof. The Clifford algebra C is central simple of dimension 2^{2m-2}. The representation $C \to \mathrm{End}(V)$ is then injective and we view C as the subalgebra of $\mathrm{End}(V)$ generated by S and T. By the Double Centralizer Theorem, $C \otimes A \cong \mathrm{End}(V)$ where $A = \mathrm{End}_C(V)$ is the centralizer of C. Then $\dim A = 4$ so that A must be a quaternion algebra, and I_q preserves C so it induces an involution K on A.

The element $z \in C$ anti-commutes with every element of $S_1 \cup T$ and $J(z) = \pm z$. If $a \in A$ and $K(a) = \pm a$ then az can be adjoined to S or T, depending on whether $I_q(za) = K(a)J(z)$ equals $-za$ or za. When $a = 1$ we have the situation of the Expansion Lemma 2.5. To adjoin more than one dimension to (S, T) we need anticommuting elements of A, so let us stick to the pure quaternions A_0. Define the eigenspaces $A_0 = A_+ \oplus A_-$ where $K(x) = \lambda x$ for $x \in A_\lambda$. Then either $(S + zA_+, T + zA_-)$ or $(S + zA_-, T + zA_+)$ forms an (s', t')-family in $\mathrm{Sim}(V, q)$. Since $\dim A_+ + \dim A_- = 3$ we see that $s' + t' = s + t + 3 = 2m + 2$.

For the uniqueness suppose $(S", T")$ is some expansion of (S, T), say $(S", T") = (S \perp R_-, T \perp R_+)$. Then $R_- + R_+ \subseteq zA$ since every element of $R_- + R_+$ anticommutes with $S_1 \cup T$. If $R_- + R_+ = Fz$ then the family $(S", T")$ was obtained just by adjoining z. Otherwise $R_- + R_+ \subseteq zA_0$. Furthermore if $f \in R_\varepsilon$ then $K(f) = \pm f$, and it follows that R_- and R_+ are contained in A_+ and A_-, in some order. Therefore $(S", T")$ is contained in (S', T'). □

Of course the exact dimension of A_+ (either 0 or 2 as in (6.8)), and whether zA_+ is adjoined to S or to T, depend on the values of s and t. We do not need to keep careful track of this in the proof above because we know from (7.3) that $s' \equiv t' \pmod 8$.

Exactly when does a given pair (σ, τ) possess a quadratic module of dimension 2^m? We can now refine Theorem 2.11 and answer this question, provided the Witt invariant is quaternion.

7.7 Theorem. *Suppose* $c(\beta) =$ quaternion. *Then there is a quadratic* (σ, τ)*-module of dimension* 2^m *if and only if one of the following holds:*

(1) $s + t \le 2m - 1$.

(2) $s+t = 2m$ *and either:* $d\beta = \langle 1 \rangle$ *and* $s \equiv t \pmod 4$, *or:* $c(\beta)$ *is split by* $F(\sqrt{d\beta})$ *and* $s \equiv t - 2, t$ *or* $t + 2 \pmod 8$.

(3) $s + t = 2m + 1$, $c(\beta) = 1$ *and* $s \equiv t + 1$ *or* $t - 1 \pmod 8$.

(4) $s + t = 2m + 2$, $d\beta = \langle 1 \rangle$, $c(\beta) = 1$ *and* $s \equiv t \pmod 8$.

Proof. Suppose $(\sigma, \tau) < \mathrm{Sim}(V, q)$ where $\dim V = 2^m$. Then we know that $s + t \le 2m + 2$. If $s + t \le 2m - 1$ then (7.5) applies and if $s + t = 2m + 2$ we use (7.3). If $s + t = 2m + 1$, then by the Expansion Lemma 2.5 we can expand (σ, τ) to a larger family (σ', τ'). By (7.3) we know that $d\beta' = \langle 1 \rangle$, $c(\beta') = 1$ and $s' \equiv t' \pmod 8$. Since $\beta' = \beta \perp \langle d \rangle$ for some $d \in F^\bullet$, it follows that $c(\beta) = c(\beta' \perp \langle -d \rangle) = c(\beta')[d\beta', d] = 1$ and either $s \equiv t + 1$ or $s + 1 \equiv t \pmod 8$.

Now suppose that $s+t = 2m$. Choose a subfamily (σ_0, τ_0) where $s_0+t_0 = 2m-1$. If the original family is non-faithful then it must be obtained from (σ_0, τ_0) by adjoining z and we conclude from the Expansion Lemma that $d\beta = \langle 1\rangle$ and $s \equiv t \pmod 4$. Otherwise by (7.6) the family (σ, τ) lies within a full expansion (σ', τ') where $s'+t' = 2m+2$. Then (s', t') must equal $(s+2, t)$, $(s+1, t+1)$ or $(s, t+2)$, and we know that $s \equiv t-2$, t or $t+2 \pmod 8$. Also $\beta' = \beta \perp \langle x, y\rangle$ for some $x, y \in F^\bullet$. Then $d\beta = d(\beta' \perp \langle -x, -y\rangle) = \langle -xy\rangle$ and $c(\beta) = c(\beta' \perp \langle -x, -y\rangle) = [-x, -y]$ which is split by the field $F(\sqrt{-xy}) = F(\sqrt{d\beta})$.

For the converse suppose (σ, τ) is given satisfying one of those conditions. If $s+t = 2m-1$ we are done by (7.5) and if $s+t = 2m+2$ we apply (7.3). Suppose $s+t = 2m+1$. Letting $d = -d\beta$ we find that $c(\beta \perp \langle d\rangle) = c(\beta)[d\beta, d] = 1$ since $c(\beta) = 1$. Let (σ', τ') equal either $(\sigma \perp \langle d\rangle, \tau)$ or $(\sigma, \tau \perp \langle -d\rangle)$, according as $s \equiv t-1$ or $t+1 \pmod 8$. Then by (7.3) we have $(\sigma, \tau) \subset (\sigma', \tau') < \mathrm{Sim}(V, q)$ where $\dim V = 2^m$.

Suppose $s+t = 2m$. In the case $d\beta = \langle 1\rangle$ and $s \equiv t \pmod 8$ we can remove one dimension from σ or τ to get a subfamily (σ_0, τ_0) having $s_0 + t_0 = 2m-1$ and $c(\beta_0) = c(\beta) =$ quaternion. Then there is a quadratic (σ_0, τ_0)-module of dimension 2^m, and the Expansion Lemma makes it a (σ, τ)-module. In the final case suppose $d\beta = \langle d\rangle$ and $c(\beta) = [d, x]$ for some $x \in F^\bullet$. Define $\beta' = \beta \perp \langle -x, xd\rangle$ and note that $d\beta' = \langle 1\rangle$ and $c(\beta') = c(\beta)[-x, xd][d, d] = 1$. Define a pair (σ', τ') by enlarging (σ, τ) appropriately to make $\beta' \simeq \sigma' \perp -\tau'$ and $s' \equiv t' \pmod 8$. Then again by (7.2) we get $(\sigma, \tau) \subset (\sigma', \tau') < \mathrm{Sim}(V, q)$ where $\dim V = 2^m$. □

The information in this theorem can be restated to provide the dimension of an unsplittable (σ, τ)-module whenever $c(\beta) =$ quaternion. We do this now, choosing the notation so that in each case the smallest possible unsplittable dimension is 2^m. That is, $m = \delta(s, t)$ in the sense of (2.15).

7.8 Theorem. *Let* (σ, τ) *be a pair of quadratic forms where* σ *represents* 1 *and* $\dim \sigma = s$ *and* $\dim \tau = t$. *Define* $\beta = \sigma \perp -\tau$ *and suppose* $c(\beta) =$ quaternion. *Let* ψ *be an unsplittable quadratic* (σ, τ)*-module.*

If $\boxed{s \equiv t \pmod 8}$ *let* $s+t = 2m+2$. *Then* $m \equiv t-1 \pmod 4$ *and:*

$\dim \psi = 2^m$ *iff* $d\beta = \langle 1\rangle$ *and* $c(\beta) = 1$.

$\dim \psi = 2^{m+1}$ *iff the first case fails and either* $d\beta = \langle 1\rangle$ or $c(\beta)$ *is split by* $F(\sqrt{d\beta})$.

$\dim \psi = 2^{m+2}$ *otherwise.*

If $\boxed{s \equiv t \pm 1 \pmod 8}$ *let* $s+t = 2m+1$. *Then* $m \equiv t$ *or* $t-1 \pmod 4$ *and:*

$\dim \psi = 2^m$ *iff* $c(\beta) = 1$.

$\dim \psi = 2^{m+1}$ *otherwise.*

If $\boxed{s \equiv t \pm 2 \pmod 8}$ *let* $s+t = 2m$. *Then* $m \equiv t \pm 1 \pmod 4$ *and:*

$\dim \psi = 2^m$ *iff* $c(\beta)$ *is split by* $F(\sqrt{d\beta})$.

$\dim \psi = 2^{m+1}$ *otherwise.*

If $\boxed{s \equiv t + 4 \pmod 8}$ *let* $s + t = 2m$. *Then* $m \equiv t + 2 \pmod 4$ *and:*

$\dim \psi = 2^m$ *iff* $d\beta = \langle 1 \rangle$.

$\dim \psi = 2^{m+1}$ *otherwise.*

If $\boxed{s \equiv t \pm 3 \pmod 8}$ *let* $s + t = 2m - 1$. *Then* $m \equiv t + 2$ *or* $t + 3 \pmod 4$ *and:*

$\dim \psi = 2^m$.

Proof. These criteria can be read off directly from (7.7). □

The pairs (σ, τ) whose unsplittable quadratic modules are as small as possible are the nicest kind. Recall from (2.15) that for given (s, t) the smallest unsplittable module that an (s, t)-family can have is $2^{\delta(s,t)}$. We define an (s, t) pair (σ, τ) to be a *minimal* pair if its unsplittable quadratic modules have this smallest possible dimension $2^{\delta(s,t)}$. Then (σ, τ) is minimal if and only if $c(\beta) =$ quaternion and (σ, τ) satisfies the conditions for $\dim \psi = 2^m$ given in (7.8).

Remark. The dimensions of unsplittables for alternating (σ, τ)-modules can be found by altering in (7.8) each of the congruences for s and t by 4 (mod 8). (See Exercise 2.6.) We can also define (σ, τ) to be a (-1)-*minimal pair* if its unsplittable alternating modules have the smallest possible dimension.

7.9 Proposition. *Suppose* (σ, τ) *is an* (s, t)*-pair where* $s \geq 1$, $t \geq 0$ *and where the dimension of a quadratic unsplittable is* 2^m. *Then* (σ, τ) *is minimal if and only if one of the following equivalent conditions holds:*

(1) $m = \delta(s, t)$.

(2) *Each unsplittable quadratic* (σ, τ)*-module remains unsplittable after any scalar extension.*

(3) $s > \rho_t(2^{m-1})$.

(4) $$s + t = \begin{cases} 2m + 1 & \text{if } m \equiv t \\ 2m & \text{if } m \equiv t + 1 \\ 2m - 1 \text{ or } 2m & \text{if } m \equiv t + 2 \\ 2m - 1, 2m, 2m + 1 \text{ or } 2m + 2 & \text{if } m \equiv t + 3 \end{cases} \pmod 4.$$

Proof. (1) $\iff$ (2) follows from the definition of "minimal".

(3) $\iff$ (4): Use the formulas in (2.13). The lower bounds in (4) come from condition (3). For the upper bounds note that there exists a (σ, τ)-module of dimension 2^m so that $s \leq \rho_t(2^m)$.

(2) $\implies$ (3): Suppose (σ, τ) is a minimal (s, t)-pair with an unsplittable module (V, q) of dimension 2^m. If $s \leq \rho_t(2^{m-1})$ then there is some (s, t)-pair (σ', τ')

having a module of dimension 2^{m-1}. Passing to an extension field K we may assume $(\sigma', \tau') \simeq (\sigma, \tau)$. But then (V_K, q_K) is not unsplittable, contrary to hypothesis.

(3) $\Longrightarrow$ (2): If $s > \rho_t(2^{m-1})$ then (σ, τ) must be minimal since no (s, t)-pair can have a module of dimension 2^{m-1}. □

For example the possible sizes of minimal (s, t) pairs with $s \geq t$ and having unsplittables of dimension 8 are: (4, 1), (4, 2), (4, 3), (4, 4), (5, 0), (5, 1), (6, 0), (7, 0), (8, 0). Every pair $(s\langle 1\rangle, t\langle 1\rangle)$ is minimal (see Exercise 4). The minimal pairs are characterized by a strong uniqueness property for their unsplittable modules. Compare Lemma 7.2.

7.10 Proposition. *An (s, t)-pair (σ, τ) is minimal if and only if there exists a (σ, τ)-module (V, q) such that I_q is the unique 1-involution on* End(V) *compatible with (C, J_S).*

Proof. Let $(\sigma, \tau) < \text{Sim}(V, q)$, view V as a C-module and recall that I_q is a 1-involution on End(V) compatible with (C, J_S). Let $A = \text{End}_C(V)$ and K the involution on A induced by I_q. Then the 1-involutions on End(V) compatible with (C, J_S) are exactly the involutions I_q^a where $a \in A^\bullet$ and $K(a) = a$. The unique involution property is equivalent to requiring that $\mathcal{S}^+(A, K)$ have dimension 1. Since this condition is independent of scalar extension we may assume F is algebraically closed.

If $s \leq \rho_t(2^{m-1})$ then there is a quadratic (C, J_S)-module (W, φ) of dimension 2^{m-1}. Let $V = W \oplus W$ and consider the forms $\varphi \perp \langle b\rangle\varphi$ on V for $b \in F^\bullet$. For different values of b these forms provide unequal 1-involutions on End(V) compatible with (C, J_S).

Conversely suppose $(\sigma, \tau) < \text{Sim}(V, q)$ where $\dim V = 2^m$ and $s > \rho_t(2^{m-1})$. We will show that I_q is unique. If $s + t \geq 2m + 1$ the uniqueness is clear since C maps surjectively onto End(V). Suppose $s + t = 2m - 1$ so that $A = \text{End}_C(V)$ is a quaternion algebra with $C \otimes A \cong \text{End}(V)$. Then I_q is unique iff K is the bar involution on A, which occurs iff K has type -1. By (6.7) this is equivalent to saying that J_S has type -1 and by (7.4) it occurs iff $s \equiv t \pm 3 \pmod 8$. Since $s + t = 2m - 1$, this congruence is the same as $m \equiv t + 2$ or $t + 3 \pmod 4$.

The remaining case is when $m \equiv t + 1 \pmod 4$ and $s + t = 2m$. Then $s \equiv t + 2 \pmod 8$. As before we have $C_0 \otimes A' \cong \text{End}(V)$ for a quaternion algebra A' having an induced involution K'. Then $A = \text{End}_C(V)$ is the centralizer of $z' = \pi(z)$ in A'. (Here π is the corresponding representation of C.) Since $s \equiv t + 2 \pmod 8$, the sign computation says that $J_S(z) = -z$ so that $K(z') = -z'$. Then z' is a pure quaternion and $A = F + Fz'$. Therefore $\mathcal{S}^+(A, K) = F$ so that I_q is unique. □

The uniqueness of the involution I_q for a minimal pair (σ, τ) implies that all (σ, τ)-unsplittables are C-similar (with the standard exception when C is not simple).

7.11 Corollary. *Suppose (σ, τ) is a minimal (s, t)-pair with unsplittable quadratic module (V, ψ). Then every unsplittable (σ, τ)-module is C-similar to (V, ψ), (up to a twist by the main automorphism when C is not simple). Consequently, $(\sigma, \tau) < \mathrm{Sim}(\alpha)$ if and only if $\psi \mid \alpha$.*

Proof. Suppose (V', ψ') is another (σ, τ)-unsplittable. Then V and V' are C-modules, $\dim V = \dim V' = 2^m$ and $s > \rho_t(2^{m-1})$.

Claim. We may assume $V' \cong V$ as C-modules. For if C is simple the modules are certainly isomorphic. Otherwise s + t is even and we know $s + t \geq 2m - 1$. If $s+t = 2m+2$ the two module structures differ only by the usual "twist" as described in (4.12), so we can arrange $V' \cong V$. Suppose $s + t = 2m$. If there exist two different C-module structures then both cases in Theorem 7.7 (2) hold true. Therefore $s \equiv t \pmod 8$, $d\beta = \langle 1 \rangle$ and $c(\beta) = 1$. But then there exists an (s, t)-family on some quadratic space of dimension 2^{m-1} contrary to the hypothesis $s > \rho_t(2^{m-1})$. This proves the claim.

The argument is completed as in the proof of (7.2). □

Suppose $(\sigma, \tau) < \mathrm{Sim}(q)$ is an (s, t)-family with $s + t \geq 2m - 1$. If $\dim q = 2^m$ then the Expansion Proposition 7.6 implies that there exists an $(m + 1, m + 1)$-family in $\mathrm{Sim}(q)$. This statement can fail if we allow $\dim q = 2^m n_0$, as seen in Exercise 10. However the assertion does generalize in some cases.

7.12 Corollary. *Suppose $(\sigma, \tau) < \mathrm{Sim}(q)$ is an (s, t)-family and $\dim q = n = 2^m n_0$ where n_0 is odd. If $s = \rho_t(n)$ is the maximal value, then (σ, τ) is minimal pair and there exists an $(m + 1, m + 1)$-family in $\mathrm{Sim}(q)$.*

Proof. Since $s = \rho_t(2^m) > \rho_t(2^{m-1})$ the pair is minimal. Let $(\sigma, \tau) < \mathrm{Sim}(\psi)$ be the unique unsplittable, so that $\dim \psi = 2^m$ and $q \simeq \psi \otimes \gamma$ where $\dim \gamma$ is odd. Since $s + t \geq 2m - 1$ the Expansion Proposition 7.6 implies that $\mathrm{Sim}(\psi)$ admits an (s', t')-family where $s' + t' = 2m + 2$. Then $s' \equiv t' \pmod 8$ and shifting produces an $(m + 1, m + 1)$-family. □

From Theorem 7.8 we can read off the criteria for an (s, t)-pair (σ, τ) to be minimal. It is interesting to display this calculation explicitly in the case of a single form σ over the real field $\mathbb{R}$.

7.13 Proposition. *Let $\sigma = p\langle 1 \rangle \perp r\langle -1 \rangle$ over $\mathbb{R}$. Then σ is not minimal if and only if there is a dot (•) in the corresponding entry of the following table, indexed by the values of p and r* (mod 8).

$p \backslash r$	0	1	2	3	4	5	6	7
0								
1				•		•	•	•
2						•	•	•
3		•				•	•	•
4								
5		•	•	•				•
6		•	•	•				
7		•	•	•		•		

Proof. Using calculations of $d\sigma$ and $c(\sigma)$ in Exercises 3.5 and 3.6 we can translate the criteria in (7.8) to congruence conditions on $\dim \sigma$ and $\operatorname{sgn} \sigma$. These yield the table.□

Remark. The proof also shows that σ is (-1)-minimal if and only if $\sigma \perp 2\mathbb{H}$ is minimal. Some of the symmetries in this table are explored in Exercise 4.

At this point we can complete the classification of (s, t)-pairs which have hyperbolic type, as defined in (4.14) and discussed in (6.16). Recall that these are the pairs (σ, τ) such that the unsplittables are not irreducible. With our usual notations, this says that an irreducible C-module does not have a symmetric bilinear form admitting (C, J). Some of the details of the proof below are left to the reader.

7.14 Proposition. *Let (σ, τ) be an (s, t)-pair such that σ represents* 1, *and $\beta = \sigma \perp -\tau$. Then (σ, τ) is of hyperbolic type if and only if one of the following conditions holds:*

$s \equiv t \pm 3 \pmod 8$ *and* $c(\beta) = 1$.

$s \equiv t \pm 2 \pmod 8$ *and* $d\beta = \langle 1 \rangle$.

$s \equiv t + 4 \pmod 8$ *and $c(\beta)$ is split by* $F(\sqrt{d\beta})$.

Proof. Let $C = C(-\sigma_1 \perp \tau)$ with involution $J = J_S$ as usual, and let V be an irreducible C-module. Then (σ, τ) has hyperbolic type iff there is no symmetric bilinear form on V which admits (C, J). Equivalently, there does not exist a 1-involution on $\operatorname{End}(V)$ compatible with (C, J).

Suppose $s + t$ is odd so that C is central simple. Then $A = \operatorname{End}_C(V)$ is a central division algebra and $C \otimes A \cong \operatorname{End}_F(V)$. By (6.13) there exists an involution K on A, and $J \otimes K$ induces an involution I on $\operatorname{End}(V)$. If $A \neq F$ then by (6.15) A has involutions of both types and one of them yields a 1-involution I. If $A = F$ then $\operatorname{type}(I) = \operatorname{type}(J)$. Then by (7.4) we see that (σ, τ) has hyperbolic type iff $c(\beta) = [A] = 1$ and $s \equiv t \pm 3 \pmod 8$.

Suppose $s+t$ is even so that $C = C_0 \otimes Z$ where $Z = F \otimes Fz$. Let $A = \mathrm{End}_{C_0}(V)$, so that A is central simple and $C_0 \otimes A \cong \mathrm{End}_F(V)$. First assume that $d\beta = \langle d \rangle \neq \langle 1 \rangle$. Then $Z \cong F(\sqrt{d})$ is a field and we may view $Z \subseteq A$.

Claim. There exist involutions K_+, K_- on A such that $K_\varepsilon(z) = \varepsilon z$.

This follows from an extension theorem for involutions due to Kneser, (see Scharlau (1985), Theorem 8.10.1). The claim is also proved below in Exercise 10.13.

Let $K = K_\varepsilon$ with ε chosen to make $K(z) = J(z)$. Define $B = \mathrm{Cent}_A(Z) = \mathrm{End}_C(V)$ so that B is a division algebra with center Z. If there exists $x \in B^\bullet$ with $K(x) = -x$ then K and K^x are involutions of both types on A and compatible with (C, J). Therefore if our 1-involution on $\mathrm{End}(V)$ fails to exist then no such x exists, and we see that $K(z) = z$ and $B = Z$. From the dimensions of centralizers we see that A must be a quaternion algebra containing the subfield Z. Furthermore, $J^+ \otimes K$ must have type -1. Since $K(z) = z$ we know that $s \equiv t \pmod 4$ and K has type 1. Then J^+ must have type -1 and $s \equiv t+4 \pmod 8$ by (7.4). Thus in this case when $s+t$ is even and $d\beta \neq \langle 1 \rangle$, we see that (σ, τ) is of hyperbolic type iff $c(\beta) = [A]$ is split by $F(\sqrt{d})$ and $s \equiv t+4 \pmod 8$.

Finally suppose $d\beta = \langle 1 \rangle$ so that $z^2 = 1$ and z acts as ± 1 on the irreducible module V. Then V is an irreducible C_0-module and A is a division algebra. If $J(z) = -z$ there can be no compatible involutions at all. This is the case $s \equiv t+2 \pmod 4$ already noted after (4.14). Otherwise $s \equiv t \pmod 4$ so that $J(z) = z$ and any involution K on A is compatible with (C, J). As before if $A \neq F$ there exist involutions of both types on A. Then (σ, τ) has hyperbolic type iff $A = F$ and the induced involution J^+ on C_0 has type -1. By (7.4) this occurs iff $c(\beta) = 1$ and $s \equiv t+4 \pmod 8$. □

Remark. The criteria for (σ, τ) to be of (-1)-hyperbolic type are obtained by cycling the congruences above by 4 (mod 8).

7.15 Corollary. *Let $\sigma = p\langle 1 \rangle \perp r\langle -1 \rangle$ over $\mathbb{R}$. Then σ is of hyperbolic type if and only if there is a dot (•) in the corresponding entry of the following table, indexed by the values of p and r* (mod 8).

$p \backslash r$	0	1	2	3	4	5	6	7
0								
1		•	•	•		•		
2		•	•	•				
3		•	•	•				•
4								
5		•				•	•	•
6						•	•	•
7				•		•	•	•

Proof. Apply the proposition and the calculations of $d\sigma$ and $c(\sigma)$. □

Remark. From the symmetries of the tables we see that σ has hyperbolic type if and only if $\sigma \perp 4\mathbb{H}$ does as well. Analysis of the proof shows that σ has (-1)-hyperbolic type if and only if $\sigma \perp 4\langle 1\rangle$ has hyperbolic type.

If $(S, T) \subseteq \mathrm{Sim}(V, q)$ is a pair of amicable subspaces, then so is $(S', T') = (fSg, fTg)$ for any $f, g \in \mathrm{Sim}^\bullet(V, q)$. Conversely if (S, T) and (S', T') are pairs of amicable subspaces in $\mathrm{Sim}(V, q)$, how can we tell whether they are equivalent in this way? One obvious necessary condition is that the induced pairs of quadratic forms (σ, τ) and (σ', τ') be similar. For minimal pairs that condition suffices.

7.16 Corollary. *Suppose (S, T) and (S', T') are pairs of amicable subspaces of* $\mathrm{Sim}(V, q)$ *which are similar as quadratic spaces: $S' \simeq \langle c\rangle S$ and $T' \simeq \langle c\rangle T$ for some $c \in F^\bullet$. If $\dim V = 2^m$ and $s > \rho_t(2^{m-1})$, then there exist $f, g \in \mathrm{Sim}^\bullet(V, q)$ such that $(S', T') = (fSg, fTg)$.*

Proof. We may assume $1_V \in S$. Then there exists $f \in S'$ with $\mu(f) = c$. We compose with f^{-1} to assume $S' \simeq S$ and $T' \simeq T$. The Clifford algebra $C = C(-\sigma_1 \perp \tau)$ with the involution $J = J_S$ then has two representations π and π' on (V, q) corresponding to these two (s, t)-families. That is, (V, q) becomes a (C, J)-module in two ways. In the notation used at the start of Chapter 4, the subspaces $\breve{S}, \breve{T} \subseteq C$ satisfy: $S = \pi(\breve{S})$, $T = \pi(\breve{T})$, and $S' = \pi'(\breve{S})$, $T' = \pi'(\breve{T})$.

Since $s > \rho_t(2^{m-1})$ the (C, J)-module structures on V must be unsplittable. By (7.11) these two unsplittables are C-similar (possibly after twisting π in the non-simple case). Let $h : V \to V$ be a C-similarity carrying the π-structure to the π'-structure. Then $h(\pi(c)x) = \pi'(c)h(x)$ for all $c \in C$ and $x \in V$. That is, $\pi'(c) = h \circ \pi(c) \circ h^{-1}$. Therefore $S' = hSh^{-1}$ and $T' = hTh^{-1}$. □

In some cases we can eliminate the restriction on dimensions in (7.16). We are given (C, J) and two quadratic (C, J)-modules (V, q) and (V', q') which are F-similar, and hope to conclude that they are C-similar. First suppose C is simple, so that V and V' are isomorphic as C-modules. They break into unsplittables

$$V = V_1 \perp \cdots \perp V_k \quad \text{and} \quad V' = V_1' \perp \cdots \perp V_k'.$$

Assuming (σ, τ) is minimal we see from (7.11) that all V_i and V_j' are C-similar. In order to glue these similarities we must find the unsplittables together with C-similarities $g_j : V_j \to V_j'$ such that the norms $\mu(g_j)$ are all equal. For example suppose $F = \mathbb{R}$ and (V, q) is positive definite. Then any C-similarity between the unsplittable components has positive norm so it can be scaled to yield a C-isometry, and the "gluing" works. The same idea goes through in a few more cases over $\mathbb{R}$ (see Exercise 8).

Suppose now that C is not simple, so that $s + t$ is even and $d\sigma = d\tau$. In order to ensure that the two C-module structures on V are isomorphic, we require that the two (s, t)-families have the same "character". Let $z = z(S_1 \perp T)$ be an element of highest degree with $z^2 = 1$. As mentioned before (4.12) there are exactly two irreducible C-modules V_+ and V_-, chosen so that z acts as $\varepsilon 1_{V_\varepsilon}$ on V_ε. Any C-module V is isomorphic to a direct sum of n_+ of copies of V_+ and n_- copies of V_-, for some integers $n_+, n_- \geq 0$. Then

$$\dim V = (n_+ + n_-) \cdot 2^m \quad \text{and} \quad \operatorname{trace}(\pi(z)) = (n_+ - n_-) \cdot 2^m,$$

where $2^m = \dim V_+ = \dim V_-$. Therefore two C-modules are isomorphic iff they have the same dimension and the same value for $\operatorname{trace}(\pi(z))$.

Since we are interested only in the spaces S, T and not in the representation π, we may "twist" π by replacing it by $\pi \circ \alpha$ where α is the canonical automorphism of C. This operation leaves the subspaces S and T unchanged but it alters the sign of $\operatorname{trace}(\pi(z))$. Therefore the non-negative integer $|\operatorname{trace}(\pi(z))|$ depends only on the given family (S, T), and not on the choice of the representation π.

7.17 Definition. If $(S, T) \subseteq \operatorname{Sim}(V, q)$ is an (s, t)-family, let z be an element of highest degree in the Clifford algebra C, chosen so that if C is not simple then $z^2 = 1$. Define $\chi(S, T) = |\operatorname{trace}(\pi(z))|$, the *character* of the family.

7.18 Lemma. *If* $\chi(S, T) \neq 0$ *then* $s \equiv t \pmod 4$, $d\sigma = d\tau$ *and* (S, T) *is maximal.*

Proof. If (S, T) can be expanded in $\operatorname{Sim}(V, q)$ then there exists $f \in \operatorname{Sim}^\bullet(V, q)$ which anticommutes with $\pi(z)$, so that $\operatorname{trace}(\pi(z)) = 0$. If $s + t$ is odd then (S, T) can be expanded. If $s \equiv t + 2 \pmod 4$ then $J(z) = -z$ so that $\operatorname{trace}(\pi(z)) = 0$. Finally suppose $s \equiv t \pmod 4$ but $d\sigma \neq d\tau$. Then $Z = F + Fz \cong F(\sqrt{d})$ is a field and the minimal polynomial for $\pi(z)$ is $x^2 - d$, which is irreducible. Then $\operatorname{trace}(\pi(z)) = 0$ since the characteristic polynomial must be a power of $x^2 - d$. □

7.19 Proposition. *Suppose* (V, q) *is positive definite over the real field* $\mathbb{R}$. *Suppose* (S, T) *and* (S', T') *are* (s, t)*-families in* $\operatorname{Sim}(V, q)$ *such that* $\chi(S, T) = \chi(S', T')$. *Then* $(S', T') = (hSh^{-1}, hTh^{-1})$ *for some* $h \in \mathrm{O}(V, q)$.

Proof. Since the forms are positive definite over $\mathbb{R}$ we have $S \simeq S' \simeq s\langle 1\rangle$ and $T \simeq T' \simeq t\langle 1\rangle$ as quadratic spaces. For C and J as usual, we see that (V, q) becomes a quadratic (C, J)-module in two ways. We may twist the representation π by α, if necessary, to assume that $\operatorname{trace}(\pi(z)) = \operatorname{trace}(\pi'(z))$. Then these two C-module structures are isomorphic. The two (C, J)-modules can then be broken into unsplittables

$$V = V_1 \perp \cdots \perp V_k \quad \text{and} \quad V' = V_1' \perp \cdots \perp V_k'$$

in such a way that V_i and V_i' are isomorphic C-modules. Since $(s\langle 1\rangle, t\langle 1\rangle)$ is a minimal pair we know as in (7.11) that V_i and V_i' are C-similar. The norm of such a similarity must be positive in $\mathbb{R}$ so we may scale it to find a C-isometry $h_i : V_i \to V_i'$. Glue these h_i's together to obtain an isometry $h : (V, q) \to (V, q)$ carrying the π-structure to the π'-structure. This completes the proof, as in (7.16). □

7.20 Corollary. (1) *Suppose* $(S, T) \subseteq \mathrm{Sim}(V, n\langle 1\rangle)$ *over* $\mathbb{R}$. *If* $\chi(S, T) = 0$ *then* (S, T) *can be enlarged to a family of maximal size. That trace condition always holds if* $s \not\equiv t \pmod 4$.

(2) *Every sum of squares formula of size* $[r, n, n]$ *over* $\mathbb{R}$ *is equivalent to one over* $\mathbb{Z}$.

Exercises for Chapter 7

1. **Maximal families.** Suppose $(S, T) \subseteq \mathrm{Sim}(V, B)$ is an (s, t)-family with associated representation $\pi : C \to \mathrm{End}(V)$.

(1) If π is non-faithful then (S, T) is maximal. More generally if $\chi(S, T) \neq 0$ (as defined in (7.17)) then (S, T) is maximal.

(2) Find examples of faithful maximal families. If $(S, T) \subseteq \mathrm{Sim}(V, B)$ is maximal and faithful, what can be said about the algebra $A = \mathrm{End}_{C_0}(V)$?

(Hint. (1) If $f \in \mathrm{Sim}^\bullet(V)$ anticommutes with $S_1 + T$ then f must anticommute with $\pi(z)$.)

2. **Why is $c(\beta)$ split by $F(\sqrt{\beta})$?** In the situation of Theorem 7.7 suppose $s + t = 2m$ and there is a quadratic module (V, q) of dimension 2^m. Let $Z \cong F(\sqrt{\beta})$ be the center of the Clifford algebra C and suppose Z is a field. Then C is a central simple Z-algebra and there is an induced Z-action on V. Then $\dim_Z C = 2^{2m-2}$, $\dim_Z V = 2^{m-1}$ and $C \cong \mathrm{End}_Z(V)$. Therefore $1 = [C]_Z = [C_0 \otimes Z]$ and $c(\beta) = [C_0]$ is split by $F(\sqrt{\beta})$. If $s \equiv t \pmod 4$ then $J(z) = z$. Compute type(J) as a Z-involution to see $s \equiv t \pmod 8$. Is there a similar argument when $d\beta = \langle 1\rangle$?

3. The following can be proved by methods of Chapter 2 or by applying (7.8).

(1) If the dimension of an unsplittable (σ, τ)-module is 2^m then the dimension of an unsplittable $(\sigma \perp \langle a\rangle, \tau \perp \langle a\rangle)$-module is 2^{m+1}.

(2) If (σ, τ) is a minimal pair and α is any quadratic form, then $(\sigma \perp \alpha, \tau \perp \alpha)$ is also minimal. If α represents 1 then (α, α) is minimal. If $(\sigma, \tau) < \mathrm{Sim}(\varphi)$ is unsplittable, what is the unsplittable quadratic module for $(\sigma \perp \alpha, \tau \perp \alpha)$?

(3) For any $s \geq 1, t \geq 0$ the pair $(s\langle 1\rangle, t\langle 1\rangle)$ is minimal with (unique) unsplittable module $2^m\langle 1\rangle$, where $m = \delta(s, t)$.

4. (1) If (σ, τ) is minimal and $\varphi = \langle\langle a, b, c\rangle\rangle$ then $(\sigma \perp \varphi, \tau)$ is also minimal.

(2) If (σ, τ) is minimal and $a \in D_F(\sigma)$ then $(\langle a\rangle\sigma, \langle a\rangle\tau)$ is also minimal.

(3) If σ is minimal then $\sigma \perp 8\langle 1\rangle$, $\sigma \perp 8\langle -1\rangle$ and $\sigma \perp \mathbb{H}$ are minimal. If σ is also isotropic then $\langle -1\rangle\sigma$ is minimal. Interpret these in terms of the symmetry of the table in (7.13).

(4) Repeat the observations above using "hyperbolic type" rather than "minimal". Observe from (7.13) and (7.15) that the entry (p, r) is marked in one chart iff $(p, -r)$ is marked in the other. Is there any deeper explanation of this coincidence?

(Hint. (1) Express $\varphi = \alpha \perp (d\alpha)\alpha$ where $\alpha = \langle 1, a, b, c\rangle$ and shift.)

5. Suppose (σ, τ) has the property that every unsplittable (σ, τ)-module is similar to a Pfister form. Then $(\sigma \perp \alpha, \tau \perp \alpha)$ has the same property.

6. Suppose (σ, τ) is a pair where σ represents 1 with unsplittables of dimension 2^m. Then there exist subforms $\sigma' \subset \sigma$ and $\tau' \subset \tau$ such that σ' represents 1 and (σ', τ')-unsplittables have dimension 2^{m-1}.

7. (1) Given an (s, t)-pair (σ, τ) where $\sigma = \langle 1\rangle \perp \sigma_1$, let $\beta = \sigma \perp -\tau$. For which $a \in F^\bullet$ is the $(s+1, t)$-pair $(\sigma \perp \langle a\rangle, \tau)$ minimal? This occurs if and only if one of the following conditions holds:

$s \equiv t$ or $t-2 \pmod 8$ and $c(\beta) = [d\beta, -a]$.

$s \equiv t+1$ or $t-3 \pmod 8$ and $c(\beta)$ is split by $F(\sqrt{-a \cdot d\beta})$.

$s \equiv t+2$ or $t+4 \pmod 8$ and $c(\beta)[d\beta, -a] =$ quaternion.

$s \equiv t+3 \pmod 8$ and $d\beta = \langle -a\rangle$ and $c(\beta) =$ quaternion.

$s \equiv t-1 \pmod 8$ and $d\beta = \langle -a\rangle$ and $c(\beta) = 1$.

(2) For what (s, t) is it possible that a non-minimal (s, t)-pair can be expanded to a minimal $(s+1, t)$-pair?

(3) Similarly analyze the cases where $(\sigma, \tau \perp \langle b\rangle)$ is minimal.

(Hint. (2) $\delta(s+1, t) = 1 + \delta(s, t)$ if and only if $s - t \equiv 0, 1, 2, 4 \pmod 8$.)

8. **Conjugate subspaces.** (1) Suppose $\{1_V, f_2, \dots, f_t\}$ is an orthogonal basis of some subspace of $\mathrm{Sim}(V, q)$. Define

$$S = \mathrm{span}\{1_V, f_2, f_3, f_4\} \quad \text{and} \quad S' = \mathrm{span}\{1_V, f_2, f_3, f_2 f_3\}.$$

Then S' cannot be expressed as fSg for any $f, g \in \mathrm{GL}(V)$.

(2) Explain Exercise 1.16 using the more abstract notions of (7.16). The strong conjugacy in that exercise seems to require a Clifford algebra C such that $\bar{c} \cdot c \in F$ for every $c \in C$.

(3) Suppose σ, q are forms over $\mathbb{R}$ such that σ is minimal and both forms represent 1. Suppose $S, S' \subseteq \mathrm{Sim}(V, q)$ with $1_V \in S \cap S'$, $S \simeq S' \simeq \sigma$, and $\chi(S) = \chi(S')$.

Question. For which σ, q does it follow that $S' = hSh^{-1}$ for some $h \in \mathrm{O}(V, q)$?

From (7.19) we know it is true when σ, q are positive definite. The same argument proves the statement when σ is positive definite and $\dim \sigma \not\equiv 0 \pmod 4$, (in those cases the algebra C is simple). If σ is of hyperbolic type the statement is certainly true. It fails in all other cases.

(Hint. If σ is definite and C is not simple let $(V_\varepsilon, \psi_\varepsilon)$ be the positive definite irreducible (C, J)-modules. Let $V = \psi_1 \perp \langle -1\rangle\psi_1 \perp \psi_{-1} \perp \langle -1\rangle\psi_{-1}$ and $V' = \psi_1 \perp \psi_1 \perp \langle -1\rangle\psi_{-1} \perp \langle -1\rangle\psi_{-1}$ to get a counterexample. If σ is indefinite and regular type, an irreducible (C, J)-module (W, ψ) admits no C-similarity of norm -1. Then $\psi \perp \psi$ and $\psi \perp \langle -1\rangle\psi$ are C-isomorphic and F-isometric, but are not (C, J)-similar. (Use the Cancellation Theorem mentioned after (4.10).)

9. **Spaces not containing 1.** Suppose $S \subseteq \mathrm{Sim}(V, q)$, choose $g \in S^\bullet$ and define the character $\chi(S) = \chi(g^{-1}S)$ following (7.17) for spaces containing 1_V.

(1) This value is independent of the choice of g.

(2) Generalize the definition and (7.19) to amicable pairs $(S, T) \subseteq \mathrm{Sim}(V, q)$.

(Hint. Recall $z(S)$ defined in Exercise 2.8. Suppose $\dim S \equiv 0 \pmod 4$ and $dS = \langle 1\rangle$. If we choose $z(S)^2 = 1$ then $\chi(S) = |\operatorname{trace}(z(S))|$.)

10. **Non-minimal behavior.** There exists an example where $(\langle 1, a\rangle, \langle x\rangle) < \mathrm{Sim}(V, q)$ where $\dim q = 12$ but such that $\mathrm{Sim}(q)$ does not admit any $(3, 3)$-family. Compare this with the assertion in (7.12). Find an explicit example over $\mathbb{R}$.

(Hint. Recall (5.7)(4) and find q such that $\langle\langle a\rangle\rangle \mid q$, $x \in G_F(q)$ but q does not have a 2-fold Pfister factor.)

11. **Unique unsplittables.** A pair (σ, τ) is defined to have *unique unsplittables* if all unsplittable quadratic (σ, τ)-modules are (C, J)-similar, possibly after twisting the associated representation in the non-simple case.

(1) If $(\sigma, \tau) < \mathrm{Sim}(\varphi)$ is unsplittable and (σ, τ) has unique unsplittables, then: $(\sigma, \tau) < \mathrm{Sim}(q)$ if and only if $\varphi \mid q$.

(2) Suppose (σ, τ) is an (s, t)-pair where $s + t$ is odd, and suppose $(\sigma, \tau) < \mathrm{Sim}(V, q)$ is unsplittable. Let C be the associated Clifford algebra with centralizer A, so that $C \otimes A \cong \mathrm{End}(V)$ and $J \otimes K \cong I_q$ as usual. Then (σ, τ) has unique unsplittables iff every $f \in A$ with $K(f) = f$ can be expressed as $f = r \cdot K(g)g$ for some $g \in A$ and $r \in F$.

12. Let (σ, τ) be an (s, t)-pair and suppose $c(\beta) = [-x, -y] \neq 1$.

If $s \equiv t \pm 3 \pmod 8$ then (σ, τ) has unique unsplittables, as defined in Exercise 11.

If $s \equiv t \pm 1 \pmod 8$ then the (C, J)-similarity classes of unsplittables are in one-to-one correspondence with $D_F(\langle x, y, xy\rangle)/F^{\bullet 2}$.

(Hint. Let (V, q) be unsplittable so that $C \otimes A \cong \mathrm{End}(V)$ where $A = (-x, -y/F)$ with induced involution K. If $s \equiv t \pm 3$ then $K = $ bar. Otherwise every (C, J)-unsplittable arises from a 1-involution on A. These are the involutions K_0^e where $K_0 = $ bar and $e \in A_0^\bullet$. Apply (6.8)(3).)

13. By Exercise 3.15(3) we know that $\langle\langle a_1\rangle\rangle \otimes \langle 1, a_2, \dots, a_m\rangle < \mathrm{Sim}(\langle\langle a_1, \dots, a_m\rangle\rangle)$. This module is unsplittable iff m is odd. That space of dimension $2m$ is minimal

iff $m \not\equiv 0 \pmod 4$. From Corollary 7.11 we find that: If $m \not\equiv 0 \pmod 4$ and if the forms $\langle\langle a_1\rangle\rangle \otimes \langle 1, a_2, \dots, a_m\rangle$ and $\langle\langle b_1\rangle\rangle \otimes \langle 1, b_2, \dots, b_m\rangle$ are similar, then $\langle\langle a_1, \dots, a_m\rangle\rangle \simeq \langle\langle b_1, \dots, b_m\rangle\rangle$.

14. **More on trace forms.** (1) **Lemma.** *Let* $C = C(-\alpha \perp \tau)$ *where* $\dim \alpha = a$, $\dim \tau = t$ *and* $a + t = 2m$ *is even. Let* $J = J_{A,T}$ *be the involution extending the map* $(-1) \perp (1)$ *on* $-\alpha \perp \tau$. *Then* J *has type* 1 *iff* $a - t \equiv 0$ *or* 6 (mod 8).

Recall the notation $P(\alpha)$ from Exercise 3.14.

(2) Suppose α and τ are forms as above and $c(-\alpha \perp \tau) = 1$.

If $a - t \equiv 2$ or 4 (mod 8), the Pfister form $P(\alpha \perp \tau)$ is hyperbolic.

If $a - t \equiv 0$ or 6 (mod 8), then $P(\alpha \perp \tau) \simeq q \otimes q$ for some form q.

(3) If $\dim q = 2^m$ and there is an $(m+1, m+1)$-family in $\mathrm{Sim}(q)$ then $q \otimes q$ is a Pfister form.

(4) **Corollary.** *If* $\dim \sigma = 2m$ *and* $\sigma \in I^3F$ *then*

$$P(\sigma) \simeq \begin{cases} 2^m\langle 1\rangle \otimes \psi & \text{if } m \equiv 0 \\ \text{hyperbolic} & \text{if } m \not\equiv 0 \end{cases} \pmod 4.$$

(Hint. (2) For C and J as above define the trace form B_J on C by $B_J(x, y) = \ell(J(x)y)$. By Exercise 3.14, $(C, B_J) \simeq P(\alpha \perp \beta)$ as quadratic spaces. Also $C \cong \mathrm{End}(V)$ where $\dim V = 2^m$ and J induces an involution I_B on $\mathrm{End}(V)$ for some λ-form B. The induced map $\ell : \mathrm{End}(V) \to F$ is the scalar multiple of the trace map having $\ell(1_V) = 1$. By Exercise 1.13 it follows that $(C, B_J) \simeq (V \otimes V, B \otimes B)$. If $a - t \equiv 2 \pmod 8$ then B is an alternating form by (1), and $B \otimes B$ is hyperbolic. Otherwise B corresponds to a quadratic form q.

(4) Let φ be a $2m$-fold Pfister form. Then $\varphi \simeq q \otimes q$ iff $\varphi \simeq 2^m\langle 1\rangle \otimes \psi$ for some m-fold Pfister form ψ. This can be proved using:

Lemma. *If* φ *and* γ *are Pfister forms and* $\gamma \subset \varphi$ *then* $\varphi \simeq \gamma \otimes \delta$ *where* δ *is a Pfister form.*

See Exercise 9.15 or Lam (1973), Chapter 10, Exer. 8.)

Notes on Chapter 7

The idea of using a chart as in (7.13) follows Gauchman and Toth (1994), §2.

The equivalence and expansion results in (7.18) and (7.19) were done over $\mathbb{R}$ by Y. C. Wong (1961) using purely matrix methods.

Exercise 13. Wadsworth and Shapiro (1977b) used a different method to prove that if φ is a round form and if $\varphi \otimes (\langle 1\rangle \perp \alpha)$ and $\varphi \otimes (\langle 1\rangle \perp \beta)$ are similar then $\varphi \otimes P(\alpha) \simeq \varphi \otimes P(\alpha)$. The main tool for this proof is Lemma 5.5 above.

Chapter 8

The Space of All Compositions

The topological space $\text{Comp}(s, n)$ of all composition formulas of type $\mathbb{R}^s \times \mathbb{R}^n \to \mathbb{R}^n$ turns out to be a smooth compact real manifold. After deriving general properties of $\text{Comp}(s, n)$, we focus on the spaces of real composition algebras. For example the space Comp(8, 8) has 8 connected components, each of dimension 56. Since these algebras have such a rich structure we compute the dimensions by another method, by considering autotopies, monotopies and the associated Triality Theorem.

The spaces $\text{Comp}(s, n)$ are accessible since they are orbits of certain group actions. This analysis requires the reader to have some familiarity with basic results from the theory of algebraic groups. For instance we use properties of orbits and stabilzers, and we assume some facts about the the orthogonal group $\text{O}(n)$ and the symplectic group $\text{Sp}(n)$ (e.g. their dimensions and number of components).

We begin with the general situation, specializing to the real case later. Let (S, σ) and (V, q) be quadratic spaces over the field F, with dimensions s, n respectively. To avoid trivialities, assume $s > 1$ so that n is even. Define the sets

$$\text{Bil}(S, V) = \{m: S \times V \to V : m \text{ is bilinear}\}$$
$$\text{Comp}(\sigma, q) = \{m \in \text{Bil}(S, V) : q(m(x, y)) = \sigma(x) \cdot q(y) \text{ for every } x \in S, y \in V\}$$

Then $\text{Bil}(S, V)$ is an F-vector space of dimension sn^2 and $\text{Comp}(\sigma, q)$ is an affine algebraic set (since it is the solution set of the Hurwitz Matrix Equations). If the base field needs some emphasis we may write $\text{Comp}_F(\sigma, q)$, etc.

The product of orthogonal groups $\text{O}(\sigma) \times \text{O}(q) \times \text{O}(q)$ acts on $\text{Comp}(\sigma, q)$ by:

$$((\alpha, \beta, \gamma) \bullet m)(x, y) = \gamma(m(\alpha^{-1}(x), \beta^{-1}(y))) \quad \text{for } x \in S \text{ and } y \in V.$$

This definition can be recast using the notation of similarities. If $m \in \text{Comp}(\sigma, q)$ define

$$\hat{m} : S \to \text{Sim}(V, q) \quad \text{by} \quad \hat{m}(x)(y) = m(x, y).$$

Then $\hat{m}$ is a linear isometry from (S, σ) to the subspace $S_m = \text{image}(\hat{m}) \subseteq \text{Sim}(V, q)$. This $\hat{m}$ determines the composition m and we think of $\hat{m}$ as an element of $\text{Comp}(\sigma, q)$. The group action becomes:

$$((\alpha, \beta, \gamma) \bullet \hat{m})(x) = \gamma \circ \hat{m}(\alpha^{-1}(x)) \circ \beta^{-1} \quad \text{for } x \in S.$$

The subspace S_m is carried to $\gamma \circ S_m \circ \beta^{-1}$ by this action.

8.1 Lemma. *If* $\langle a\rangle q \simeq q$ *then* $\text{Comp}(\sigma, q) \cong \text{Comp}(\langle a\rangle\sigma, q)$.

Proof. Given $h \in \text{Sim}^\bullet(q)$ with $\mu(h) = a$. If $m \in \text{Comp}(\sigma, q)$ then sending m to $h \circ m$ provides the isomorphism. □

We view (S, σ) as a quadratic space *with a given orthogonal basis* $\{e_1, e_2, \dots, e_s\}$. In the applications it will be $\mathbb{R}^s$ or $\mathbb{C}^s$ with the standard orthonormal basis. We may assume that σ represents 1. For if $\text{Comp}(\sigma, q) \neq \emptyset$, choose $a \in D_F(\sigma) \subseteq G_F(q)$ and apply (8.1). Then we may assume that the given basis was chosen so that $\sigma(e_1) = 1$.

8.2 Definition. $\text{Comp}^1(\sigma, q) = \{m \in \text{Comp}(\sigma, q) : \hat{m}(e_1) = 1_V\}$.

We define $\text{Bil}^1(S, V)$ similarly and note that it is a coset of a linear subspace of dimension $(s-1)n^2$ in the vector space $\text{Bil}(S, V)$.

8.3 Lemma. $\text{Comp}(\sigma, q) \cong \text{O}(q) \times \text{Comp}^1(\sigma, q)$, *an isomorphism of algebraic sets.*

Proof. Define $\varphi : \text{O}(q) \times \text{Comp}^1(\sigma, q) \to \text{Comp}(\sigma, q)$ by $\varphi(g, m_0) = g \circ m_0$. The inverse map is given by $\varphi^{-1}(m) = (\hat{m}(e_1), \hat{m}(e_1)^{-1} \circ m)$. Note that φ and φ^{-1} are polynomial maps since $\hat{m}(e_1)^{-1} = I_q(\hat{m}(e_1))$. □

The action of $\text{O}(q) \times \text{O}(q)$ on $\text{Comp}(\sigma, q)$ becomes the following action on $\text{O}(q) \times \text{Comp}^1(\sigma, q)$:

$$(\beta, \gamma) \bullet (g, \hat{m}_0) = (\gamma g \beta^{-1}, \beta \diamond \hat{m}),$$

where $\beta \diamond \hat{m}$ denotes the conjugation action of $\text{O}(q)$ on $\text{Comp}^1(\sigma, q)$ given by:

$$(\beta \diamond \hat{m})(x) = \beta \circ \hat{m} \circ \beta^{-1} \quad \text{for } x \in S.$$

To analyze this conjugation action we introduce the "character" of $m \in \text{Comp}^1(\sigma, q)$, as mentioned in the discussion before (7.17).

The map $\hat{m} : S \to \text{Sim}(V)$ sends $e_1 \mapsto 1_V$. The associated Clifford algebra $C = C(-\sigma_1)$ is generated by $\{e_2, \dots, e_s\}$, and $\hat{m}$ induces a similarity representation $\pi_m : C \to \text{End}(V)$ where $\pi_m(e_i) = \hat{m}(e_i)$. This makes V into a C-module which we denote by V_m. Define the element $z = e_2 \dots e_s \in C$ as usual. When $s \equiv 0 \pmod 4$ and $d\sigma = \langle 1\rangle$ then C is not simple and admits an irreducible unsplittable module. In that case we normalize our choice of basis to ensure that $z^2 = 1$. That normalization is automatic if $\sigma \simeq s\langle 1\rangle$ and an orthonormal basis is chosen.

8.4 Definition. If $m \in \text{Comp}^1(\sigma, q)$ define the character $\chi(m) = \text{trace}(\pi_m(z))$. Define

$$\text{Comp}^1(\sigma, q; k) = \{m \in \text{Comp}^1(\sigma, q) : \chi(m) = k\}.$$

If $s \not\equiv 0 \pmod 4$ or if $d\sigma \neq \langle 1 \rangle$ we know that $\chi(m) = 0$. Generally, $\chi(m)$ is an even integer between $-n$ and n. As we mentioned in the discussion before (7.17):

$$\chi(m) = \chi(m') \text{ if and only if } V_m \cong V_{m'} \text{ as C-modules.}$$

It easily follows that $\chi(m) = \chi(\beta \diamond m)$, so that the O($q$)-orbit of m is inside $\text{Comp}^1(\sigma, q; \chi(m))$.

This character can be extended to the whole set Comp(σ, q) by using the isomorphism φ in (8.3). Then the $\text{O}(q) \times \text{O}(q)$-orbit of m is contained in $\text{Comp}(\sigma, q; \chi(m))$. See Exercise 1 for more details.

8.5 Lemma. *Suppose* $\sigma = s\langle 1 \rangle$, $q = n\langle 1 \rangle$ *and* F *is* $\mathbb{R}$ *or* $\mathbb{C}$. *Then* O(q) *acts transitively on* $\text{Comp}^1(\sigma, q; k)$, *and* $\text{O}(q) \times \text{O}(q)$ *acts transitively on* $\text{Comp}(\sigma, q; k)$.

Proof. If $m, m' \in \text{Comp}^1(\sigma, q; k)$ then $V_m \cong V_{m'}$ as C-modules. As in (7.19), these two structures are C-isometric, so there exists $\beta \in \text{O}(V, q)$ such that $\beta \circ \pi(c) = \pi'(c) \circ \beta$ for every $c \in C$. Then $\beta \circ \hat{m}(x) = \hat{m}'(x) \circ \beta$ for every $x \in F^s$ and hence $\beta \diamond \hat{m} = \hat{m}'$. The second transitivity follows using (8.3). □

To analyze the O(q)-orbit $\text{Comp}^1(\sigma, q; k)$ we gather information about the stabilizer subgroup. Let us return briefly to the more general situation with $\sigma = \langle 1 \rangle \perp \sigma_1$ and q over F. For $m \in \text{Comp}^1(\sigma, q)$, define an automorphism group

$$\begin{aligned}\text{Aut}(m) &= \{\beta \in \text{O}(q) : \beta \diamond m = m\} \\ &= \{\beta \in \text{O}(q) : \beta \circ f \circ \beta^{-1} = f \text{ for every } f \in S_m\}.\end{aligned}$$

Since the C-module structure V_m is determined by the elements of S_m,

$$\text{Aut}(m) = \text{O}(V, q) \cap \text{End}_C(V).$$

8.6 Lemma. (1) *Suppose s is odd and let* $A = \text{End}_C(V)$. *Then A is central simple,* $C \otimes A \cong \text{End}(V)$, *and* I_q *induces an involution "$\sim$" on A, which has type* 1 *if and only if* $s \equiv \pm 1 \pmod 8$. *Then*

$$\text{Aut}(m) \cong \{a \in A : \tilde{a} \cdot a = 1\}.$$

(2) *Suppose s is even and let* $A = \text{End}_{C_0}(V)$. *Then A is central simple,* $C_0 \otimes A \cong \text{End}(V)$, *and* I_q *induces an involution "$\sim$" on A, which has type* 1 *if and only if* $s \equiv 0, 2 \pmod 8$. *Let* $y = \pi_m(z) \in A$, *where* $z = z(S) \in C$. *Then* $y^2 \in F^\bullet$, $\tilde{y} = (-1)^{s/2} \cdot y$ *and*

$$\text{Aut}(m) \cong \{a \in A : ay = ya \text{ and } \tilde{a} \cdot a = 1\}.$$

Proof. The properties of A have been mentioned earlier, the type calculation follows from (7.4) and (6.9), and the description of Aut(m) is a restatement of the definition. □

The group $\mathrm{Aut}(m)$ is an algebraic group (it is an algebraic set defined over F and the multiplication and inverse maps are defined by polynomials). We can determine the dimension of $\mathrm{Aut}(m)$ by extending scalars and computing that dimension in the case F is algebraically closed.

Since we are primarily concerned with the sums-of-squares forms over $\mathbb{R}$ and $\mathbb{C}$, let us simplify the notations a little and define:

$$\mathrm{Comp}^1(s,n) = \mathrm{Comp}^1(s\langle 1\rangle, n\langle 1\rangle),$$

and similarly for $\mathrm{Comp}^1(s,n;k)$, $\mathrm{Bil}(s,n)$, etc. We also use the standard notation $\mathrm{O}(n)$ in place of $\mathrm{O}(n\langle 1\rangle)$. The stabilizer $\mathrm{Aut}(m) \subseteq \mathrm{O}(n)$ changes only by conjugation in $\mathrm{O}(n)$ as m varies in the orbit $\mathrm{Comp}^1(s,n;k)$. Then as an abstract algebraic group, $\mathrm{Aut}(m)$ depends only on s, n and k and we sometimes write it as $\mathrm{Aut}(s,n;k)$.

8.7 Proposition. *Let* $m \in \mathrm{Comp}^1(s,n;k)$.

(1) *If s is odd let* $\varepsilon = \mathrm{type}(\sim) = (-1)^{(s^2-1)/8}$. *Then:* $\dim \mathrm{Aut}(s,n;k) = \frac{n^2}{2^s} - \frac{\varepsilon n}{2^{(s+1)/2}}$.

(2) *If* $s \equiv 2 \pmod 4$ *then:* $\dim \mathrm{Aut}(s,n;k) = \frac{n^2}{2^s}$.

(3) *If* $s \equiv 0 \pmod 4$ *let* $\varepsilon = \mathrm{type}(\sim) = (-1)^{s/4}$. *Then:* $\dim \mathrm{Aut}(s,n;k) = \frac{n^2+k^2}{2^s} - \frac{\varepsilon n}{2^{s/2}}$.

Proof. We may assume F is algebraically closed. Choose $m \in \mathrm{Comp}(s,n;k)$.

(1) From (8.6) we know that $A \cong \mathbb{M}_r(F)$ where: $r \cdot 2^{(s-1)/2} = n$. If $\varepsilon = 1$ then $\mathrm{Aut}(m) \cong \mathrm{O}(r)$ has dimension $\frac{1}{2} \cdot r(r-1)$. If $\varepsilon = -1$ then $\mathrm{Aut}(m) \cong \mathrm{Sp}(r)$ has dimension $\frac{1}{2} \cdot r(r+1)$.

(2) We have $A \cong \mathbb{M}_r(F)$ where: $r \cdot 2^{\frac{s}{2}-1} = n$, and we may assume $y \in A$ satisfies $y^2 = 1$ and $\tilde{y} = -y$. Let W be an irreducible A-module so that $\dim W = r$, $A \cong \mathrm{End}(W)$, and tilde induces an ε-symmetric form $b : W \times W \to F$. The (± 1)-eigenspaces of y are then totally isotropic subspaces of W, each of dimension $r/2$. Using dual bases for these eigenspaces the Gram matrix of b is $\begin{pmatrix} 0 & 1 \\ \varepsilon 1 & 0 \end{pmatrix}$. Representing $a \in A$ as a block matrix $\begin{pmatrix} a_1 & a_2 \\ a_3 & a_4 \end{pmatrix}$ we have $\tilde{a} = \begin{pmatrix} a_4^\top & \varepsilon a_2^\top \\ \varepsilon a_3^\top & a_1^\top \end{pmatrix}$. If $a \in \mathrm{Aut}(m)$ then $ay = ya$ implies that $a = \begin{pmatrix} a_1 & 0 \\ 0 & a_4 \end{pmatrix}$. Therefore $\mathrm{Aut}(m) \cong \left\{ \begin{pmatrix} c & 0 \\ 0 & c^{-\top} \end{pmatrix} : c \in \mathrm{GL}_{r/2}(F) \right\}$ and the dimension result follows.

(3) We have A and r as above, and $y \in A$ satisfies $y^2 = 1$ and $\tilde{y} = y$. Then V is a direct sum of r (isomorphic) irreducible C_0-modules

$$V = V_1 \oplus \cdots \oplus V_r \quad \text{where } \dim V_i = 2^{\frac{s}{2}-1} = \frac{n}{r}.$$

Therefore $A = \mathrm{End}_{C_o}(V) \cong \mathbb{M}_r(F)$, since the only C_0-linear maps from V_i to V_j are scalars. Each V_i is an irreducible C-module, and these come in two non-isomorphic versions: V_+ and V_-, depending on the action of $\pi(z)$. Suppose there are p_ε copies of V_ε, so that $p_+ + p_- = r$. We may replace z by $-z$ if necessary (adjusting via the automorphism α of C) to assume $p_+ \geq p_-$. Then $k = \chi(m) = \mathrm{trace}(\pi(z)) = (p_+ - p_-) \cdot \frac{n}{r}$. In the representation $A \cong \mathbb{M}_r(F)$ the element $y = \pi(z) \in A$ has matrix $\begin{pmatrix} 1_{p_+} & 0 \\ 0 & -1_{p_-} \end{pmatrix}$. If $a \in A$ commutes with y then $a = \begin{pmatrix} a_+ & 0 \\ 0 & a_- \end{pmatrix}$ where $a_\varepsilon \in \mathbb{M}_{p_\varepsilon}(F)$. As before $(A, \sim) \cong (\mathrm{End}(W), I_b)$ for some ε-symmetric space (W, b). Since $\tilde{y} = y$ the eigenspaces of y are orthogonal, and b induces regular forms on them. If $\varepsilon = 1$ then $\mathrm{Aut}(m) \cong \mathrm{O}(p_+) \times \mathrm{O}(p_-)$ while if $\varepsilon = -1$ then $\mathrm{Aut}(m) \cong \mathrm{Sp}(p_+) \times \mathrm{Sp}(p_-)$. Therefore $\dim \mathrm{Aut}(m) = \frac{1}{2} \cdot (p_+(p_+ - \varepsilon) + p_-(p_- - \varepsilon)) = \frac{1}{4} \cdot ((p_+ + p_- - \varepsilon)^2 + (p_+ - p_-)^2 - 1)$ and a calculation completes the proof. □

Let us review some of the properties of group actions. If G is a group acting on a set W and $x \in W$ we write $G \cdot x = \{gx : g \in G\}$ for the orbit of x and $G_x = \{g \in G : gx = x\}$ for the stabilizer (isotropy subgroup) of x. The map $G \to G \cdot x$ induces a bijection between the left cosets of G_x and the orbit $G \cdot x$:

$$G/G_x \leftrightarrow G \cdot x.$$

At this point we assume that the reader knows some of the basic theory of algebraic groups as presented, for example, in Humphreys (1975). Suppose that G is an algebraic group, W is a (nonempty) algebraic variety over $\mathbb{C}$ and G acts morphically on W (i.e. the map $G \times W \to W$ is a morphism of varieties). In general an orbit $G \cdot x$ might be embedded in W in some complicated way, but it can still be viewed as a variety.

8.8 Lemma. *Suppose H is a closed subgroup of an algebraic group G. Then G/H is a nonsingular variety with $\dim(G/H) = \dim(G) - \dim(H)$, and with all irreducible components of this dimension. If G acts morphically on a variety W, then G_x is a closed subgroup of G, the orbit $G \cdot x$ is a nonsingular, locally closed subset of W, and the boundary of $G \cdot x$ is a union of orbits of strictly lower dimension. Furthermore,*

$$\dim G \cdot x = \dim G - \dim G_x.$$

Proof. See Humphreys (1975), §8, §4.3, and §12. □

A set Y is "locally closed" if it is the intersection of an open set and a closed set, in the Zariski topology. Equivalently, Y is an open subset of its closure $\bar{Y}$. The boundary of Y is the closed set $\bar{Y} - Y$. As one consequence, the closure $\overline{G \cdot x}$ is a subvariety of W with the same dimension as the orbit $G \cdot x$.

These ideas from algebraic geometry require the base field to be algebraically closed. In some cases we can extract geometric information about the *real* part of a complex variety.

8.9 Lemma. *Suppose W is a nonsingular algebraic variety over $\mathbb{C}$, which is defined over $\mathbb{R}$. If the set of real points $W(\mathbb{R})$ is nonempty then it is a smooth real manifold and* $\dim W(\mathbb{R})$ *as a manifold coincides with* $\dim W$ *as a variety.*

Proof outline. These statements about $W(\mathbb{R})$ are well-known to the experts, but I found no convenient reference. The ideal of W is $\mathcal{I}(W) = \{f \in \mathbb{C}[X] : f(\zeta) = 0 \text{ for every } \zeta \text{ in } W\}$. Here $X = (x_1, \dots, x_n)$ is the set of indeterminates. Let $f_1, \dots, f_t$ be a set of generators for $\mathcal{I}(W)$ and consider the $t \times n$ Jacobian matrix $J = \left(\frac{\partial f_i}{\partial x_j}\right)$. Recall the classical Jacobian criterion for nonsingularity: W is nonsingular if and only if for every $\zeta \in W$, $\operatorname{rank}(J(\zeta)) = n - \dim W$. (See e.g. Hartshorne (1977), p. 31.)

Since W is defined over $\mathbb{R}$ we can arrange $f_i \in \mathbb{R}[X]$, (see Exercise 2). Now view f_i as a real valued C^∞-function on $\mathbb{R}^n$ and $W(\mathbb{R})$ as a "level surface" of $\{f_1, \dots, f_t\}$. By the Implicit Function Theorem the constant rank of the Jacobian matrix J at points $\zeta \in W(\mathbb{R})$ implies that $W(\mathbb{R})$ is a smooth real manifold whose dimension equals $\dim W$. □

8.10 Proposition. *Suppose $1 < s \le \rho(n)$. Then $\mathrm{Comp}^1_{\mathbb{C}}(s, n)$ is a nonempty, nonsingular algebraic variety. Each nonempty $\mathrm{Comp}^1_{\mathbb{C}}(s, n; k)$ is a variety with two irreducible components both of dimension equal to*

$$\frac{1}{2}n(n-1) - \dim \mathrm{Aut}(s, n; k).$$

Moreover each nonempty $\mathrm{Comp}^1_{\mathbb{R}}(s, n; k)$ is a smooth compact, real manifold with two connected components. The dimension of each component equals the value displayed above. Similar statements hold for $\mathrm{Comp}_{\mathbb{C}}(s, n; k)$ and $\mathrm{Comp}_{\mathbb{R}}(s, n; k)$.

Proof. The set is nonempty by the basic Hurwitz–Radon Theorem, and it is certainly an affine algebraic set, hence a closed subvariety of $\mathrm{Bil}(s, n)$. Most of the remaining statements follow using (8.3), (8.5), (8.8) and (8.9). Since $\mathrm{O}(n)$ has two components given by the cosets of $\mathrm{O}^+(n)$, the statement that there are two components is equivalent to:

$$\text{If } m \in \mathrm{Comp}^1(s, n; k) \text{ then } \mathrm{Aut}(m) \text{is contained in } \mathrm{O}^+(n).$$

Since $\mathrm{Aut}(m) = \mathrm{O}(n) \cap \mathrm{End}_C(V)$, every $f \in \mathrm{Aut}(m)$ centralizes the algebra C and hence commutes with every element of the subspace $S_m \subseteq \mathrm{Sim}(V, n\langle 1 \rangle)$. This implies $f \in \mathrm{O}^+(n)$, by the result of Wonenburger (1962b) mentioned in Exercise 1.17(4). The compactness follows since $\mathrm{O}(n, \mathbb{R})$ is a compact group acting transitively on the set of real points, as in (8.5). □

Here is a list of these dimensions in a few small cases. If $s \equiv 0 \pmod 4$ the maximality of s forces the representation to be non-faithful so that $\chi(m) = \pm n$. In those cases $\mathrm{Comp}^1(s,n) = \mathrm{Comp}^1(s,n;n) \cup \mathrm{Comp}^1(s,n;-n)$.

(s,n)	$\dim \mathrm{Aut}(s,n)$	$\dim \mathrm{Comp}^1(s,n)$	$\dim \mathrm{Bil}^1(s,n)$
$(2,2)$	1	0	4
$(4,4)$	3	3	48
$(8,8)$	0	28	448
$(9,16)$	0	120	2048

The values in the first two columns follow from (8.7) and (8.10).

The dimension of $\mathrm{Comp}(s,n)$ can be determined using (8.3) (see Exercise 4). For example

$$\dim \mathrm{Comp}(4,4) = 9 \quad \text{and} \quad \dim \mathrm{Comp}(8,8) = 56.$$

The proposition also determines the number of connected components. For example $\mathrm{Comp}_{\mathbb{R}}(4,4) = \mathrm{O}(4) \times \mathrm{Comp}^1_{\mathbb{R}}(4,4)$ and $\mathrm{Comp}^1_{\mathbb{R}}(4,4) = \mathrm{Comp}^1_{\mathbb{R}}(4,4;4) \cup \mathrm{Comp}^1_{\mathbb{R}}(4,4;-4)$. Since $\mathrm{O}(4)$ and $\mathrm{Comp}^1(4,4;\pm 4)$ each have two components, $\mathrm{Comp}_{\mathbb{R}}(4,4)$ has eight connected components, each of dimension 9. Similarly $\mathrm{Comp}_{\mathbb{R}}(8,8)$ has eight components each of dimension 56.

Let us now consider the set of all subspaces of similarities, as a subset of the Grassmann variety of all s-planes in n-space. Recall that the character $\chi(S)$ was defined in (7.17) and if $S' = \gamma \cdot S \cdot \beta^{-1}$ then $\chi(S') = \chi(S)$. Some information is lost in passing from $\chi(m)$ to $\chi(S)$. In fact, if $S = S_m$, then $\chi(S) = |\chi(m)|$.

8.11 Definition. $\mathrm{Sub}(s,n)$ is the set of all linear subspaces $S \subseteq \mathrm{Sim}(n\langle 1\rangle)$ such that $\dim S = s$ and the induced quadratic form on S is regular.

$$\mathrm{Sub}(s,n;k) = \{S \in \mathrm{Sub}(s,n) : \chi(S) = k\},$$
$$\mathrm{Sub}^1(s,n) = \{S \in \mathrm{Sub}(s,n) : 1_V \in S\}$$

and $\mathrm{Sub}^1(s,n;k)$ is defined similarly.

As usual, $\mathrm{Sub}(s,n) = \mathrm{Sub}(s,n;0)$ when $s \not\equiv 0 \pmod 4$. If $F = \mathbb{R}$ the regularity condition on the induced quadratic form is automatic. We may view $\mathrm{Sub}(s,n;k)$ and $\mathrm{Sub}^1(s,n;k)$ as nonsingular algebraic varieties, since they are orbits of algebraic group actions. Note that sending m to $S_m = \mathrm{image}(\hat{m})$ provides a surjection

$$\varphi : \mathrm{Comp}(s,n;k) \to \mathrm{Sub}(s,n;|k|).$$

The action of $\mathrm{O}(n) \times \mathrm{O}(n)$ on $\mathrm{Comp}(s,n;k)$ descends to the action $(\beta,\gamma) \bullet S = \gamma S \beta^{-1}$ on $\mathrm{Sub}(s,n;|k|)$.

8.12 Lemma. *Suppose* $k \geq 0$ *and* $\text{Comp}(s, n, k)$ *is nonempty.*

$$\dim \text{Sub}(s, n; k) = \dim \text{Comp}(s, n; k) - \frac{s(s-1)}{2},$$
$$\dim \text{Sub}^1(s, n; k) = \dim \text{Comp}^1(s, n; k) - \frac{(s-1)(s-2)}{2}.$$

Proof. Given $S \in \text{Sub}(s, n; k)$, the fiber $\varphi^{-1}(S) = \{m \in \text{Comp}(s, n; k) : S_m = S\} \cong \{\hat{m} : \mathbb{R}^s \to S \text{ an isometry}\} \cong \text{O}(s)$, and the first dimension formula follows. For the second formula, restrict φ to $\varphi_1 : \text{Comp}^1 \to \text{Sub}^1$ and compute the fiber $\varphi_1^{-1}(S) \cong \{\hat{m} : \mathbb{R}^s \to S \text{ an isometry with } \hat{m}(e_1) = 1_V\} \cong \text{O}(s-1)$. □

For example, $\dim \text{Sub}^1(4, 4) = 0$. In fact we have already seen (in Exercise 1.4) that $\text{Sub}^1(4, 4)$ contains exactly two elements.

8.13 Proposition. $\text{Sub}^1_{\mathbb{R}}(s, n; k)$ *and* $\text{Sub}_{\mathbb{R}}(s, n; k)$ *are smooth real manifolds. If* $\text{Sub}^1_{\mathbb{R}}(s, n; k)$ *is nonempty, then it has two connected components and* $\text{Sub}_{\mathbb{R}}(s, n; k)$ *has four connected components.*

Proof. The fact that these spaces are manifolds follows from the general theory as before. Since the components of $\text{O}(n)$ are the cosets of $\text{O}^+(n)$, the $\text{O}(n) \times \text{O}(n)$ orbit $\text{Sub}_{\mathbb{R}}(s, n; k)$ breaks into four $\text{O}^+(n) \times \text{O}^+(n)$ orbits, each of which is connected. Given $S \in \text{Sub}^1_{\mathbb{R}}(s, n; k)$, these four orbits are represented by:

$$\begin{array}{cc} S & \beta S \beta^{-1} \\ \beta S & S\beta \end{array}$$

where $\beta \in \text{O}^-(n)$, i.e. $\det(\beta) = -1$. We must show that these four orbits are disjoint. For if that is done certainly $\text{Sub}_{\mathbb{R}}(s, n; k)$ has those four components. Moreover the two orbits of $\text{O}^+(n)$ acting (by conjugation) on $\text{Sub}^1_{\mathbb{R}}(s, n; k)$ are contained in the larger orbits represented by the first row above, and hence are also disjoint.

Recall from Exercise 1.17 that if $f \in S$ or if $f \in \beta S \beta^{-1}$ then f is proper, and hence if $f \in \beta S$ or $f \in S\beta$ then f is not proper. Therefore the orbits in the top row above are disjoint from the orbits in the bottom row. To complete the argument we invoke the following lemma, whose proof is surprisingly tricky. □

8.14 Lemma. *Suppose* $1_V \in S \subseteq \text{Sim}(V, q)$ *and* $s = \dim S > 2$. *If* $\beta, \gamma \in \text{Sim}^\bullet(V, q)$ *and* $\gamma S \beta^{-1} = S$ *then* β *and* γ *are proper.*

See Exercise 12 for an outline of the proof.

Finally we turn to a case of particular interest: real division algebras. Recall that a *real division algebra* is defined to be a finite dimensional $\mathbb{R}$-vector space D together with an $\mathbb{R}$-bilinear mapping $m : D \times D \to D$ such that: $m(x, y) = 0$ only when

$x = 0$ or $y = 0$. No associativity or commutativity is assumed; an identity element is not assumed to exist. Each of the classical composition algebras $\mathbb{R}$, $\mathbb{C}$, $\mathbb{H}$, $\mathbb{O}$ is a real division algebra satisfying many algebraic properties. There are several classification results, each assuming that the division algebra satisfies some algebraic property and then listing all the possibilities up to isomorphism. Here are some classical examples when A is a real division algebra with 1:

- If A is associative then $A \cong \mathbb{R}$, $\mathbb{C}$ or $\mathbb{H}$ (Frobenius 1877).
- If A is a composition algebra then $A \cong \mathbb{R}$, $\mathbb{C}$, $\mathbb{H}$ or $\mathbb{O}$ (Hurwitz 1898).
- If A is alternative then $A \cong \mathbb{R}$, $\mathbb{C}$, $\mathbb{H}$ or $\mathbb{O}$ (Zorn 1933).
- If A is commutative then $A \cong \mathbb{R}$ or $\mathbb{C}$ (Hopf 1940).

Actually in 1898 Hurwitz proved that $\dim A = 1, 2, 4, 8$ and only stated the uniqueness of the solutions. This uniqueness was worked out by his student E. Robert (1912). The classification results mentioned above are described further in Koecher and Remmert (1991), §8.2, §8.3, §9.3. The Hopf theorem was proved using topology, as outlined in Exercise 12.12.

More recent work in this direction has been done with quadratic division algebras, with flexible ones (satisfying the flexible law: $xy \cdot x = x \cdot yx$), with algebras having a large derivation algebra, and with various other types. Flexible real division algebras were classified by Benkart, Britten and Osborn (1982). A survey of such results appears in Benkart and Osborn (1981).

Can general real division algebras be classified is some reasonable way? Even determining the possible dimensions for such algebras is a deep question. In 1940 Stiefel and Hopf used algebraic topology to prove that if D is an n-dimensional real division algebra then $n = 2^m$ for some m. (See (12.4) below.) Finally in 1958 Bott's Periodicity Theorem was used to prove that n must be $1, 2, 4$ or 8. This theorem later became an corollary of topological K-theory. (See Exercise 0.8 and (12.20).)

Let $\mathrm{Div}(n)$ be the set of n-dimensional real division algebras. Then $\mathrm{Div}(n)$ is nonempty only when $n = 1, 2, 4$ or 8. It is fairly easy to describe $\mathrm{Div}(1)$ and $\mathrm{Div}(2)$ explicitly. The challenge is to describe the sets $\mathrm{Div}(4)$ and $\mathrm{Div}(8)$, and possibly to find some general algebraic classifications. Useful results in this direction remain elusive. Let us consider four algebraic methods for constructing examples of division algebras.

(1) *Isotopes*. Two F-algebras D, D' are isotopic if there exist bijective linear maps $f, g, h : D \to D'$ such that

$$f(xy) = g(x) \cdot h(y) \quad \text{for every } x, y \in D.$$

If D is a division algebra then any isotope of D is also a division algebra. Then isotopy is an equivalence relation on $\mathrm{Div}(n)$. This concept was introduced in Steenrod's work on homotopy groups and was formalized by Albert (1942b). Every division algebra is isotopic to one with an identity element (see Exercise 0.8). Then $\mathrm{Div}(1)$ and $\mathrm{Div}(2)$ each have only one isotopy class, but $\mathrm{Div}(4)$ and $\mathrm{Div}(8)$ are much more complicated. The concept of isotopy (or isotopism) arises naturally in several contexts. For example, two division rings are isotopic if and only if they coordinatize isomorphic projective planes. See Hughes and Piper (1973), p. 177.

(2) *Mutations.* A mutation of an F-algebra D with parameters $r, s \in F$ is given by altering the multiplication of D to $m_{r,s} : D \times D \to D$ defined by

$$m_{r,s}(x, y) = rxy + syx.$$

If D is a composition division algebra over $\mathbb{R}$ then this mutation is also a division algebra with identity, provided $r \neq \pm s$. The "bar" map is still an involution for the mutation and if $r + s = 1$ the elements of the mutation have the same inverses as they do in D. (Compare Lex (1973).)

(3) *Bilinear perturbations.* Suppose D is a composition algebra over $\mathbb{R}$ and $\beta : D \times D \to \mathbb{R}$ is a bilinear form. Define $m_\beta : D \times D \to D$ by $m_\beta(x, y) = xy + \beta(x, y) \cdot 1$. This furnishes a division algebra if and only if the quadratic form $Q(x) = x \cdot \bar{x} + \beta(x, \bar{x})$ is anisotropic. For example let $\ell : D \to \mathbb{R}$ be the trace map $\ell(x) = \frac{1}{2} \cdot (x + \bar{x})$. If D is a division algebra, then $\beta(x, y) = \ell(xy)$ or $\ell(x)\ell(y)$ yield division algebras. We also get division algebras from $\beta(x, y) = t_1 \cdot \ell(xy) + t_2 \cdot \ell(x\bar{y}) + t_3 \cdot \ell(x)\ell(y)$ for certain values of the real parameters t_1, t_2, t_3. The examples in Hähl (1975) are of this type.

(4) *Twisted quaternions.* Choosing $b \in \mathbb{C}$, define an algebra $\mathbb{H}_b = \mathbb{C} \oplus \mathbb{C}j$, with multiplication given as follows. For $r, s, u, v \in \mathbb{C}$ define

$$(r + sj) \cdot (u + vj) = (ru + bs\bar{v}) + (rv + s\bar{u})j.$$

Then $\mathbb{H}_b$ is a 4-dimensional $\mathbb{R}$-vector space with basis $\{1, i, j, ij\}$, $1 \in \mathbb{C}$ is the identity element, $jx = \bar{x}j$ for every $x \in \mathbb{C}$ and $j^2 = b$. If $b < 0$ then $\mathbb{H}_b \cong \mathbb{H}$, the associative quaternion algebra. If $b \notin \mathbb{R}$ then $\mathbb{H}_b$ is a division algebra (use the formula to analyze zero-divisors) and $\mathbb{H}_b$ is not associative: in fact, $j \cdot j^2 \neq j^2 \cdot j$. Even though every non-zero element of $\mathbb{H}_b$ has a left inverse and a right inverse, those inverses can differ. For example, $(b^{-1}j) \cdot j = 1$ but $j \cdot (b^{-1}j) \neq 1$. The twisted quaternion algebras discussed by Bruck (1944) are of this type. Such algebras are studied more generally by Waterhouse (1987).

If the entries of the multiplication table of a real division algebra are altered by small amounts then the result yields another division algebra. That is, the collection $\mathrm{Div}(n)$ of n-dimensional real division algebras is an open set. Generally, let $\mathrm{Bil}(r, s, n)$ be the set of all bilinear maps $f : \mathbb{R}^r \times \mathbb{R}^s \to \mathbb{R}^n$. It is a vector space of dimension rsn. Such a map f is defined to be *nonsingular* if $f(x, y) \neq 0$ whenever $x \neq 0$ in $\mathbb{R}^r$ and $y \neq 0$ in $\mathbb{R}^s$. Let $\mathrm{Nsing}(r, s, n)$ be the set of all nonsingular elements in $\mathrm{Bil}(r, s, n)$. Then $\mathrm{Div}(n) = \mathrm{Nsing}(n, n, n)$.

8.15 Lemma. $\mathrm{Nsing}(r, s, n) \subseteq \mathrm{Bil}(r, s, n)$ *is an open set.*

Proof. If $f \in \mathrm{Bil}(r, s, n)$ then $f(S^{r-1}, S^{s-1}) \subseteq \mathbb{R}^n$ is a compact subset, since the spheres S^k are compact. Define $\omega(f)$ to be the distance between 0 and this compact subset. The map $\omega : \mathrm{Bil}(r, s, n) \to [0, \infty)$ is continuous and $\mathrm{Nsing}(r, s, n)$ is the complement of $\omega^{-1}(0)$. □

It is usually difficult to determine whether $\mathrm{Nsing}(r, s, n)$ is nonempty. (See Chapter 12.) But if it is nonempty, then $\mathrm{Nsing}(r, s, n)$ is an open set of dimension rsn. For the classical cases of $\mathrm{Div}(n)$ we obtain:

$$\dim \mathrm{Comp}(4,4) = 9 \qquad \dim \mathrm{Div}(4) = 64$$
$$\dim \mathrm{Comp}(8,8) = 56 \qquad \dim \mathrm{Div}(8) = 512.$$

Therefore the algebraic constructions of division algebras (e.g. by isotopy or mutation) cannot produce all the possible division algebras of dimension 4 or 8. For example the set of algebra multiplications which are isotopic to a fixed octonion algebra forms one orbit of an action of $\mathrm{GL}(8) \times \mathrm{GL}(8) \times \mathrm{GL}(8)$. This orbit has dimension at most $3 \cdot 8^2 = 192$ inside $\mathrm{Div}(8)$. Compare Exercise 18.

Let us now consider real division algebras with a (2-sided) identity element. To facilitate the discussion we simplify and extend some of the notations. As before let $e = e_1 = (1, 0, \ldots, 0)$ be the first element of the standard basis of $\mathbb{R}^n$. Define:

$$\mathrm{Bil}(n) = \{m : \mathbb{R}^n \times \mathbb{R}^n \to \mathbb{R}^n \text{ such that } m \text{ is bilinear}\};$$
$$\mathrm{Bil}^1(n) = \{m \in \mathrm{Bil}(n) : e \text{ is a left identity element for } m\};$$
$$\mathrm{Bil}^{11}(n) = \{m \in \mathrm{Bil}(n) : e \text{ is a 2-sided identity element for } m\}.$$

Then $m \in \mathrm{Bil}(n)$ is a multiplication on $\mathbb{R}^n$ (setting $x * y = m(x, y)$). It is a *division algebra* if: $m(x, y) = 0$ implies $x = 0$ or $y = 0$. It is a *composition algebra* if it satisfies the norm property: $|m(x, y)| = |x| \cdot |y|$ for every $x, y \in \mathbb{R}^n$. Let us use similar notations for the sets of division algebras and composition algebras:

$$\mathrm{Div}(n) \qquad \mathrm{Div}^1(n) \qquad \mathrm{Div}^{11}(n)$$
$$\mathrm{Comp}(n) \qquad \mathrm{Comp}^1(n) \qquad \mathrm{Comp}^{11}(n).$$

Of course these are nonempty only when $n = 1, 2, 4$ or 8. Note that $\mathrm{Bil}(n)$ is a vector space of dimension n^3; $\mathrm{Bil}^1(n)$ is a coset of a linear subspace with dimension $(n-1)n^2$; and $\mathrm{Bil}^{11}(n)$ is a coset of a linear subspace of dimension $(n-1)^2 n$. We know that $\mathrm{Div}(n)$ is an open subset of $\mathrm{Bil}(n)$. Similarly, $\mathrm{Div}^1(n) \subseteq \mathrm{Bil}^1(n)$ and $\mathrm{Div}^{11}(n) \subseteq \mathrm{Bil}^{11}(n)$ are open subsets.

What is the dimension of $\mathrm{Comp}^{11}(n)$ and how many connected components does it have? We present the answer to this question twice, using different methods. The first uses the direct group action ideas mentioned above. The second approach employs the Triality Theorem.

8.16 Propositon. $\mathrm{Comp}^{11}(4)$ *is a set of two points.*

$\mathrm{Comp}^{11}(8)$ *is a nonsingular algebraic variety with two components each isomorphic to 7-dimensional projective space.*

Proof. Let $\mathbb{R}$ be the base field (although more general fields F will work here as well). Suppose $m \in \mathrm{Comp}^{11}(4)$. Then $xy = m(x, y)$ makes $\mathbb{R}^4$ into a composition algebra with identity element $e = e_1$. The mapping m is determined by the values $e_i e_j$ where $\{e_1, \dots, e_4\}$ is the given orthonormal basis. We know that $e_2^2 = e_3^2 = e_4^2 = -1$ and $e_2 e_3 = \pm e_4$. The other values $e_i e_j$ are determined by that choice of sign since m is associative. Then either m is the standard quaternion multiplication, $m(x, y) = xy$, or else m comes from the opposite algebra: $m(x, y) = yx$. These are the two points in $\mathrm{Comp}^{11}(4)$. (Exercise 1.4 is relevant here.)

If $m \in \mathrm{Comp}^{11}(8) \subseteq \mathrm{Comp}^1(8, 8)$ we defined the character $\chi(m)$ as $\mathrm{trace}(\pi(z))$, using the associated representation $\pi : C \to \mathrm{End}(V)$ and the central element z satisfying $z^2 = 1$. Then $\pi(z) = \pm 1$ and $\chi(m) = \pm 8$. Then $\mathrm{Comp}^{11}(8)$ is a union of two disjoint components $\mathrm{Comp}^{11}(8, +) \subseteq \mathrm{Comp}^1(8, 8; 8)$ and $\mathrm{Comp}^{11}(8, -) \subseteq \mathrm{Comp}^1(8, 8; -8)$. Any $m \in \mathrm{Comp}^{11}(8)$ has an associated operation m' defined: $m'(x, y) = m(y, x)$. Since $\chi(m') = -\chi(m)$, those two spaces are isomorphic.

Recall that $\mathrm{O}(8)$ acts transitively on the space $\mathrm{Comp}^1(8, 8; 8)$ as in (8.5). Let $m_0(x, y) = x \cdot y = xy$ be the standard octonion multiplication. If $m(x, y) = x * y$ lies in $\mathrm{Comp}^1(8, 8; 8)$ then m arises from m_0 by the action of some $\beta \in \mathrm{O}(8)$. Working through the defnitions, we find:

$$\beta(x * y) = x \cdot \beta(y) \quad \text{for every } x, y.$$

Certainly this operation $*$ admits e as a left-identity element. If m lies in $\mathrm{Comp}^{11}(8, +)$ then e is also a right-identity: $x * e = x$. This occurs if and only if $\beta(x) = x \cdot \beta(e)$ for every x. Thus $\beta = \mathcal{R}_b$ is a right multiplication map on the octonions, for some $b = \beta(e)$ with $|b| = 1$. This provides a surjective map from the sphere S^7 of unit octonions to the space $\mathrm{Comp}^{11}(8, +)$, sending b to the operation $*$ determined by:

$$(x * y) \cdot b = x \cdot (y \cdot b).$$

To examine the fibers of this map, suppose $b, c \in S^7$ both go to the same operation. Then

$$(x \cdot yb)b^{-1} = (x \cdot yc)c^{-1} \quad \text{for every } x, y.$$

Setting $x = b$ and using the Moufang identity (as in (1.A.10)) we find: $by = (byb)b^{-1} = (b \cdot yc)c^{-1}$ so that $by \cdot c = b \cdot yc$. Exercise 1.27 implies that $1, b, c$ must be linearly dependent. Interchanging b and c if necessary we may write $c = r + sb$ for some $r, s \in \mathbb{R}$. The alternative law then implies that $(w \cdot b^{-1}) \cdot c = w \cdot (b^{-1} \cdot c)$. In particular $x \cdot yc = (x \cdot yb)(b^{-1}c)$ and plugging in $c = r + sb$ yields: $rxy = r(x \cdot yb)b^{-1}$. Suppose c is not a scalar multiple of b, so that b is not scalar and $r \neq 0$. Then $xy \cdot b = x \cdot yb$ for every x, y and Exercise 1.27 implies b is a scalar, a contradiction. Hence $c = \pm b$.

Consequently $\mathrm{Comp}^{11}(8, +)$ is exactly the sphere S^7 with antipodal points identified, so it is 7-dimensional projective space. □

We can also analyze the space Comp(8) by using the action of the full group $O(8)\times O(8)\times O(8)$. This approach yields another proof of (8.16) but more importantly it leads to a consideration of the interesting phenomenon of "triality".

The group $O(8) \times O(8) \times O(8)$ acts transitively on Comp(8). This fact follows from the Clifford algebra theory (see (8.5)), but more direct proofs can be given for this case. What is the stabilizer of the standard octonion algebra D ? From the definition of the action, this stabilizer is related to the group of autotopies defined below. The next results are valid over general fields F (where $2 \neq 0$), provided D is a division algebra.

The results here are well known but the terminology follows ideas of J. H. Conway. As in the appendix of Chapter 1, we use $[x] = \bar{x}x$ to denote the norm form in the octonion algebra and we write $O(D)$ for the orthogonal group of this norm form. For the usual case over $\mathbb{R}$ this group becomes $O(8)$.

8.17 Definition. Let D be an octonion division algebra over F. If $\alpha, \beta, \gamma \in GL(D)$ the triple (α, β, γ) is an *autotopy* of D if $\gamma(xy) = \alpha(x) \cdot \beta(y)$ for every $x, y \in D$. If (α, β, γ) is an autotopy define γ to be a *monotopy*. Let Autot(D) and Mon(D) the groups of all autotopies and monotopies of D, respectively. Define Autot$^{\text{o}}(D)$ and Mon$^{\text{o}}(D)$ to be the corresponding groups of isometries (restricting to the case $\alpha, \beta, \gamma \in O(D)$).

It is easy to see that Autot(D) is a group under componentwise composition and Mon(D) is the image of the projection π : Autot$(D) \to GL(D)$ sending (α, β, γ) to γ. Similarly Mon$^{\text{o}}(D)$ is the image of Autot$^{\text{o}}(D)$. If $\varphi \in$ Aut(D) is an algebra automorphism then $(\varphi, \varphi, \varphi)$ is an autotopy and φ is an isometry. Hence Aut$(D) \subseteq$ Mon$^{\text{o}}(D)$.

8.18 Lemma. ker(Autot$^{\text{o}}(D) \to$ Mon$^{\text{o}}(D)) \cong \{\pm 1\}$.

Proof. An element of the kernel is $(\alpha, \beta, 1)$ where $\alpha(x)\beta(y) = xy$. Then $\alpha(x) = xa$ and $\beta(y) = by$ for every x, y (where $a = \beta(1)^{-1}$ and $b = \alpha(1)^{-1}$). Then $xa \cdot by = xy$ and consequently $xa \cdot z = x \cdot az$ for every x, z. This says that a is in the nucleus $\mathcal{N}(D) = F$ (as in Exercise 1.27). □

If (α, β, γ) is an autotopy then each of α, β, γ is a monotopy. To see this suppose $z = xy$ and consider the resulting "braiding sequence": $xy = z, x = zy^{-1}, z^{-1}x = y^{-1}, z^{-1} = y^{-1}x^{-1}, yz^{-1} = x^{-1}, y = x^{-1}z$. Each of these six expressions leads to another autotopy. For example from $x = zy^{-1}$ we find $\alpha(zy^{-1}) = \alpha(x) = \gamma(z)\beta(y)^{-1}$ so that $(\gamma, \iota\beta\iota, \alpha)$ is also an autotopy. (Here ι denotes the inverse map:

$\iota(x) = x^{-1}$). The six associated autotopies are best displayed in a hexagon:

$$\begin{array}{ccc} & (\alpha, \beta, \gamma) & \\ (\iota\alpha\iota, \gamma, \beta) & & (\gamma, \iota\beta\iota, \alpha) \\ (\beta, \iota\gamma\iota, \iota\alpha\iota) & & (\iota\gamma\iota, \alpha, \iota\beta\iota) \\ & (\iota\beta\iota, \iota\alpha\iota, \iota\gamma\iota) & \end{array}$$

Therefore α, β, γ are monotopies.

Recall from (1.A.10) that D satisfies various weak forms of associativity, including:

$a \cdot ab = a^2b$ and $ba \cdot a = ba^2$	(the alternative laws)
$ax \cdot a = a \cdot xa$	(flexible law)
$a(xy)a = ax \cdot ya$	(Moufang identity)

Setting $L_a(x) = ax$, $R_a(x) = xa$ and $B_a(x) = axa$, Moufang says that (L_a, R_a, B_a) is an autotopy for every $a \in D^\bullet$. Therefore each L_a, R_a and B_a is a monotopy. It is clear that these maps are similarities, relative to the norm form. In fact, $L_a, R_a, B_a \in \mathrm{Sim}^+(D)$ by Exercise 1.17.

The bi-multiplication map B_a is closely related to the hyperplane reflection τ_a on D, relative to the norm form $[x] = \bar{x}x$. Recall that $\tau_a(x) = x - \frac{2[x,a]}{[a]} \cdot a$. Since $x\bar{a} + a\bar{x} = 2[x, a]$ we find $\tau_a(x) = -[a]^{-1} \cdot a\bar{x}a$. Then $\tau_1(x) = -\bar{x}$ and

$$B_a = [a] \cdot \tau_a\tau_1.$$

This proves again that $B_a \in F^\bullet \mathrm{O}^+(D) \subseteq \mathrm{Sim}^+(D)$.

8.19 Triality Theorem. $\mathrm{Mon}(D) = \mathrm{Sim}^+(D)$ *and* $\mathrm{Mon}^{\mathrm{o}}(D) = \mathrm{O}^+(D)$

Consequently every $\gamma \in \mathrm{O}^+(D)$ has associated maps $\alpha, \beta \in \mathrm{O}^+(D)$ making (α, β, γ) an autotopy, and these α, β are unique up to sign. This three-fold symmetry among α, β, γ is a version of the Triality Principle studied in Lie theory and elsewhere.

For the usual cases over $\mathbb{R}$ we find that $\mathrm{Mon}(8) = \mathrm{Sim}^+(8) = \mathbb{R}^\bullet \cdot \mathrm{O}^+(8)$ has dimension 29, and using (8.18): $\dim \mathrm{Autot}(8) = 30$. Similarly $\dim \mathrm{Mon}^{\mathrm{o}}(8) = \dim \mathrm{Autot}^{\mathrm{o}}(8) = 28$.

As a step toward the proof of this theorem we show that monotopies are similarities.

8.20 Lemma. $\mathrm{Mon}(D) \subseteq \mathrm{Sim}^\bullet(D)$.

Proof. If (α, β, γ) is an autotopy, $\gamma(xy) = \alpha(x) \cdot \beta(y)$. Then $\gamma(x) = \alpha(x) \cdot \beta(1)$. Since $\alpha, \gamma \in \mathrm{GL}(D)$, $\beta(1)$ must be invertible and we may set $a = \beta(1)^{-1}$ and conclude: $\alpha(x) = \gamma(x) \cdot a$. Similarly $\beta(y) = b \cdot \gamma(y)$ where $b = \alpha(1)^{-1}$ and

$$\gamma(xy) = \gamma(x)a \cdot b\gamma(y) \quad \text{for every } x, y \in D.$$

The elements a, b are called the "companions" of γ. Take norms to find $[\gamma(xy)] = r \cdot [\gamma(x)] \cdot [\gamma(y)]$, where $r = [ab]$. Then the form $q(x) = r \cdot [\gamma(x)]$ satisfies

$q(xy) = q(x)q(y)$ and D is a composition algebra relative to q. It follows (Exercise 13) that the forms $q(x)$ and $[x]$ coincide, and γ is a similarity. □

Proof of the Triality Theorem. The "bar" map $J(x) = \bar{x}$ is an anti-monotopy and an improper similarity. (Define (α, β, γ) to be an anti-autotopy if $\gamma(xy) = \alpha(y)\beta(x)$, etc.) If $g \in \mathrm{Sim}^{\bullet}(D)$ then $g = L_{g(1)} \circ h$ where $h \in \mathrm{O}(D)$. This h can be expressed as a product of hyperplane reflections τ_a (by a weak form of the Cartan–Dieudonné Theorem). As mentioned before (8.19), each τ_a is a scalar multiple of $B_a \circ J$ so it is an anti-monotopy. Therefore g is in the group generated by maps L_u, B_a and J so that g is a monotopy or an anti-monotopy. Moreover, g is a monotopy if and only if an even number of τ_a's are involved, if and only if g is a proper similarity. Conversely if $g \in \mathrm{Mon}(D)$ then $g \in \mathrm{Sim}^{\bullet}(D)$ and the same parity argument shows that g is proper. □

We can use this theorem to analyze the spaces of composition algebras over $\mathbb{R}$ or $\mathbb{C}$. These numbers, summarized in the next corollary, agree with the earlier computations.

8.21 Corollary. *The table below lists the number of components and the dimensions of the spaces under discussion.*

	# of components	dimension
Comp(8)	8	56
$\mathrm{Comp}^1(8)$	4	28
$\mathrm{Comp}^{11}(8)$	2	7

Proof. The group $\mathrm{O}(8)^3$ has 8 components and acts transitively on Comp(8). Since $\mathrm{Comp}(8) \cong \mathrm{O}(D)^3/\,\mathrm{Autot}^{\mathrm{o}}(D)$ we find $\dim \mathrm{Comp}(8) = 3 \cdot 28 - 28 = 56$. Since $\mathrm{Autot}^{\mathrm{o}}(D) \subseteq \mathrm{O}^+(D)^3$, which is one component of $\mathrm{O}(D)^3$, there are still 8 components in Comp(8).

Using (8.5) we know that O(8) acting on $\mathrm{Comp}^1(8)$ has two orbits and $\mathrm{Stab}(D) = \{\beta \in \mathrm{O}(8) : \beta(xy) = x\beta(y)$ for every $x, y \in D\}$. If $\beta \in \mathrm{Stab}(D)$ then $\beta = R_b$ where $b = \beta(1)$ and $xy \cdot b = x \cdot yb$ (compare the proof of (8.16)). Then b is scalar (as in Exercise 1.27) and $\mathrm{Stab}(D) = \{\pm 1\}$, so that $\mathrm{Comp}^1(8) \cong \mathrm{O}(8)/\{\pm 1\}$ has 4 components and dimension 28.

Finally if $*$ is in $\mathrm{Comp}^1(8)$ define a new multiplication $\heartsuit$ by: $x \heartsuit y = R_e^{-1}(x) * y$. That is, $\heartsuit$ is defined by the formula: $(x * e) \heartsuit y = x * y$. Then e is a 2-sided identity element for $\heartsuit$ (see Exercise 0.8). The projection map $\pi : \mathrm{Comp}^1(8) \to \mathrm{Comp}^{11}(8)$, defined by $\pi(*) = \heartsuit$, acts as the identity map on $\mathrm{Comp}^{11}(8)$. O(8) acts on Comp(8) by: $(\alpha \bullet m)(x, y) = m(\alpha(x), y)$, and the subgroup $\mathrm{O}(7) = \{\alpha \in \mathrm{O}(8) : \alpha(e) = e\}$ acts on $\mathrm{Comp}^1(8)$. The point is that every O(7)-orbit in $\mathrm{Comp}^1(8)$ contains exactly one element in $\mathrm{Comp}^{11}(8)$. The uniqueness is easy and the existence follows since $\pi(m) = R_e^{-1} \bullet m$. Then $\mathrm{Comp}^{11}(8)$ becomes the orbit space $\mathrm{Comp}^1(8)/\,\mathrm{O}(7)$. Therefore $\mathrm{Comp}^{11}(8)$ has half as many components as $\mathrm{Comp}^1(8)$ and $\dim \mathrm{Comp}^{11}(8) = \dim \mathrm{Comp}^1(8) - \dim \mathrm{O}(7) = 28 - 21 = 7$. $\square$

For $n = 4$ or 8, the eight components of $\mathrm{Comp}(n, n)$ are represented by the eight standard multiplications:

$$\begin{array}{cc} xy & yx \\ \bar{x}y & y\bar{x} \\ x\bar{y} & \bar{y}x \\ \bar{x}\bar{y} & \bar{y}\bar{x} \end{array}$$

Since $e * y = y$ for every multiplication in $\mathrm{Comp}^1(n)$, all the $\bar{y}$ terms are eliminated and the four components of $\mathrm{Comp}^1(n)$ are represented by the top four multiplications in the list. Similarly the two components of $\mathrm{Comp}^{11}(n)$ are represented by the first two cases: xy and yx.

We compared the dimensions of $\mathrm{Comp}(n)$ and $\mathrm{Div}(n)$ in the table after (8.10). For algebras with identity we see that $\mathrm{Comp}^{11}(8)$ is a compact 7-dimensional space inside $\mathrm{Div}^{11}(8)$ which is an open subset of the flat space $\mathrm{Bil}^{11}(8)$ of dimension 392. Actually the set of composition algebras inside $\mathrm{Div}^{11}(8)$ is a little larger, because $\mathrm{Comp}^{11}(8)$ uses a fixed norm form on $\mathbb{R}^8$. See Exercise 20.

Exercises for Chapter 8

1. **Defining $\chi(m)$.** Suppose $m \in \mathrm{Comp}(\sigma, q)$. Then $\hat{m} : S \to \mathrm{End}(V)$ and the space (S, σ) has a given orthogonal basis $\{e_1, \dots, e_s\}$. In the case $s \equiv 0 \pmod 4$ and $d\sigma = \langle 1 \rangle$ we also assume that $\sigma(e_1) \dots \sigma(e_s) = 1$. We defined $\chi(m)$ to equal $\chi(m_0)$ where $m_0 = \varphi^{-1}(m)$ as in (8.3).

(1) Setting $f_i = \hat{m}(e_i) \in S_m$ we have $\chi(m) = \mathrm{trace}(\tilde{f}_1 f_2 \tilde{f}_3 f_4 \dots)$.

(2) If $s \not\equiv 0 \pmod 4$ or if $d\sigma \neq \langle 1 \rangle$ then $\chi(m) = 0$. In any case, $\chi(m)$ is an even integer between $-n$ and n. (See (7.18).)

(3) If $(\alpha, \beta, \gamma) \in \mathrm{O}(\sigma) \times \mathrm{O}(q) \times \mathrm{O}(q)$ then $\chi((\alpha, \beta, \gamma) \bullet m) = (\det \alpha) \cdot \chi(m)$.

(Hint. (1) $\mathrm{image}(\hat{m}_0)$ has orthogonal basis $1_V, \tilde{f}_1 f_2, \dots, \tilde{f}_1 f_s$ so the element " z" equals $(\tilde{f}_1 f_2)(\tilde{f}_1 f_3) \dots (\tilde{f}_1 f_s)$. Then $z = \tilde{f}_1 f_2 \tilde{f}_3 f_4 \dots$, at least up to some scale factor. If $s \equiv 0 \pmod 4$ and $d\sigma = \langle 1 \rangle$ then $\tilde{z} = z$ and $z^2 = \mu(z) = 1$, and the scale factor needed was 1. Compare Exercise 2.8.

(3) See Exercise 2.8 (3), (4).)

2. **Generation of radical ideals.** In (8.9) the variety W is defined over $\mathbb{R}$. This means that $W = \mathcal{Z}(g_1, \dots, g_k)$, the zero set for a list of polynomials $g_j \in \mathbb{R}[X]$. The Jacobian criterion might not work directly for these g_i. (Provide an example where it fails.) In the proof of (8.9) we need to know that $\mathcal{I}(W)$ is generated by elements of $\mathbb{R}[X]$. If $\mathcal{A} = (g_1, \dots, g_k)\mathbb{C}[X]$ the Nullstellensatz says that $\mathcal{I}(W) = \sqrt{\mathcal{A}}$, the radical of the ideal $\mathcal{A}$. Our claim follows from a more general result:

Lemma. *Suppose K/F is a separable algebraic extension of fields. If $\mathcal{B} \subseteq F[X]$ is an ideal, then $\sqrt{\mathcal{B} \otimes K} = \sqrt{\mathcal{B}} \otimes K$.*

(Hint. It is enough to show that the ring $K[X]/(\sqrt{\mathcal{B}} \otimes K)$ is reduced (i.e. has no non-zero nilpotent elements). Since $\sqrt{\mathcal{B}}$ is an intersection of primes, $F[X]/\sqrt{\mathcal{B}}$ embeds into some direct product of fields. It suffices to show: if L/F is a field extension then $L \otimes K$ is reduced. We may assume here that K/F is finite and separable.)

3. Suppose W is an algebraic variety and G is an algebraic group which acts morphically on W, all defined over $\mathbb{C}$. If $G(\mathbb{C})$ acts transitively on $W(\mathbb{C})$ then W is a nonsingular variety and all the irreducible components of W have the same dimension. Moreover if G, W and the G-action are defined over $\mathbb{R}$ then $W(\mathbb{R})$ is a smooth real manifold. Does it follow that $G(\mathbb{R})$ acts transitively on $W(\mathbb{R})$?

(Hint. Let $G = W = \mathbb{C}^\bullet$ (an affine variety embedded in $\mathbb{C}^2$), with action $g \bullet w = g^2 w$.)

4. **Connected components.** $\text{Comp}^1_{\mathbb{R}}(s, n)$ has $2c$ components and $\text{Comp}_{\mathbb{R}}(s, n)$ has $4c$ components, where c is the number of k for which $\text{Comp}^1_{\mathbb{R}}(s, n; k) \neq \emptyset$. If $C = C((s-1)\langle -1 \rangle)$ then c is the number of non-isomorphic n-dimensional C-modules. Hence

$$c = \begin{cases} 1 & \text{if } s \not\equiv 0 \pmod 4 \\ 1 + \frac{n}{2^m} & \text{otherwise.} \end{cases}$$

The irreducible dimension 2^m can be computed directly from the structure of C.

5. **Characters.** Let D be a quaternion or octonion algebra with left representation $\mathcal{L} : D \to \text{Sim}(D)$ and compute the character of $\mathcal{L}$. Similarly compute the character of the right representation. Are the left and right characters equal?

6. **More division algebras.** Let D be a real composition algebra with $n = \dim D = 2$, 4 or 8. Suppose $b : D \times D \to D$ is an $\mathbb{R}$-bilinear map with the property that $|b(x, y)| < 1$ whenever $|x| = |y| = 1$. Define

$$m_b : D \times D \to D \quad \text{by} \quad m_b(x, y) = xy + b(x, y).$$

Then (D, m_b) is a division algebra. If we assume only $|b(x, y)| \leq 1$ what further conditions are needed to ensure that m_b is a division algebra? This construction provides a space of division algebras of dimension n^3. Does it equal the whole space $\text{Div}(n) = \text{Nsing}(n, n)$?

7. **Inverses.** Suppose A is an F-algebra, where F is a field.

(1) If A is an alternative division algebra and $\dim A$ is finite then A must have an identity element.

(2) Suppose A has an identity element and every non-zero element of A has an inverse (that is: if $0 \neq a \in A$, there exists $b \in A$ with $ab = ba = 1$). Does it follow that A is a division algebra? Find a counterexample where $F = \mathbb{R}$, $\dim A = 3$.

(3) A *strong inverse* for $a \in A$ is $a^{-1} \in A$ such that $a^{-1} \cdot ax = x = xa \cdot a^{-1}$ for every $x \in A$. Suppose every non-zero element of A has a strong inverse. Check that $(a^{-1})^{-1} = a$ and deduce that A is a division algebra. In a composition algebra every a with $[a] \neq 0$ has a strong inverse.

Theorem. *Suppose A is a ring with* 1. *Then: every non-zero element of A has a strong inverse if and only if A is an alternative division ring.*

(Hint. (1) If $a \neq 0$, find e with $ae = a$. Prove $e^2 = e$.

(2) Let $A = \mathbb{R}^3$ with basis $\{1, i, j\}$, choose $\delta \in A$ and define $i^2 = j^2 = -1$ and $ij = -ji = \delta$. Then every non-zero element has an inverse. (Define "bar" and compute $u \cdot \bar{u} = \bar{u} \cdot u$.) Moreover, if $\delta \neq 0$ then $uv = vu = 0$ implies $u = 0$ or $v = 0$. Hence inverses are unique, but clearly A cannot be a division algebra.

(3) The proof of the theorem is elementary but not easy. See Hughes and Piper (1973), pp. 137–138 and p. 151, or see Mal'cev (1973), pp. 91–94.)

8. **Components.** $\mathrm{Div}(n)$ is a nonempty open subset of $\mathrm{Bil}(n)$, provided $n = 1, 2, 4, 8$.

(1) Describe the topological spaces $\mathrm{Div}^{11}(2)$ and $\mathrm{Div}(2)$ explicitly. Check that $\mathrm{Div}^{11}(2)$ is the "interior" of a certain parabola in $\mathrm{Bil}^{11}(2) \cong \mathbb{R}^2$. Then $\mathrm{Div}(2)$ is 8-dimensional with 4 components. Everything in $\mathrm{Div}(2)$ is isotopic to $\mathbb{C}$ and $\mathrm{Autot}(\mathbb{C}) \cong \mathbb{C}^* \times \mathbb{C}^* \times \{1, -1\}$.

(2) If $m \in \mathrm{Div}(n)$ and $0 \neq x \in \mathbb{R}^n$, define $\lambda(m) = \mathrm{sgn}(\det(m_x))$, where m_x is the left multiplication map. This $\lambda(m)$ is independent of x. Let $\rho(m)$ be the sign for the right multiplications. For signs ε, η let $\mathrm{Div}^{\varepsilon\eta} = \{m \in \mathrm{Div}(n) : \lambda(m) = \varepsilon \text{ and } \rho(m) = \eta\}$. These four subsets are represented by $xy, \bar{x}y, x\bar{y}, \overline{xy}$.

(3) If $m \in \mathrm{Div}^{++}(n)$ there is a path in $\mathrm{Div}^{++}(n)$ from m to some $m_1 \in \mathrm{Div}^{11}(n)$.

(4) Buchanan (1979) used homotopy theory to prove:

Theorem. *If $n = 4$ or 8 then $\mathrm{Div}^{11}(n)$ has two connected components, represented by the multiplications xy and yx.*

Corollary. $\mathrm{Div}(4)$ *and* $\mathrm{Div}(8)$ *each have* 8 *connected components, represented by the eight standard multiplications.*

(Hint. (3) By Exercise 0.8, there exist $f, g \in \mathrm{GL}^+(n)$ with $(f, g) * m \in \mathrm{Div}^{11}(n)$. Choose paths in $\mathrm{GL}^+(n)$ from 1_n to f and from 1_n to g.)

9. **Sub$(s, n; k)$.** Define $\varphi : \mathrm{O}(n) \times \mathrm{Sub}^1(s, n; k) \to \mathrm{Sub}(s, n; k)$ by: $\varphi(g, T) = gT$. Then φ is surjective with fiber $\varphi^{-1}(S) \cong S \cap \mathrm{O}(n)$, which is the unit sphere in S. Hence $\dim \mathrm{Sub}(s, n; k) = \frac{n(n-1)}{2} + \dim \mathrm{Sub}^1(s, n; k) - (s - 1)$. Is this consistent with (8.12)?

10. How does the $\mathrm{O}(s)$ action relate to the $\mathrm{O}(n) \times \mathrm{O}(n)$ action?

(1) Let $\alpha \in \mathrm{O}(s)$ and $m \in \mathrm{Comp}^1(s, n)$. Then $\chi(\alpha \bullet m) = (\det \alpha) \cdot \chi(m)$. Each orbit of the group $\mathrm{O}(s) \times \mathrm{O}(n)$ equals $\mathrm{Comp}^1(s, n; k) \cup \mathrm{Comp}^1(s, n; -k)$ for some k.

(2) Let $S \in \text{Sub}^1(s, n; k)$. Define $\text{Aut}^{\&}(S) = \{(\beta, \gamma) \in \text{O}(n) \times \text{O}(n) : \gamma S \beta^{-1} = S\}$ and consider the induced group homomorphism $\text{Aut}^{\&}(S) \to \text{O}(S)$. The image is $\text{O}^+(n)$ if $\chi(S) \neq 0$, and it is $\text{O}(n)$ if $\chi(S) = 0$.

(3) There is an exact sequence

$$1 \to \text{Aut}(m) \to \text{Aut}^{\&}(S) \to \begin{Bmatrix} \text{O}(S) & \text{if } \chi(S) = 0 \\ \text{O}^+(S) & \text{if } \chi(S) \neq 0 \end{Bmatrix} \to 1.$$

Compute $\dim \text{Aut}^{\&}(S)$ and use this to give another computation of $\dim \text{Sub}(s, n)$.

(4) Similarly analyze $\text{Aut}(S) = \{\beta \in \text{O}(n) : \beta S \beta^{-1} = S\}$.

(Hint. (2) If $\alpha \in \text{O}^+(S)$ then α is in the image, using C-isometries as in (8.5) or (7.19). Conversely suppose α is in that image. If $\chi(m) \neq 0$ apply part (1).)

11. **Automorphism groups.** There are several reasonable definitions for "the" automorphism group of a composition $m \in \text{Comp}(s, n)$. For example,

$$\text{Aut}(m) = \{\beta \in \text{O}(n) : \beta \diamond m = m\} = \{\beta \in \text{O}(n) : (1, \beta, \beta) \bullet m = m\},$$

as defined above.

$$\text{Aut}^{\&}(m) = \{(\beta, \gamma) \in \text{O}(n) \times \text{O}(n) : (1, \beta, \gamma) \bullet m = m\}.$$

$$\text{Aut}^{\%}(m) = \{(\alpha, \beta) \in \text{O}(s) \times \text{O}(n) : (\alpha, \beta, \beta) \bullet m = m\}.$$

$$\text{Autot}(m) = \{(\alpha, \beta, \gamma) : (\alpha, \beta, \gamma) \bullet m = m\}.$$

These are related to the groups $\text{Aut}(S)$ and $\text{Aut}^{\&}(S)$ defined in Exercise 10. What are the dimensions of these algebraic groups?

12. **Proper similarities.** Here is a sketch of the proof of (8.14). Suppose $1_V \in S \subseteq \text{Sim}(V, q)$ and $s = \dim S > 2$.

First Step. If $g \in \text{Sim}^\bullet(V, q)$ and $gSg^{-1} = S$ then g is proper.

(1) Find a counterexample when $\dim S = \dim V = 2$. If $\dim S = 2$ and $4 \mid \dim V$ then g is proper.

(2) Suppose $C = C(W, \varphi)$ is a Clifford algebra with center Z. If $x \in W$ is anisotropic then $xWx^{-1} = W$. (In fact the map $w \mapsto xwx^{-1}$ is the reflection through the line Fx.)

Lemma. *If $u \in C^\bullet$ and $uWu^{-1} = W$ then $u = y \cdot x_1 \cdot x_2 \dots x_k$ for some $y \in Z$ and $x_i \in W$.*

(3) *Proof of First Step when s is odd.* $C = C(-S_1)$ is central simple, and $C \otimes A = \text{End}(V)$ where $A = \text{End}_C(V)$. The involution I_q = "$\sim$" preserves C and A. Then $gCg^{-1} = C$ so there exists $u \in C^\bullet$ such that $a = u^{-1}g \in A$. Since $\tilde{g}g = \mu(g)$ conclude that $\tilde{a}a$ and $\tilde{u}u$ are scalars. Since a commutes with elements of S it is proper, by Exercise 1.17. Since $uS_1u^{-1} = S_1$ and $Z = F$, the lemma implies $u = x_1 \cdot x_2 \dots x_k$ for some $x_i \in S_1$. Hence $g = ua$ is proper.

(4) *Proof of First Step when s is even.* C_0 is central simple, and $C_0 \otimes A = \text{End}(V)$ where $A = \text{End}_{C_0}(V)$. As before, there exists $u \in C_0^\bullet$ such that $a = u^{-1}g \in A$

and $\tilde{a}a$ and $\tilde{u}u = \beta$ are scalars. Then a is proper since it commutes with $f_2 f_3$. Since $Z = F \oplus Fz$ is the center of C, $gzg^{-1} = \varepsilon z$ for some $\varepsilon = \pm 1$. Then $aza^{-1} = u^{-1}gzg^{-1}u = \varepsilon z$ and hence $aS_1a^{-1} = S_1$ since $S_1 \subseteq zC_0$. Therefore $uS_1u^{-1} = S_1$ and the lemma applies as before to show that u and $g = ua$ are proper.

(5) $h \in S$ implies $hSh \subseteq S$.

(6) Suppose F is algebraically closed. If $f \in S$ there exists $h \in S$ with $h^2 = f$.

(7) Suppose $\gamma S \beta^{-1} = S$ as in (8.14). Assume F is algebraically closed and use (6) to find $h \in S$ such that $h^2 = \gamma^{-1}\beta$. Let $g = \gamma h$ so that $g^{-1} = h\beta^{-1}$. By (5), $gSg^{-1} = \gamma hSh\beta^{-1} = S$ and the First Step implies that g is proper. Since h is proper conclude that both β and γ are proper.

(Hint. (1) Exercise 1.17.

(2) The map $w \mapsto uwu^{-1}$ is in $\mathrm{O}(W)$, and hence is a product of hyperplane reflections. Compare Cassels (1978), pp. 175–177 or Scharlau (1985), pp. 334–336.

(5) Choose the basis of S so that $h = a + bf_2$ and compute hf_jh.

(6) If $f = r + sf_2$ let $h = x + yf_2$ and solve for x and y.)

13. **Norm form uniqueness.** Suppose D is a composition algebra (with identity) relative to two quadratic forms $q(x)$ and $q'(x)$. These forms must coincide.

(Hint. The theory in Chapter 1 provides associated involutions $\bar{x}$ and $\tilde{x}$ so that $q(x) = x \cdot \bar{x}$ and $q'(x) = x \cdot \tilde{x}$. Show that these involutions coincide.)

14. **Trilinear map.** (1) For euclidean spaces U, V, W the following are equivalent:

(a) There is a bilinear $f : U \times V \to W$ with the norm property $|f(u, v)| = |u| \cdot |v|$.

(b) There is a trilinear map $g : U \times V \times W \to \mathbb{R}$ such that $|g(u, v, w)| \le |u| \cdot |v| \cdot |w|$ and moreover for every u, v there exists a non-zero w such that equality holds.

(2) If $\dim U = \dim V = \dim W$ then condition (b) is symmetric in U, V, W.

(Hint. (2) f, g are related by $g(u, v, w) = \langle f(u, v) | w \rangle$, where $\langle x | y \rangle$ is the dot product.)

15. The Triality Theorem implies that for every $\gamma \in \mathrm{O}^+(8)$, there exist $\alpha, \beta \in \mathrm{O}^+(8)$ such that (α, β, γ) is an autotopy, relative to the standard octonion multiplication. Moreover α, β are uniquely determined up to sign.

(1) Every $\gamma \in \mathrm{O}^+(8)$ equals $B_{\bar{a}}B_bB_{\bar{c}} \dots B_{\bar{g}}$, a product of (at most) 7 bi-multiplication maps.

(2) Then $\alpha = L_{\bar{a}}L_bL_{\bar{c}} \dots L_{\bar{g}}$ and $\beta = R_{\bar{a}}R_bR_{\bar{c}} \dots R_{\bar{g}}$, up to sign.

(3) Every $\alpha \in \mathrm{O}^+(8)$ can be expressed as a product of 7 of the maps L_a and also as a product of 7 of the maps R_a.

(4) If $(\alpha, \beta, \gamma) \in \mathrm{Autot}(D)$ then: $\alpha = \beta = \gamma$ is an automorphism $\iff \alpha(1) = \beta(1) = 1$. Compare Exercise 1.24.

(5) How much of this theory goes through for octonion algebras over a general field?

(Hint. (1) The Cartan–Dieudonné Theorem (proved in Artin (1957) or Lam (1973)) implies that $\gamma = \tau_1 \cdot \tau_a \dots \tau_g$ for some 7 unit vectors $a, \dots, g \in \mathbb{R}^8 = D$. Then $\gamma = B_{\bar{a}} B_b B_{\bar{c}} \dots B_{\bar{g}}$.

(2) Use the explicit autotopies (L_u, R_u, B_u) and the uniqueness of α, β.

(5) There are some difficulties with scalars over a general field F. For instance the group $\mathcal{B}$ generated by the B_a's consists of all $\theta(\sigma) \cdot \sigma$ where $\sigma \in O^+(D)$ and $\theta(\sigma)$ denotes the spinor norm of σ. The group $F^\bullet \cdot \mathcal{B}$ can be a proper subgroup of $\mathrm{Sim}^+(D)$. Does the group generated by the L_a's equal $\mathrm{Sim}^+(D)$?)

16. **Automorphism and autotopy.** (1) If D is the octonion division algebra over $\mathbb{R}$ determine $\dim \mathrm{Aut}(D)$.

(2) The "companion" map $\mathrm{Autot}^o(8) \to S^7 \times S^7$ sends $(\alpha, \beta, \gamma) \in \mathrm{Autot}^o(8)$ to $(a, b) = (\beta(1)^{-1}, \alpha(1)^{-1})$. Then $\alpha = R_a \circ \gamma$ and $\beta = L_b \circ \gamma$. The nonempty fibers of this companion map are the cosets $(\alpha, \beta, \gamma) \cdot \mathrm{Aut}(D)$.

(3) $\mathrm{Autot}^o(8)$ is a connected 2-fold covering group of $\mathrm{Mon}^o(8) = O^+(8)$.

(4) The companion map is surjective.

How does composition of autotopies corresponding to an operation on the associated companion pairs in $S^7 \times S^7$?

(Hint. (1) $\dim \mathrm{Aut}(D) = 14$. For D is generated by unit vectors i, j, v such that $D = H \perp Hv$ where H is the quaternion algebra generated by i, j. If $\varphi \in \mathrm{Aut}(D)$ then $\varphi(i)$ can be any unit vector in $\{1\}^\perp$, a choice in S^6. Given $\varphi(i)$, then $\varphi(j)$ can be any unit vector in $\{1, i\}^\perp$, etc.

(3) $\pi : \mathrm{Autot}^o(8) \to \mathrm{Mon}^o(8)$ is a homomorphism with kernel $\{(1, 1, 1), (-1, -1, 1)\}$. Find a path between those two points in $\mathrm{Autot}^o(8)$ by using autotopies (L_a, R_a, B_a).

(4) Compute dimensions.)

17. **Dimension 1, 2, 4.** (1) Analyze the spaces $\mathrm{Comp}^{11}(1)$ and $\mathrm{Comp}^{11}(2)$.

(2) Work out the parallels of (8.17) through (8.21) for quaternion algebras. Deduce that $\dim \mathrm{Autot}(4) = 11$ and $\dim \mathrm{Autot}^o(4) = 9$. What is the analog of the companion map of Exercise 16?

18. **Isotopy and isomorphism.** Let D be a quaternion or octonion division algebra over $\mathbb{R}$.

(1) If $n = 4$ or 8 let $\mathrm{Isotop}(n)$ be the set of all multiplications on $\mathbb{R}^n$ which are isotopic to the multiplication of D. (Why is this independent of the choice of D?) Then $\mathrm{Isotop}(n)$ can be viewed as an algebraic variety. What is its dimension?

(2) Similarly analyze $\mathrm{Isomor}(n)$, the set of algebras isomorphic to D.

(Hint. (1) $\mathrm{Isotop}(n)$ is an orbit of $\mathrm{GL}(n)^3$ with stabilizer $\mathrm{Autot}(D)$.

(2) $\mathrm{Isomor}(n)$ is an orbit of $\mathrm{GL}(n)$ with stabilizer $\mathrm{Aut}(D)$.)

19. **Loops.** An *inverse loop* is a set G with a binary operation such that (i) there is an identity element $1 \in G$; and (ii) for every $x \in G$ there exists $x^{-1} \in G$ satisfying: $x^{-1} \cdot xy = y = yx \cdot x^{-1}$ for every $y \in G$. An *autotopy* on G is a triple (α, β, γ) of invertible maps on G such that: $\gamma(xy) = \alpha(x)\beta(y)$ for every x, y.

(1) If $xy = z$ then $x = zy^{-1}, z^{-1}x = y^{-1}, \ldots$, and we get the associated hexagon of six autotopies of G. Define a monotopy and deduce that α, β, γ are monotopies.

(2) γ is a monotopy if and only if there exist $a, b \in G$ such that $\gamma(xy) = \gamma(x)a \cdot b\gamma(y)$ for every x, y. The elements a, b are the *companions* of γ. Note that $\alpha = R_a \circ \gamma$ and $\beta = L_b \circ \gamma$ provide the autotopy, and $a = \beta(1)^{-1}$ and $b = \alpha(1)^{-1}$.

(3) For a as above, $B_a(x) = axa$ is unambiguously defined and (L_a, R_a, B_a) is an autotopy. Similarly we find autotopies (B_a, L_a^{-1}, L_a), (R_a^{-1}, B_a, R_a) etc. These imply the Moufang identities: $ax \cdot ya = a(xy)a$; $axa \cdot a^{-1}y = a \cdot xy$; $xa^{-1} \cdot aya = xy \cdot a$.

(4) If $a \in G$ the following are equivalent:

(i) a is the image of 1 under a monotopy;
(ii) (L_a, R_a, B_a) is an autotopy;
(iii) the Moufang identities hold for a.

A Moufang loop (or "Moup") is an inverse loop in which every a, x, y satisfies the Moufang identities. Then G is a Moufang loop if and only if the monotopies act transitively on G.

(Hint. (3) $(\beta, \iota\gamma\iota, \iota\alpha\iota) \circ (\gamma, \iota\beta\iota, \alpha)^{-1} = (\beta\gamma^{-1}, \iota\gamma\beta^{-1}\iota, \iota\alpha\iota\alpha^{-1}) = (L_a, R_a, \iota\alpha\iota\alpha^{-1})$ is an autotopy, so that $\iota\alpha\iota\alpha^{-1}(xy) = ax \cdot ya$ for every x, y. Then $B_a(x) = ax \cdot a = a \cdot xa$ and (L_a, R_a, B_a) works. The six autotopies derived from this one provide other examples.)

20. **Other norm forms.** Fix $e \neq 0$ in $\mathbb{R}^8$ and let $\mathrm{Comp}^e(8)$ be the set of all multiplications $m \in \mathrm{Bil}(8)$ which make $\mathbb{R}^8$ into a composition division algebra with identity element e. Here we do not assume that the standard inner product is the norm form. Then $\mathrm{Comp}^e(8) \subseteq \mathrm{Div}^{11}(8)$. Is $\mathrm{Comp}^e(8)$ a nice topological space? What is its dimension?

(Hint. If $\mathrm{PD}(n) = \{$positive definite quadratic forms on $\mathbb{R}^n\}$, then $\dim \mathrm{PD}(n) = n(n+1)/2$ since $\mathrm{PD}(n) \cong \mathrm{GL}(n)/\mathrm{O}(n)$. Then $\mathrm{PD}^1(8) = \{q \in \mathrm{PD}(8) : q(e) = 1\}$ has dimension 35. Is there a bijection: $\mathrm{Comp}^e(8) \leftrightarrow \mathrm{Comp}^{11}(8) \times \mathrm{PD}^1(8)$?)

21. For which $\alpha, \beta, \gamma \in \mathrm{GL}(8)$ does the action of (α, β, γ) on $\mathrm{Bil}(8)$ preserve the subset $\mathrm{Div}^{11}(8)$?

(*Idea.* Let $m_1(x, y) = xy$ be the octonion multiplication with identity e. If $\varphi \in \mathrm{GL}(8)$ with $\varphi(e) = e$ then $m_\varphi = (\varphi, \varphi, \varphi) \bullet m_1$ is in $\mathrm{Div}^{11}(8)$. Then $(r\varphi, s\varphi, rs\varphi)$ preserves $\mathrm{Div}^{11}(8)$ when $r, s \in \mathbb{R}^\bullet$. Conversely if (α, β, γ) preserves it then $\varphi^{-1}\gamma^{-1}(x) = \varphi^{-1}\alpha^{-1}(x) \cdot \varphi^{-1}\beta^{-1}(e)$ and $\varphi^{-1}\gamma^{-1}(y) = \varphi^{-1}\alpha^{-1}(e) \cdot \varphi^{-1}\beta^{-1}(y)$ for every $x, y \in D$ and every such φ. Must $\alpha^{-1}(e)$ and $\beta^{-1}(e)$ be scalar multiples of e?)

22. **Split octonion algebras.** In (8.17) through (8.21) we assumed that the octonion algebra D is a division algebra (so the norm form $[x]$ is anisotropic). Are the same results true when D is a "split" octonion algebra, that is, when the norm form $[x]$ on D is hyperbolic?

(Note. If F is infinite the non-invertible elements form the zero set of a polynomial function. Therefore almost all elements of D are invertible.)

23. **Robert's Thesis (1912).** Let $\mathcal{A}$ be the set of all $n \times n$ matrices A whose entries are $\mathbb{C}$-linear forms in $X = (x_1, \dots, x_s)$ and which satisfy $A^\top \cdot A = (x_1^2 + \cdots + x_s^2) \cdot I_n$.

(1) Each $A \in \mathcal{A}$ corresponds to a unique $m \in \mathrm{Comp}_{\mathbb{C}}(s, n)$.

(2) $\mathrm{O}(n) \times \mathrm{O}(n)$ acts on $\mathcal{A}$ by: $(P, Q) * A = P \cdot A \cdot Q^\top$. This corresponds to the action on Comp described above. Consequently $\mathcal{A}$ is an algebraic variety, and we know the number of components and their dimensions.

(Hint. (1) Recall the original treatment by Hurwitz as described in Chapter 0.)

Notes on Chapter 8

The ideas presented in the first part of the Chapter are based on results of Petersson (1971) and of Bier and Schwardmann (1982). The homology and stable homotopy groups of the topological spaces $\mathrm{Comp}^1(s, n)$ were computed by Bier and Schwardmann.

Zorn characterized finite dimensional alternative division algebras over any base field. One proof appears in Schafer (1966), p. 56. The result has a remarkable generalization, due to Kleinfeld, Bruck and Skornyakov:

Theorem. *Any simple alternative ring, which is not a nilring and which is not associative, must be an octonion algebra over its center.*

This theorem is proved in Kleinfeld (1953) and in Zhevlakov et al. (1982), §7.3. An easier proof, assuming characteristic $\neq 2$, is given in Kleinfeld (1963).

There are more constructions of real division algebras, usually done by "twisting" the standard algebras in various ways. For example see Althoen, Hansen and Kugler (1994). Further information on real division algebras is contained in Myung (1986). Certain "pseudo-octonion" algebras are 8-dimensional division algebras (without an identitiy element) which are especially symmetric. See also Elduque and Myung (1993).

The dimension argument after (8.15) showing that not all division algebra multiplications are isotopic to a composition algebra is due to Petersson.

Dimension counts show how hard it might be to get a useful classification of real division algebras. However, there is a positive result about general elements of $\mathrm{Div}^{11}(n)$.

Theorem. *If D is a real division algebra with identity and* $\dim D > 1$, *then D contains a subalgebra isomorphic to $\mathbb{C}$. That is, there exists $a \in D$ with $a^2 = -1$.*

Proofs appear in Yang (1981) and Petro (1987). Both proofs use topological properties to prove that the map $x \mapsto x^2$ is surjective on D.

Following (8.15), $\text{Div}(n) = \text{Nsing}(n, n, n)$. Bier (1979) showed that $\text{Nsing}(r, s, n)$ is a semi-algebraic set and hence has a finite number of connected components. He also proved that if $n \geq r + s - 1$ then $\text{Nsing}(r, s, n)$ is dense in $\text{Bil}(r, s, n)$ and if moreover $n > (r \# s) + r + s - 1$ then $\text{Nsing}(r, s, n)$ is connected. (This notation $r \# s$ is defined in Chapter 12.)

The viewpoint and terminology of autotopies and monotopies, as defined in (8.17), was explained to me in 1980 by J. H. Conway. Versions of Conway's approach are also seen in Exercises 16 and 20, as well as in the appendix to Chapter 1.

Our presentation of the Triality Principle 8.19 basically follows van der Blij and Springer (1960), who prove it without restrictions on the characteristic of the ground field. Some simplifications in the proof use Conway's approach. Other authors use the terms autotopism and isotopism. See Hughes and Piper (1973), Chapter VIII.

Over any field F (with characteristic $\neq 2$), every $m \in \text{Comp}(4, 4)$ is isotopic to a quaternion algebra H. Letting xy be the multiplication in H, then the multiplication $m(x, y)$ is expressible as one of four types:

$$(1)\ axcyb \qquad (2)\ axb\bar{y}c \qquad (3)\ c\bar{x}ayb \qquad (4)\ a\bar{x}c\bar{y}b$$

where $a, b, c \in H$ and $N(abc) = 1$. For what choices of a, b, c are two of these algebras isomorphic? This question is analyzed by Stampfli-Rollier (1983).

Kuz'min (1967) discusses the topological space of all isomorphism classes of n-dimensional real division algebras (with identity). He considers the subspaces of power-associative algebras, quadratic algebras, etc., and determines their dimensions.

Exercise 4. Bier and Schwardmann (1982) discuss this number of components.

Exercise 7. (2) A similar remark is made in Althoen and Weidner (1978).

(3) Stronger theorem: If every non-zero element of A has a strong right inverse, then A is alternative. This result is related to the geometry of projective planes. See Hughes and Piper (1973), pp. 140–149.

Exercise 8. Buchanan's proof uses homotopy theory.

Define $\mathcal{A}(n) = \{A \in \text{GL}_n(\mathbb{R}) : A \text{ has no real eigenvalues}\}$ and $\mathcal{W}(n) = \{W \in \text{O}(n) : W \text{is skew-symmetric}\}$. Buchanan proves $\mathcal{W}(n)$ is a strong deformation retract of $\mathcal{A}(n)$. The space $\mathcal{W}(n)$ has two connected components, separated by the Pfaffian (see (10.8)). Any $m \in \text{Div}^{11}(n)$ induces $\hat{m} : \mathbb{R}^n - \{0\} \to \mathcal{A}(n)$ and this maps to $\mathcal{W}(n)$. The standard composition algebras yield multiplications xy and yx with unequal Pfaffians. Hence $\text{Div}^{11}(n)$ has at least two components. A computation of $\pi_{n-2}(\mathcal{A}(n))$ leads to a proof that there are only two components.

A somewhat simpler proof in the case $n = 4$ is given by Gluck, Warner and Yang (1983), §8. The components are separated by their "handedness".

Exercise 11. The group $\mathrm{Aut}^{\%}(m)$ was studied by Riehm (1982), a work motivated by a question of A. Kaplan (1981). Those ideas were extended in Riehm (1984).

Exercise 15–16. This threefold symmetry for α, β, γ in $\mathrm{O}^+(8)$ is one aspect of triality. The sign ambiguities can be removed if we work with the covering group Spin(8) instead. From Exercise 16(3) it follows that $\mathrm{Autot}^{\circ}(8) \cong \mathrm{Spin}(8)$. Many aspects of triality have appeared in the mathematical literature. For example see Knus et al. (1998), §35, and Chapter 10.

Exercise 19. This approach to Moufang loops is due to J. H. Conway. Connections between Moufang loops and geometry are described in Bruck (1963).

Exercise 23. E. Robert, in his 1912 thesis, analyzed these matrices A in the cases $r = n = 4, 8$. He showed essentially that $\mathrm{Comp}_{\mathbb{C}}(n, n)$ consists of two orbits of $\mathrm{O}(n) \times \mathrm{O}(n)$, distinguished by the "character".

Chapter 9

The Pfister Factor Conjecture

We focus now on the form q rather than on (σ, τ). Suppose F is a field (in which $2 \neq 0$). Given n, which n- dimensional forms q over F admit the largest possible families in $\text{Sim}(q)$? We stated the following conjecture in (2.17).

9.1 Pfister Factor Conjecture. Let q be a quadratic form over F with $\dim q = n = 2^m n_0$ where n_0 is odd. If there is an $(m+1, m+1)$-family in $\text{Sim}(q)$ then $q \simeq \varphi \otimes \omega$ where φ is an m-fold Pfister form and $\dim \omega$ is odd.

One attraction of this conjecture is that it relates the forms involved in the Hurwitz–Radon type of "multiplication" of quadratic forms with the multiplicative quadratic forms studied by Pfister. We will reduce the question to the case $n = 2^m$ and to prove it whenever $m \leq 5$. The difficulties in extending our proof seem closely related to the difficulties in extending Pfister's result (3.21) for forms in I^3F. For certain special classes of fields we can prove the conjecture. For example, it is true for every global field. In the appendix we describe (without proofs) some results about function fields of quadratic forms and use that theory to provide another proof of the cases $m \leq 5$.

This conjecture can be restated in terms of the original sort of composition defined in Chapter 1. For as noted in the (7.12), if $\dim q = 2^m \cdot (\text{odd})$, then there exists $\sigma < \text{Sim}(q)$ with $\dim \sigma = \rho(n)$ if and only if there exists an $(m+1, m+1)$-family in $\text{Sim}(q)$.

If either σ or τ is isotropic then $(\sigma, \tau) < \text{Sim}(q)$ implies that q is hyperbolic, by (1.9). In this case the conjecture is trivial so we may assume that σ and τ are anisotropic.

9.2 Conjecture PC(m). Suppose q is a quadratic form over F with $\dim q = 2^m$. If there exists an $(m+1, m+1)$-family in $\text{Sim}(q)$, then q is similar to a Pfister form.

Proof that PC(m) *is equivalent to the Pfister Factor Conjecture* 9.1. Certainly (9.1) implies PC(m). Conversely assume PC(m) and suppose q is given with $\dim q = n = 2^m n_0$ and with an $(m+1, m+1)$-family $(\sigma, \tau) < \text{Sim}(q)$. The Decomposition Theorem 4.1 implies that all the (σ, τ)-unsplittables have the same dimension 2^k. Since q is a sum of unsplittables, $2^k \mid n$ so that $k \leq m$. If φ is an unsplittable then $s + t = 2m + 2$ implies $\dim \varphi = 2^m$. Then the uniqueness in (7.2) implies that all (σ, τ)-unsplittables are similar to φ. Therefore $q \simeq \varphi \otimes \omega$ for some form ω of dimension n_0, which is odd. The form φ is similar to a Pfister form by PC(m). Absorbing the scale factor into ω, we may assume φ is a Pfister form. □

9.3 Lemma. PC(m) *is true for* $m \leq 3$.

Proof. The cases $m = 1, 2$ are vacuous. Suppose $m = 3$ and $\dim q = 8$ and q admits a $(4, 4)$-family. By (1.10) a $(3, 0)$-family $\langle 1, a, b\rangle < \mathrm{Sim}(q)$ already implies that $\langle\langle a, b\rangle\rangle \mid q$, forcing q to be similar to a Pfister form. □

Suppose $\dim q = 2^m$ and $(\sigma, \tau) < \mathrm{Sim}(q)$ is an $(m + 1, m + 1)$-family. As mentioned after (7.1) we have $d\sigma = d\tau$, $c(\sigma) = c(\tau)$, so that $\sigma \equiv \tau \pmod{J_3(F)}$. If applications of the Shift Lemma can transform the pair (σ, τ) into some pair (δ, δ), then (2.16) implies the Conjecture PC(m). To state this idea more formally we introduce the set $\mathcal{P}_m$ of all (s, t)- pairs of quadratic forms over F where $s + t = 2m + 2$. Define the relation $\approx$ on $\mathcal{P}_m$ to be the equivalence relation generated by three "elementary" relations motivated by the ideas in Chapter 2:

(1) $(\sigma, \tau) \approx (\tau, \sigma)$.

(2) $(\sigma, \tau) \approx (\langle a\rangle\sigma, \langle a\rangle\tau)$ whenever $a \in D_F(\sigma)D_F(\tau)$.

(3) $(\sigma \perp \varphi, \tau \perp \psi) \approx (\sigma \perp \langle d\rangle\psi, \tau \perp \langle d\rangle\varphi)$ whenever $\dim\varphi \equiv \dim\psi \pmod 4$ and $\langle d\rangle = (\det\varphi)(\det\psi)$.

The motivation for this definition arises from the following basic observation:

If $(\sigma, \tau) \approx (\sigma', \tau')$ then: $(\sigma, \tau) < \mathrm{Sim}(V, B)$ if and only if $(\sigma', \tau') < \mathrm{Sim}(V, B)$.

9.4 Definition. Let $\mathcal{P}_m^\circ$ be the set of all (s, t)-pairs (σ, τ) such that $s + t = 2m + 2$, $d\sigma = d\tau$, $c(\sigma) = c(\tau)$ and $s \equiv t \pmod 8$. Equivalently, $\mathcal{P}_m^\circ$ is the set of all $(\sigma, \tau) \in \mathcal{P}_m$ such that $\sigma \equiv \tau \pmod{J_3(F)}$ and $s \equiv t \pmod 8$.

We first observe that $\mathcal{P}_m^\circ$ is a subset of $\mathcal{P}_m$ preserved by the equivalence relation.

9.5 Lemma. *Suppose* $(\sigma, \tau) \in \mathcal{P}_m$.

(1) $(\sigma, \tau) \in \mathcal{P}_m^\circ$ *if and only if* $(\sigma, \tau) < \mathrm{Sim}(q)$ *for some* q *with* $\dim q = 2^m$.

(2) *If* $(\sigma, \tau) \approx (\sigma', \tau')$ *and* $(\sigma, \tau) \in \mathcal{P}_m^\circ$, *then* $(\sigma', \tau') \in \mathcal{P}_m^\circ$.

Proof. (1) Apply (7.3).

(2) This follows from (1) and ideas from Chapter 2. Here is a more direct proof. We may assume that the (s', t')-pair (σ', τ') is obtained from the (s, t)-pair (σ, τ) by applying one of the three elementary relations. Since $s \equiv t \pmod 8$ is easily follows that $s' \equiv t' \pmod 8$. Let $\beta = \sigma - \tau$ and $\beta' = \sigma' - \tau'$. The elementary relations

imply the following equations in the Witt ring:

$$\begin{aligned}\beta' &= -\beta && \text{if type 1.}\\ \beta' &= \langle a\rangle\beta && \text{if type 2.}\\ \beta' &= \beta + \langle\langle d\rangle\rangle \otimes (\varphi \perp -\psi) && \text{if type 3.}\end{aligned}$$

Now $\beta \in I^2F$ so that $\beta \equiv \langle x\rangle\beta \pmod{I^3F}$ for every $x \in F^\bullet$. Also since $d(\varphi \perp -\psi) = (d\varphi)(d\psi) = \langle d\rangle$, Exercise 3.7 (4) implies $\langle\langle d\rangle\rangle \otimes (\varphi \perp -\psi) \in I^3F$. Then in each case $\beta' \equiv \beta \pmod{I^3F}$. Since $I^3F \subseteq J_3(F)$ we have $\beta' \in J_3(F)$ so that $(\sigma', \tau') \in \mathcal{P}_m^\circ$. □

In trying to prove PC(m) by induction we are led to a related question.

9.6 The Shift Conjecture SC(m). If $(\sigma, \tau) \in \mathcal{P}_m^\circ$ then $(\sigma, \tau) \approx (\sigma', \tau')$ where σ' and τ' represent a common value.

Of course σ' and τ' represent a common value if and only if the form $\beta' = \sigma' \perp -\tau'$ is isotropic. If SC(m') is true for every $m' \le m$ then PC(m) follows. Here is a more formal statement of this idea.

9.7 Lemma. *If* SC(m) *and* PC$(m-1)$ *are true over F then* PC(m) *is also true over F.*

Proof. Suppose $(\sigma, \tau) < \mathrm{Sim}(q)$ is an $(m+1, m+1)$-family where $\dim q = 2^m$. Then (7.3) implies $(\sigma, \tau) \in \mathcal{P}_m^\circ$. By SC$(m)$ we may alter σ, τ to assume $\sigma \simeq \sigma' \perp \langle a\rangle$ and $\tau \simeq \tau' \perp \langle a\rangle$. The Eigenspace Lemma 2.10 implies that $q \simeq q' \otimes \langle\langle a\rangle\rangle$ and $(\sigma', \tau') < \mathrm{Sim}(q')$. By PC$(m-1)$ this q' is similar to a Pfister form and therefore so is q. □

If SC(m) is true over F for all m then the Pfister Factor Conjecture holds over F. In nearly every case where PC(m) has been proved for a field F, the condition SC(m) can be proved as well.

Before discussing small cases of this conjecture we note that: if F satisfies SC(m) for all m then $I^3F = J_3(F)$, which is a major part of Merkurjev's Theorem. Therefore it seems unlikely that an easy proof of the Shift Conjecture will arise.

9.8 Proposition. *Suppose* SC(m') *is true over F for all $m' \le m$. If $\beta \in J_3(F)$ and $\dim \beta = 2m+2$ then $\beta \in I^3F$.*

Proof. Write $\beta = \sigma \perp -\tau$ for some forms σ, τ of dimension $m+1$. Then $(\sigma, \tau) \in \mathcal{P}_m^\circ$ and application of SC(m') for $m' = m, m-1, m-2, \dots$ implies that $(\sigma, \tau) \approx (\delta, \delta)$ for some form δ. By the proof of (9.5), $\beta = \sigma - \tau \equiv \delta - \delta \equiv 0 \pmod{I^3F}$. □

9.9 Proposition. SC(m) *is true for $m \le 4$.*

Proof. Let $(\sigma, \tau) \in \mathcal{P}_m^\circ$. If $m \leq 2$ the equal invariants imply that $\sigma \simeq \tau$ (see Exercise 3.5). If $m = 3$ then $(\sigma, \tau) \approx (\varphi, 0)$ where $\varphi \simeq \sigma \perp (d\tau)\tau$. Then $\dim \varphi = 8$ and $\varphi \in J_3(F)$ so that φ is similar to a Pfister form by (3.21). If $\varphi \simeq \langle a\rangle\langle\langle x, y, z\rangle\rangle$ then $(\varphi, 0) \approx (\delta, \delta)$ where $\delta \simeq \langle a\rangle\langle 1, x, y, z\rangle$. Now suppose $m = 4$. Then $\beta = \sigma \perp -\tau \in J_3(F)$ and $\dim \beta = 10$. This β must be isotropic by Pfister's Theorem (3.21), and σ and τ represent a common value. □

It seems difficult to know whether a general pair (σ, τ) can be shifted to some better (σ', τ'). In some cases knowledge of certain types of subforms yields the result.

9.10 Lemma. *Suppose* (σ, τ) *is an* (s, t)*-pair. Then* $(\sigma, \tau) \approx (\sigma', \tau')$ *for some* σ' *and* τ' *which represent a common value, provided there exist subforms* $\varphi \subset \sigma$ *and* $\psi \subset \tau$ *such that* $\varphi \neq 0, \sigma$*;* $\dim \varphi \equiv \dim \psi \pmod 4$ *and* $\det \varphi = \det \psi$.

For example, this definition holds if $s > 2$ and σ and τ contain 2-dimensional subforms of equal determinant. The condition also holds if $s > 4$ and σ contains a 4-dimensional subform of determinant $\langle 1\rangle$.

Proof. Express $\sigma = \sigma_1 \perp \varphi$ and $\tau = \tau_1 \perp \psi$. Since $\varphi \neq 0, \sigma$, we may express $\sigma_1 = \langle x\rangle \perp \sigma_2$ and $\varphi = \langle a\rangle \perp \varphi_1$. Use (2.6) to shift $\langle x\rangle \perp \varphi_1$ and ψ. Since $\det(\langle x\rangle \perp \varphi_1))(\det \psi) = \langle ax\rangle$ we obtain $(\sigma', \tau') = (\sigma_2 \perp \langle a\rangle \perp \langle ax\rangle\psi, \tau_1 \perp \langle ax\rangle(\langle x\rangle \perp \varphi_1))$. Both σ' and τ' represent a. □

9.11 Proposition. SC(5) *is true.*

Proof. Suppose $(\sigma, \tau) \in \mathcal{P}_5^\circ$. We may shift (σ, τ) to a (10, 2)-pair (σ_0, τ_0). Then $\beta = \sigma_0 \perp -\tau_0$ is a 12-dimensional element of $J_3(F)$. If β is isotropic then σ_0 and τ_0 represent a common value and we are done. Assume β is anisotropic and write $\tau_0 \simeq \langle -a\rangle\langle 1, -b\rangle$, for some $a, b \in F^\bullet$. Then $\beta \simeq \langle a\rangle\langle\langle -b\rangle\rangle \perp \sigma_0$. Pfister's Theorem 3.21 implies that $\beta \simeq \varphi_1 \perp \varphi_2 \perp \varphi_3$, where $\langle a\rangle\langle\langle -b\rangle\rangle \subset \varphi_1$ and each φ_i is 4-dimensional of determinant $\langle 1\rangle$. Then $\varphi_2 \subset \sigma$ and (9.10) applies. □

We have been unable to prove SC(6) over an arbitrary field because we lack information about 14-dimensional forms in I^3F. Rost (1994) proved that any such form β is a transfer of the pure part of some 3-fold Pfister form over a quadratic extension of F. Hoffmann and Tignol (1998) deduced from this that β must contain an Albert subform. (Recall that an Albert form is a 6-dimensional form in I^2F.) This information leads to a possible approach to SC(6).

9.12 Lemma. *If the following hypothesis holds true over* F*, then* SC(6) *is true over* F*. Hypothesis: Whenever* β *is an anisotropic* 14*-dimensional form in* I^3F *and* $\gamma \subset \beta$ *is a given* 3*-dimensional subform, then there exists an Albert form* α *such that* $\gamma \subset \alpha \subset \beta$.

Proof. Suppose $(\sigma, \tau) \in \mathcal{P}_6^\circ$ is an (11, 3)-family. Then $\beta = \sigma \perp -\tau$ is a 14-dimensional form in $J_3(F)$. By Merkurjev's Theorem, $\beta \in I^3F$. If β is isotropic the conclusion of SC(6) is clear. If β is anisotropic the hypothesis provides an Albert form α with $-\tau \subset \alpha \subset \beta$. Expressing $\alpha = \alpha' \perp -\tau$ we have $\dim \alpha' = \dim \tau = 3$, $\alpha' \subset \sigma$ and $\det \alpha' = \det \tau$. Then (9.10) applies. □

It is not at all clear whether the strong condition in (9.12) is always true. Finding a counterexample to it would be interesting. But it might be much more interesting to construct a non-Pfister form of dimension 64 admitting a (7, 7)-family!

If the field F satisfies some nice properties, then the conjecture SC(m) is true for all m. Recall that the u-invariant $u(F)$ of a non-real field F is the maximal dimension of an anisotropic quadratic form over F.

9.13 Corollary. *If F satisfies one of the properties below then* SC(m) *is true over F for all m.*

(1) *F is nonreal and $u(F) < 14$.*

(2) *Every anisotropic form σ over F with* $\dim \sigma \geq 11$ *contains a 4-dimensional subform of determinant $\langle 1 \rangle$.*

Proof. We may assume $m \geq 6$ and $(\sigma, \tau) \in \mathcal{P}_m^\circ$.

(1) By hypothesis, every quadratic form over F of dimension ≥ 14 is isotropic. Since $\dim(\sigma \perp -\tau) = 2m + 2 \geq 14$, σ and τ must represent a common value.

(2) We can shift the given (σ, τ) to assume $\dim \sigma \geq 11$. The claim then follows from (9.10). □

Every algebraic number field satisfies condition (2) above. More generally, every "linked" field satisfies (2). Recall that two 2-fold Pfister forms φ and ψ are said to be linked if they can be written with a "common slot": $\varphi \simeq \langle\langle a, x \rangle\rangle$ and $\psi \simeq \langle\langle a, y \rangle\rangle$ for some $a, x, y \in F^\bullet$. The field F is said to be *linked* if every pair of 2-fold Pfister forms is linked.

9.14 Lemma. *The following conditions are equivalent for a field F.*

(1) *F is linked.*

(2) *The quaternion algebras form a subgroup of the Brauer group.*

(3) *For every form q over F, $c(q) =$ quaternion.*

(4) *Every 6-dimensional form α over F with $d\alpha = \langle 1 \rangle$ is isotropic.*

(5) *Every 5-dimensional form over F contains a 4-dimensional subform of determinant $\langle 1 \rangle$.*

We omit the details of the proof. Most of the work appears in Exercise 3.10.

The standard examples of linked fields are finite fields, local fields, global fields, fields of transcendence degree ≤ 2 over $\mathbb{C}$, fields of transcendence degree 1 over $\mathbb{R}$. Of course by (9.13) we know that $SC(m)$, and hence the Pfister Factor Conjecture, is true over any linked field.

We digress for a moment to discuss the Pfister behavior of general unsplittable modules over linked fields. If (σ, τ) is an (s, t)-pair over a linked field, is every unsplittable (σ, τ)-module necessarily similar to a Pfister form? The exceptions are called "special" pairs.

9.15 Definition. A pair (σ, τ) is *special* if $s \equiv t \pmod 8$ and the form $\beta = \sigma \perp -\tau$ satisfies: $d\beta \neq \langle 1 \rangle$ and $c(\beta)$ is a quaternion algebra not split by $F(\sqrt{d\beta})$.

In the notation of Theorem 7.8 the special pairs are exactly the ones having unsplittables of dimension 2^{m+2}. We are assuming throughout that σ represents 1.

9.16 Proposition. *Suppose F is a linked field and (σ, τ) is a pair which is not special. Then every unsplittable (σ, τ)-module is similar to a Pfister form.*

Proof. Theorem 7.8 applies here since F is linked so that $c(\beta)$ must be quaternion. Let $m = \delta(s, t)$ and suppose α is an unsplittable (σ, τ)-module. If $\dim \alpha = 2^m$ then $s + t \geq 2m - 1$ and the Expansion Proposition 7.6 implies that there is an $(m + 1, m + 1)$-family in $\text{Sim}(\alpha)$. Then $PC(m)$ implies that α is similar to a Pfister form. Suppose $\dim \alpha = 2^{m+1}$. If $s + t \geq 2m + 1 = 2(m + 1) - 1$, we are done as before using $PC(m + 1)$. The remaining cases have $s + t = 2m$ and $s \equiv t \pm 2$ or $t + 4 \pmod 8$. Dropping one dimension from σ or from τ we can find an (s', t')-pair $(\sigma', \tau') \subset (\sigma, \tau)$ where $s' + t' = 2m - 1$ and $s' \equiv t' \pm 3 \pmod 8$. Again since F is linked we may use Theorem 7.8 to get an unsplittable (σ', τ')-module ψ of dimension 2^m. Then $PC(m - 1)$ implies ψ is similar to a Pfister form. Furthermore (σ', τ') is a minimal pair and (7.18) implies that ψ is the unique (σ', τ')-unsplittable. Therefore $\alpha \simeq \psi \otimes \langle a, b \rangle$ for some $a, b \in F^\bullet$ and α is also similar to a Pfister form. The last case when $\dim \alpha = 2^{m+2}$ occurs only when (σ, τ) is special. □

The special pairs really do behave differently. Using (5.11) we gave examples of special $(2, 2)$-pairs over the rational field $\mathbb{Q}$ which have 8-dimensional unsplittable modules which are not similar to Pfister forms.

The Pfister Factor Conjecture can be reformulated purely in terms of algebras with involution. (Compare (6.12).) This version is interesting but seems harder to work with than the original conjecture.

9.17 Conjecture. In the category of F-algebras with involution, suppose $(A, K) \cong (Q_1, J_1) \otimes \cdots \otimes (Q_m, J_m)$ where each (Q_k, J_k) is a quaternion algebra with involution. If the algebra A is split, then there is a decomposition

$$(A, K) \cong (Q_1', J_1') \otimes \cdots \otimes (Q_1', J_m')$$

where each (Q'_1, J'_k) is a split quaternion algebra with involution.

Claim. (9.17) *is equivalent to* PC(m).

Proof. Assume (9.17) and suppose $(\sigma, \tau) \in \mathcal{P}^\circ_m$ with associated Clifford algebra C. By hypothesis, $C \cong C_0 \times C_0$ and $C_0 \cong \mathrm{End}(V)$ for a space V of dimension 2^m. Since $s \equiv t \pmod 8$ the involution $J = J_S$ on C induces an involution J_0 of type 1 on C_0 as in (7.4). This provides the involution I_q on $\mathrm{End}(V)$ corresponding to a quadratic form q on V. The conjecture PC(m) says exactly that this q must be a Pfister form. The algebra C_0 can be decomposed as a tensor product of quaternion subalgebras, each preserved by the involution J_0 (compare Exercise 3.14). Therefore we may apply (9.17) to conclude that (C_0, J_0) is a product of split quaternion algebras with involution, (Q'_1, J'_k). Expressing $Q'_k \cong \mathrm{End}(U_k)$ where $\dim U_k = 2$, the involution J'_k induces a λ_k-form B_k on U_k. It follows that $V \cong U_1 \otimes \cdots \otimes U_m$ and $q \simeq B_1 \otimes \cdots \otimes B_m$. If all the types λ_k are 1 then q is a product of binary quadratic forms, so it is similar to a Pfister form. Otherwise some skew forms occur in the product (necessarily an even number of them) and q is hyperbolic, so again it is Pfister.

Conversely, assume PC(m) and let (A, K) be a split algebra with a decomposition as in (9.17). Then $A \cong \mathrm{End}(V)$ where $\dim V = 2^m$ and the involution K induces a regular λ-form B on V.

Claim. It suffices to decompose $(V, B) \simeq \bigotimes(U_j, B_j)$ for some λ_j-spaces with $\dim U_j = 2$. For if such a factorization exists we can use (6.10) to see that $(A, K) \cong \bigotimes(\mathrm{End}(U_j), I_{B_j})$ as required.

Since A is a product of quaternions we may reverse the procedure in (3.14) to view A as some Clifford algebra: $A \cong C(W, q)$. Since K preserves each quaternion algebra it also preserves the generating space W. Then K is an (s, t)-involution on $C(W, q)$, for some (s, t) where $s + t = 2m + 1$. The isomorphism $(A, K) \cong (\mathrm{End}(V), I_B)$ then provides an (s, t)-family in $\mathrm{Sim}(V, B)$. If $\lambda = 1$, PC(m) implies that (V, B) is similar to a Pfister form, so it has a decomposition into binary forms, as in the claim. If $\lambda = -1$, then $(V, B) \simeq \begin{pmatrix} 0 & 1 \\ -1 & 0 \end{pmatrix} \otimes 2^{m-1}\langle 1\rangle$ (compare Exercise 1.7) and again a (V, B) is a product of binary forms. □

A direct proof of the Conjecture 9.17 does not seem obvious even for the cases $m \le 3$.

One tiny bit of evidence for the truth of PC(m) is the observation that if $\dim q = 2^m$ and there is an $(m + 1, m + 1)$-family in $\mathrm{Sim}(q)$, then $q \otimes q$ is a Pfister form. This follows from Exercise 7.14 (3). Of course this condition is far weaker that saying that q itself is a Pfister form.

Appendix to Chapter 9. Pfister forms and function fields

In this appendix we discuss, without proofs, the notion of the function field $F(q)$ of a quadratic form q over F. That theory yields another proof of PC(m) for $m \leq 5$. These "transcendental methods" in quadratic form theory were clarified in the Generic Splitting papers of Knebusch (1976, 1977a). Expositions of this theory appear in Lam's lectures (1977), in Scharlau's text (1985) and in the booklet by Knebusch and Scharlau (1980).

As usual, all quadratic forms considered here are regular and F is a field of characteristic not 2. If q is a form over F and K is an extension field of F we write q_K for $q \otimes K$. We use the notation $q \sim 0$ to mean that q is hyperbolic. (This "$\sim$" stands for Witt equivalence.)

A quadratic form φ of dimension n over F can be considered from two viewpoints. It can be viewed geometrically as an inner product space (V, φ) or it can be viewed algebraically as a polynomial $\varphi(X) = \varphi(x_1, \dots, x_n)$ homogeneous of degree 2 in n variables. Over the field $F(X)$ of rational functions it is clear that the form $\varphi \otimes F(X)$ represents the value $\varphi(X)$. For example the form $\langle a, b\rangle$ represents the value $ax_1^2 + bx_2^2$ over $F(x_1, x_2)$. Furthermore if $\varphi \subset q$ (i.e. φ is isometric to a subform of q) then $q \otimes F(X)$ represents the value $\varphi(X)$.

A.1 Subform Theorem. *Let φ, q be quadratic forms over F such that q is anisotropic. The following statements are equivalent.*

(1) $\varphi \subset q$.

(2) *For every field extension K of F, $D_K(\varphi_K) \subseteq D_K(q_K)$.*

(3) *$q \otimes F(X)$ represents $\varphi(X)$, where $X = (x_1, \dots, x_n)$ is a system of $n = \dim \varphi$ indeterminates.*

This theorem, due to Cassels and Pfister, has many corollaries. Among them is the following characterization of Pfister forms as the forms which are "generically multiplicative".

A.2 Corollary. *Let φ be an anisotropic form over F with* $\dim \varphi = n$. *Let X, Y be systems of n indeterminates. The following statements are equivalent.*

(1) *φ is a Pfister form.*

(2) *For every field extension K of F, $D_K(\varphi_K)$ is a group.*

(3) *$\varphi \otimes F(X, Y)$ represents the value $\varphi(X) \cdot \varphi(Y)$.*

(4) $\varphi(X) \in G_{F(X)}(\varphi_{F(X)})$.

Suppose φ is a quadratic form of dimension n over F and X is a system of n indeterminates. If $n \geq 2$ and $\varphi \not\cong \mathbb{H}$ then $\varphi(X)$ is an irreducible polynomial and we define the *function field*

$$F(\varphi) = \text{the field of fractions of } F[X]/(\varphi(X)).$$

Certainly φ becomes isotropic over $F(\varphi)$, for if $\xi_i \in F(\varphi)$ is the image of x_i, then $\varphi(\xi_1, \dots, \xi_n) = 0$. In fact $F(\varphi)$ is a "generic zero field" for φ in the sense of Knebusch (1976). Changing the variables in φ or multiplying φ by a non-zero scalar alters the function field $F(\varphi)$ only by an isomorphism.

If $\varphi \simeq \langle 1 \rangle \perp \psi$ then $\varphi(X) = x_1^2 + \psi(X')$ where $X' = (x_2, \dots, x_n)$ and we calculate that

$$F(\varphi) \cong F(X')(\sqrt{\psi(X')}).$$

For example if $\varphi \simeq \langle 1, a \rangle$ then $F(\varphi) \cong F(x)(\sqrt{-a})$, a purely transcendental extension of $F(\sqrt{-a})$. If φ is isotropic then $F(\varphi)$ is a purely transcendental extension of F. (See Exercise 12.) To simplify later statements let us define $F(\mathbb{H}) = F(x)$, where x is an indeterminate.

Using results about quadratic forms over valuation rings Knebusch proved the following result about norms of similarities.

A.3 Norm Theorem. *Let φ, q be quadratic forms over F such that φ represents* 1 *and* $\dim \varphi = m \geq 2$. *Let X be a system of m indeterminates. The following are equivalent.*

(1) $q \otimes F(\varphi) \sim 0$.

(2) $\varphi(X) \in G_{F(X)}(q_{F(X)})$.

The condition (2) here is equivalent to the existence of a "rational composition formula"

$$\varphi(X) \cdot q(Y) = q(Z)$$

where $X = (x_1, \dots, x_m)$ and $Y = (y_1, \dots, y_n)$ are systems of independent indeterminates and each entry z_k of Z is a linear form in Y with coefficients in $F(X)$. If each z_k is actually bilinear in X, Y then we have $\varphi < \text{Sim}(q)$, as in (1.9) (3).

A.4 Corollary. *Let φ be an anisotropic form which represents* 1 *and* $\dim \varphi \geq 2$ *over F. Then φ is a Pfister form if and only if $\varphi \otimes F(\varphi) \sim 0$.*

Proof. If φ is a Pfister form then since $\varphi \otimes F(\varphi)$ is isotropic it must be hyperbolic by (5.2) (2). The converse follows from (A.3) and (A.2). □

A.5 Corollary. *Suppose q is an anisotropic form and $q \otimes F(\varphi) \sim 0$. Then φ is similar to a subform of q. In particular* $\dim \varphi \leq \dim q$.

Proof. Let $b \in D_F(\varphi)$ so that $\langle b \rangle \varphi$ represents 1. The Norm Theorem then implies that $b \cdot \varphi(X) \in G_{F(X)}(q_{F(X)})$. For any $a \in D_F(q)$ it follows that $q_{F(X)}$ represents $ab \cdot \varphi(X)$ and the Subform Theorem implies that $\langle ab \rangle \varphi \subset q$. □

A.6 Corollary. *Let φ be a Pfister form and q and anisotropic form over F. Then $q \otimes F(\varphi) \sim 0$ if and only if $\varphi \mid q$.*

Proof. If $\varphi \mid q$ apply (A.4). Conversely suppose $q \otimes F(\varphi) \sim 0$. Then (A.5) implies $q \simeq \langle a_1 \rangle \varphi \perp q_1$ for some $a_1 \in F^\bullet$ and some form q_1. But then $q_1 \otimes F(\varphi) \sim 0$, since φ is a Pfister form, and we may proceed by induction. □

This corollary is a direct generalization of Lemma 3.20(2) since if $\varphi = \langle\langle b \rangle\rangle$ then $F(\varphi)$ is a purely transcendental extension of $F(\sqrt{-b})$. Now let us apply these results to our questions about spaces of similarities.

A.7 Lemma. *If $\sigma < \mathrm{Sim}(q)$ where $\dim \sigma \geq 2$ then $q \otimes F(\sigma) \sim 0$.*

Proof. For any field extension K of F, $\sigma_K < \mathrm{Sim}(q_K)$. Since $\sigma \otimes F(\sigma)$ is isotropic the claim follows from (1.4). Here is another proof: We may assume σ represents 1. Let X be a system of $s = \dim \sigma$ indeterminates. Since $\sigma_{F(X)}$ represents $\sigma(X)$ and $\sigma_{F(X)} < \mathrm{Sim}(q_{F(X)})$ we conclude that $\sigma(X) \in G_{F(X)}(q_{F(X)})$. The Norm Theorem applies. □

The anisotropic cases of (1.10) follow as corollaries. For example, suppose $\langle 1, a, b \rangle < \mathrm{Sim}(q)$ where q is anisotropic. Let $\varphi = \langle\langle a, b \rangle\rangle$ and note that $\langle 1, a, b \rangle \otimes F(\varphi)$ is isotropic. Then the argument in (A.7) implies that $q \otimes F(\varphi) \sim 0$ and (A.6) implies that $\varphi \mid q$.

By the Expansion Proposition 7.6 the following statement of the conjecture is equivalent to "PC(m) over all fields":

Pfister Factor Conjecture. If $\sigma < \mathrm{Sim}(q)$ where $\dim q = 2^m$ and $\dim \sigma = \rho(2^m)$ then q is similar to a Pfister form.

A.8 Lemma. *The following statement is equivalent to the Pfister Factor Conjecture. Suppose $\sigma < \mathrm{Sim}(q)$ where $\dim q = 2^m$ and $\dim \sigma = \rho(2^m)$. If q is isotropic then q is hyperbolic.*

Proof. If q is similar to a Pfister form and is isotropic then it is hyperbolic by (5.2). Conversely suppose the statement here is true and $\sigma < \mathrm{Sim}(q)$ over F where $\dim q = 2^m$ and $\dim \sigma = \rho(2^m)$. Then $\sigma \otimes F(q) < \mathrm{Sim}(q \otimes F(q))$ and the assumed statement implies that $q \otimes F(q)$ is hyperbolic. By (A.4) it follows that q is similar to a Pfister form. □

In trying to prove this conjecture we suppose that $\sigma < \mathrm{Sim}(q)$ as above. Assuming that q is isotropic but not hyperbolic we try to derive a contradiction. Express $q = q_a \perp k\mathbb{H}$ where q_a is anisotropic and non-zero. Then $q_a \otimes F(\sigma) \sim 0$ by (A.7) and therefore $\dim q_a \geq \dim \sigma = \rho(2^m)$, by (A.5). If $m \leq 3$ this already provides a contradiction since $\rho(2^m) = 2^m = \dim q$ in those cases. The case $m = 4$ is settled by the next lemma which we could have proved after (1.4).

A.9 Lemma. *Suppose $S \subseteq \mathrm{Sim}(V, q)$ is a (regular) subspace where* $\dim S = s$. *Suppose q is isotropic but not hyperbolic and $v \in V$ is an isotropic vector. Then $S \cdot v$ is a totally isotropic subspace of V of dimension s.*

Proof. If $f \in S$ then $q(f \cdot v) = \mu(f)q(v) = 0$. Therefore $S \cdot v$ is totally isotropic. Suppose f is in the kernel of the evaluation map $\varepsilon : S \to S \cdot v$. Then $f(v) = 0$ so that f is not injective and it follows that $\mu(f) = 0$. However (1.4) implies that S is anisotropic and consequently $f = 0$. Therefore ε is a bijection. □

Now suppose $m = 4$, so that $\dim q = 16$ and $\dim \sigma = 9$. The lemma implies that q has a totally isotropic subspace of dimension 9 which is certainly impossible since $9\mathbb{H}$ cannot fit inside q. If $m = 5$ then $\dim q = 32$ and $\dim \sigma = 10$ and the lemma shows that $10\mathbb{H} \subset q$. Therefore $10 \le \dim q_a \le 12$, since the earlier argument implies that $\dim q_a \ge \dim \sigma = 10$. The next idea is to observe that these inequalities hold over any extension field K such that $q \otimes K$ is not hyperbolic.

A.10 Proposition. *Suppose q is a form of even dimension which is not hyperbolic over F. Then there exists an extension field K such that $q \otimes K \simeq \psi \perp k\mathbb{H}$ and ψ is similar to an anisotropic (non-zero) Pfister form.*

Proof. Suppose $q \simeq q_0 \perp i_0\mathbb{H}$ where q_0 is anisotropic. Let $F_1 = F(q_0)$ be the function field so that $q_0 \otimes F_1 \simeq q_1 \perp i_1\mathbb{H}$ for some anisotropic form q_1 and some $i_1 \ge 1$. If $q_1 \ne 0$ let $F_2 = F_1(q_1)$ and express $q_1 \otimes F_2 \simeq q_2 \perp i_2\mathbb{H}$ for some anisotropic form q_2 and some $i_2 \ge 1$. Repeat this process to get a tower of fields $F \subseteq F_1 \subseteq F_2 \subseteq \cdots \subseteq F_h$ where $q \otimes F_h \sim 0$ but $q \otimes F_{h-1} \not\sim 0$. Let $K = F_{h-1}$ and express $q \otimes K \simeq \psi \perp k\mathbb{H}$ where $\psi = q_{h-1}$ is anisotropic. By construction $\psi \ne 0$ and $\psi \otimes K(\psi) \sim 0$. Therefore ψ is similar to a Pfister form by (A.4). □

A.11 Proposition. *The Pfister Factor Conjecture is true if $m \le 5$.*

Proof. We already settled the cases $m \le 4$ and showed that if $m = 5$ then $10 \le \dim q_a \le 12$. Replacing F by the field K of (A.10) we get the extra information that $\dim q_a$ is a power of 2. This contradiction completes the proof. □

Exercises for Chapter 9

1. ***u*-invariants.** For a nonreal field F, $u(F)$ is defined to be the maximal dimension of an anisotropic quadratic form over F.

(1) Suppose $u = u(F(\sqrt{a}))$ is finite. Then every anisotropic form σ over F with $\dim \sigma \ge u + 3$ contains a 4-dimensional subform of determinant $\langle 1 \rangle$.

(2) If $u(F(\sqrt{a})) \leq 8$ then for every m, SC(m) is true over F. For example this condition holds if F is an extension of $\mathbb{R}$ of transcendence degree ≤ 3.

(Hint. (1) Use Lemma 3.20.

(2) The theory of C_i-fields shows that if $K/\mathbb{C}$ has transcendence degree ≤ 3 then $u(K) \leq 8$. See e.g. §2.15 of Scharlau (1985).)

2. If every form over F of dimension 12 contains a 4-dimensional subform of determinant $\langle 1\rangle$, then PC(m) is true for all m.

(Hint. If $m = 6$ suppose $(\sigma, \tau) < \mathrm{Sim}(q)$ is an $(11, 3)$-family where $\dim q = 64$. Find a related $(12, 0)$-family and apply (9.11) and (2.10) to find that $q \simeq \langle\langle a\rangle\rangle \otimes q'$ where $\dim q' = 32$ and q' admits a $(7, 3)$-family. From $\rho_3(32) = 7$ use (7.12) to find a $(6, 6)$-family in $\mathrm{Sim}(q')$.)

3. (1) Suppose (σ, τ) is a pair such that $\dim \sigma \geq 8$ and σ contains an Albert subform. Then $(\sigma, \tau) \approxeq (\sigma', \tau')$ where τ' is isotropic. Consequently, if $(\sigma, \tau) < \mathrm{Sim}(q)$ then q must be hyperbolic. Compare Exercise 6.4 (4).

(2) Extend the definition of the equivalence $\approxeq$ to include cases as mentioned in Exercise 2.4 (1). Will this change the validity of results in Chapter 9?

(Hint. (1) Scale to assume $\sigma \simeq \langle a, b, ab\rangle \perp \langle -x, -y, -xy\rangle \perp \langle u, v, \ldots\rangle$. Shift twice.)

4. Let $F((t))$ be the field of formal Laurent series over F. Then PC(m) over $F((t))$ implies PC(m) and PC($m-1$) over F.

(Hint. Use Springer's Theorem about quadratic forms over valued fields.) Compare Exercise 10.

5. Suppose q is a form of dimension 2^m over F and there is an $(m+1, m+1)$-family in $\mathrm{Sim}(q)$. Then $q \in I^3F$. What are the possible values of the signature $\mathrm{sgn}_P(q)$ when P is an ordering of F?

6. **PC(6).** Suppose $\sigma < \mathrm{Sim}(V, q)$ over F where $\dim \sigma = 11$ and $\dim q = 2^6 = 64$. As usual, let $C = C(-\sigma_1)$ and $A = \mathrm{End}_C(V)$, so that $C \otimes A \cong \mathrm{End}(V)$. Then A is a quaternion algebra with induced involution "bar". If there is a quadratic extension L/F such that $\sigma \otimes L$ is isotropic and $c(\sigma) = [A]$ is split by L, then q must be similar to a Pfister form?

7. **Pfister unsplittables.** Suppose (C, J) is the Clifford algebra with involution associated to an (s, t)-pair (σ, τ) where $s + t = 2m + 1$. Then $(C, J) \cong (Q_1, J_1) \otimes \cdots \otimes (Q_m, J_m)$ where each (Q_k, J_k) is a quaternion algebra with involution as in Exercise 6.4. Suppose $Q_k \cong (a_k, b_k)$ corresponding to generators e_k, f_k where $J(e_k) = \pm e_k$ and $J(f_k) = \pm f_k$.

(1) Suppose all a_k belong to a two element set $\{1, d\}$. Then every (σ, τ)-unsplittable is similar to a Pfister form.

(2) For what (s, t)-pairs does the condition in (1) apply? We can use Exercise 3.14 to get explicit quaternion algebras in C. For example (1) applies when $\sigma = \langle 1\rangle \perp \langle\langle -c\rangle\rangle \otimes \alpha$ and $\tau = 0$. It also applies when $(\sigma, \tau) = (\langle 1\rangle \perp \alpha, \langle 1\rangle \perp \alpha)$.

(Hint. (1) Note that $(d, u) \otimes (d, v) \cong (1, u) \otimes (d, uv)$ and the involution preserves the factors. Then $(C, J) \cong (C', J') \otimes (Q, J'')$ where Q is quaternion and $C' \cong \mathrm{End}(U)$ is a tensor product of split quaternions. Suppose (V, q) is unsplittable for (C, J) and apply (6.11) to find that $(V, q) \simeq (U, \varphi) \otimes (W, \omega)$, where (W, ω) is an unsplittable (Q, J'')-module. Show that φ and ω are Pfister.)

8. **Definition.** $I^n F$ is *linked* if every pair φ, ψ of n-fold Pfister forms is linked. That is, $\varphi \simeq \langle\langle a\rangle\rangle \otimes \alpha$ and $\psi \simeq \langle\langle b\rangle\rangle \otimes \alpha$ for some $(n-1)$-fold Pfister form α. The linked fields mentioned above are the ones where $I^2 F$ is linked.

Proposition. *If $I^3 F$ is linked then for every m,* $\mathrm{SC}(m)$ *is true over F.*

(Hint. If $I^n F$ is linked then every anisotropic $q \in I^n F$ has a "simple decomposition": $q \simeq \varphi_1 \perp \cdots \perp \varphi_k$ where each φ_j is similar to an n-fold Pfister form. (See Elman, Lam and Wadsworth (1979), Corollary 3.6.) Given $(\sigma, \tau) \in \mathcal{P}^\circ_m$ let $\beta = \sigma \perp -\tau$. By Merkurjev's Theorem $\beta \in I^3 F$. We may assume β is anisotropic. A simple decomposition implies $t \equiv 0 \pmod 4$. Shift to assume $\tau = 0$, use the decomposition and (9.10).)

9. **Adjusting signatures.** If P is an ordering of F then $\mathrm{sgn}_P(\sigma)$ denotes the signature of the form σ relative to P.

(1) Suppose P is an ordering of F. If $(\sigma, \tau) \in \mathcal{P}^\circ_m$ then

$$\mathrm{sgn}_P(\sigma) \equiv \mathrm{sgn}_P(\tau) \pmod 8.$$

(2) **Signature Shift Conjecture.** If $(\sigma, \tau) \in \mathcal{P}^\circ_m$ then $(\sigma, \tau) \approx (\sigma', \tau')$ for some pair (σ', τ') where $\dim \sigma' = \dim \tau'$ and $\mathrm{sgn}_P(\sigma') = \mathrm{sgn}_P(\tau')$ for all orderings P.

Definition. F has the property ED if for every $b \in F^\bullet$ and every form q over F such that $q \perp \langle -b\rangle$ is totally indefinite, q represents bt for some totally positive $t \in F^\bullet$.

Lemma. *If the field F satisfies ED then the Signature Shift Conjecture holds. This applies, for example, if F is an algebraic extension of a uniquely ordered field.*

Remark. It might be possible to find a counterexample to $\mathrm{SC}(m)$ by finding a field for which the Signature Shift Conjecture fails.

(Hint. (1) If $\beta \in I^3\mathbb{R}$ then $\dim \beta \equiv 0 \pmod 8$.

(2) Mimic the idea in (9.10).)

10. **Laurent series fields.** Let F be a complete discrete valued field with valuation ring $\mathcal{O}$, maximal ideal $\mathfrak{m} = \pi\mathcal{O}$, and non-dyadic residue field $k = \mathcal{O}/\mathfrak{m}$ (i.e. char $k \neq 2$).

A quadratic form q over F has "good reduction" if there exists an orthogonal basis $\{e_1, \dots, e_n\}$ such that $q(e_i) \in \mathcal{O}^\bullet$. In this case let $L = \mathcal{O}e_1 + \dots + \mathcal{O}e_n$, a free $\mathcal{O}$-module. There is a corresponding "reduced" form $\bar{q}$ over k obtained from $L/\mathfrak{m}L$. By Springer's Theorem the isometry class of $\bar{q}$ is independent of the choice of basis and $\bar{q}$ is isotropic iff q is isotropic. Any quadratic form q over F can be expressed as $q = q_1 \perp \langle\pi\rangle q_2$ where q_1 and q_2 have good reduction. These reduced forms $\bar{q}_1$ and $\bar{q}_2$ are uniquely determined up to Witt equivalence. (For more details see the texts of Lam or Scharlau.)

(1) **Lemma.** *Suppose (V, q) is anisotropic with good reduction, and $L \subseteq V$ as above. If $f \in \mathrm{Sim}(V, q)$ with norm $\mu(f) \in \mathcal{O}$ then $f(L) \subseteq L$.*

(2) **Corollary.** *Suppose q, σ, τ are anisotropic forms with good reduction over F. If $(\sigma, \tau) < \mathrm{Sim}(q)$ over F then $(\bar{\sigma}, \bar{\tau}) < \mathrm{Sim}(\bar{q})$ over k.*

(3) Suppose $F = k((t))$ is a Laurent series field. If (V, q) is a quadratic space over F then $(V, q) = (V_1, q_1) \perp (V_2, \langle t\rangle q_2)$ where q_1, q_2 are forms with good reduction. If q is anisotropic then the subspaces V_i are uniquely determined. E.g. $V_1 = \{v \in V : q(v) \in k\}$.

(4) **Corollary.** *Suppose σ, τ, q_1, q_2 are anisotropic forms over k. If $(\sigma, \tau) < \mathrm{Sim}(q_1 \perp \langle t\rangle q_2)$ over $k((t))$ then $(\sigma, \tau) < \mathrm{Sim}(q_1)$ and $(\sigma, \tau) < \mathrm{Sim}(q_2)$ over k.*

(5) **Corollary.** *Suppose σ, τ, q are anisotropic forms over k. Then $(\sigma \perp \langle t\rangle, \tau \perp \langle t\rangle) < \mathrm{Sim}(q \otimes \langle\langle t\rangle\rangle)$ over $k((t))$ iff $(\sigma, \tau) < \mathrm{Sim}(q)$ over k.*

(Hint. (1) Suppose $v \in L$ and let r be the smallest non-negative integer with $\pi^r \cdot f(v) \in L$. If $r > 0$ then $q(\pi^r \cdot f(v)) \in \mathfrak{m}$ and the anisotropy implies $\pi^r \cdot f(v) \in \mathfrak{m}L = \pi L$, contrary to the minimality.)

11. **History.** (1) The following result of Cassels (1964) was a major motivation for Pfister's theory: $1 + x_1^2 + \dots + x_n^2$ is not expressible as a sum of n squares in $\mathbb{R}(x_1, \dots, x_n)$.

(2) The level $s(F)$ was defined in Exercise 5.5. Given m there exists a field of level 2^m. In fact let $X = (x_1, \dots, x_n)$ be a system of indeterminates, let $d = x_1^2 + \dots + x_n^2$ and define $K_n = \mathbb{R}(X)(\sqrt{-d})$. If $2^m \le n < 2^{m+1}$ Pfister proved: $s(K_n) = 2^m$.

(3) The function field methods in quadratic form theory began with the "Hauptsatz" of Arason and Pfister:

Theorem. *If q is a non-zero anisotropic form in I^nF then* $\dim q \ge 2^n$.

(Hint. (1) Use $q = n\langle 1\rangle$ and $\varphi(x) = x_0^2 + \dots + x_n^2$ over $\mathbb{R}(x_0, \dots, x_n)$ in the Subform Theorem (A.1).

(2) Apply Exercise 5.5. Alternatively, K_n is equivalent to the function field $\mathbb{R}((n+1)\langle 1\rangle)$. Certainly $s(K_n) \le n$ hence $s(K_n) \le 2^m$. If not equal then $2^m\langle 1\rangle$ is isotropic, hence hyperbolic, over K_n. Get a contradiction using (A.5).

(3) Given $q \sim \langle c_1\rangle\varphi_1 \perp \dots \perp \langle c_k\rangle\varphi_k$ where each φ_j is an n-fold Pfister form. Suppose $k > 1$ and assume the result for any such sum of fewer than k terms (over any field). If $q \otimes F(\varphi_1) \sim 0$ apply (A.5). Otherwise apply the induction hypothesis to the anisotropic part of $q \otimes F(\varphi_1)$.)

12. (1) Suppose $K = F(x_1, \ldots, x_n)$ is a purely transcendental extension of F and q is a form over F. If $q \otimes K$ is isotropic then q must be isotropic over F.

(2) Let φ be a form over F with $\dim\varphi \geq 2$. Then $F(\varphi)/F$ is purely transcendental if and only if φ is isotropic.

(Hint. (2) Suppose φ is isotropic with $\dim\varphi > 2$. Changing variables we may assume that $\varphi(X) = x_1x_2 + \alpha$ where $\alpha = \alpha(X')$ is a non-zero quadratic form in $X' = (x_3, \ldots, x_n)$.)

13. **More versions of PC(m).** The following statements are equivalent to PC(m) (over all fields):

(1) If $\dim\varphi = 2^m$, φ represents 1, and there is an $(m+1, m+1)$-family in $\operatorname{Sim}(V, \varphi)$, then φ is round. That is: for every $c \in D_F(\varphi)$ there exists $f \in \operatorname{Sim}^\bullet(\varphi)$ with $\mu(f) = c$.

(2) Suppose (A, K) is a tensor product of m quaternion algebras with involution, as in (9.17). Suppose A is split and there exists $0 \neq h \in A$ with $J(h) \cdot h = 0$. Then for every $c \in F$ there exists $f \in A$ such that $J(f) \cdot f = c$.

(Hint. (1) Use (A.2).

(2) Let $A \cong \operatorname{End}(V)$ where $\dim V = 2^m$ with J corresponding to I_φ, for a quadratic form φ on V. Equivalently $\operatorname{Sim}(V, \varphi)$ admits an $(m+1, m+1)$-family. The condition $I_\varphi(h) \cdot h = 0$ implies φ is isotropic. The conclusion says that φ is round.)

14. **Pfister neighbors.** (1) If φ is a hyperbolic form and $\alpha \subset \varphi$ with $\dim\alpha > \frac{1}{2}\dim\varphi$ then α must be isotropic.

(2) A form α is called a *Pfister neighbor* if there is a Pfister form ρ such that $\alpha \subset \langle a\rangle\rho$ for some $a \in F^\bullet$ and $\dim\alpha > \frac{1}{2}\dim\rho$. In this case: α is isotropic iff ρ is hyperbolic. Every form of dimension ≤ 3 is a Pfister neighbor.

(3) If α is a Pfister neighbor then the associated Pfister form is unique.

(4) Suppose α is Pfister neighbor associated to ρ. If $\alpha < \operatorname{Sim}(q)$ and q is anisotropic then $\rho \mid q$. In fact, if $\alpha < \operatorname{Sim}(q)$ and $q \simeq q_0 \perp m\mathbb{H}$ where q_0 is anisotropic, then $\rho \mid q_0$.

(Hint. (1) Viewed geometrically, the space (V, φ) of dimension $2m$ has a totally isotropic subspace S with $\dim S = m$. The subspace (A, α) has $\dim A > m$. Then $A \cap S \neq \{0\}$.

(3) If α is associated to ρ and to ψ then $\psi \otimes F(\rho)$ is isotropic, hence hyperbolic.)

15. **More on Pfister neighbors.** If φ is an m-fold Pfister form and $\langle 1, a, b\rangle \subset \varphi$ then $\varphi \cong \langle\langle a, b, c_3, \ldots, c_m\rangle\rangle$ for some $c_j \in F^\bullet$. (Compare (5.2)(3) and Exercise 5.23.) More generally:

Proposition. *Suppose φ is a Pfister form and α is a Pfister neighbor with associated Pfister form ρ. If $\alpha \subset \varphi$ then $\varphi \cong \rho \otimes \delta$ for some Pfister form δ.*

(Hint. Assume φ is anisotropic. Exercise 14(1) and (A.6) imply that $\rho \mid \varphi$. Then $\varphi \cong \rho \perp \gamma$ for some form γ. If $\dim\gamma > 0$ choose $c \in D_F(\gamma)$ and let $\rho_1 := \rho \otimes \langle\langle c\rangle\rangle$.

Since $\rho \perp \langle c\rangle$ is a subform of φ and is a Pfister neighbor associated to ρ_1 we have $\rho_1 \mid \varphi$. Iterate the argument.)

16. If there is a counterexample to the Pfister Factor Conjecture when $m = 6$, then there exists a field F and $\sigma < \mathrm{Sim}(q)$ where $\dim \sigma = 12$, $\dim q = 64$ and $q \simeq \psi \perp k\mathbb{H}$ where ψ is an anisotropic Pfister form of dimension 16 or 32.

Notes on Chapter 9

Several of the ideas used in the proof of $\mathrm{SC}(m)$ for $m \leq 5$ are due to Wadsworth. In particular he had the idea of examining 4-dimensional subforms of determinant $\langle 1\rangle$. The approach to the Pfister Factor Conjecture given in the appendix follows Wadsworth and Shapiro (1977a).

The property $\mathrm{SC}(m)$ was proved in (9.13) for certain classes of fields. However there exist fields not satisfying any of these properties. For example there is a field F and a quadratic form β such that $\beta \in I^3F$, $\dim \beta = 14$ and β contains no 4-dimensional subform of determinant $\langle 1\rangle$. If fact, if k is a field and $F = k((t_1))((t_2))((t_3))$ is the iterated Laurent series field then there are examples of such β over F. This is proved in Hoffmann and Tignol (1998), where the stated property is called D(14).

The class of linked fields as defined in Lemma 9.14 was first examined by Elman and Lam (1973b). Some of their proofs were simplified by Elman (1977), Elman, Lam and Wadsworth (1979) and Gentile (1985).

(A.10) is due to Knebusch (1976). The Pfister form ψ there is called the "leading form" of q. For further information see Knebusch and Scharlau (1980) or Scharlau (1985), p. 163–165.

Exercise 7. See Yuzvinsky (1985).

Exercise 9. This property ED (for "effective diagonalization") was introduced by Ware and studied by Prestel and Ware (1979).

Exercise 10 follows a communication from A. Wadsworth (1976).

Exercise 14–15. For Pfister neighbors see Knebusch (1977a) or Knebusch and Scharlau (1980).

Chapter 10

Central Simple Algebras and an Expansion Theorem

Our previous expansion result (7.6) followed from an explicit analysis of the possible involutions on a quaternion algebra. The Expansion Theorem in this chapter depends on similar information about involutions on a central simple algebra of degree 4. Albert (1932) proved that any such algebra A is a tensor product of two quaternion algebras. However there can exist involutions J on A which do not arise from quaternion subalgebras. It is the analysis of these "indecomposable involutions" which provides the necessary information for the Expansion Theorem. The principal ingredient is Rowen's observation that a symplectic involution on a central simple algebra of degree 4 must be decomposable.

The chapter begins with a discussion of maximal (s, t)-families and a characterization of those dimensions for which expansions are always possible. The Expansion Theorem requires knowledge of involutions on algebras of degree 4. We derive the needed results from a general theory of Pfaffians. This theory is first described for matrix rings, then lifted to central simple algebras, and finally specialized to algebras of degree 4. The exposition would be considerably shortened if we restrict attention to the degree 4 case from the start. (Most of the results needed here appear in Knus et al. (1998), Ch. IV.) Our long digression about general Pfaffians is included here since it is a novel approach and it helps clarify some of the difficulties of generalizing the theory to larger algebras.

Suppose $(S, T) \subseteq \mathrm{Sim}(V, q)$ is an (s, t)-family. If $\dim V = 2^m$ and $s+t = 2m-1$ the Expansion Proposition (7.6) says that (S, T) can be enlarged to some family of maximal size. We will sharpen this result by showing families of certain smaller sizes can also be enlarged. For example let us consider the case $\dim q = 16$. If $S \subseteq \mathrm{Sim}(V, q)$ where $\dim S = 5$ then there exists T such that $(S, T) \subseteq \mathrm{Sim}(V, q)$ is a $(5, 5)$-family. On the other hand there exist quadratic forms q with $\dim q = 16$ such that $\mathrm{Sim}(q)$ has $(3, 3)$-families but admits no (s, t)-families of larger size. See Exercise 1.

The Expansion Lemma (2.5) provides examples of maximal families. For instance if $S_0 \subseteq \mathrm{Sim}(V, q)$ is a 3-dimensional subspace with orthogonal basis $\{1_V, f, g\}$ then it can be expanded by adjoining fg. The expanded space $S = \mathrm{span}\{1_V, f, g, fg\}$ is maximal family because no non-zero map can anticommute with f, g and fg.

If $\mu(f) = a$ and $\mu(g) = b$ the quadratic form on S is $\sigma = \langle 1, a, b, ab\rangle$ and the associated Clifford algebra is $C = C(-\sigma_1) = C(\langle -a, -b, -ab\rangle)$. If $\{e_1, e_2, e_3\}$ is the set of generators of C then $z = e_1e_2e_3$ is the element of highest degree, generating the center of C. If $\pi : C \to \mathrm{End}(V)$ is the representation corresponding to S we see that $\pi(z) = (f)(g)(fg) = -ab \cdot 1_V$, a scalar. Then π is not faithful (i.e. not injective). This sort of behavior always occurs when a family arises from the Expansion Lemma. More generally recall the properties of the character $\chi(S, T)$ defined in (7.17). The next lemma is a repetition of (7.18).

10.1 Lemma. *Suppose $(S, T) \subseteq \mathrm{Sim}(V, q)$ is an (s, t)-family with forms (σ, τ). If $\chi(S, T) \neq 0$ then $s \equiv t \pmod 4$, $d\sigma = d\tau$ and (S, T) is maximal.*

We call this sort of family "trivially maximal". If $s + t$ is odd then no (s, t)-family can be maximal since we can always expand by one dimension to get a non-faithful (maximal) family. To avoid this sort of triviality we will investigate when (S, T) can be expanded by 2 (or more) dimensions.

We have already considered some expansion results. For example Proposition 7.6 states that if $(S, T) \subseteq \mathrm{Sim}(V, q)$ is an (s, t)-family such that $\dim q = 2^m$ and $s + t = 2m - 1$, then (S, T) can be expanded by 3 dimensions. As another example, recall that $\langle 1, a\rangle < \mathrm{Sim}(q)$ if and only if $(\langle 1, a\rangle, \langle 1, a\rangle) < \mathrm{Sim}(q)$, and similarly for $\langle 1, a, b\rangle < \mathrm{Sim}(q)$. These results are be generalized in the next proposition, which is a mild refinement of (7.12).

10.2 Proposition. *Let (σ, τ) be a minimal pair with unsplittable (σ, τ)-modules of dimension 2^m. Suppose $(S, T) \subseteq \mathrm{Sim}(V, q)$ is an (s, t)-family with forms (σ, τ). Then there is an associated (s', t')-family in $\mathrm{Sim}(V, q)$ with $s' + t' = 2m + 2$.*

Proof. If (S, T) is trivially maximal, this associated family cannot be an actual expansion of (S, T). Let $C = C(-\sigma_1 \perp \tau)$ with the usual involution J, and let (W, ψ) be an unsplittable (C, J)-module. If C does not act faithfully on W, we replace (S, T) by a smaller family obtained by deleting one dimension. This smaller family is still minimal. By (7.11) we know that every unsplittable module is (C, J)-similar to (W, ψ). The Decomposition Theorem 4.1 then yields a (C, J) isometry $(V, q) \simeq (W, \psi) \otimes_F \langle a_1, \dots, a_r\rangle$ for some $a_i \in F^\bullet$. Now the Expansion Proposition 7.6 can be applied to (W, ψ) to produce the larger family as desired. □

Suppose $(S, T) \subseteq \mathrm{Sim}(V, q)$ is an (s, t)-family with $s + t$ odd. Let (σ, τ) be the corresponding forms and $C = C(-\sigma_1 \perp \tau)$ the associated Clifford algebra. Then C is a central simple F-algebra of dimension 2^{s+t-1} and the given representation $\pi : C \to \mathrm{End}(V)$ induces an isomorphism

$$C \otimes A \cong \mathrm{End}(V)$$

where $A = \mathrm{End}_C(V)$ is the centralizer of C in $\mathrm{End}(V)$. Then A is also a central simple F-algebra and since the involution J on C and I_q on $\mathrm{End}(V)$ are compatible, there is an induced involution K on A.

10.3 Lemma. *(S, T) can be expanded by 2 dimensions if and only if there is a quaternion subalgebra $Q \subseteq A$ which is preserved by the involution K.*

Proof. Such Q exists if and only if there exist $a, b \in A$ such that a^2 and b^2 are in $F^\bullet$, a, b anticommute, $K(a) = \pm a$ and $K(b) = \pm b$. Let z be an element of highest degree in C so that z anticommutes with $S_1 + T$, $z^2 \in F^\bullet$ and $J(z) = \pm z$. Let $f = za$ and $g = zb$. Then Q exists if and only if there exist $f, g \in \mathrm{End}(V)$ which anticommute with $S_1 + T$, f^2 and g^2 are in $F^\bullet$, $I_q(f) = \pm f$ and $I_q(g) = \pm g$. This occurs if and only if (S, T) can be expanded by 2 dimensions. □

Of course this lemma is just a slight generalization of the Expansion Proposition 7.6. In order to go further we need information about quaternion subalgebras of larger algebras with involution. Recall that if A is a central simple F-algebra then $\dim_F A = n^2$ is a perfect square (since over some splitting field E, $A \otimes E \cong \mathbb{M}_n(E)$ for some n). Define the *degree* of the algebra A to be this integer n. Then a quaternion algebra has degree 2.

The basic examples of central simple F-algebras with involution are tensor products of split algebras and quaternion algebras. For instance if $A \cong Q_1 \otimes Q_2$ where Q_1 and Q_2 are quaternion algebras, then A is a central simple algebra of degree 4. Certainly this A has an involution, since we can use $J = J_1 \otimes J_2$ where J_i is an involution on Q_i. We consider the converse.

10.4 Definition. Let A be a central simple F-algebra. Then A is *decomposable* if

$$A \cong A_1 \otimes A_2$$

for some central simple F-algebras A_i with $\deg A_i > 1$.

If J is an involution on A then (A, J) is *decomposable* if $(A, J) \cong (A_1, J_1) \otimes (A_2, J_2)$ for some central simple F-algebras A_i with involutions J_i and with $\deg A_i > 1$. When the algebra A is understood we say that the involution J is decomposable.

Note that J is decomposable if and only if there exists a proper J-invariant central simple subalgebra A_1 of A. For A_2 can be recovered as the centralizer of A_1.

Every algebra of prime degree is certainly indecomposable. In particular, quaternion algebras are indecomposable. If $A \cong \mathrm{End}(V)$ is split and J is any involution of symplectic type on A then J is decomposable if and only if $\deg A > 2$. Similarly if $J = I_q$ is the adjoint involution of a quadratic form q on V and if $q \simeq \alpha \otimes \beta$ for some quadratic forms α, β of dimension > 1, then J is decomposable. (See (6.10).)

Let us now concentrate on algebras of degree 4. Albert (1932) proved that if A has degree 4 and possesses an involution then A is decomposable as a tensor product of quaternion subalgebras. Rowen (1978) used Pfaffians to prove that symplectic involutions on a division algebra of degree 4 are always decomposable. The next theorem is a refinement of these results.

10.5 Theorem. *Let A be a central simple F-algebra of degree 4 with involution. Then $A \simeq Q_1 \otimes Q_2$ for some quaternion algebras Q_i.*

(1) *If J is an involution on A of symplectic type then J is decomposable. Furthermore if $y \in A - F$ such that $y^2 \in F^\bullet$ and $J(y) = \pm y$, then there exists a J-invariant quaternion subalgebra Q with $y \in Q$.*

(2) *Suppose J is an involution on A of orthogonal type. Then J is decomposable if and only if there exists $y \in A$ such that $y^2 \in F^\bullet$ and $J(y) = -y$. Furthermore if such y is given, then there exists a J-invariant quaternion subalgebra Q with $y \in Q$.*

Certainly there exist indecomposable involutions on split algebras of degree 4, provided F is not quadratically closed. (Just use I_q on $\text{End}(V)$ where $(V, q) \simeq \langle 1, 1, 1, c\rangle$ for some non-square $c \in F$.) Indecomposable involutions on division algebras of degree 4 were first exhibited by Amitsur, Rowen, Tignol (1979). These examples were clarified by work of Knus, Parimala, Sridharan on the "discriminant" of an involution. We present an exposition of the theory of Pfaffians, the characterization of indecomposable involutions on algebras of degree 4, and a proof of Theorem 10.5.

Before beginning those tasks, we mention an easy lemma and then apply that theorem to deduce another expansion result for (s, t)-families.

10.6 Lemma. *Suppose A is a central simple F-algebra of degree 4 with involution J. Then (A, J) is decomposable if and only if $(A, J) \cong (C(U, \alpha), J')$ for some 4-dimensional quadratic space (U, α) and some involution J' which preserves U.*

Proof. Suppose A is a product of two invariant quaternion algebras. Choose generators which are J-invariant (i.e. $J(x) = \pm x$). Alter the two quaternion algebras to a Clifford algebra as in (3.14), and note that the Clifford generators are still J-invariant. □

Suppose $(S, T) \subseteq \text{Sim}(V, q)$ is an (s, t)-family where $\dim q = 2^m$ and $s + t = 2m - 3$. Then $\dim C = 2^{2m-4}$ and the centralizer A will be central simple of degree 4. If the induced involution K on A has symplectic type then (10.3) and (10.5) imply that (S, T) can be expanded to a family of maximal size. This is the situation mentioned at the start, when $S \subseteq \text{Sim}(q)$ where $\dim q = 16$ and $\dim S = 5$.

For exactly which dimensions s, t and 2^m are we guaranteed that a family will expand to one of maximal size? One necessary condition is easily verified: if $s = \rho_t(2^{m-2})$ then there exists some (s, t)-family on 2^m-space (over some field) which cannot be expanded by 2 dimensions.

In fact we can construct one over the real field $\mathbb{R}$. For such s, t, m there is a family $(s\langle 1\rangle, t\langle 1\rangle) < \mathrm{Sim}(2^{m-2}\langle 1\rangle)$. Therefore $(s\langle 1\rangle, t\langle 1\rangle) < \mathrm{Sim}(q)$ where $q = 2^{m-2}\langle 1, 1, 1, -1\rangle$. Then $\dim q = 2^m$ but q is not a Pfister form. Then $\mathrm{Sim}(q)$ admits no family of maximal size because $\mathrm{PC}(m)$ holds over $\mathbb{R}$. We prove that this necessary condition is also sufficient.

10.7 Expansion Theorem. *Suppose $(S, T) \subseteq \mathrm{Sim}(V, q)$ is an (s, t)-family and $\dim q = 2^m$. If $s > \rho_t(2^{m-2})$ then there is an associated (s', t')-family $(S', T') \subseteq \mathrm{Sim}(V, q)$ where $s' + t' = 2m + 2$.*

Here the family (S', T') might not be an expansion of (S, T), since (S, T) could be trivially maximal. For such cases $s+t$ is even and the representation is not faithful. Then we first pass to a subfamily of (S, T) of codimension 1 and expand that to the family (S', T').

Note. That inequality is equivalent to:

$$s + t \geq \begin{cases} 2m - 3 & \text{if } m \equiv t \\ 2m - 1 & \text{if } m \equiv t + 1 \\ 2m - 2 & \text{if } m \equiv t + 2 \\ 2m - 3 & \text{if } m \equiv t + 3 \end{cases} \quad (\text{mod } 4).$$

Of course this condition is related to the condition for minimal pairs given in (7.9). In this situation an unsplittable (σ, τ)-module must have dimension 2^{m-1} or 2^m. In the former case we find that (σ, τ) is a minimal pair and the unsplittable module $(S, T) \subseteq \mathrm{Sim}(W, \varphi)$ is unique by (7.11). Since (V, q) is a sum of unsplittable components, it follows that $(S, T) \subseteq \mathrm{Sim}(V, q)$ expands uniquely to a family of maximal size. Therefore the new content of the theorem occurs when unsplittables have dimension 2^m.

Proof. If $s + t = 2m - 1$ then (7.6) implies that the family always expands by 3 dimensions. Suppose $s + t = 2m - 3$ and $m \equiv t$ or $t + 3 \pmod 4$. Then $C \otimes A \cong \mathrm{End}(V)$ with involutions $J \otimes K \cong I_q$, where (A, K) is an algebra of degree 4 with involution. Since $s = 2m - 3 - t \equiv t \pm 3 \pmod 8$ we see from (7.4) that the involution J on C has type -1. Then (6.9) implies that K has type -1 on A. Now (10.5) and (10.6) imply that

$$(A, K) \cong (C(U, \alpha), J')$$

where $\dim U = 4$ and J' preserves U. Then there exists an orthogonal basis $h_1, \dots, h_4$ of U such that $J'(h_i) = \pm h_i$. Then the elements zh_i, along with $zh_1h_2h_3h_4$, can be adjoined to (S, T) to provide a family of maximal size ($s' + t' = 2m + 2$).

Finally suppose that $s+t = 2m-2$ and $m \equiv t+2 \pmod 4$. Then $s \equiv t+2 \pmod 8$, the involution K has type 1 and $J(z) = -z$. Then the representation $\pi : C \to \mathrm{End}(V)$ cannot send z to a scalar, and therefore π must be faithful. We may identify C with its image $\pi(C) \subseteq \mathrm{End}(V)$. Since C_0 is central simple of dimension 2^{2m-4} its centralizer

A is central simple of degree 4 and

$$C_0 \otimes A \cong \mathrm{End}(V).$$

Since the involutions J and I_q are compatible, I_q restricts to an involution K on A. Since z commutes with C_0 we find that $z \in A$ and $K(z) = J(z) = -z$. By Theorem 10.5(2) the involution K is decomposable, so that $(A, K) \cong (C(U, \alpha), J')$ as above. Furthermore in this isomorphism the element z corresponds to an element of U. We choose an orthogonal basis $\{z, h_1, h_2, h_3\}$ of U with $I_q(h_i) = \pm h_i$ and expand the family (S, T) by adjoining $\{h_1, h_2, h_3, zh_1h_2h_3\}$. □

There is a fine point to be made here about "maximal" families. Suppose $s+t$ is odd and an (s,t)-family $(S, T) \subseteq \mathrm{Sim}(V, q)$ is given. Let the corresponding forms be σ, τ and suppose that there exists $(\sigma, \tau) \subset (\sigma', \tau') < \mathrm{Sim}(q)$ where $s' + t' = s + t + 2$. It does not necessarily follow that the original family (S, T) can be expanded by 2 dimensions. The explanation is that a given (s, t)-pair (σ, τ) can have different realizations as an (s, t)-family in $\mathrm{Sim}(q)$. (See Exercise 2(2).)

We now begin our analysis of Pfaffians and central simple algebras, ultimately leading to a proof of Theorem 10.5. Few of the results here are new, but the properties of the set $\mathcal{D}(A)$ provide an interesting approach. As usual in this book we assume that F is a field of *characteristic not* 2. This restriction simplifies the exposition. The results have analogs in characteristic 2 and there exist treatments of the subject which unify both cases.

If A is an F-algebra (always assumed finite dimensional, associative and with 1) then $A^\bullet$ denotes the group of invertible elements in A. If $\mathcal{S} \subseteq A$ is a subset we write $\mathcal{S}^\bullet$ for the set $\mathcal{S} \cap A^\bullet$.

10.8 Classical Definition. Let S be a skew-symmetric $n \times n$ matrix over F such that n is even. Then the *Pfaffian* $\mathrm{Pf}(S) \in F$ is defined with the following properties:

(1) $\mathrm{Pf}(S)$ is a form (homogeneous polynomial) of degree $n/2$ in the entries of S. In particular $\mathrm{Pf}(cS) = c^{n/2}\,\mathrm{Pf}(S)$ for any $c \in F$.

(2) $\mathrm{Pf}(S)^2 = \det S$.

(3) $\mathrm{Pf}(P^\top \cdot S \cdot P) = \mathrm{Pf}(S) \cdot \det P$.

(4) $\mathrm{Pf}(S_n) = 1$ where $S_n = \begin{pmatrix} 0 & 1 \\ -1 & 0 \end{pmatrix} \oplus \cdots \oplus \begin{pmatrix} 0 & 1 \\ -1 & 0 \end{pmatrix}$, with $n/2$ summands.

There are several proofs that $\mathrm{Pf}(S)$ is well defined. One way is to use the theory of alternating spaces to show that if S is skew-symmetric then $S = P^\top \cdot S_n \cdot P$ for some P. Then $\det S = (\det P)^2$. We could define $\mathrm{Pf}(S) = \det P$ and then prove that this value is independent of the choice of P (using the lemma: $Q \in \mathrm{Sp}_n$ implies $\det Q = 1$).

Alternatively we could use a "generic" skew-symmetric S over $\mathbb{Z}[s_{ij}]$, argue as above that $\det S$ is a square in $\mathbb{Q}(s_{ij})$. Then it is also a square in $\mathbb{Z}[s_{ij}]$. Choose a square root, $\mathrm{Pf}(S)$, for this generic case, with the sign chosen so that the specialization to S_n yields the value 1.

Another method avoids alternating spaces, using induction to prove directly that the generic S has a square determinant (see Jacobson (1968)). One can also define Pfaffians using exterior algebras and multilinear algebra. For example see Chevalley (1954) or (1955), Bourbaki (1959), §5, n° 2.

Remark. There exists a "Pfaffian adjoint" $\mathrm{Pfadj}(S)$ which is a $n \times n$ skew- symmetric matrix satisfying

$$S \cdot \mathrm{Pfadj}(S) = \mathrm{Pfadj}(S) \cdot S = \mathrm{Pf}(S) \cdot I_n$$

The entries of $\mathrm{Pfadj}(S)$ are forms of degree $n/2 - 1$ in the entries of S. Consequently there exists a "Pfaffian expansion by minors" as well. The existence of Pfadj can be proved using the generic Pfaffian. Each cofactor S_{ij} in the matrix S must be a multiple of the (irreducible) polynomial $\mathrm{Pf}(S)$. Cancel $\mathrm{Pf}(S)$ from the equation $S \cdot \mathrm{adj}(S) = (\det S) \cdot I_n$ to obtain $\mathrm{Pfadj}(S)$. This approach appears in Jacobson (1968).

10.9 Corollary. (1) *If A, B are skew symmetric then*

$$\mathrm{Pf}\begin{pmatrix} A & 0 \\ 0 & B \end{pmatrix} = (\mathrm{Pf}\, A) \cdot (\mathrm{Pf}\, B).$$

(2) *If S is invertible and skew-symmetric $n \times n$ then* $\mathrm{Pf}(S^{-1}) = (-1)^{n/2}(\mathrm{Pf}\, S)^{-1}$.
(3) *For any $m \times m$ matrix C and an $m \times m$ skew-symmetric matrix S,*

$$\mathrm{Pf}\begin{pmatrix} S & C \\ -C^{\mathsf{T}} & 0 \end{pmatrix} = (-1)^{\frac{m(m-1)}{2}} \cdot \det C.$$

These properties are easy to derive from the definition. In particular, $\mathrm{Pf}\begin{pmatrix} 0 & 1_m \\ -1_m & 0 \end{pmatrix} = (-1)^{\frac{m(m-1)}{2}}$. In the 4×4 case let

$$S = \begin{pmatrix} 0 & a_{12} & a_{13} & a_{14} \\ & 0 & a_{23} & a_{24} \\ & & 0 & a_{34} \\ & & & 0 \end{pmatrix}$$

where we omit writing the lower half. Then

$$\mathrm{Pfadj}(S) = \begin{pmatrix} 0 & -a_{34} & a_{24} & -a_{23} \\ & 0 & -a_{14} & a_{13} \\ & & 0 & -a_{12} \\ & & & 0 \end{pmatrix},$$

and $\mathrm{Pf}(S) = a_{12}a_{34} - a_{13}a_{24} + a_{14}a_{23}$.

It is convenient to introduce a new notation for the eigenspaces of an involution J. If J has type λ on $\mathrm{End}(V)$ define

$$\mathrm{Sym}(J) = \{f \in \mathrm{End}(V) : J(f) = \lambda f\},$$
$$\mathrm{Alt}(J) = \{f \in \mathrm{End}(V) : J(f) = -\lambda f\}.$$

Then for any J, if $\dim V = n$ then $\dim \mathrm{Alt}(J) = \frac{n(n-1)}{2}$. The classical Pfaffian map on matrices is defined on $\mathrm{Alt}(^\top)$. Note also that $\mathrm{Alt}(J) = \mathrm{image}(1 - \lambda J) = \{g - \lambda J(g) : g \in \mathrm{End}(V)\}$.

When J has symplectic type, there is a natural notion of "Pfaffian" for elements of $\mathrm{Alt}(J)$, defined independently of the matrix Pfaffian mentioned above. If $f \in \mathrm{Alt}(J)$ then $J(f) = f$ so the matrix B of f satisfies: $M^{-1} \cdot B^\top \cdot M = B$. Then the matrix $T = MB$ is skew-symmetric. Such a matrix B can also be characterized by: $B = ST$ for some skew-symmetric matrices S, T such that S is nonsingular. It quickly follows that the characteristic polynomial $\chi_f(x)$ is the square of another polynomial. (For $\chi_f(x) = \det(x1 - B) = \det M^{-1} \cdot \det(xM - T)$. Since M^{-1} and $xM - T$ are skew-symmetric over the field $F(x)$, $\chi_f(x)$ is a square in $F(x)$ and hence is a square in $F[x]$.) With a little more work we get a stronger result.

10.10 Lemma. *For f as above, every elementary divisor of f has even multiplicity.*

Proof of Theorem A.7. Here the elementary divisors are the polynomials which appear as the characteristic polynomials of blocks in the Rational Canonical Form for f. (Each of them is a power of an irreducible polynomial.) First assume that F contains all the eigenvalues of f. If λ is an eigenvalue the elementary divisors $(x - \lambda)^m$ are determined by the numbers $d_j = \dim\ker(\lambda 1 - f)^j$ for $j = 1, 2, \ldots$ Since MB is skew symmetric and hence has even rank we know that $\mathrm{rank}\, f = \mathrm{rank}(MB) = \text{even}$. Similarly since $(\lambda 1 - f)^j \in \mathrm{Alt}(J)$ we conclude that $d_j = n - \mathrm{rank}(\lambda 1 - f)^j = \text{even}$. It follows that $(x - \lambda)^m$ occurs with even multiplicity.

In general if K/F is a field extension, the elementary divisors of $f \otimes K$ over K determine the elementary divisors of f over F. Passing to a field K containing all the eigenvalues of f the result follows. □

Proof #2, following Kaplansky (1983). We are given $B = M^{-1}T$ where M, T are skew-symmetric and M is invertible. Then $xI - B = M^{-1}(xM - T)$. The matrix $xM - T$ is skew-symmetric over the principal ideal domain $F[x]$. Applying the theory of alternating spaces over $F[x]$, (e.g. see Kaplansky (1949), p. 475 or Bourbaki (1959), §5, n° 1) there exists some invertible matrix R over $F[x]$ such that

$$R \cdot (aM - T) \cdot R^\top = \begin{pmatrix} 0 & p_1 \\ -p_1 & 0 \end{pmatrix} \oplus \begin{pmatrix} 0 & p_2 \\ -p_2 & 0 \end{pmatrix} \oplus \cdots$$

where $p_i \in F[x]$ and each p_i divides p_{i+1}. Absorbing the factor M^{-1} and applying some elementary column operations, we find that there exist invertible matrices P, Q over $F[x]$ such that $P \cdot (xI - B) \cdot Q = \mathrm{diag}(p_1, p_1, p_2, p_2, \ldots)$. Therefore the invariant factors of B are $p_1, p_1, p_2, p_2, \ldots$ This shows that the invariant factors, and hence the elementary divisors, of B have even multiplicities. □

Proof #3. There is a more geometric proof due to Tignol (1991). Suppose (V, b) is a (regular) alternating space over F and $f \in \mathrm{End}(V)$ is self-adjoint (i.e. $I_b(f) = f$). Then there exists a decomposition $V = U \oplus U'$ such that U and U' are totally isotropic and f-invariant. The action of f on U' is dual to the action of f on U so that there exists a basis for which the matrix of f is $\begin{pmatrix} C & 0 \\ 0 & C^{\top} \end{pmatrix}$. The proof uses the "primary decomposition" of V relative to f but does not employ more complicated linear algebra. □

For a ring A and $a, b \in A$ define the relation $a \sim b$ to mean that $b = pap^{-1}$ for some $p \in A^{\bullet}$. If $A \cong \mathbb{M}_n(F)$ then $a \sim b$ if and only if a and b are "similar" matrices, or equivalently, they have exactly the same elementary divisors.

10.11 Proposition. *For $f \in \mathrm{End}(V)$ with $n \times n$ matrix B over F, the following are equivalent:*

(1) $J(f) = f$ *for some symplectic involution J on* $\mathrm{End}(V)$.

(2) $B = ST$ *for some skew-symmetric S, T such that S is nonsingular.*

(2′) $B = S'T'$ *for some skew-symmetric S', T' such that T' is nonsingular.*

(3) *All elementary divisors of f have even multiplicity.*

(4) *n is even and* $B \sim \begin{pmatrix} C & 0 \\ 0 & C \end{pmatrix}$ *for some $n/2 \times n/2$ matrix C.*

Proof. (1) $\Longleftrightarrow$ (2) is clear using $S = M^{-1}$. For (2) $\Longleftrightarrow$ (2′) note that $ST = (STS) \cdot S^{-1}$. The implication (1) $\Longrightarrow$ (3) is done in Lemma 10.10. (3) $\Longrightarrow$ (4) is standard linear algebra. (4) $\Longrightarrow$ (2): Since $C \sim C^{\top}$ we find that $B \sim \begin{pmatrix} C & 0 \\ 0 & C^{\top} \end{pmatrix} = ST$ where $S = \begin{pmatrix} 0 & I \\ -I & 0 \end{pmatrix}$ and $T = \begin{pmatrix} 0 & -C^{\top} \\ C & 0 \end{pmatrix}$. Then there is an invertible matrix P such that $B = P \cdot ST \cdot P^{-1} = (PSP^{\top}) \cdot (P^{-\top}TP^{-1})$, verifying statement (2). □

We define $\mathcal{D} = \mathcal{D}(\mathrm{End}(V))$ to be the set of all $f \in \mathrm{End}(V)$ satisfying these equivalent conditions. When we consider $\mathbb{M}_n(F)$ rather than $\mathrm{End}(V)$, we write $\mathcal{D}_n$. Here are some basic properties of this set $\mathcal{D}$:

$\mathcal{D}$ is closed under polynomials. ($p \in F[x]$ and $f \in \mathcal{D}$ imply $p(f) \in \mathcal{D}$.)
$\mathcal{D}$ is closed under inverses. ($f \in \mathcal{D}^{\bullet}$ implies $f^{-1} \in \mathcal{D}$.)
$\mathcal{D}$ is closed under conjugation. ($f \in \mathcal{D}$ and $g \in \mathrm{GL}(V)$ imply $gfg^{-1} \in \mathcal{D}$.)
Let J be any involution on $\mathrm{End}(V)$.

If $f, g \in \mathrm{Alt}(J)$ and f or g is invertible, then $fg \in \mathcal{D}$.

If J has symplectic type then $\mathrm{Alt}(J) \subseteq \mathcal{D}$, a linear subspace of dimension $n(n-1)/2$.

We can now define Pfaffians on $\mathcal{D}$ by using that matrix C.

10.12 Definitions. Suppose $f \in \mathcal{D}(\mathrm{End}(V))$ where $n = \dim V$. Choose a basis of V such that the matrix of f is $\begin{pmatrix} C & 0 \\ 0 & C \end{pmatrix}$, as in Proposition 10.11.

Define $\mathrm{pf}(f) = \det C$, the Pfaffian of f. Define

$$\mathrm{pf}\,\chi_f(x) = \chi_C(x) = \det(xI_{n/2} - C),$$

the Pfaffian characteristic polynomial.

Define $\pi(f) \in \mathcal{D}(\mathrm{End}(V))$ to be the map with matrix $\begin{pmatrix} \mathrm{adj}\, C & 0 \\ 0 & \mathrm{adj}\, C \end{pmatrix}$.

Here we have used a lower case "p" to distinguish this Pfaffian from the previous "matrix Pfaffian" $\mathrm{Pf}(S)$. Of course we must verify that these definitions do not depend on the choice of the basis. Suppose f has matrix $\begin{pmatrix} C & 0 \\ 0 & C \end{pmatrix}$ with respect to one basis of V and has matrix $\begin{pmatrix} D & 0 \\ 0 & D \end{pmatrix}$ with respect to another basis. Then C and D have the same elementary divisors, so that $C \sim D$. Consequently $\mathrm{pf}(f)$ and $\mathrm{pf}\,\chi_f(x)$ are well defined. One way to prove that this adjoint map is well defined is to recall the following fact about the classical adjoint:

Let $p(x) = x^m + a_{m-1}x^{m-1} + \cdots + a_0$ be the characteristic polynomial of C (and of D). If $p^*(x) = (-1)^{m-1} \cdot \frac{p(x)-p(0)}{x} = (-1)^{m-1}(x^{m-1} + a_{m-1}x^{m-2} + \cdots + a_1)$, then $\mathrm{adj}\, C = p^*(C)$. (See Exercise 7.) Since $\begin{pmatrix} C & 0 \\ 0 & C \end{pmatrix} = Q \cdot \begin{pmatrix} D & 0 \\ 0 & D \end{pmatrix} \cdot Q^{-1}$ for some matrix Q, we find that $Q \cdot \begin{pmatrix} \mathrm{adj}\, D & 0 \\ 0 & \mathrm{adj}\, D \end{pmatrix} \cdot Q^{-1} = Q \cdot p^*\left\{\begin{pmatrix} D & 0 \\ 0 & D \end{pmatrix}\right\} \cdot Q^{-1} = p^*\left\{\begin{pmatrix} C & 0 \\ 0 & C \end{pmatrix}\right\} = \begin{pmatrix} \mathrm{adj}\, C & 0 \\ 0 & \mathrm{adj}\, C \end{pmatrix}$. Therefore $\pi(f)$ is well defined (and $\pi(f) = p^*(f)$).

10.13 Lemma. *Suppose $n = \dim V$ is even and let $\mathcal{D} = \mathcal{D}(\mathrm{End}(V))$.*

(1) $\mathrm{pf} : \mathcal{D} \to F$ *is a polynomial map of degree $n/2$. If $f \in \mathcal{D} = \mathcal{D}(\mathrm{End}(V))$ then:*
$\mathrm{pf}(f)^2 = \det f$.
$\mathrm{pf}(g^{-1}fg) = \mathrm{pf}(f)$ *for any $g \in \mathrm{GL}(V)$.*
$\mathrm{pf}(f^k) = \mathrm{pf}(f)^k$. *In particular,* $\mathrm{pf}(1_V) = 1$ *and if $f \in \mathcal{D}^\bullet$ then* $\mathrm{pf}(f^{-1}) = \mathrm{pf}(f)^{-1}$.
If $f \in \mathcal{D}(\mathrm{End}(V))$ and $g \in \mathcal{D}(\mathrm{End}(W))$ then $\mathrm{pf}(f \oplus g) = \mathrm{pf}(f) \cdot \mathrm{pf}(g)$.

(2) $\operatorname{pf}\chi_f(x)$ *is a monic polynomial of degree* $n/2$ *and* $\operatorname{pf}\chi_f(f) = 0$.

(3) $\pi : \mathcal{D} \to \mathcal{D}$ *is a polynomial map of degree* $n/2$, *satisfying*

$$f \cdot \pi(f) = \pi(f) \cdot f = \operatorname{pf}(f) \cdot 1_V$$
$$\pi(g \cdot f \cdot g^{-1}) = g \cdot \pi(f) \cdot g^{-1}$$
$$\pi(\pi(f)) = \operatorname{pf}(f)^{\frac{n}{2}-2} \cdot f \text{ and } \operatorname{pf}(\pi(f)) = \operatorname{pf}(f)^{\frac{n}{2}-2}.$$

Proof. (1) Clear from the definitions.

(2) Apply the Cayley–Hamilton Theorem.

(3) Use standard properties of the classical adjoint adj C. The second statement follows from the fact that adj f is well defined, independent of the basis chosen. For the final equations recall that $\operatorname{adj}(\operatorname{adj} C)) = (\det C)^{m-2} \cdot C$ for any $m \times m$ matrix C. (See Exercise 7.) Note that the situation needs some special interpretation when $n = 2$ and $f = 0_V$. □

This version of the Pfaffian on $\mathcal{D}$ is related to the classical version for skew-symmetric matrices.

10.14 Lemma. (1) *Suppose* M, T *are skew-symmetric* $n \times n$ *matrices and* M *is invertible. Then* $M^{-1} \cdot T \in \mathcal{D}_n$ *and* $\operatorname{pf}(M^{-1} \cdot T) = (\operatorname{Pf} M)^{-1} \cdot (\operatorname{Pf} T)$.

(2) *Suppose* $J(f) = f$ *for a symplectic involution* J. *Then for any* $g \in \mathrm{GL}(V)$, $\operatorname{pf}(J(g)fg) = \operatorname{pf}(f) \cdot \det g$.

(3) *Suppose* J *is a symplectic involution on* $\operatorname{End}(V)$. *If* $f, g \in \operatorname{Alt}(J)$ *and either* f *or* g *is invertible then* $fg \in \mathcal{D}$. *In this case*

$$\operatorname{pf}(fg) = \operatorname{pf}(f) \cdot \operatorname{pf}(g) \quad \textit{and} \quad \pi(fg) = \pi(g) \cdot \pi(f).$$

In particular if $f \in \mathcal{D}$ *then* $\pi(f^k) = \pi(f)^k$.

Proof. (1) Choose independent generic skew-symmetric $n \times n$ matrices S_0, T_0 and use determinants to see that $\operatorname{pf}(S_0T_0) = \varepsilon \cdot \operatorname{Pf}(S_0) \cdot \operatorname{Pf}(T_0)$ for some $\varepsilon = \pm 1$. This formula specializes to all $n \times n$ skew-symmetric S, T over F, with the same sign ε. Evaluate ε by computing one special case.

(2) Pick a basis and let B be the matrix of f and P the matrix of g. Represent J as $J(X) = M^{-1} \cdot X^\top \cdot M$ where M is nonsingular skew-symmetric. Then MB is skew-symmetric and $J(P)BP = M^{-1} \cdot (P^\top \cdot MB \cdot P)$ so that $\operatorname{pf}(J(P)BP) = (\operatorname{Pf} M)^{-1} \cdot \operatorname{Pf}(P^\top \cdot MB \cdot P) = (\operatorname{Pf} M)^{-1} \cdot \operatorname{Pf}(MB) \cdot \det P = \operatorname{pf}(B) \cdot \det P$.

(3) Let B, C be the matrices of f, g and M is given as in (2). Since $J(f) = f$ we know that MB and BM^{-1} are skew-symmetric. Similarly MC and CM^{-1} are skew-symmetric. Suppose f is invertible. Then $\operatorname{pf}(f) \cdot \operatorname{pf}(g) = \operatorname{pf}(B) \cdot \operatorname{pf}(C) = \operatorname{pf}(BM^{-1} \cdot M) \cdot \operatorname{pf}(M^{-1} \cdot MC) = \operatorname{Pf}(MB^{-1})^{-1} \cdot \operatorname{Pf}(M) \cdot \operatorname{Pf}(M)^{-1} \cdot \operatorname{Pf}(MC) = \operatorname{pf}((MB^{-1})^{-1} \cdot MC) = \operatorname{pf}(BC) = \operatorname{pf}(fg)$, using several applications of part (1).

From (10.13)(3) we derive $\pi(fg) \cdot fg = \operatorname{pf}(fg) = \operatorname{pf}(f) \cdot \operatorname{pf}(g) = \operatorname{pf}(f) \cdot \pi(g)g = \pi(g)(\operatorname{pf}(f)1_V)g = \pi(g)\pi(f) \cdot fg$. Then if $f, g \in \operatorname{Alt}(J)^\bullet$ we have

$\pi(fg) = \pi(g) \cdot \pi(f)$. Now for fixed $f \in \mathrm{Alt}(J)^\bullet$ we need to verify that formula for all $g \in \mathrm{Alt}(J)$. (The case when g is invertible is similar). If $|F|$ is infinite this follows since $\mathrm{Alt}(J)^\bullet$ is Zariski dense in $\mathrm{Alt}(J)$. For the general case we use a generic argument. Let $S = (s_{ij})$ be a generic skew-symmetric matrix and set $\hat{C} = M^{-1}S$. Then the given matrix B and this $\hat{C}$ are in $\mathrm{Alt}(J)^\bullet$ over the field $F(s_{ij})$ so that $\pi(B\hat{C}) = \pi(\hat{C})\pi(B)$. This equation holds over the ring $F[s_{ij}]$ (since $\pi(\hat{C}) = \sum_{j=0}^{n} a_j \hat{C}^j$ for some $a_j \in F[s_{ij}]$, as in Exercise 10). Therefore it can be specialized to any $C \in \mathrm{Alt}(J)$. □

Suppose $n = \dim V = 4$. We will analyze $\mathcal{D} = \mathcal{D}_4 = \mathcal{D}(\mathrm{End}(V))$ in further detail. The results above show that $\mathrm{pf} : \mathcal{D} \to F$ is a quadratic form and $\pi : \mathcal{D} \to \mathcal{D}$ is a linear form. These maps have natural extensions to the whole space $\mathrm{End}(V)$. To describe these extensions we use the trace map $\mathrm{tr}(f) = \mathrm{trace}(f)$. Note that $\mathrm{tr}(1_V) = n$.

10.15 Example. Suppose $n = 4$. Define $Q : \mathrm{End}(V) \to F$ by $Q(f) = \frac{1}{8} \cdot \mathrm{tr}(f)^2 - \frac{1}{4} \cdot \mathrm{tr}(f^2)$. Define $\pi' : \mathrm{End}(V) \to \mathrm{End}(V)$ by $\pi'(f) = \frac{1}{2} \cdot \mathrm{tr}(f) \cdot 1_V - f$.

(1) Then Q is a regular quadratic form extending $\mathrm{pf} : \mathcal{D} \to F$ and π' is a linear form extending $\pi : \mathcal{D} \to \mathcal{D}$. Also $Q(f) = \frac{1}{2} \cdot \mathrm{tr}(\pi'(f) \cdot f)$ and $Q(fg) = Q(gf)$ so that $Q(s^{-1}fs) = Q(f)$. Furthermore

$$\pi'(\pi'(f)) = f \quad \text{and} \quad Q(\pi'(f)) = Q(f).$$

Any $f \in \mathrm{End}(V)$ is expressed as $f = \alpha 1_V + f_0$ where $\alpha = \frac{1}{4} \cdot \mathrm{tr}(f)$ is a scalar and $\mathrm{tr}(f_0) = 0$. Then $\pi'(f) = \alpha 1_V - f_0$.

(2) If $f \in \mathcal{D}$ then f has minimal polynomial $m_f(x)$ of degree ≤ 2. The following are equivalent for any $f \in \mathrm{End}(V)$ which is not a scalar:

$m_f(x) = x^2 - \frac{1}{2} \cdot \mathrm{tr}(f) \cdot x + \beta$ for some $\beta \in F$

$f = \alpha 1_V + f_0$ such that $\mathrm{tr}(f_0) = 0$ and $f_0^2 \in F$.

$f \cdot \pi'(f) \in F$.

These conditions imply $f \in \mathcal{D}$, except in the case $f_0^2 = 0$ and rank $f_0 = 1$. In particular if $m_f(x)$ is irreducible of degree 2 then $f \in \mathcal{D}$.

Proof. (1) If $f \in \mathcal{D}$ then the matrix of f is $\begin{pmatrix} C & 0 \\ 0 & C \end{pmatrix}$ for some 2×2 matrix C. The characteristic polynomial of C is $p(x) = x^2 - (\mathrm{tr}\, C)x + (\det C)$ so that $p^*(x) = (\mathrm{tr}\, C) - x$. Then $\pi(f) = p^*(f) = \frac{1}{2}\mathrm{tr}(f) - f$. Also since $\mathrm{pf}(f)$ is a scalar we find that $\mathrm{pf}(f) = \frac{1}{4} \cdot \mathrm{tr}(\mathrm{pf}(f) 1_V) = \frac{1}{4} \cdot \mathrm{tr}(\pi(f) \cdot f) = \frac{1}{4} \cdot \mathrm{tr}((\frac{1}{2} \cdot \mathrm{tr}(f) \cdot 1_V - f) \cdot f) = \frac{1}{8} \cdot \mathrm{tr}(f)^2 - \frac{1}{4} \cdot \mathrm{tr}(f^2)$. Therefore π' extends π and Q extends pf. The remaining properties are easily checked. (Compare Exercise 10.)

(2) If $m_f(x) = x^2 - \frac{1}{2} \cdot \mathrm{tr}(f) \cdot x + \beta$ then $f_0^2 = (f - \frac{1}{4}\mathrm{tr}(f))^2 \in F$. If $f_0^2 \in F$ then $f \cdot \pi'(f) = (\alpha 1_V + f_0) \cdot (\alpha 1_V - f_0) = \alpha^2 1_V - f_0^2$ is a scalar. If $f \cdot \pi'(f) \in F$ then $(f - \frac{1}{4} \cdot \mathrm{tr}(f))^2 = f_0^2$ is a scalar, so that $f^2 - \frac{1}{2} \cdot \mathrm{tr}(f) \cdot f + \beta = 0_V$ for some

$\beta \in F$. Then $m_f(x) = x^2 - \frac{1}{2} \cdot \mathrm{tr}(f) \cdot x + \beta$. Suppose these conditions hold but $f \notin \mathcal{D}$. Then $m_f(x)$ must be reducible (why?) so the minimal polynomial of f_0 must be $(x - \alpha)(x + \alpha)$ for some $\alpha \in F$. If $\alpha \neq 0$ each elementary divisor must equal $x \pm \alpha$, and f_0 is similar to a diagonal matrix. But then tr $f_0 = 0$ implies $f \in \mathcal{D}$. Therefore $\alpha = 0$ and $f_0^2 = 0_V$. Since $f_0 \notin \mathcal{D}$ the elementary divisors of f_0 must be $\{x, x, x^2\}$ so that f_0 has rank 1. □

Now let us turn to the main topic of this chapter: central simple algebras. We assume the standard facts about central simple F-algebras with involution, as presented in Scharlau's book, for example. We continue to assume all involutions here are of the "first kind", unless explicitly stated otherwise.

If J is a λ-involution on the central simple F-algebra A, we define

$$\mathrm{Alt}(A, J) = \mathrm{Alt}(J) = \{a \in A : J(a) = -\lambda a\}.$$

If A is an algebra of degree n then $\dim \mathrm{Alt}(A, J) = \frac{n(n-1)}{2}$.

10.16 Proposition. *Let A be a central simple F-algebra with involution. Suppose $n = \deg A$ is even. Define*

$$\mathcal{D}(A) = \{a \in A : J(a) = a \text{ for some } (-1)\text{-involution } J \text{ on } A\}.$$

For any involution J_0 on A,

$$\mathcal{D}(A) = \{bc : b \in \mathrm{Alt}(J_0)^\bullet \text{ and } c \in \mathrm{Alt}(J_0)\} = \{a \in A : \mathrm{Alt}(J_0)^\bullet \cdot a \cap \mathrm{Alt}(J_0) \neq \emptyset\}.$$

This set $\mathcal{D}(A)$ is closed under polynomials, under inverses and under conjugation.

(1) *There is a "reduced Pfaffian" map* $\mathrm{pf}_A : \mathcal{D}(A) \to F$ *which is a polynomial map of degree $n/2$ satisfying*

$\mathrm{pf}_A(a)^2 = \mathrm{nrd}(a)$

$\mathrm{pf}_A(p^{-1}ap) = \mathrm{pf}_A(a)$

$\mathrm{pf}_A(a^k) = \mathrm{pf}_A(a)^k$ *(In particular,* $\mathrm{pf}_A(1) = 1$ *and* $\mathrm{pf}_A(a^{-1}) = \mathrm{pf}_A(a)^{-1}$ *if* $a \in \mathcal{D}(A)^\bullet$.*)*

If $J(a) = a$ for a (-1)-involution J and if $b \in A^\bullet$ then $\mathrm{pf}_A(J(b)ab) = \mathrm{pf}_A(a) \cdot \mathrm{nrd}(b)$.

(2) *If $a \in \mathcal{D}(A)$ define the polynomial* $p_a(x) = \mathrm{pf}_{A(x)}(x1 - a) \in F[x]$. *Then $p_a(x)$ is monic of degree $n/2$ and $p_a(a) = 0$.*

(3) *There is a polynomial map* $\pi_A : \mathcal{D}(A) \to \mathcal{D}(A)$ *of degree $n/2 - 1$ satisfying*

$a \cdot \pi_A(a) = \pi_A(a) \cdot a = \mathrm{pf}_A(a) \cdot 1$

$\pi_A(bab^{-1}) = b \cdot \pi_A(a) \cdot b^{-1}$ *for any* $b \in A^\bullet$

$\pi_A(\pi_A(a)) = \mathrm{pf}_A(a)^{\frac{n}{2}-2} \cdot a$ *and* $\mathrm{pf}_A(\pi_A(a)) = \mathrm{pf}_A(a)^{\frac{n}{2}-1}$.

If J is a (-1)-involution $a, b \in \mathrm{Alt}(J)$ and either a or b is invertible then $ab \in \mathcal{D}(A)$ and

$$\mathrm{pf}_A(ab) = \mathrm{pf}_A(a) \cdot \mathrm{pf}_A(b) \quad \text{and} \quad \pi_A(ab) = \pi_A(b) \cdot \pi_A(a).$$

Proof. The equivalence of the two descriptions of $\mathcal{D}(A)$ and the various closure properties follow as before. To define pf_A we use "descent", following the standard definition of the reduced norm, nrd. Let K be a splitting field for A and choose an algebra isomorphism $\varphi : A \otimes_F K \xrightarrow{\cong} \mathbb{M}_n(K)$. Given the (-1)-involution J on A define the involution I on $\mathbb{M}_n(K)$ by requiring it to be K-linear and $I(\varphi(a \otimes 1)) = \varphi(J(a) \otimes 1)$ for every $a \in A$. That is, the diagram

$$\begin{array}{ccc} A \otimes K & \xrightarrow{\varphi} & \mathbb{M}_n(K) \\ {\scriptstyle J\otimes 1}\downarrow & & \downarrow{\scriptstyle I} \\ A \otimes K & \xrightarrow{\varphi} & \mathbb{M}_n(K) \end{array}$$

commutes. Then I has symplectic type on $\mathbb{M}_n(K)$ and it follows that if $a \in \mathcal{D}(A)$ then $\varphi(a \otimes 1) \in \mathcal{D}_n$. Define $\mathrm{pf}_A(a) = \mathrm{pf}(\varphi(a \otimes 1)) \in K$.

First note that this value does not depend on the choice of K (for we may pass to an algebraic closure of F and note that the matrix is unchanged). Furthermore it is independent of the choice of the isomorphism φ. (Another isomorphism ψ differs from φ by an inner automorphism: there exists $p \in \mathrm{GL}_n(K)$ such that $\psi(x) = p^{-1}\varphi(x)p$ for all $x \in A \otimes K$. Recall that $\mathrm{pf}(p^{-1}xp) = \mathrm{pf}(x)$ for matrices.) Finally suppose that K/F is a Galois extension (using the theorem that there exists a separable splitting field). The standard "descent" argument (as in Scharlau (1985), pp. 296–297) used to prove that the reduced norm has values in F also applies here to show that $\mathrm{pf}_A(a) \in F$.

The stated properties of pf_A follow from the corresponding properties for the matrix Pfaffian. The polynomial $p_a(x)$ is the analog of the Pfaffian characteristic polynomial defined in (10.12) above.

The map π_A arises from the Pfaffian adjoint map discussed in (10.12) and (10.13). Defining $\pi_A(a) = \varphi^{-1}(\pi(\varphi(a \otimes 1))) \in \mathcal{D}(A \otimes K)$, the usual descent argument shows that this value lies in $\mathcal{D}(A)$. The stated formulas follow from Lemmas 10.13 and 10.14. □

A question about a central simple algebra can often be reduced to the split case after an extension to a splitting field. In order to exploit this idea we need a technical lemma.

10.17 Lemma. *Let K/F be an extension of infinite fields.*

(1) *Suppose U is a K-vector space and $p : U \to K$ is a polynomial function. If $U = V \otimes_F K$ for some F-vector space V and if p vanishes on $V \otimes 1$, then $p = 0$.*

(2) *If A is a finite dimensional F-algebra and $W \subseteq A$ is an F-linear subspace such that $(W \otimes K) \cap (A \otimes K)^{\bullet} \neq \emptyset$ then $W \cap A^{\bullet} \neq \emptyset$.*

Proof. (1) Choosing an F-basis of V this statement becomes: if $X = (x_1, \dots, x_n)$ is a system of indeterminates and $p(X) \in K[X]$ vanishes on F^n then $p(X) = 0$. This follows by induction on n and the fact that a non-zero polynomial in one variable has finitely many roots.

(2) Let $\mathcal{L} : A \to \mathrm{End}_F(A)$ be the representation defined by: $\mathcal{L}(a)(x) = ax$. Define $N : A \to F$ by $N(c) = \det(\mathcal{L}(c))$. Then $p = N \otimes 1$ is a polynomial function on $A \otimes K$ and c is a unit in $A \otimes K$ if and only if $p(c) \neq 0$. Apply part (1). □

Note that these assertions are false over finite fields (see Exercise 11). The next result is related to (6.15) but is proved independently here.

10.18 Corollary. *Let A be a central simple F-algebra with involution J. There exists $a \in A^\bullet$ such that $J(a) = -a$, except when A is (split) of odd degree and J has orthogonal type. Consequently A admits a 1-involution, and it admits a (-1)-involution provided* $\deg A$ *is even.*

Proof. That exception is necessary since a skew-symmetric matrix must have even rank. Also recall that a division algebra with involution must have 2-power degree. (This was mentioned earlier in (6.17).) Then an algebra of odd degree with involution must be split.

Suppose $A \cong \mathbb{M}_n(F)$ is split and express $J(X) = M^{-1} \cdot X^\top \cdot M$ for some λ-symmetric matrix M. If J has symplectic type then $J(M) = -M$. If J has orthogonal type choose a nonsingular skew-symmetric matrix S, which exists since we assume that n is even. Then $J(M^{-1}S) = -(M^{-1}S)$.

Now suppose A is not split. As mentioned above this implies that $n = \deg A$ is even. In addition, Wedderburn's Theorem on finite division rings implies that F is infinite. Let $W = \{a \in A : J(a) = -a\}$. Let K be a splitting field, $\varphi : A \otimes K \xrightarrow{\cong} \mathbb{M}_n(K)$ and I the involution on $\mathbb{M}_n(K)$ corresponding to J. Since $W \otimes K$ contains units, by the split case analyzed above, (10.17) implies that W contains a unit of A. □

10.19 Corollary. *Let A be a central simple F-algebra with involution and let K be a splitting field with $\varphi : A \otimes K \xrightarrow{\cong} \mathbb{M}_n(K)$. Let $a, b \in A$ and $f = \varphi(a \otimes 1)$, $g = \varphi(b \otimes 1)$.*

(1) *$a \in \mathcal{D}(A)$ if and only if $f \in \mathcal{D}_n$.*

(2) *$a \sim b$ in A if and only if $f \sim g$ in $\mathbb{M}_n(K)$.*

(3) *For any involution J on A, $a \sim J(a)$.*

Proof. If $A \cong \mathbb{M}_n(F)$ is split, we may alter φ by an inner automorphism to assume that φ induces the inclusion $\mathbb{M}_n(F) \subseteq \mathbb{M}_n(K)$. Since the elementary divisors of $a \in \mathbb{M}_n(F)$ are determined by its elementary divisors over K, the assertions (1) and (2) follow. For (3) express J as $J(a) = M^{-1} \cdot a^\top \cdot M$. Then $a \sim a^\top \sim J(a)$ holds for every $a \in A$.

Suppose A is not split so that F is infinite by Wedderburn.

(1) Let J be a 1-involution on A and let $W = \{c \in A : J(c) = -c \text{ and } J(ca) = -ca\}$. If $c \in W \cap A^\bullet$ then $a = c^{-1} \cdot ca \in \mathcal{D}(A)$. The statement follows by applying Lemma 10.17 to this space W.

(2) Use $W = \{c \in A : ac = cb\}$.

(3) Use $W = \{c \in A : ac = cJ(a)\}$. □

We begin our discussion of algebras of degree 4 with a preliminary lemma.

10.20 Lemma. *Let A be a central simple F-algebra of degree 4 with a (-1)-involution J. Then the restriction of* pf *to the 6-dimensional space* $\mathrm{Alt}(J)$ *is a regular quadratic form.*

Proof. We may extend scalars to assume $A \cong \mathrm{End}(V)$ is split. Then $J = I_b$ is the adjoint involution for some regular alternating form b on V. Choosing a symplectic basis for (V, b) we get the matrix of the form is $M = \begin{pmatrix} 0 & 1 \\ -1 & 0 \end{pmatrix}$ in 2×2 blocks. Then B is the matrix of some $f \in \mathrm{Alt}(J)$ if and only if MB is skew-symmetric if and only if

$$B = \begin{pmatrix} x & y & 0 & r \\ z & w & -r & 0 \\ 0 & -s & x & z \\ s & 0 & y & w \end{pmatrix} \quad \text{for some } x, y, z, w, r, s \in F.$$

Then the formulas in Lemma 10.14(1) and after Corollary 10.9 show that $\mathrm{pf}(B) = -rs + xw - yz$. This is a regular quadratic form in 6 variables. □

10.21 Proposition (Albert, Rowen). *Suppose A is a central simple F-algebra of degree 4 with involution. Then any (-1)-involution on A is decomposable. In particular A is decomposable as an algebra.*

Proof. By (10.16) there is a linear map $\pi : \mathrm{Alt}(J) \to \mathrm{Alt}(J)$ such that $a \cdot \pi(a) = \pi(a) \cdot a = \mathrm{pf}(a)$ for every $a \in \mathrm{Alt}(J)$. Furthermore $\pi(\pi(a)) = a$. In fact, as in Example 10.15, π is the restriction of the linear map $\pi' : A \to A$ defined by $\pi'(x) = \frac{1}{2} \cdot \mathrm{trd}(x) - x$. Therefore $\mathrm{Alt}(J) = F \oplus W$ where W is the (-1)-eigenspace of π and $\dim W = 5$.

The quadratic form pf_J on $\mathrm{Alt}(J)$ has associated bilinear form B_J given by $2B_J(x, y) = \mathrm{pf}_J(x+y) - \mathrm{pf}_J(x) - \mathrm{pf}_J(y) = (x+y)\cdot\pi(x+y) - x\cdot\pi(x) - y\cdot\pi(y) = x\cdot\pi(y) + y\cdot\pi(x)$. If $y \in W$ then $2B_J(1, y) = (-y) + y = 0$. Hence $\mathrm{Alt}(J) \simeq F \perp W$ relative to the quadratic form pf_J and consequently the induced form on W is regular (using Lemma 10.20). Choose x, y as part of an orthogonal basis of W relative to pf_J. Then $x^2 = -x \cdot \pi(x) = -\mathrm{pf}(x) \in F^\bullet$ and similarly $y^2 \in F^\bullet$. Also $xy + yx = -2B_J(x, y) = 0$ and we conclude that $\{x, y\}$ generates a quaternion subalgebra Q of A. Since $W \subseteq \mathrm{Alt}(J)$ this Q is J-invariant. □

Although we are interested mainly in the case A has degree 4, we will define the Pfaffian associated to an orthogonal involution in the general case of a central simple algebra of degree n. Suppose that J is an involution of *orthogonal* type on A. We define a Pfaffian on $\mathrm{Alt}(J)$ in analogy to the classical Pfaffian on skew symmetric matrices. Since $\mathrm{Alt}(J)^\bullet \cdot \mathrm{Alt}(J) \subseteq \mathcal{D}(A)$, as mentioned in (10.16), we obtain a "Pfaffian" map and a "Pfaffian adjoint" associated to a fixed $s \in \mathrm{Alt}(J)^\bullet$:

$$\mathrm{Pf}_s : \mathrm{Alt}(J) \to F \text{ is defined by } \mathrm{Pf}_s(a) = \mathrm{pf}(sa).$$
$$\pi_s : \mathrm{Alt}(J) \to \mathrm{Alt}(J) \text{ is defined by } \pi_s(a) = \pi(sa)s.$$

Some aspects of these maps are independent of the choice of s.

10.22 Lemma. *Let J be a 1-involution on a central simple algebra A of even degree n. Let $s \in \mathrm{Alt}(J)^\bullet$.*

(1) *If $a, b \in \mathrm{Alt}(J)^\bullet$ then $\mathrm{pf}(a^{-1}b) = \mathrm{Pf}_s(a)^{-1} \cdot \mathrm{Pf}_s(b)$.*
If $s, t \in \mathrm{Alt}(J)^\bullet$ let $\lambda = \mathrm{pf}(ts^{-1})$. Then for every $a \in \mathrm{Alt}(J)$

$$\mathrm{Pf}_t(a) = \lambda \cdot \mathrm{Pf}_s(a) \quad \textit{and} \quad \pi_t(a) = \lambda \cdot \pi_s(a).$$

(2) $\mathrm{Pf}_s(a)^2 = \mathrm{nrd}(s) \cdot \mathrm{nrd}(a)$ *for every $a \in \mathrm{Alt}(J)$.*
$\mathrm{Pf}_s(J(b) \cdot a \cdot b) = \mathrm{Pf}_s(a) \cdot \mathrm{nrd}(b)$ *for every $a \in \mathrm{Alt}(J)$ and $b \in A^\bullet$.*

(3) *If $a \in \mathrm{Alt}(J)$ then $\pi_s(a) \cdot a = a \cdot \pi_s(a) = \mathrm{Pf}_s(a)$.*

(4) *If $a \in \mathrm{Alt}(J)$ then $\pi_s(\pi_s(a)) = (\mathrm{nrd}\, s) \cdot (-1)^{\frac{n}{2}} \cdot \mathrm{Pf}_s(a)^{\frac{n}{2}-2} \cdot a$.*

Proof. (1) This generalizes Lemma 10.14(1). Define another involution J_0, by setting $J_0(x) = s \cdot J(x) \cdot s^{-1}$. Then J_0 is a (-1)-involution (since $J(s) = -s$), $J_0(s) = -s$ and $\mathrm{Alt}(J_0) = s \cdot \mathrm{Alt}(J) = \mathrm{Alt}(J) \cdot s^{-1}$. Since sa and $sb \in \mathrm{Alt}(J_0)$ the last statement in (10.16) implies $\mathrm{pf}(a^{-1}b) = \mathrm{pf}((sa)^{-1} \cdot sb) = \mathrm{pf}(sa)^{-1}\,\mathrm{pf}(sb)$, as claimed. For the second statement, note that $ts^{-1} \in \mathrm{Alt}(J) \cdot s^{-1} = \mathrm{Alt}(J_0)$ and $sa \in s \cdot \mathrm{Alt}(J) = \mathrm{Alt}(J_0)$. Then (10.16)(3) implies: $\mathrm{Pf}_t(a) = \mathrm{pf}(ta) = \mathrm{pf}(ts^{-1} \cdot sa) = \mathrm{pf}(ts^{-1}) \cdot \mathrm{pf}(sa) = \lambda \cdot \mathrm{Pf}_s(a)$. The second equality is proved later.

(2) The first statement is clear. The second follows from (10.16)(1) since $\mathrm{pf}(sJ(b)ab) = \mathrm{pf}(J_0(b) \cdot sa \cdot b) = \mathrm{pf}(sa) \cdot \mathrm{nrd}(b)$.

(3) Certainly $\pi_s(a) \cdot a = \pi(sa) \cdot sa = \mathrm{pf}(sa) = \mathrm{Pf}_s(a)$. For the second equality recall that $sa \cdot \pi(sa) = \mathrm{Pf}_s(a)$ is a scalar so that $\mathrm{Pf}_s(a) = s^{-1} \cdot sa\pi(sa) \cdot s = a \cdot \pi_s(a)$. Now to finish the proof of (1): using (3) the equation $\pi_t(a) = \lambda \cdot \pi_s(a)$ holds whenever $a \in A^\bullet$. The standard "generic" argument now applies.

(4) This follows from the definition in terms of π and the properties of π stated in (10.16) (after noting that $s^2, sa \in \mathrm{Alt}(J_0)$ and $\mathrm{pf}(s^2) = (-1)^{\frac{n}{2}} \cdot (\mathrm{nrd}\, s)$.) Alternatively we note that if $a \in \mathrm{Alt}(J)^\bullet$ then $\pi_s(a) = \mathrm{Pf}_s(a) \cdot a^{-1}$. Then $\pi_s(\pi_s(a)) = \mathrm{Pf}_s(a)^{\frac{n}{2}-1} \cdot \mathrm{Pf}_s(a^{-1}) \cdot a$. Since $\mathrm{Pf}_s(a^{-1}) = (-1)^{\frac{n}{2}} \cdot \mathrm{Pf}_s(a)^{-1} \cdot \mathrm{nrd}(s)$ the claim holds. Since this claim is a polynomial equation valid for every $a \in A^\bullet$ the standard generic argument applies again. □

Let us now specialize again to the case of main interest: A has degree $n = 4$. Then $(\mathrm{Alt}(J), \mathrm{Pf}_s)$ is a quadratic space of dimension 6 whose similarity class is independent of the choice of s, depending only on the algebra A.

10.23 Corollary. *Let A be a central simple F-algebra with involution and degree 4. If J is an involution on A define the form $\varphi_J : \mathrm{Alt}(J) \to F$ as follows:*

If J has type -1 let $\varphi_J(a) = \mathrm{pf}(a)$.

If J has type 1 choose $s \in \mathrm{Alt}(J)^\bullet$ and define $\varphi_J(a) = \mathrm{Pf}_s(a)$.

Then $(\mathrm{Alt}(J), \varphi_J)$ is a regular 6-dimensional quadratic space, and all these spaces are similar.

Proof. First we prove the similarity. Let J be any 1-involution on A and choose $s \in \mathrm{Alt}(J)^\bullet$. Let J_1 be any (-1)-involution on A. Then there exists $t \in \mathrm{Alt}(J)^\bullet$ such that $J_1(x) = t \cdot J(x) \cdot t^{-1}$ so that $\mathrm{Alt}(J_1) = t \cdot \mathrm{Alt}(J)$. The left-multiplication map $\mathcal{L}_t : \mathrm{Alt}(J) \to \mathrm{Alt}(J_1)$ provides the desired similarity, since for any $a \in \mathrm{Alt}(J)$ we have $\varphi_{J_1}(\mathcal{L}_t(a)) = \mathrm{pf}(ta) = \mathrm{Pf}_t(a) = \lambda \cdot \mathrm{Pf}_s(a) = \lambda \cdot \varphi_J(a)$, where $\lambda = \mathrm{pf}(ts^{-1})$ as in (10.22). The regularity of φ_J now follows from (10.20). □

Define the *Albert form* α_A to be this 6-dimensional quadratic form associated to A. To calculate α_A note that A is decomposable (by (10.21)) so that $A \cong C(V, q)$ for some 4-dimensional quadratic space (V, q). Use the involution J_0 which is the identity on V, so that J_0 has type (-1) and $\mathrm{Alt}(J_0) = F \oplus V \oplus Fz$. Here $z = z(V, q)$ so that $z^2 = \delta$ where $dq = \langle \delta \rangle$. From Example 10.15 we know that $\pi(\alpha + v + \beta z) = \alpha - v - \beta z$. Therefore $\mathrm{pf}(\alpha + v + \beta z) = \alpha^2 - q(v) - \beta^2\delta$ and α_A is similar to $(\mathrm{Alt}(J_0), \mathrm{pf}_{J_0}) \simeq \langle 1, -dq\rangle \perp -q$.

It is this form for which Albert proved: A is a division algebra if and only if the form α_A is anisotropic. (See Exercises 3.10 (5) and 3.17.) This Albert form can also be expressed nicely in terms of a decomposition $A \cong Q_1 \otimes Q_2$ for quaternion algebras Q_i. Let φ_i be the norm form of Q_i with pure parts φ_i' (so that $\varphi_i \simeq \langle 1 \rangle \perp \varphi_i'$). Then α_A is similar to the form $\varphi_1' \perp -\varphi_2'$. It is easy to recover the algebra A from the Albert form α_A since $c(\alpha_A) = c(\varphi_1 \perp -\varphi_2) = c(\varphi_1)c(\varphi_2) = [Q_1] \cdot [Q_2] = [A]$. If these formulas for the Albert form α_A are taken as the definition, the uniqueness properties do not seem clear. (See Exercise 3.17.)

10.24 Lemma. *Suppose A is a central simple F-algebra with involution J of orthogonal type. If A has even degree then $\mathrm{Alt}(J)^\bullet \neq \emptyset$ and all values of $\mathrm{nrd}(b)$ for $b \in \mathrm{Alt}(J)^\bullet$ lie in the same square class in $F^\bullet/F^{\bullet 2}$.*

Proof. We proved the first statement in Corollary 10.18. Now suppose $b, c \in \mathrm{Alt}(J)^\bullet$. Then $bc \in \mathcal{D}(A)$ and therefore $\mathrm{nrd}(b) \cdot \mathrm{nrd}(c) = \mathrm{nrd}(bc) = \mathrm{pf}(bc)^2 \in F^{\bullet 2}$. □

Define the *determinant* $\det(J) \in F^\bullet/F^{\bullet 2}$ to be that common square class. That is, if J is a 1-involution on the central simple algebra A and $\deg A$ is even, then $\det(J) = \langle \mathrm{nrd}(b) \rangle \in F^\bullet/F^{\bullet 2}$ for any $b \in \mathrm{Alt}(J)^\bullet$.

10.25 Lemma. (1) *Let (V, q) be a quadratic space of even dimension. Then* $\det(I_q) = \det q$ *in* $F^\bullet/F^{\bullet 2}$.

(2) *Suppose (A_i, J_i) are central simple F-algebras with involutions of orthogonal type and with even degrees. Then* $\det(J_1 \otimes J_2) = \langle 1 \rangle$.

Proof. (1) Pick a basis and let M be the symmetric matrix of the form q. Let B be the matrix of $b \in \mathrm{End}(V)$. Then $I_q(B) = M^{-1} \cdot B^\top \cdot M$. If $b \in \mathrm{Alt}(I_q)$ we find that MB is skew- symmetric so that $\det(MB)$ is a square. Therefore $\det(I_q) = \langle \det(B) \rangle = \langle \det(M) \rangle = \det q$ in $F^\bullet/F^{\bullet 2}$.

(2) If $\deg(A_i) = n_i$ and $a_i \in A_i$ recall that $\mathrm{nrd}(a_1 \otimes a_2) = (\mathrm{nrd}\, a_1)^{n_2}(\mathrm{nrd}\, a_2)^{n_1}$ where the reduced norms are computed in the appropriate algebras. Now simply choose $b \in \mathrm{Alt}(J_1)^\bullet$, which exists in A_1 by Corollary 10.18, note that $b \otimes 1 \in \mathrm{Alt}(J_1 \otimes J_2)^\bullet$ and compute $\mathrm{nrd}(b \otimes 1) = \mathrm{nrd}(b)^{n_2}$ is a square. □

Thus one necessary condition that a 1-involution J be decomposable (relative to subalgebras of even degree) is that $\det(J) = \langle 1 \rangle$. In the case A has degree 4 this was proved by Knus, Parimala and Sridharan to be a sufficient condition as well. The key idea is the linear map π_s discussed in (10.22).

10.26 Proposition. *Let A be a central simple F-algebra of degree* 4 *with involution J. Then J is indecomposable if and only if J has orthogonal type and* $\det(J) \neq \langle 1 \rangle$.

Proof. The "if" part is in (10.25). We proved in (10.21) that symplectic involutions are decomposable. Therefore we assume that J is an involution of orthogonal type with $\det(J) = \langle 1 \rangle$ and search for a J-invariant quaternion subalgebra. By definition there exists $b \in \mathrm{Alt}(J)^\bullet$ such that $\mathrm{nrd}(b) = \lambda^2$ for some $\lambda \in F^\bullet$. Then by (10.22) $\pi_s \circ \pi_s = \lambda^2 \cdot 1_{\mathrm{Alt}(J)}$ so the 6-dimensional space $\mathrm{Alt}(J)$ breaks into $\pm\lambda$-eigenspaces: $\mathrm{Alt}(J) = U^+ \oplus U^-$. Let B_s be the bilinear form associated to the quadratic form Pf_s. Then $2B_s(x, y) = \mathrm{Pf}_s(x + y) - \mathrm{Pf}_s(x) - \mathrm{Pf}_s(y) = x \cdot \pi_s(y) + y \cdot \pi_s(x)$. Similarly we argue that this quantity equals $\pi_s(x) \cdot y + \pi_s(y) \cdot x$.

If $x \in U^+$ and $y \in U^-$ then $2B_s(x, y) = x \cdot (-\lambda y) + (\lambda x) \cdot y = -\lambda \cdot (xy - yx)$ and it also equals $(\lambda x) \cdot y + (-\lambda y) \cdot x = \lambda \cdot (xy - yx)$. Therefore $xy - yx = 0$ and we conclude that U^+ centralizes U^- and that $\mathrm{Alt}(J) = U^+ \perp U^-$ relative to the quadratic form Pf_s. Consequently the restrictions of Pf_s to the subspaces U^+ and U^- are regular.

We may assume $\dim U^+ \geq 3$ (otherwise interchange λ and $-\lambda$). If $x, y \in U^+$ then $2B_s(x, y) = \lambda \cdot (xy + yx)$ and in particular $x^2, y^2 \in F$. Choose $x, y \in U^+$ to be part of an orthogonal basis relative to B_s. Then x, y are units and $xy + yx = 0$, so they generate a quaternion subalgebra $Q \subseteq A$. Since $x, y \in \mathrm{Alt}(J)$ this Q is certainly J-invariant. (In fact, the induced involution on Q is the standard "bar".) □

Now we are in a position to prove Theorem 10.5.

Proof of Theorem 10.5. There are three cases to be considered. If y is given with $y^2 = d \in F^{\bullet 2}$, it suffices to find some $a \in \mathcal{D}(A)^{\bullet}$ which anti-commutes with y and with $J(a) = \pm a$. For with such a we know that $a^2 = \alpha a + \beta$ for some $\alpha, \beta \in F$, since $\deg(a) \leq 2$, by (10.16)(2). Conjugating by y and subtracting, we find that $\alpha = 0$ so that $a^2 = \beta \in F^{\bullet}$. Then y and a generate a J-invariant quaternion subalgebra Q.

Let K be a splitting field of A with $\sqrt{d} \in K$, let $\varphi : A \otimes K \xrightarrow{\cong} \mathrm{End}_K(V)$ and $f = \varphi(y \otimes 1)$. Then $f^2 = d \cdot 1_V$ so that f provides an eigenspace decomposition $V = V^+ \oplus V^-$ with dimensions $4 = n^+ + n^-$. The matrix of f relative to a compatible basis is

$$\begin{pmatrix} \sqrt{d} \cdot I_{n^+} & 0 \\ 0 & -\sqrt{d} \cdot I_{n^-} \end{pmatrix}.$$

(1) We know J is decomposable from (10.21). Suppose first that $J(y) = y$. Then $y \in \mathcal{D}(A)$. Since $f \in \mathcal{D}$ the dimensions n^+ and n^- are even. Then $n^+ = n^- = 2$, since $y \notin F$. Following the notations in the proof of (10.21) we see that $\mathrm{trd}(y) = 0$ so that $y \in W$. Extending $\{y\}$ to an orthogonal basis $\{y, a, \dots\}$ of W, we see that $a \in \mathrm{Alt}(J)^{\bullet} \subseteq \mathcal{D}(A)^{\bullet}$ and a, y anti-commute.

Suppose y is given with $J(y) = -y$. Then $J(f) = -f$ in $\mathrm{End}(V)$ so that $f \sim -f$. Therefore $n^+ = n^- = 2$ and hence $f \in \mathcal{D}(\mathrm{End}(V))$. Then $y \in \mathcal{D}(A)$ by (10.19) so there exists some (-1)-involution J_1 on A with $J_1(y) = y$. Express $J_1 = J^a$ so that $J(a) = a$ and $y = J^a(y) = a^{-1} \cdot J(y) \cdot a = -a^{-1} y a$. Then $a \in \mathcal{D}(A)$ and a, y anti-commute.

(2) If J is decomposable we can certainly find such an element y inside a J-invariant quaternion subalgebra. Conversely suppose J is a 1-involution with $J(y) = -y$. As before we find that $f \sim -f$ so that $n^+ = n^- = 2$. Then $\mathrm{nrd}(y) = \det(f) = (\sqrt{d})^2(-\sqrt{d})^2 = d^2$. Then $\det(J) = \langle 1 \rangle$ and (10.26) implies that J is decomposable. As above $y \in \mathcal{D}(A)$ so there exists some (-1)-involution J_1 with $J_1(y) = y$. Express $J_1 = J^a$ and note that $J(a) = -a$ and a, y anti-commute. Since $ay \in \mathcal{D}(A)$ and y, ay anticommute, the claim follows. □

The existence of an indecomposable involution on a degree 4 division algebra was first proved by Amitsur, Rowen and Tignol (1979). The Knus, Parimala and Sridharan Theorem (10.26) shows that the determinant $\det(J)$ determines whether J is indecomposable. This criterion is made clearer by the following result of Knus, Lam, Shapiro, Tignol (1992).

Proposition. *Let A be a central simple F-algebra of degree 4, with involution. The following subsets of $F^{\bullet}$ are equal.*

$\{d : \langle d \rangle = \det(J)$ *for some* 1*-involution* J *on* $A\}$.

$G_F(\alpha_A)$, *the group of similarity factors of an Albert form of* A.

$\mathrm{nrd}(A^{\bullet}) \cdot F^{\bullet 2}$, *the group of square classes of reduced norms.*

Consequently the algebra A admits an indecomposable involution if and only if the Albert form α_A has a similarity factor which is not a square.

Analogous decomposition results fail for algebras of larger degree. Any tensor product of three quaternion algebras is a central simple algebra of degree 8. However Amitsur, Rowen and Tignol (1979) found an example of a division algebra D of degree 8 over its center and such that D has an involution but is indecomposable (i.e. D has no quaternion subalgebras).

Several standard properties of quadratic forms have analogs for orthogonal involutions of central simple algebras. We end this chapter with some remarks about this correspondence. An orthogonal involution on $\mathrm{End}(V)$ must equal the adjoint involution I_q for some quadratic form q on V, unique up to scalar multiple. Any invariant of q which remains unchanged if q is altered by a similarity should be definable entirely in terms of the involution I_q. For example:

$\det q \in F^\bullet/F^{\bullet 2}$, in the case $n = \dim q$ is even.

$|\operatorname{sgn}_P(q)|$, the absolute value of the signature of q at an ordering P of F.

$C_0(q)$, the even Clifford algebra.

The Witt index of q.

$G_F(q)$, the group of similarity factors (or norms) of the form q.

Are there analogous invariants for orthogonal involutions on arbitrary central simple algebras, coinciding with the given invariants in the split case? Of course we hope that the newly defined invariant will be useful in the theory of involutions.

We have already seen one example of this program: the determinant $\det(J)$ is the analog of $\det q$. Lewis and Tignol (1993) have investigated the signature of an involution. The analog of the even Clifford algebra was done long ago by Jacobson (1964) and discussed further by Tits (1968). The determinant $\det(J)$ also arises naturally out of Jacobson's theory. This even Clifford algebra of an algebra with involution (A, J) is investigated extensively in Knus et al. (1998).

The Pfister Factor Conjecture provides another example of this theme. A quadratic space (V, q) is similar to a Pfister form when q is a tensor product of some binary forms. Equivalently, the algebra $(\mathrm{End}(V), I_q)$ is a tensor product of split quaternion algebras with involution. Motivated by this, let (A, J) be a central simple algebra with 1-involution and define it to be a "Pfister algebra" if it is a tensor product of some quaternion algebras with involution. The Pfister Factor Conjecture says: When A is split then these two notions coincide. A precise statement appears in (9.17).

Exercises for Chapter 10

1. **Maximal examples.** (1) If $\dim q = 16$ and $(\sigma, \tau) < \mathrm{Sim}(q)$ is an (s, t)- familiy where $s + t \geq 7$, then q is similar to a Pfister form. Find an example of q over $\mathbb{R}$ such that $\dim q = 16$ and $\mathrm{Sim}(q)$ has a (3, 3)-family but admits no families of larger size.

(2) There exists $(\langle 1, a\rangle, \langle x\rangle) < \text{Sim}(V, q)$ where $\dim q = 12$ but such that $(\langle 1, a\rangle, \langle x\rangle)$ does not admit any expansion by 2 dimensions. (See Exercise 7.10.) Find similar examples $(\sigma, \tau) < \text{Sim}(q)$ of an (s, t)-family where $s + t = 2m - 1$ and $\dim q = 2^m \cdot 3$, but σ admits no expansion by 2 dimensions.

(3) **Open question.** Are there similar examples in other dimensions? For instance, is there some $\sigma < \text{Sim}(q)$ where $\dim \sigma = 5$, $\dim q = 48$, but the 5-plane does not expand by 2 dimensions? That involves a degree 4 Clifford algebra D (which must be a division algebra) and a (-1)-involution on $\mathbb{M}_3(D)$ having no invariant quaternion subalgebras. Does such an involution exist?

(4) When can $\langle 1, a\rangle < \text{Sim}(q)$ be maximal as a subspace? Certainly if $\langle\langle a\rangle\rangle \mid q$ but q has no 2-fold Pfister factor then this occurs. The converse is unknown.

Open question. If $\langle\langle a\rangle\rangle \mid q$ and $\langle\langle x, y\rangle\rangle \mid q$ then must there exist $b \in F^\bullet$ with $\langle\langle a, b\rangle\rangle \mid q$?

(Hint. (1) If $s + t \geq 7$ then (10.7) shows that there is a $(5, 5)$-family and q is Pfister by PC(4). Find a proof that does not invoke Theorem 10.7.)

2. **Non-uniqueness.** (1) Suppose (σ, τ) is an (s, t)-pair where $s + t$ is odd, and let (C, J) be the corresponding Clifford algebra with involution. Then $(\sigma, \tau) < \text{Sim}(V, q)$ if and only if there is a central simple F-algebra with involution (A, K) such that $(C \otimes A, J \otimes K) \cong (\text{End}(V), I_q)$. However this (A, K) need not be unique.

(2) The two representations π_α and π_β of $C \to \text{End}(V)$ arising from the two choices above yield two $(2, 1)$-families on the 8-dimensional space (V, q). One of them expands to a $(4, 4)$-family and the other does not admit any expansion of 2 or more dimensions.

(Hint. (1) Let $(\sigma, \tau) = (\langle 1, 1\rangle, \langle 1\rangle)$ so that $(C, J) \cong (\mathbb{M}_2(\mathbb{Q}), I_{\langle\langle 1\rangle\rangle})$. Let $q \simeq \langle\langle 1, 1, 1\rangle\rangle$, $\alpha = \langle 1, 1, 1, 1\rangle$ and $\beta = \langle 1, 1, 1, 2\rangle$. Then $\langle\langle 1\rangle\rangle \otimes \alpha \simeq \langle\langle 1\rangle\rangle \otimes \beta$ but α, β are not similar.)

3. **Matrix Pfaffians.** (1) If S, T are skew-symmetric $n \times n$ matrices which anticommute then ST is also skew-symmetric and $\text{Pf}(ST) = \pm \text{Pf}(S) \cdot \text{Pf}(T)$. Is this sign independent of S, T?

(2) Suppose R commutes with some nonsingular skew-symmetric S. Then $R^\top \cdot R \in \mathcal{D}$ and $\text{pf}(R^\top \cdot R) = \det R$.

(3) If $S, T \in \text{GL}_n$ are skew-symmetric then $ST \in \mathcal{D}_n$ and $\text{pf}(ST) = (-1)^{\frac{n}{2}} \text{Pf}(S) \cdot \text{Pf}(T)$. Consequently if $S_1 S_2 S_3 S_4 = I_n$ where each S_i is skew-symmetric then $\text{Pf}(S_1) \cdot \text{Pf}(S_2) \cdot \text{Pf}(S_3) \cdot \text{Pf}(S_4) = 1$. Are there analogous results when I_n equals a product of some k skew-symmetric matrices?

(4) If S is skew-symmetric $n \times n$ then $\text{Pfadj}(\text{Pfadj}(S)) = (-1)^{\frac{n}{2}} \cdot (\text{Pf } S)^{\frac{n}{2}-2} \cdot S$.

4. **Properties of π.** (1) Let M, T be given as in 10.14. Then $\pi(M^{-1} \cdot T) = \text{Pfadj}(M)^{-1} \cdot \text{Pfadj}(T)$.

(2) If $f \in \text{Alt}(J)$ and $g \in \text{GL}(V)$ then $\pi(J(g) f g) = (\det g) \cdot g^{-1} \cdot \pi(f) \cdot J(g)^{-1}$.

5. Let A be a central simple F-algebra with involution. Suppose $\deg A = n$ is even and $n > 2$.

(1) **Lemma.** *$\mathcal{D}(A)$ contains an F-basis of A.*

(2) If J is an involution on A then $\mathrm{Alt}(J)$ generates A as an F-algebra. Does $\mathrm{Sym}(J)$ generate A as well?

Corollary. (i) *If J, J' are involutions on A then $J = J'$ if and only if* $\mathrm{Alt}(J) = \mathrm{Alt}(J')$.

(ii) $(A, J) \cong (A, J')$ *if and only if* $\mathrm{Alt}(J') = x \cdot \mathrm{Alt}(J) \cdot x^{-1}$ *for some* $x \in A^\bullet$.

(Note. This assertion is also true when A is quaternion.)

(3) Given the subspace $S = \mathrm{Alt}(J) \subseteq A$, express the subspace $\mathrm{Sym}(J)$ somehow directly in terms of S.

(Hint. (1) It suffices to settle the split case. An *ad hoc* proof can be given, but the claim follows immediately from a theorem of Kasch (1953). Further references appear in Leep, Shapiro, Wadsworth (1985), §4.

(3) $\mathrm{Sym}(J) = (\mathrm{Alt}(J))^\perp$ relative to the trace form $\tau : A \times A \to F$ defined by $\tau(x, y) = \mathrm{trd}(xy)$.)

6. (1) Let J be a λ-involution on $\mathrm{End}(V)$ and fix $s_0 \in \mathrm{Alt}(J)^\bullet$. Then $f \in \mathrm{Alt}(J)$ iff $f = J(g) \cdot s_0 \cdot g$ for some $g \in \mathrm{End}(V)$.

(2) Does (1) remain valid for involutions on a central simple algebra A?

(Hint. Let B be the λ-form on V corresponding to J, and B_0 the alternating form for J^{s_0}. Then (V, B_0) has a symplectic basis and the regular part of B^f has a symplectic basis. Choose a (not necessarily injective) isometry $g : (V, B^f) \to (V, B_0)$.)

7. Let C be an $m \times m$ matrix over F.

(1) If $p(x) = \det(xI_m - C)$ is the characteristic polynomial, define $p^*(x) = (-1)^{m+1} \cdot \frac{p(x)-p(0)}{x}$. Then $\mathrm{adj}\, C = p^*(C)$.

(2) $\mathrm{adj}(\mathrm{adj}\, C)) = (\det C)^{m-2} \cdot C$.

(3) If $\dim V = 2$, then $\mathcal{D}(\mathrm{End}(V)) = F \cdot 1_V$. If $f = \alpha \cdot 1_V$ for $\alpha \in F$, then $\mathrm{pf}(f) = \alpha$, $\mathrm{pf}\chi_f(x) = x - \alpha$ and $\pi(f) = 1_V$. Explain the difficulty in the definition when $f = 0_V$.

(Hint. (1) Verify first that $C \cdot p^*(C) = (\det C) \cdot I_m$. The claim follows for nonsingular C. Apply this case to a generic matrix C, or to the matrix $C + x \cdot I_m$ in $F(x)$, and then specialize to deduce it for arbitrary C.

(2) Apply the equation $X \cdot \mathrm{adj}\, X = (\det X) I_m$ to $X = C$ and $X = \mathrm{adj}\, C$ and deduce the claim when C is nonsingular. Complete the argument as before.)

8. **Subspaces of $\mathcal{D}$.** Let A be a degree 4 algebra with involution. If $S \subseteq \mathcal{D}(A)$ is a linear subspace with $\dim S = 6$ and $1_V \in S$, then $S = \mathrm{Alt}(J)$ for some (-1)-involution J.

(Hint. Let S_0 be the subspace of trace 0 elements. Then $(S, \mathrm{pf}) \simeq \langle 1\rangle \perp -\psi$ as a quadratic space, where $\psi(c) = c^2$ for $c \in S_0$. There is an induced algebra homomorphism $\pi : C(\psi) \to A$. If ψ is regular then π is surjective and the involution J_0 on $C(\psi)$ induces the desired J on A. Otherwise, pass to the split case and find $T \subseteq S_0$ with $\dim T = 3$ and $t^2 = 0$ for every $t \in T$. Get a contradiction using Jordan forms and the fact that every such t has even rank.)

9. **Albert forms.** Let A be a central simple algebra of degree 4, with involution. Then the Albert form α_A is uniquely defined up to a scale factor. If J is a (-1)-involution on A let $\mathrm{Alt}_0(J)$ be the subspace of trace 0 elements of $\mathrm{Alt}(J)$. Then α_A has a special presentation: $(\mathrm{Alt}(J), \mathrm{pf}) \simeq \langle 1\rangle \perp -\psi$ where $\psi(c) = c^2$ for $c \in \mathrm{Alt}_0(J)$. Conversely, if there is a realization of α_A which represents 1, then there is a corresponding (-1)-involution J. Consequently, if α is one choice for the Albert form, then there is a bijective correspondence:

$$\{\text{isomorphism classes of } (-1)\text{-involutions on } A\} \leftrightarrow D_F(\alpha)/G_F(\alpha).$$

10. If $f \in \mathcal{D}(\mathrm{End}(V))$ then $\pi(f)$ is a polynomial in f. For example, when $n = \dim V$:

$$\text{if } n = 4 \text{ then } \pi(f) = \tfrac{1}{2}\cdot(\mathrm{tr}\, f)1_V - f;$$

$$\text{if } n = 6 \text{ then } \pi(f) = f^2 - \tfrac{1}{2}\cdot(\mathrm{tr}\, f)\cdot f + \left(\tfrac{1}{8}\cdot(\mathrm{tr}\, f)^2 - \tfrac{1}{4}\cdot(\mathrm{tr}\, f^2)\right)1_V.$$

(Hint. If $n = 6$ then $\chi_f(x) = x^6 - c_1x^5 + c_2x^4 - \cdots = p(x)^2$ where $p(x) = x^3 + ax^2 + bx + c$. Then $\pi(f) = p^*(f)$ where $p^*(x) = x^2 + ax + b$. Then $a = -\frac{1}{2}c_1$ and $b = \frac{1}{2}c_2 - \frac{1}{8}c_1^2$. For the eigenvalues λ_i, $c_1 = \sum \lambda_i = \mathrm{tr}(f)$ and $c_2 = \sum \lambda_i\lambda_j = \frac{1}{2}((\mathrm{tr}\, f)^2 - \mathrm{tr}(f^2))$.)

11. **Finite field examples.** (1) Suppose $\mathcal{S} \subseteq \mathbb{M}_n(F)$ is a linear subspace of singular matrices, but that for some extension field K/F the space $\mathcal{S} \otimes K \subseteq \mathbb{M}_n(K)$ contains a nonsingular matrix. Then F must be finite and $n > |F|$.

(2) The set of all $\begin{pmatrix} x & * & * \\ 0 & y & * \\ 0 & 0 & x+y \end{pmatrix}$ provides a 5-dimensional example in $\mathbb{M}_3(\mathbb{F}_2)$. Find a similar example of $\mathcal{S} \subseteq \mathbb{M}_4(\mathbb{F}_3)$ with $\dim \mathcal{S} = 9$.

12. Suppose A is a central simple F-algebra.

(1) If J is an involution on A and $a \in A$ then $a \sim J(a)$, by Corollary 10.19. In fact $J(a) = bab^{-1}$ for some b such that $J(b) = \lambda b$, where $\lambda = \mathrm{type}(J)$.

(2) If $a \in A$ is nilpotent then $a \sim -a$.

13. **Linear algebra.** (1) **Lemma.** *If $C \in \mathbb{M}_n(F)$ then there exists some symmetric $S \in \mathrm{GL}_n(F)$ such that $S \cdot C \cdot S^{-1} = C^{\top}$.*

(2) **Corollary.** *Let A be a central simple F-algebra with involution and $a \in A$. Then there exists a 1-involution J on A such that $J(a) = a$.*

(3) **Proposition.** *Let A be as before and suppose $\varepsilon = \pm 1$ is given. If $a \in A^{\bullet}$ with $a \sim -a$ then there exists an ε-involution J such that $J(a) = -a$.*

(Hint. (1) Use the rational canonical form to reduce to the case C is a companion matrix. Now S can be exhibited explicitly. It can also be derived as the Gram matrix of a trace form on the algebra $F[x]/(p(x))$ where $p(x)$ is the characteristic polynomial of A.

(2) Suppose $A = \mathrm{End}(V)$ is split, choose a basis, apply (1) and define $J(X) = S^{-1} \cdot X^{\top} \cdot S$. If A is not split then F is infinite. Fix a 1-involution J_0, consider the linear subspace $W = \{c \in A : J_0(c) = c \text{ and } J_0(ca) = ca\}$, and apply (10.17).

(3) The same steps work, but the split case is harder. References appear in the Notes below.)

14. **Generalizing $\mathcal{D}$.** Define

$$\mathcal{D}_n^0 = \{B \in \mathrm{M}_n(F) : B = ST \text{ for some skew-symmetric } S, T\}.$$

(1) $\mathcal{D}_n \subseteq \mathcal{D}_n^0$ with strict containment if $n \geq 3$.

(2) If $B \in \mathcal{D}_n^0$ then every elementary divisor of B not of the form x^k occurs with even multiplicity.

(3) Find some $B \in \mathcal{D}_3^0$ with $\mathrm{rank}(B) = 1$. What conditions on the elementary divisors characterize elements of $\mathcal{D}_n^0$? (See the Notes for references.)

(Hint. (1) Find 4×4 skew-symmetric S, T such that ST has rank 1.

(2) Note that $QBQ^{-1} = (QSQ^{\top})(Q^{-\top}TQ^{-1})$ and choose Q so that $QSQ^{\top} = \begin{pmatrix} H & 0 \\ 0 & 0 \end{pmatrix}$ for some nonsingular skew-symmetric H. Then $B \sim \begin{pmatrix} B_0 & B_1 \\ 0 & 0 \end{pmatrix}$ where $B_0 \in \mathcal{D}$. The multiplicity of a non-zero eigenvalue of B equals that of B_0 and (10.10) applies.)

15. (1) Let $f \in \mathrm{End}(V)$. Then f lies in $\mathcal{D} \iff f \sim \begin{pmatrix} C & 0 \\ 0 & C \end{pmatrix}$. Here is a "basis-free" version: $f \in \mathcal{D} \iff f$ centralizes some split quaternion subalgebra of $\mathrm{End}(V)$.

(2) **Proposition.** *Let A be a central simple F-algebra with involution and suppose $Q \subseteq A$ is a quaternion subalgebra. Then $C_A(Q) \subseteq \mathcal{D}(A)$. The converse is true if A is split of even degree or if A has degree 4.*

(Hint: (1) If $f \in \mathcal{D}$ then $V = U \oplus W$ with bases $\{u_1, \dots\}$ and $\{w_1, \dots\}$ such that $f(u_j) = \sum_i c_{ij} u_i$ and $f(w_j) = \sum_i c_{ij} w_i$. Define $g, h \in \mathrm{End}(V)$ by $g = \begin{pmatrix} 0 & 1 \\ 1 & 0 \end{pmatrix}$ and $h = \begin{pmatrix} 1 & 0 \\ 0 & -1 \end{pmatrix}$. Then f centralizes the algebra generated by g and h.

(2) Let $C = C_A(Q)$ so that $A \cong Q \otimes C$. If $c \in C$ then by Exercise 13(2) there is a 1-involution J_0 with $J_0(c) = c$. If $J = (\text{bar}) \otimes J_0$ then J is a (-1)-involution and $J(c) = c$. Then $c \in \mathcal{D}(A)$. The converse is in (1) when A is split. What if $\deg A = 4$?)

16. **Characteristic polynomial.** Let A be a central simple F-algebra of degree n and let x be an indeterminate. If $a \in A$, define $p_a(x) = \text{nrd}(x - a)$, the reduced norm computed in $A \otimes F(x)$. (We abuse notation here, writing $x - a$ rather than $1 \otimes x - a \otimes 1$.) If K is a splitting field for A and $\varphi : A \otimes K \xrightarrow{\cong} \text{End}_K(V)$ with $f = \varphi(a \otimes 1)$ as usual, then $p_a(x)$ is the characteristic polynomial of f. Therefore $p_a(x) \in F[x]$ is monic of degree n.

(1) $p_a(x) = x^n - \text{trd}(a) \cdot x^{n-1} + \cdots + (-1)^n \text{nrd}(a)$ and $p_a(a) = 0$.

(2) Let $m_a(x) \in F[x]$ be the minimal polynomial of a over F in the usual sense. Then $m_a(x)$ divides $p_a(x)$ and those two polynomials involve the same irreducible factors in $F[x]$.

(Hint. (2) Define $\mathcal{L}_a : A \to A$ by $\mathcal{L}_a(x) = ax$ and show $\det(\mathcal{L}_a) = \text{nrd}(a)^n$. (*Proof idea.* Pass to K, and prove: $\det(\mathcal{L}_f) = (\det f)^n$.) Then $\mathcal{L}_a$ has minimal polynomial $m_a(x)$ and characteristic polynomial $\det(\mathcal{L}_{x-a}) = \text{nrd}(x - a)^n = p_a(x)^n$.)

17. Let A be a central simple F-algebra of even degree n and with involution J.

(1) If $a \in A$ has minimal polynomial $m_a(x)$ which is irreducible of degree k then $k \mid n$.

(2) If $m_a(x)$ is separable irreducible of degree k and $k \mid n/2$ then $a \in \mathcal{D}(A)$. In this case, $\text{pf}_A(a) = (-1)^{\frac{n}{2}} \cdot m_a(0)^{\frac{n}{2k}}$.

(3) Is the result still true if $m_a(x)$ is inseparable?

18. **Decomposability.** A quadratic form q is defined to be decomposable if $q \simeq \alpha \otimes \beta$ for some smaller forms α, β. If (V, q) is decomposable then the algebra with involution $(\text{End}(V), I_q)$ is decomposable. For the converse, suppose (V, q) is a quadratic space and $A \subseteq \text{End}(V)$ is a proper I_q-invariant central simple subalgebra.

(1) If A is split then q is decomposable.

(2) If A is quaternion then q is decomposable.

Open question. If $(\text{End}(V), I_q)$ is decomposable, must q be decomposable?

(Hint. (1) Compare (6.11).)

19. **Albert forms of higher degree.** Define a *degree d space* to be a pair (V, φ) where V is an F-vector space and $\varphi : V \to F$ is a form of degree d. Two degree d spaces (V, φ) and (W, ψ) are *similar* if there exists a bijective linear map $f : V \to W$ and a scalar $\lambda \in F^\bullet$ such that $\psi(f(v)) = \lambda \cdot \varphi(v)$ for every $v \in V$.

(1) Let (A, J) be a central simple algebra of even degree n, with an involution. Define the *Albert form* α_A to be $(\text{Alt}(J), \text{Pf}_s)$ if $\text{type}(J) = 1$, and to be $(\text{Alt}(J), \text{pf})$

is type$(J) = -1$. This is a degree $n/2$ space of dimension $\frac{n(n-1)}{2}$. Generalize (10.23) to show that the similarity class of α_A is independent of the choice of the involution J.

(2) If A has degree 8, then α_A is a quartic form in 28 variables.

Open Question. Can this α_A be used somehow? If $A = C(V, q)$ where $\dim V = 6$, how can the pfaffian of $a \in \mathcal{D}(A)$ be computed?

20. **Decomposable involutions. Proposition.** *Suppose (A, J) is a central simple F-algebra of degree $n > 2$, with a symplectic involution. If $[A] = [D]$ for a quaternion algebra D then (A, J) is decomposable. In fact, $(D, \text{bar}) \subset (A, J)$.*

(Hint. We may assume D is a division algebra. Let V be an irreducible right A-module, so that $A \otimes D \cong \text{End}(V)$ and $A = \text{End}_D(V)$. Here D acts naturally on the left, so that $d \cdot va = dv \cdot a$. The involution $J \otimes$ (bar) on $A \otimes D$ yields some I_B for a symmetric bilinear form B on V. This B admits D (as in the appendix to Chapter 4) so it lifts to a hermitian form $h : V \times V \to D$. The adjoint involution I_h on $\text{End}_D(V)$ coincides with the original J on A. There are *right* actions of D on V which commute with the given left action: Choose an orthogonal D-basis $\{v_1, \ldots, v_s\}$ of V. Every $v \in V$ has a unique expression $v = \sum d_i v_i$ for $d_i \in D$. For $x \in D$ define $\mathcal{R}_x(v) = v * x = \sum d_i x v_i$. Then $dv * x = d \cdot (v * x)$ so that $\mathcal{R} : D \to \text{End}_D(V) = A$. Furthermore, $h(v * x, w) = h(v, w * \bar{x})$ since $h(v_i, v_j) \in F$ for every i, j. Then $\mathcal{R}$ provides the desired embedding.)

Open Question. Is there a similar result when D is a product of 2 quaternion algebras?

Notes on Chapter 10

Skew-determinants are called Pfaffians, based on an 1815 paper of Pfaff which dealt with systems of linear differential equations. Pfaff's method was extended by works of Jacobi. Cayley (1847) was the first to prove that the determinant of any skew-symmetric matrix of even order is the square of a Pfaffian. Further details on these historical developments appear in Muir (1906).

Many of the results on Pfaffians also appear in Knus et al. (1998), §2. Some information and references concerning division algebras with involution are mentioned in (6.17).

Lemma 10.10 on the multiplicities of elementary divisors is well known. For example see Bennett (1919); Stenzel (1922); Hodge, Pedoe (1947), pp. 383–384; Freudenthal (1952); Kaplansky (1983) and Gow, Laffey (1984). Proofs that the characteristic polynomial of such a map f must be a perfect square appear in Voss (1896), Jacobson (1939), Drazin (1952).

Some authors (e.g. Fröhlich (1984)) use (10.14)(2) as the definition of $\text{pf}_J(f)$ for $f \in \text{Alt}(J)^{\bullet}$. First show that any such f is expressible as $f = J(g)g$ for some $g \in \text{GL}(V)$, using the fact that all regular alternating forms on V are isometric. Define

$\mathrm{pf}_J(f) = \det g$. To prove it is well defined use the lemma: If $J(h) \cdot h = 1_V$ then $\det h = 1$.

The notion of a reduced Pfaffian on a central simple algebra, parallel to the reduced norm, was introduced independently by Fröhlich (1984), Jacobson (1983) and Tamagawa (1977). Also Jančevskii (1974) proved that if J is a symplectic involution on an F-division algebra of degree 4 then $\mathrm{nrd}(a) \in F^2$ for every $a \in \mathrm{Alt}(J)$. See Knus (1988) and Knus et al. (1998), §2 for a discussion of the reduced Pfaffian, done uniformly for fields of any characteristic.

Exercise 13. These results have appeared in various forms in the literature. Part (1) goes back at least to Voss (1896) (over the complex field $\mathbb{C}$) and has been re-proved by a number of authors since then, including Frobenius (1910), Taussky, Zassenhaus (1959), Kaplansky (1969), Theorem 66. Part (3) in the split case characterizes those nonsingular matrices which are expressible as a product ST where S is symmetric and T is skew-symmetric. Such results were proved over $\mathbb{C}$ by Stenzel (1922) and over $\mathbb{R}$ by Freudenthal (1952). More general statements were proved for arbitrary fields in Hodge, Pedoe (1947) (p. 376, pp. 389–390), in Gow and Laffey (1984), and in Shapiro (1992).

Exercise 13 is also related to the following extension result due to Kneser (stated here only for involutions of the first kind), and proved in Scharlau (1985), §8.10 and in Knus et al. (1998), (4.14).

Theorem. *Suppose A is a central simple F-algebra with involution and $B \subseteq A$ is a simple subalgebra. Any involution on B can be extended to an involution on A.*

Exercise 14. $B \in \mathcal{D}_n^0$ if and only if every elementary divisor of B not of the form x^k occurs with even multiplicity, and the remaining elementary divisors are arrangeable in pairs x^k, x^k or x^{k+1}, x^k. This calculation was done over $\mathbb{C}$ by Stenzel (1922), over $\mathbb{R}$ by Freudenthal (1952), over a general field by Gow, Laffey (1984).

Exercise 16. These results on the "reduced characteristic polynomial" also follow from the theory of the "generic minimum polynomial" described in Jacobson (1968), pp. 222–226. For a central simple (associative) F-algebra A, the generic minimum polynomial of $a \in A$ coincides with the reduced characteristic polynomial of Exercise 16. If J is a (-1)- involution on $\mathrm{End}(V)$, then $\mathrm{Alt}(J)$ can be viewed as a Jordan algebra. If $a \in \mathrm{Alt}(J)$ the generic norm $n(a)$ is exactly the Pfaffian $\mathrm{pf}(a)$ as we defined above. See pp. 230–232 of Jacobson (1968). Also compare Knus et al. (1998), §32.

Exercise 20. See Bayer, Shapiro, Tignol (1992).

Chapter 11

Hasse Principles

In this chapter we determine when there is a "Hasse Principle" for (s, t)-families. Before discussing definitions and details, we can get a rough idea of this topic by considering the field $\mathbb{Q}$ of rational numbers. The completions of $\mathbb{Q}$ with respect to various absolute values are well known. They are $\mathbb{R}$ (the field of real numbers) and $\mathbb{Q}_p$ (the field of p-adic numbers) where p is a prime number. To unify the notation let $\mathbb{Q}_\infty = \mathbb{R}$. If q is a quadratic form over $\mathbb{Q}$, write $q_p = q \otimes \mathbb{Q}_p$ for the extension of q to $\mathbb{Q}_p$. The Hasse Principle for "$\sigma < \operatorname{Sim}(q)$" is the implication:

$$\sigma_p < \operatorname{Sim}(q_p) \text{ for every } p \text{ implies } \sigma < \operatorname{Sim}(q) \text{ over } \mathbb{Q}.$$

Here "every p" means p is either ∞ or a prime number. The main result of the chapter is that this Hasse Principle does hold in most cases. In fact it fails if and only if σ is special (in the sense of Definition 9.15). We prove this in the more general context of (s, t)-families over an arbitrary global field. We also obtain a version of the Hasse Principle valid for special pairs.

This chapter is fairly specialized and the results here are not used later in the book. One goal here is to establish a new theorem, the Modified Hasse Principle (Theorem 11.17). This was conjectured in 1978 and has not previously appeared in the literature.

Throughout this chapter F is a global field. We assume that the reader is familiar with the basic results about quadratic forms over local fields and global fields (e.g. the Approximation Theorem for Valuations, the Hasse–Minkowski Theorem). This background is described (sometimes without complete proofs) in several texts. For example, see O'Meara (1963), Lam (1973), Cassels (1978) or Scharlau (1985). Before beginning the discussion of (s, t)-families we review the notations and results that will be used.

We concentrate on the case F is an algebraic number field. That is, F is a finite algebraic extension of the field $\mathbb{Q}$ of rational numbers. There is another type of global field, namely the finite algebraic extensions of $\mathbb{F}_p(t)$, where $\mathbb{F}_p$ is the field of p elements and t is an indeterminate. These "algebraic function fields" are often easier to handle than the algebraic number fields. To simplify the exposition we omit the function field case (see Exercise 2). Throughout this chapter, F is an algebraic number field unless specifically stated otherwise.

A *prime* $\mathfrak{p}$ of F is an equivalence class of valuations on F. Other authors use the terms "prime spot" or "place" for $\mathfrak{p}$. Let $F_\mathfrak{p}$ denote the completion of F at $\mathfrak{p}$.

Each prime is either "finite" or "infinite". A *finite prime* is one arising from a P-adic valuation relative to a prime ideal P in the ring of integers of F. In this case $\mathfrak{p}$ lies over a unique rational prime p and $F_\mathfrak{p}$ is a finite extension of the field $\mathbb{Q}_p$. An infinite prime is one arising from an archimedean valuation. In that case either $F_\mathfrak{p} \cong \mathbb{R}$ and $\mathfrak{p}$ is called a *real prime*, or $F_\mathfrak{p} \cong \mathbb{C}$ and $\mathfrak{p}$ is called a *complex prime*. The real primes correspond to embeddings of F into $\mathbb{R}$, or equivalently they correspond to orderings of F. There are finitely many infinite primes of F.

Quadratic forms over the completed field $F_\mathfrak{p}$ are fairly easy to work with. If $\mathfrak{p}$ is a finite prime then every quadratic form of dimension ≥ 5 over $F_\mathfrak{p}$ is isotropic. In fact, there exists a unique quaternion division algebra over $F_\mathfrak{p}$ and its norm form is the unique 4-dimensional anisotropic quadratic form over $F_\mathfrak{p}$. The isometry class of a quadratic form α over $F_\mathfrak{p}$ is determined by its invariants: $\dim \alpha$, $d\alpha$ and $c(\alpha)$. Note that $c(\alpha)$ can take on only 2 values since there is only one non-split quaternion algebra.

If $\mathfrak{p}$ is complex the isometry class of a quadratic form over $F_\mathfrak{p} \cong \mathbb{C}$ is determined by its dimension. If $\mathfrak{p}$ is real the isometry class of a quadratic form over $F_\mathfrak{p} \cong \mathbb{R}$ is determined by its dimension and its signature.

If q is a form over F then information about q over F is said to be "global", while information about $q_\mathfrak{p} = q \otimes F_\mathfrak{p}$ is called "local". The idea of a "local-global principle" or "Hasse Principle" is a central concept in this theory. A property $\mathcal{L}$ is said to satisfy a Hasse Principle if $\mathcal{L}$ can be checked over F by verifying it at all the completions $F_\mathfrak{p}$. The next theorem is the classic example of a "Hasse Principle".

11.1 Hasse–Minkowski Theorem. *Suppose q is a quadratic form over F and $q_\mathfrak{p}$ is isotropic for all primes $\mathfrak{p}$ of F. Then q is isotropic over F.*

Consequently if α and β are two quadratic forms over F then: $\alpha \simeq \beta$ over F if and only if $\alpha \otimes F_\mathfrak{p} \simeq \beta \otimes F_\mathfrak{p}$ for every prime $\mathfrak{p}$. Therefore isometry of quadratic forms is decided by the invariants $\dim \alpha$, $d\alpha$, $\text{sgn}_\mathfrak{p}(\alpha) = \text{sgn}(\alpha \otimes F_\mathfrak{p})$ at the real primes $\mathfrak{p}$, and $c_\mathfrak{p}(\alpha) = c(\alpha \otimes F_\mathfrak{p})$ at the finite primes $\mathfrak{p}$.

11.2 Definition. Let (σ, τ) be an (s, t)-pair over the global field F. The *Hasse Principle* for (σ, τ) is the following statement: If q is a regular quadratic form over F and if $(\sigma_\mathfrak{p}, \tau_\mathfrak{p}) < \text{Sim}(q_\mathfrak{p})$ over $F_\mathfrak{p}$ for every prime $\mathfrak{p}$ of F, then $(\sigma, \tau) < \text{Sim}(q)$ over F.

Our first goal is to prove that the Hasse Principle for (σ, τ) holds if and only if (σ, τ) is not special (Theorem 11.13). We will then modify the Hasse Principle to get a positive result for special pairs as well (Theorem 11.17).

Since the global field F is linked, (9.13) and (9.16) imply that the Pfister Factor Conjecture is true over F and that every unsplittable (σ, τ)-module is similar to a Pfister form, provided (σ, τ) is not special. Recall that the special (s, t)-pairs (as described in (9.15)) are exactly the ones whose unsplittables have dimension 2^{m+2} where $m = \delta(s, t)$ in the notation of Theorem 7.8.

11.3 Lemma. *In proving the Hasse principle for* (σ, τ), *we may assume* σ *represents* 1.

Proof. Suppose $(\sigma_{\mathfrak{p}}, \tau_{\mathfrak{p}}) < \mathrm{Sim}(q_{\mathfrak{p}})$ for all primes $\mathfrak{p}$ of F. We assume $\dim \sigma \geq 1$ after switching σ and τ if necessary. For any $a \in D_F(\sigma)$ the Hasse–Minkowski Theorem implies $\langle a \rangle q \simeq q$. Let $(\sigma', \tau') = (\langle a \rangle \sigma, \langle a \rangle \tau)$ so that σ' represents 1 and for every $\mathfrak{p}$, $(\sigma'_{\mathfrak{p}}, \tau'_{\mathfrak{p}}) < \mathrm{Sim}(q_{\mathfrak{p}})$. If the Hasse principle for (σ', τ') is true we conclude that $(\sigma', \tau') < \mathrm{Sim}(q)$ and therefore $(\sigma, \tau) < \mathrm{Sim}(q)$. □

Since (s, t)-families are closely related to "divisibility" of quadratic forms we first consider a Hasse principle for such division. Recall that $\varphi \mid q$ means that $q \simeq \varphi \otimes \omega$ for some quadratic form ω.

11.4 Proposition. *Let* φ *and* q *be quadratic forms over a global field* F. *If* $\varphi_{\mathfrak{p}} \mid q_{\mathfrak{p}}$ *over* $F_{\mathfrak{p}}$ *for all primes* $\mathfrak{p}$ *of* F, *then* $\varphi \mid q$ *over* F.

Proof of a special case. We first give the short proof in the case φ is a Pfister form. This is the only case which we need later. The full proof of (11.4) is presented in the appendix.

Suppose φ is a Pfister form and $c \in D_F(q)$. Since $\varphi_{\mathfrak{p}} \mid q_{\mathfrak{p}}$ Lemma 5.5(1) shows that $\langle c \rangle \varphi_{\mathfrak{p}} \subset q_{\mathfrak{p}}$ for every prime $\mathfrak{p}$. Hasse–Minkowski then implies that $q \simeq \langle c \rangle \varphi \perp q'$ for some form q' over F. Then Lemma 5.5(2) implies that $\varphi_{\mathfrak{p}} \mid q'_{\mathfrak{p}}$ for every p. The result now follows by induction. □

11.5 Corollary. *The Hasse Principle is true for every minimal pair* (σ, τ) *over* F. *It is also true for all pairs* (σ, τ) *having unsplittables of dimension* ≤ 4.

Proof. Suppose (σ, τ) is a minimal pair and q is a quadratic form over F such that $(\sigma_{\mathfrak{p}}, \tau_{\mathfrak{p}}) < \mathrm{Sim}(q_{\mathfrak{p}})$ for every $\mathfrak{p}$. Let ψ be the quadratic unsplittable for (σ, τ) as in (7.11). For any prime $\mathfrak{p}$ the pair $(\sigma_{\mathfrak{p}}, \tau_{\mathfrak{p}})$ is again minimal (see (7.9)) and has unsplittable module $\psi_{\mathfrak{p}}$. Then $\psi_{\mathfrak{p}} \mid q_{\mathfrak{p}}$ by (7.11), so that $\psi \mid q$ over F by (11.4). Another application of (7.11) shows that $(\sigma, \tau) < \mathrm{Sim}(q)$ over F. Note that ψ is similar to a Pfister form here, by (9.16), so we used only the special case of (11.4) proved above.

If the unsplittables for (σ, τ) have dimension ≤ 4 then (σ, τ) can be replaced by one of the examples listed in Theorem 5.7. (See Exercise 5.8.) If (σ, τ) is one of the pairs listed in (5.7) then for a form q over F the relationship $(\sigma, \tau) < \mathrm{Sim}(q)$ is characterized by certain "factors" of the form q, or by certain terms in $G_F(q)$. Then (11.4) implies the Hasse Principle for (σ, τ). □

Here is one more useful comment about minimal pairs.

11.6 Lemma. *If* (σ, τ) *is an* (s, t)*-pair over* F *and if* $(\sigma_{\mathfrak{p}}, \tau_{\mathfrak{p}})$ *is a minimal pair over* $F_{\mathfrak{p}}$ *for every* $\mathfrak{p}$ *then* (σ, τ) *is a minimal pair over* F.

Proof. Check the criteria in Theorem 7.8. □

We pause in the exposition of (s, t)-families to recall some notations and results about quadratic forms representing values with "prescribed signatures".

11.7 Definition. If $\mathfrak{p}$ is a real prime of F (i.e. an ordering) and q is a quadratic form over F then $\mathrm{sgn}_{\mathfrak{p}}(q) = \mathrm{sgn}(q_{\mathfrak{p}})$ denotes the signature of q relative to this ordering. Let X_F be the set of all orderings (i.e. real primes) of F.

Various "Approximation Theorems" are useful tools in number theory. We need to know some special cases of approximation involving the real primes of F. First we quote the standard "Weak Approximation Theorem" which is a consequence of a general result about the independence of valuations on fields. Here $|\cdot|_{\mathfrak{p}}$ denotes a fixed absolute value corresponding to the prime $\mathfrak{p}$.

11.8 Lemma. *Let S be a finite set of primes of F. Let $a_{\mathfrak{p}} \in F_{\mathfrak{p}}$ be given for $\mathfrak{p} \in S$ and let $\varepsilon > 0$ be a given real number, arbitrarily small. Then there exists an $a \in F$ such that $|a - a_{\mathfrak{p}}|_{\mathfrak{p}} < \varepsilon$ for every $\mathfrak{p} \in S$.*

Let us now consider the signs of values represented by a form.
If $w = (c_1, c_2, \dots, c_n) \in F_{\mathfrak{p}}^n$ define the norm $||w||_{\mathfrak{p}} = \max_j\{|c_j|_{\mathfrak{p}}\}$.

11.9 Corollary. *Let q be a quadratic form over F. For each real prime $\mathfrak{p}$ let $\delta_{\mathfrak{p}} \in \{\pm 1\}$ be a value represented by $q_{\mathfrak{p}}$. Then there exists $a \in D_F(q)$ such that $\mathrm{sgn}_{\mathfrak{p}}(a) = \delta_{\mathfrak{p}}$ for every $\mathfrak{p}$.*

Proof. Let $q \simeq \langle c_1, \dots, c_n\rangle$. For each $\mathfrak{p}$ choose a vector $x_{\mathfrak{p}} = (x_{1\mathfrak{p}}, \dots, x_{n\mathfrak{p}}) \in F_{\mathfrak{p}}^n$ such that $q(x_{\mathfrak{p}}) = \delta_{\mathfrak{p}}$. By (11.8), there exists a vector $x \in F^n$ which is very close to $x_{\mathfrak{p}}$ for every $\mathfrak{p}$. (That is, for given $\varepsilon > 0$ there exists x such that $||x - x_{\mathfrak{p}}||_{\mathfrak{p}} < \varepsilon$ for every $\mathfrak{p}$.) Let $a = q(x)$ for this vector x. Then a is close to $q(x_{\mathfrak{p}}) = \delta_{\mathfrak{p}}$ in $F_{\mathfrak{p}}$ for every $\mathfrak{p}$, so that a has the same sign as $\delta_{\mathfrak{p}}$. □

Suppose $\mathfrak{p} \in X_F$. A quadratic space (V, q) is called "positive definite at $\mathfrak{p}$" if for every $0 \neq v \in V$ the value $q(v)$ is positive relative to the ordering $\mathfrak{p}$. If $q \simeq \langle a_1, \dots, a_n\rangle$ then q is positive definite at $\mathfrak{p}$ if all the a_i are positive at $\mathfrak{p}$. Similarly we define "negative definite". A form is called "indefinite" at $\mathfrak{p}$ if it is neither positive nor negative definite at $\mathfrak{p}$.

11.10 Definition. If γ is a form over F let $H(\gamma) = \{\mathfrak{p} \in X_F : \gamma$ is positive definite at $\mathfrak{p}\}$. If $a_j \in F^\bullet$ let $H(a_1, a_2, \dots, a_n) = H(\langle a_1, a_2, \dots, a_n\rangle)$. If $\Omega \subseteq X_F$ let ε_Ω denote any element of $F^\bullet$ with the property that $H(\varepsilon_\Omega) = \Omega$.

Then ε_Ω is positive at an ordering $\mathfrak{p}$ if and only if $\mathfrak{p} \in \Omega$. The Approximation Theorem (11.8) implies that for every Ω there does exist such an element ε_Ω.

The Hasse–Minkowski Theorem implies that the isometry of quadratic forms over F is determined by the classical invariants $\dim q$, dq, $c(q_{\mathfrak{p}})$ and $\operatorname{sgn}_{\mathfrak{p}}(q)$. If q lies in I^3F (the ideal in the Witt ring generated by all 3-fold Pfister forms) then $dq = \langle 1\rangle$ and $c(q_{\mathfrak{p}}) = 1$ for all $\mathfrak{p}$. In this case the isometry class of q is determined by its dimension and its signatures. For instance if ψ is a 3-fold Pfister form then $\psi \simeq \langle\langle 1, 1, \varepsilon_\Omega\rangle\rangle = 4\langle\langle \varepsilon_\Omega\rangle\rangle$ where $\Omega = H(\psi)$. Similarly if ψ is an $(m+1)$-fold Pfister form where $m \geq 2$ and $\Omega = H(\psi)$, then $\psi \simeq \langle\langle 1, \dots, 1, \varepsilon_\Omega\rangle\rangle \simeq 2^m\langle\langle \varepsilon_\Omega\rangle\rangle$.

We now return to the discussion of the Hasse Principle for an (s,t)-pair (σ, τ).

11.11 Lemma. *If $(\sigma, \tau) < \operatorname{Sim}(q)$ and $\mathfrak{p}$ is a real prime with $\mathfrak{p} \notin H(\sigma \perp \tau)$ then $\operatorname{sgn}_{\mathfrak{p}}(q) = 0$. Consequently, $H(q) \subseteq H(\sigma \perp \tau)$. Furthermore if (σ, τ) is a minimal pair and q is an unsplittable which represents 1, then $H(q) = H(\sigma \perp \tau)$.*

Proof. For such $\mathfrak{p}$ either σ or τ represents a value a which is negative at $\mathfrak{p}$. Then $\langle 1, -a\rangle \otimes q$ is hyperbolic (since $\langle a\rangle q \simeq q$) so that $2 \cdot \operatorname{sgn}_{\mathfrak{p}}(q) = 0$. In the other direction, suppose (σ, τ) is minimal and q is an unsplittable which represents 1. Then $\dim q = 2^m$ where $m = \delta(s,t)$. If $\mathfrak{p} \in H(\sigma \perp \tau)$ then $(\sigma_{\mathfrak{p}}, \tau_{\mathfrak{p}}) \simeq (s\langle 1\rangle, t\langle 1\rangle)$ and Exercise 7.3 (3) implies that the unique unsplittable for $(\sigma_{\mathfrak{p}}, \tau_{\mathfrak{p}})$ is $2^m\langle 1\rangle$. Then $q_{\mathfrak{p}}$ is similar to $2^m\langle 1\rangle$ and hence $q_{\mathfrak{p}} \simeq 2^m\langle 1\rangle$ since q represents 1. Therefore $\mathfrak{p} \in H(q)$. □

If (σ, τ) is not minimal there is more freedom to prescribe the signatures of unsplittables.

11.12 Proposition. *Let (σ, τ) be an (s,t)-pair over F such that σ represents 1 and the (σ, τ)-unsplittables have dimension 2^{m+1} where $m = \delta(s,t) \geq 2$. Then $(\sigma, \tau) < \operatorname{Sim}(2^m\langle\langle \varepsilon_\Omega\rangle\rangle)$ for every $\Omega \subseteq H(\sigma \perp \tau)$.*

Proof. The strategy is to expand the non-minimal pair (σ, τ) to some minimal pair $(\sigma \perp \langle a\rangle, \tau)$ having an unsplittable of dimension 2^{m+1}. The criteria for doing this are listed in Exercise 7.7. Suppose such $a \in F^\bullet$ can be chosen so that $H(\sigma \perp \tau) \cap H(a) = \Omega$. To prove the proposition let φ be any $(\sigma \perp \langle a\rangle, \tau)$-unsplittable which represents 1. By (11.11) $H(\varphi) = H(\sigma \perp \langle a\rangle \perp \tau) = \Omega$. By (9.16) this φ is an $(m+1)$-fold Pfister form. Since $m \geq 2$ we conclude that $\varphi \simeq 2^m\langle\langle \varepsilon_\Omega\rangle\rangle$, as hoped.

In order to construct such an element a, first note that if $s \equiv t \pm 3 \pmod 8$ then (σ, τ) is minimal and the hypotheses cannot occur. Next suppose $s \equiv t + 2$ or $t + 4 \pmod 8$. Then $(\sigma \perp \langle a\rangle, \tau)$ is minimal for every $a \in F^\bullet$. Choosing $a = \varepsilon_\Omega$ we are done.

Suppose $s \equiv t \pmod 8$. If $d\beta = \langle 1\rangle$ we may shift to assume $s \geq 2$, replace (σ, τ) by some $(s-1, t)$-pair (σ', τ) and apply the case $s \equiv t - 1$ to that pair. Suppose $d\beta = \langle d\rangle \not\simeq \langle 1\rangle$. Then (7.8) implies that $c(\beta)$ is split by $F(\sqrt{d})$, so that $c(\beta) = [d, x]$ for some $x \in F^\bullet$. The pair $(\sigma \perp \langle a\rangle, \tau)$ is minimal if and only if $c(\beta) = [d, -a]$, by Exercise 7.7. This occurs when $[d, -ax] = 1$, or equivalently when $a \in D_F(\langle -x\rangle\langle 1, -d\rangle)$. By (11.9) we can choose a in that set with prescribed signs at every ordering $\mathfrak{p}$ where d is positive. A calculation shows that $\langle d\rangle = d\beta = (\det\sigma)(\det\tau)$ so that d is positive at every $\mathfrak{p} \in H(\sigma \perp \tau)$. Then we may choose a so that $H(\sigma \perp \tau) \cap H(a) = \Omega$.

Suppose $s \equiv t+1 \pmod 8$. Then $c(\beta) \neq 1$ by (7.8), so suppose $c(\beta) = [-x, -y]$. By Exercise 7.7 we see that $(\sigma \perp \langle a\rangle, \tau)$ is minimal if and only if $c(\beta)$ is split by $F(\sqrt{-ad})$ where $d\beta = \langle d\rangle$. This occurs if and only if $a \in D_F(\langle -d\rangle\langle x, y, xy\rangle)$. By (11.9) we can choose a in that set with prescribed signs at every ordering $\mathfrak{p}$ where $\langle x, y, xy\rangle$ is indefinite. If $\mathfrak{p} \in H(\sigma \perp \tau)$ then $\operatorname{sgn}_{\mathfrak{p}}(\beta) \equiv 1 \pmod 8$ and Exercise 3.5(5) implies that $c_{\mathfrak{p}}(\beta) = 1$. Therefore $[-x, -y]_{\mathfrak{p}} = 1$ and $\langle x, y, xy\rangle_{\mathfrak{p}}$ must be indefinite. Then we may choose a so that $H(\sigma \perp \tau) \cap H(a) = \Omega$.

Finally if $s \equiv t+6$ or $t+7 \pmod 8$ there is no expansion $(\sigma \perp \langle a\rangle, \tau)$ which is minimal. In these cases we may shift part of σ to the right to assume $t > 0$, scale to assume τ represents 1 and consider the (t, s)-family (τ, σ). One of the earlier cases now applies. □

11.13 Theorem. *The Hasse Principle is true for every non-special pair over F.*

Proof. Let (σ, τ) be a non-special (s, t)-pair. By (11.3) we may assume σ represents 1. By (11.5) the result is true if (σ, τ) is minimal so let us assume it is non-minimal. Then the dimension of an unsplittable is 2^{m+1} where $m = \delta(s, t)$. We proceed by induction on m. Since the case $m = 1$ was proved in (11.5) we assume $m \geq 2$. Suppose q is a form over F such that $(\sigma_{\mathfrak{p}}, \tau_{\mathfrak{p}}) < \operatorname{Sim}(q_{\mathfrak{p}})$ for every prime $\mathfrak{p}$. We may replace q by $\langle\varepsilon_\Delta\rangle q$ where $\Delta = \{\mathfrak{p} : \operatorname{sgn}_{\mathfrak{p}}(q) \geq 0\}$ in order to assume that $\operatorname{sgn}_{\mathfrak{p}}(q) \geq 0$ for every real prime $\mathfrak{p}$.

Lemma 11.6 implies that there exists some $\mathfrak{p}$ where $(\sigma_{\mathfrak{p}}, \tau_{\mathfrak{p}})$ is not minimal. At that prime $\mathfrak{p}$ the dimension of an unsplittable $(\sigma_{\mathfrak{p}}, \tau_{\mathfrak{p}})$-module is 2^{m+1}. Therefore $\dim q = 2^{m+1} \cdot r$ for some $r \geq 1$.

Now choose subforms $\sigma' \subset \sigma$ and $\tau' \subset \tau$ such that σ' represents 1, $\dim\sigma' \not\equiv \dim\tau' \pmod 8$ and unsplittable (σ', τ')-modules have dimension 2^m. (See Exercise 7.6.) By the induction hypothesis we know q is a (σ', τ')-module over F, so that $q \simeq q_1 \perp \cdots \perp q_k$ where each q_j is an unsplittable (σ', τ')-module. By (9.16) each q_j is similar to an m-fold Pfister form. Since $m \geq 2$ we have $dq_j = \langle 1\rangle$ and therefore $dq = \langle 1\rangle$.

Claim. $c(q) = 1$. If $m \geq 3$, then $c(q_j) = 1$ for each j and therefore $c(q) = 1$. Suppose $m = 2$. We prove the claim by checking that $c_{\mathfrak{p}}(q) = 1$ for every prime $\mathfrak{p}$. If $\mathfrak{p}$ is a prime where $(\sigma_{\mathfrak{p}}, \tau_{\mathfrak{p}})$ is not minimal then each of its unsplittables is similar to a 3-fold Pfister forms and it follows that $c_{\mathfrak{p}}(q) = 1$. If $\mathfrak{p}$ is a prime where $(\sigma_{\mathfrak{p}}, \tau_{\mathfrak{p}})$ is minimal then $(\sigma_{\mathfrak{p}}, \tau_{\mathfrak{p}})$ has a unique unsplittable module φ which is a 2-fold Pfister form. Then $q_{\mathfrak{p}} \simeq \varphi \otimes \alpha$ for some form α where $\dim\alpha$ is even since $8 \mid \dim q$. It again follows that $c_{\mathfrak{p}}(q) = 1$, proving the claim.

Since $dq = \langle 1\rangle$ and $c(q) = 1$ the isometry class of q is determined by its signatures. Define $\Omega_j = \{\mathfrak{p} \in X_F : \operatorname{sgn}_{\mathfrak{p}}(q) \geq 2^{m+1} \cdot j\}$ for $1 \leq j \leq r$. Then $H(\sigma \perp \tau) \supseteq \Omega_1 \supseteq \cdots \supseteq \Omega_r = H(q)$. Let $\varepsilon_j = \varepsilon_{\Omega_j}$ and define

$q' = 2^m\langle\langle\varepsilon_1\rangle\rangle \perp \cdots \perp 2^m\langle\langle\varepsilon_r\rangle\rangle$. Then the dimension, discriminant, Witt invariant and all signatures for q' match those of q. Therefore $q \simeq q'$. Finally by Proposition 11.12 we conclude that $(\sigma, \tau) < \mathrm{Sim}(q)$ over F. □

The special pairs are more difficult to work with. Suppose (σ, τ) is a special (s, t)-pair so that $s \equiv t \pmod 8$, $d\beta \neq \langle 1\rangle$ and $c(\beta)$ is not split by $F(\sqrt{d\beta})$.

11.14 Lemma. *If F is an algebraically closed, real closed or p-adic field, no special pairs can exist over F.*

Proof. Suppose (σ, τ) is a special pair over F and $\beta = \sigma \perp -\tau$. Then $d\beta = \langle d\rangle \neq \langle 1\rangle$ and $c(\beta) = [-x, -y]$ is not split by $F(\sqrt{d})$. Since d is not a square, F is not algebraically closed. If F is real closed we must have $\langle d\rangle = \langle -1\rangle$. But then $F(\sqrt{d})$ is algebraically closed and hence splits $c(\beta)$, contrary to hypothesis. If F is p-adic there is a unique anisotropic 4-dimensional quadratic form, and it has determinant $\langle 1\rangle$. Since $[-x, -y]$ is not split by $F(\sqrt{d})$ it follows that $\langle d, x, y, xy\rangle$ is anisotropic, and the uniqueness implies $\langle d\rangle = \langle 1\rangle$. □

11.15 Proposition. *Let (σ, τ) be a special pair over the global field F, where $s + t = 2m + 2$. then $(\sigma_\mathfrak{p}, \tau_\mathfrak{p}) < \mathrm{Sim}(2^m\mathbb{H}_\mathfrak{p})$ for every prime $\mathfrak{p}$.*

Proof. For each $\mathfrak{p}$ the lemma implies that there is a $(\sigma_\mathfrak{p}, \tau_\mathfrak{p})$-module φ of dimension 2^{m+1}. Scaling φ if necessary we know that it is a Pfister form over $F_\mathfrak{p}$, by (9.16). If $\mathfrak{p}$ is a real prime and $\mathfrak{p} \notin H(\sigma \perp \tau)$ then φ admits a (-1)-similarity and hence $\varphi \simeq 2^m\mathbb{H}_\mathfrak{p}$. If $\mathfrak{p} \in H(\sigma \perp \tau)$ then $(\sigma_\mathfrak{p}, \tau_\mathfrak{p}) = (s\langle 1\rangle, t\langle 1\rangle)$ has unsplittable module $2^m\langle 1\rangle$ over $F_\mathfrak{p}$ by Exercise 7.3. Therefore $2^m\mathbb{H}_\mathfrak{p} \simeq 2^m\langle 1\rangle \otimes \mathbb{H}_\mathfrak{p}$ is a $(\sigma_\mathfrak{p}, \tau_\mathfrak{p})$-module. Finally suppose $\mathfrak{p}$ is a finite prime. If $m \geq 2$ then every $(m+1)$-fold Pfister form over the p-adic field $F_\mathfrak{p}$ is $2^m\mathbb{H}_\mathfrak{p}$ and we are done. No special pair exists when $m = 0$. The remaining case when $s = t = 2$ is settled by the following lemma, when $q = 2\mathbb{H}$. □

11.16 Lemma. *Let F be a global field, $(\langle 1, a\rangle, \langle x, y\rangle)$ be a $(2,2)$-pair and q be a form over F. Then $\langle\langle a\rangle\rangle \mid q$, $\langle\langle xy\rangle\rangle \mid q$ and $x \in G_F(q)$ if and only if $(\langle 1, a\rangle_\mathfrak{p}, \langle x, y\rangle_\mathfrak{p}) < \mathrm{Sim}(q_\mathfrak{p})$ for every prime $\mathfrak{p}$ of F.*

Proof. The "if" part follows from (1.10) and from (11.4), the Hasse Principle for divisibility. For the converse suppose $\langle\langle a\rangle\rangle \mid q$, $\langle\langle xy\rangle\rangle \mid q$ and $x \in G_F(q)$. If $\langle axy\rangle_\mathfrak{p} \simeq \langle 1\rangle_\mathfrak{p}$ we use (5.7) (4) and the Expansion Lemma 2.5. Otherwise $\langle 1, a, -x, -y\rangle_\mathfrak{p}$ is isotropic (since there is a unique anisotropic 4-dimensional form) and (5.7) (5) applies. □

Consequently the Hasse Principle for (σ, τ) is false whenever (σ, τ) is special. This is clear from Proposition 11.12 since $(\sigma, \tau) < \mathrm{Sim}(2^m\mathbb{H})$ is impossible over F (unsplittable (σ, τ)-modules have dimension 2^{m+2} while $2^m\mathbb{H}$ has dimension 2^{m+1}).

However the obstruction to the Hasse Principle for a special pair seems to be in the dimension of q. We are led to a modification of the original principle:

11.17 Modified Hasse Principle. *Suppose (σ, τ) is a special pair over a global field F, with $s + t = 2m + 2$. If q is a form over F such that $(\sigma_\mathfrak{p}, \tau_\mathfrak{p}) < \mathrm{Sim}(q_\mathfrak{p})$ for every prime $\mathfrak{p}$ of F and such that $2^{m+2} \mid \dim q$ then $(\sigma, \tau) < \mathrm{Sim}(q)$ over F.*

The rest of this chapter is concerned with the proof of this principle. The most difficult part of this result is the case $m = 1$, when (σ, τ) is a special $(2, 2)$-pair. This case is settled by the following theorem, which will be proved later in the chapter using the "trace-form" technique developed in Chapter 5. For $a, b, x \in F^\bullet$ define the set

$$\mathcal{M} = \mathcal{M}(\langle\langle a\rangle\rangle, \langle\langle b\rangle\rangle, \langle x\rangle) = \{q : \langle\langle a\rangle\rangle \mid q,\ \langle\langle b\rangle\rangle \mid q \text{ and } x \in G_F(q)\}.$$

Of course every $q \in \mathcal{M}$ is an orthogonal sum of certain $\mathcal{M}$-indecomposables, as defined in Chapter 5. Here is our main result, valid for any global field F.

11.18 Theorem. *Suppose $\mathcal{M} = \mathcal{M}(\langle\langle a\rangle\rangle, \langle\langle b\rangle\rangle, \langle x\rangle)$ over F as above and $\langle a\rangle \not\simeq \langle b\rangle$.*

(1) *Every $\mathcal{M}$-indecomposable form has dimension 4.*

(2) *If q_1 and q_2 are $\mathcal{M}$-indecomposables then $(\langle 1, a\rangle, \langle x, bx\rangle) < \mathrm{Sim}(q_1 \perp q_2)$.*

We will keep these notations for the rest of the chapter, using $b = xy$. If the pair $(\langle 1, a\rangle, \langle x, y\rangle)$ is not special then this theorem follows from the work above. For in this case the form $\langle 1, a, -x, -y\rangle$ is isotropic and (5.7)(5) implies that $(\langle 1, a\rangle, \langle x, bx\rangle) < \mathrm{Sim}(q)$ if and only if $q \in \mathcal{M}_0 = \mathcal{M}(\langle\langle a\rangle\rangle, \langle\langle b\rangle\rangle)$. From (5.6) we know that every $\mathcal{M}_0$-indecomposable has dimension 4. It remains to check that $\mathcal{M}_0 = \mathcal{M}$ in this case. Since $\langle 1, a, -x, -xb\rangle$ is isotropic, the forms $\langle 1, a\rangle$ and $\langle x\rangle\langle 1, b\rangle$ represent a common value and therefore $x = uv$ for some $u \in D_F(\langle\langle a\rangle\rangle)$ and $v \in D_F(\langle\langle b\rangle\rangle)$. If $q \in \mathcal{M}_0$ then $\langle\langle a\rangle\rangle \mid q$ so that $\langle u\rangle q \simeq q$. Similarly $\langle\langle b\rangle\rangle \mid q$ implies that $\langle v\rangle q \simeq q$. Therefore $\langle x\rangle q \simeq q$ and $q \in \mathcal{M}$.

The proof of this theorem when the pair is special is quite involved. Before embarking on the proof we use the theorem to prove the Modified Hasse Principle 11.17. This argument requires several steps, involving judicious use of the Shift and Eigenspace Lemmas and some analysis of the possible signatures of $\mathcal{M}$-indecomposables.

11.19 Corollary. *The Modified Hasse Principle is true when $m = 1$.*

Proof. Suppose $(\sigma, \tau) = (\langle 1, a\rangle, \langle x, y\rangle)$ is a special $(2, 2)$-pair and suppose that $(\sigma_\mathfrak{p}, \tau_\mathfrak{p}) < \mathrm{Sim}(q_\mathfrak{p})$ for every prime $\mathfrak{p}$ of F and $8 \mid \dim q$. Let $\mathcal{M}$ be as above, using $b = xy$. Then Lemma 11.16 states that $q \in \mathcal{M}$. Part (1) of the theorem implies that $q \simeq q_1 \perp \cdots \perp q_k$ where $q_j \in \mathcal{M}$ and $\dim q_j = 4$. Since $8 \mid \dim q$ we know that k is even and part (2) of the theorem implies that $(\langle 1, a\rangle, \langle x, y\rangle) < \mathrm{Sim}(q)$. □

The second step in the proof of the Modified Hasse Principle is to reduce it to the construction of (σ, τ)-modules with prescribed signatures.

11.20 Proposition. *Suppose* (σ, τ) *is a special pair over* F *with* $s + t = 2m + 2$ *and* $m \geq 2$. *The Modified Hasse Principle is true for* (σ, τ) *provided* $(\sigma, \tau) < \mathrm{Sim}(2^m \langle\langle \varepsilon_\Delta \rangle\rangle \perp 2^m \langle\langle \varepsilon_\Omega \rangle\rangle)$ *for every* $\Delta \subseteq \Omega \subseteq H(\sigma \perp \tau)$.

Proof. We may assume $s \geq 2$ and express $\sigma = \sigma' \perp \langle a \rangle$ where σ' represents 1. Then unsplittable (σ', τ)-modules have dimension 2^{m+1}. Suppose q is a form over F such that $2^{m+2} \mid \dim q$ and $(\sigma_\mathfrak{p}, \tau_\mathfrak{p}) < \mathrm{Sim}(q_\mathfrak{p})$ for every prime $\mathfrak{p}$. We may scale q to assume that $\mathrm{sgn}_\mathfrak{p}(q) \geq 0$ for every real prime $\mathfrak{p}$. By Theorem 11.13 we know that $(\sigma', \tau) < \mathrm{Sim}(q)$. Therefore $q \simeq q_1 \perp \cdots \perp q_k$ where each q_i is an unsplittable (σ', τ)-module, and hence is similar to an $(m+1)$-fold Pfister form. Then $\dim q = 2^{m+1} \cdot r$ for some r. Since $m \geq 2$ we have $dq = 1$ and $c(q) = 1$, so that the isometry class of q is determined by its signatures. As in the proof of Theorem 11.13 we find elements ε_j such that $q \simeq 2^m \langle\langle \varepsilon_1 \rangle\rangle \perp \cdots \perp 2^m \langle\langle \varepsilon_r \rangle\rangle$ where $H(\sigma \perp \tau) \supseteq H(\varepsilon_1) \supseteq \cdots \supseteq H(\varepsilon_r) = H(q)$. By hypothesis r is even and $(\sigma, \tau) < \mathrm{Sim}(2^m \langle\langle \varepsilon_j \rangle\rangle \perp 2^m \langle\langle \varepsilon_{j+1} \rangle\rangle)$ for each $j = 1, 3, \ldots, r-1$. Therefore $(\sigma, \tau) < \mathrm{Sim}(q)$. □

The next two lemmas involve shifting the given (s, t)-pair to arrange σ and τ to represent many common values. These lemmas are valid in the more general setting of linked fields. Recall from (9.14) that F is "linked" if any two 2-fold Pfister forms can be written with a common slot. The Hasse–Minkowski Theorem implies that every global field is linked. The relation $\approx$ was defined before (9.4) above.

11.21 Lemma. *Let* (σ, τ) *be a* $(4, 4)$*-pair over a linked field* F *and suppose* σ *represents* 1. *Then* $(\sigma, \tau) \approx (\langle 1, a, b, c \rangle, \langle x, y, b, c \rangle)$ *for some* $a, b, c, x, y \in F^\bullet$.

Proof. We begin by verifying a claim about 7-dimensional forms.

Claim. If φ is a form over a linked field F and $\dim \varphi = 7$ then

$$\varphi \simeq \langle a \rangle \perp \langle\langle r \rangle\rangle \otimes \langle b, c, d \rangle \text{ for some } a, b, c, d, r \in F^\bullet.$$

Proof of claim. Replacing φ by $\langle \det \varphi \rangle \cdot \varphi$ we may assume $\det \varphi = \langle 1 \rangle$. By (9.14)(5) we can express $\varphi \simeq \alpha \perp \beta$ where β is a 4-dimensional form of determinant $\langle 1 \rangle$. Then $\dim \alpha = 3$ and $\det \alpha = \langle 1 \rangle$ so that $\varphi \simeq \langle x, y, xy \rangle \perp \langle w \rangle \langle\langle u, v \rangle\rangle$ for some $x, y, u, v, w \in F^\bullet$. Since F is linked, $\langle x, y, xy \rangle$ and $\langle u, v, uv \rangle$ represent some common value b. Then $\langle x, y, xy \rangle \simeq \langle b, b', bb' \rangle$ and $\langle u, v, uv \rangle \simeq \langle b, b'', bb'' \rangle$ for some $b', b'' \in F^\bullet$. Therefore $\varphi \simeq \langle b, b', bb' \rangle \perp \langle w \rangle \langle\langle b, b'' \rangle\rangle \simeq \langle b \rangle \perp \langle\langle b \rangle\rangle \otimes \langle b', w, wb'' \rangle$. This proves the claim.

Now starting from the given $(4, 4)$-pair, shift τ to the left to get a form φ with $\dim \varphi = 8$. Since σ represents 1 we express $\varphi \simeq \langle 1 \rangle \perp \varphi'$ and apply the claim to

φ'. Therefore $\varphi \simeq \langle 1, a, b, c, d, rb, rc, rd\rangle$ and we shift the 4-plane $\langle d, rb, rc, rd\rangle$ to the right to get the desired (4, 4)- pair, where $x = bcd$ and $y = rbcd$. □

11.22 Corollary. *Let (σ, τ) be an (s, t)-pair over a linked field F and suppose σ represents 1. If $s \equiv t \pmod 8$ and $s + t \geq 8$ then $(\sigma, \tau) \approx (\langle 1, a\rangle \perp \alpha, \langle x, y\rangle \perp \alpha)$ for some $a, x, y \in F^\bullet$ and some form α.*

Proof. Since $s \equiv t \pmod 8$ and $s + t \geq 8$ we find that $s, t \geq 4$. If $s \geq 5$, σ contains a 4-dimensional subform of determinant $\langle 1\rangle$ (since F is linked) and the shifting method of (9.10) shows that $(\sigma, \tau) \approx (\sigma' \perp \langle c\rangle, \tau' \perp \langle c\rangle)$ for some $c \in F^\bullet$ and some forms σ', τ' where σ' represents 1. Such reductions can be continued until we reach the case $s = 4$. Similarly we may reduce to smaller cases if $t \geq 5$. The remaining case is when $s = t = 4$ and the lemma applies. □

To apply (11.20) we must construct $\mathcal{M}$-indecomposables with prescribed signatures. If ψ is a 4-dimensional form in $\mathcal{M}$ then $\langle\langle a\rangle\rangle \mid \psi$ so that $\psi \simeq \langle s\rangle\langle\langle a, u\rangle\rangle$ for some $s, u \in F^\bullet$. Since $\mathcal{M}$ is closed under scaling, we concentrate on the case ψ is a 2-fold Pfister form.

11.23 Lemma. *Suppose $\langle a\rangle \not\simeq \langle b\rangle$ and φ is a 2-fold Pfister form over F. Then $\varphi \in \mathcal{M}$ if and only if $\varphi \simeq \langle\langle a, -w\rangle\rangle$ for some $w \in D_F(\langle\langle -ab\rangle\rangle)$ such that $H(a, b, -x) \subseteq H(w)$.*

Proof. First let $\mathcal{M}_0 = \mathcal{M}(\langle\langle a\rangle\rangle, \langle\langle b\rangle\rangle)$. If $w \in D_F(\langle\langle -ab\rangle\rangle)$ then $ab \in D_F(\langle\langle -w\rangle\rangle)$ and it follows that $\langle\langle a, -w\rangle\rangle \simeq \langle\langle b, -w\rangle\rangle \in \mathcal{M}_0$. Conversely, if $\varphi \in \mathcal{M}_0$ is a 2-fold Pfister form then $\varphi \simeq \langle\langle a, -v\rangle\rangle$ for some $v \in F^\bullet$ such that $\varphi' = \langle a\rangle \perp \langle -v\rangle\langle\langle a\rangle\rangle$ represents b. Express $b = ax^2 - vt$ where $t \in D_F(\langle\langle a\rangle\rangle) \cup \{0\}$. Then $t \neq 0$ since $\langle a\rangle \not\simeq \langle b\rangle$ and we define $w = avt$. Then $vw \in D_F(\langle\langle a\rangle\rangle)$ so that $\varphi \simeq \langle\langle a, -w\rangle\rangle$, and $w = (ax)^2 - ab \in D_F(\langle\langle -ab\rangle\rangle)$.

Now if $\varphi \in \mathcal{M}$ we must show that $w > 0$ at every $\mathfrak{p} \in H(a, b, -x)$. To do this note that $x < 0$ at such $\mathfrak{p}$ so that $\mathrm{sgn}_\mathfrak{p}\, \varphi = 0$ (since $\varphi \simeq \langle x\rangle\varphi$). Since $a > 0$ and $\varphi \simeq \langle\langle a, -w\rangle\rangle$ we find that $w > 0$ at $\mathfrak{p}$. Conversely, suppose $\varphi \in \mathcal{M}_0$ as above and $w > 0$ at every $\mathfrak{p} \in H(a, b, -x)$. To show that $\langle x\rangle\varphi \simeq \varphi$ it suffices to show that φ represents x. By Hasse–Minkowski, we need only check that $\varphi \perp \langle -x\rangle$ is indefinite at every real prime $\mathfrak{p}$. If this is false that form is positive definite at some $\mathfrak{p}$. Then $x < 0$ at $\mathfrak{p}$ and since $\langle\langle a\rangle\rangle$, $\langle\langle b\rangle\rangle$ and $\langle\langle -w\rangle\rangle$ are factors of φ we also know that $a, b > 0$ and $w < 0$ at that $\mathfrak{p}$. This contradicts the hypothesis on w. □

If $\varphi \in \mathcal{M} = \mathcal{M}(\langle\langle a\rangle\rangle, \langle\langle b\rangle\rangle, \langle x\rangle)$ and if any of a, b, x is negative at an ordering $\mathfrak{p}$, then $\mathrm{sgn}_\mathfrak{p}\, \varphi = 0$. That is: $H(\varphi) \subseteq H(a, b, x)$. We show that the signatures of φ can be arbitrarily prescribed, subject to that condition. Then Theorem 11.18 tells us how to prescribe signatures for unsplittable $(\langle 1, a\rangle, \langle x, bx\rangle)$-modules.

11.24 Lemma. (1) *For any* $\Gamma \subseteq H(a,b,x)$ *there exists* $\varphi = \langle\langle a, -w\rangle\rangle \in \mathcal{M}$ *with* $H(\varphi) = \Gamma$.

(2) *For any subsets* $\Delta \subseteq \Omega \subseteq H(a,b,x)$ *there exists a form* q *such that* $\dim q = 8$, $(\langle 1, a\rangle, \langle x, bx\rangle) < \mathrm{Sim}(q)$, *and:*

$$\mathrm{sgn}_{\mathfrak{p}}\, q = \begin{cases} 8 & \text{if } \mathfrak{p} \in \Delta \\ 4 & \text{if } \mathfrak{p} \in \Omega - \Delta \\ 0 & \text{if } \mathfrak{p} \notin \Omega. \end{cases}$$

Proof. (1) Since $ab > 0$ at every $\mathfrak{p} \in \Gamma$, (11.9) implies that there exists $w \in D_F(\langle\langle -ab\rangle\rangle)$ such that $H(-w) = \Gamma$. Then $w > 0$ at every $\mathfrak{p} \in H(a,b,-x)$ since $\mathfrak{p} \notin \Gamma$. The previous lemma then implies that $\varphi = \langle\langle a, -w\rangle\rangle$ is in $\mathcal{M}$ and the result follows.

(2) By part (1) there exist forms $\varphi_i = \langle\langle a, -w_i\rangle\rangle \in \mathcal{M}$ where $H(\varphi_1) = \Delta$ and $H(\varphi_2) = \Omega$. Then $q = \varphi_1 \perp \varphi_2$ has the required signatures, and $(\langle 1, a\rangle, \langle x, bx\rangle) < \mathrm{Sim}(q)$ by Theorem 11.18. □

Proof of the Modified Hasse Principle 11.17. The case $m = 1$ is settled in (11.19). Suppose $m \geq 2$ and suppose (σ, τ) is the given special (s,t)-pair where $s+t = 2m+2$. It suffices to check the criterion in (11.20) for given subsets $\Delta \subseteq \Omega \subseteq H(\sigma \perp \tau)$. First suppose $m \geq 3$ so that $s + t \geq 8$. Then (11.22) implies that $(\sigma, \tau) \approx (\langle 1, a\rangle \perp \alpha, \langle x, y\rangle \perp \alpha)$ for some $a, x, y \in F^\bullet$ and some form $\alpha \simeq \langle c_1, \dots, c_{m-1}\rangle$. Let q be the form given in (11.24), where $b = xy$. The Construction Lemma 2.7 and the equivalence above imply that

$$(\sigma, \tau) < \mathrm{Sim}(q \otimes \langle\langle c_1, \dots, c_{m-1}\rangle\rangle).$$

Since $m \geq 2$ we may check signatures to see that $q \otimes \langle\langle c_1, \dots, c_{m-1}\rangle\rangle \simeq 2^m\langle\langle \varepsilon_\Delta\rangle\rangle \perp 2^m\langle\langle \varepsilon_\Omega\rangle\rangle$. (Each c_j is positive at every $\mathfrak{p} \in H(\sigma \perp \tau)$.)

Finally suppose $m = 2$, so that (σ, τ) is a $(3,3)$-pair. The shifting approach fails here, but we can settle this case by enlarging it. Applying the known result for $m = 3$ we find that

$$(\sigma \perp \langle 1\rangle, \tau \perp \langle 1\rangle) < \mathrm{Sim}(2^3\langle\langle \varepsilon_\Delta\rangle\rangle \perp 2^3\langle\langle \varepsilon_\Omega\rangle\rangle).$$

The Eigenspace Lemma 2.10 then implies that $(\sigma, \tau) < \mathrm{Sim}(\varphi)$ for some form φ such that $2^3\langle\langle \varepsilon_\Delta\rangle\rangle \perp 2^3\langle\langle \varepsilon_\Omega\rangle\rangle \simeq \varphi \otimes \langle\langle 1\rangle\rangle$. Since φ is a multiple of a 2-fold Pfister form (since $\dim \sigma = 3$) it is determined by its signatures, and we conclude that $\varphi \simeq 2^2\langle\langle \varepsilon_\Delta\rangle\rangle \perp 2^2\langle\langle \varepsilon_\Omega\rangle\rangle$. □

We are now ready to discuss the proof of Theorem 11.18. The first step toward part (1) is to restrict the dimensions of the indecomposables.

11.25 Lemma. *Suppose* $\langle a\rangle \not\simeq \langle b\rangle$. *If* φ *is* $\mathcal{M}$*-indecomposable then* $\dim \varphi = 4$ *or* 8.

Proof. Since the indecomposables for $\mathcal{M}_0 = \mathcal{M}(\langle\langle a\rangle\rangle, \langle\langle b\rangle\rangle)$ all have dimension 4, $\dim\varphi$ must be a multiple of 4. Suppose $k = \dim\varphi > 8$. We may scale φ to assume that $\mathrm{sgn}_{\mathfrak{p}}(\varphi) \geq 0$ for every real prime $\mathfrak{p}$. We will show that $\langle\langle a, b, x\rangle\rangle$ is a subform of φ, contrary to the "indecomposable" hypothesis.

Let $\varphi' = \varphi \perp \langle -1\rangle\langle\langle a, b, x\rangle\rangle$ so that $\dim\varphi' = k+8$. If $\mathrm{sgn}_{\mathfrak{p}}(\varphi) > 0$ the divisibility conditions imply that a, b, and x are positive at $\mathfrak{p}$ so that $\mathrm{sgn}_{\mathfrak{p}}(\varphi') = \mathrm{sgn}_{\mathfrak{p}}(\varphi) - 8$. Hence $|\mathrm{sgn}_{\mathfrak{p}}(\varphi')| \leq k - 8 = \dim\varphi' - 16$ for every real $\mathfrak{p}$. The Hasse–Minkowski Theorem then implies that $8\mathbb{H} \subset \varphi'$. By cancellation we conclude that $\langle\langle a, b, x\rangle\rangle$ is a subform of φ. □

The proof of Theorem 11.18 will be done by considering transfers of hermitian forms. Recall that $\mathcal{M} = \mathcal{M}(\langle\langle a\rangle\rangle, \langle\langle b\rangle\rangle, x)$ where $\langle ab\rangle \not\simeq \langle 1\rangle$ and $\langle 1, a, -x, -bx\rangle$ is anisotropic. As in Chapter 5, let $E = F(\sqrt{ab})$ and $K = F(\sqrt{-a}, \sqrt{-b})$. Suppose q is an 8-dimensional form in $\mathcal{M}$. Since $\langle\langle a\rangle\rangle \mid q$ and $\langle\langle b\rangle\rangle \mid q$, (5.16) implies that q is the transfer of some 2-dimensional hermitian form $\langle\theta_1, \theta_2\rangle_K$ for some $\theta_1, \theta_2 \in E^\bullet$. Thus, $\theta_i = r_i + s_i\sqrt{ab}$ and

$$q \simeq \langle s_1\rangle\langle\langle a, -N\theta_1\rangle\rangle \perp \langle s_2\rangle\langle\langle a, -N\theta_2\rangle\rangle. \qquad (*)$$

Here we may assume $s_i \neq 0$. (For if $s_i = 0$, that term in $(*)$ is $2\mathbb{H}$. Apply Exercise 1.15 (1) to re-choose θ_i with $N\theta_i = 1$.) There are many ways to choose these θ_i.

11.26 Proposition. *To prove Theorem* 11.18 *it suffices to show that for every* $q \in \mathcal{M}$ *with* $\dim q = 8$, *there exist* θ_1, θ_2 *as in* $(*)$ *satisfying:*

(M1) $N\theta_i > 0$ *at every* $\mathfrak{p} \in H(a, b, -x)$.

(M2) $\theta = \theta_1\theta_2 < 0$ *at every* $\mathfrak{P} \in H_E(a, b, -x)$.

Proof. By (11.25), to prove the theorem it suffices to show that if $q \in \mathcal{M}$ with $\dim q = 8$, then $(\langle 1, a\rangle, \langle x, bx\rangle) < \mathrm{Sim}(q)$ and q is $\mathcal{M}$-decomposable. Suppose q is given and the θ_i satisfy (M1) and (M2). Then $\langle 1, a, -x\rangle \perp \langle\theta\rangle\langle 1, a\rangle$ is indefinite at every $\mathfrak{P} \in X_E$. For if it is definite then $a > 0$, $x < 0$, $\theta > 0$, and $b = \frac{1}{a} \cdot (\sqrt{ab})^2 > 0$, contrary to (M2). Then Hasse–Minkowski and (5.11) imply that $(\langle 1, a\rangle, \langle x, bx\rangle) < \mathrm{Sim}(q)$. Condition (M1) says $H(a, b, -x) \subseteq H(N\theta_i)$ and (11.23) implies $\langle s_i\rangle\langle\langle a, -N\theta_i\rangle\rangle \in \mathcal{M}$ so that q is decomposable in $\mathcal{M}$. □

The rest of the chapter is devoted to choosing θ_1 and θ_2.

11.27 Lemma. *Suppose* δ *is a form with* $\dim\delta = 2$ *and* $d\delta \in D_F(\langle\langle -ab\rangle\rangle)$. *Then* $\delta \simeq \langle s\rangle\langle\langle -N\theta\rangle\rangle$ *for some* $\theta = r + s\sqrt{ab} \in E$ *such that* $s \neq 0$.

Proof. Suppose $\delta \simeq \langle s\rangle\langle 1, -d\rangle$ for some $s, d \in F^\bullet$ where $d = u^2 - abv^2$. If $v \neq 0$ let $\theta = \frac{s}{v}(u + v\sqrt{ab}) = r + s\sqrt{ab}$ where $r = su/v$. Then $N\theta = (\frac{s}{v})^2 d$ so that $\delta \simeq \langle s\rangle\langle\langle -N\theta\rangle\rangle$ as required. If $v = 0$ then $d = u^2$ and $\delta \simeq \mathbb{H}$. In this case recall that $\langle 1, -ab\rangle$ represents 1 "transversally" since F is an infinite field. That is, there exist *non-zero* $r, s \in F$ with $r^2 - abs^2 = 1$. (See Exercise 1.15.) Let $\theta = r + s\sqrt{ab}$ so that $N\theta = 1$ and $\delta \simeq \mathbb{H} \simeq \langle s\rangle\langle\langle -N\theta\rangle\rangle$. □

In the proof below we need to prescribe signatures for the common values represented by a pair of quadratic forms. Compare Corollary 11.9 for the case of a single form. Recall that if α and β are quadratic forms over F then they represent some common value (that is, $D_F(\alpha) \cap D_F(\beta) \neq \emptyset$) if and only if the form $\alpha \perp -\beta$ is isotropic.

11.28 Lemma. *Suppose α, β are quadratic forms over F such that α and β represent some common value in $F^\bullet$. For each real prime $\mathfrak{p}$ suppose $\delta_\mathfrak{p} \in \{\pm 1\}$ is a value represented by both $\alpha_\mathfrak{p}$ and $\beta_\mathfrak{p}$. Then there exists $a \in D_F(\alpha) \cap D_F(\beta)$ such that* $\mathrm{sgn}_\mathfrak{p}(a) = \delta_\mathfrak{p}$ *for every* $\mathfrak{p}$.

Proof. We employ an Approximation Lemma stated in Exercise 4 below. Let $n = \dim\alpha$ and $m = \dim\beta$ and for each $\mathfrak{p}$ choose vectors $x_\mathfrak{p} \in F_\mathfrak{p}^n$ and $y_\mathfrak{p} \in F_\mathfrak{p}^m$ such that $\alpha(x_\mathfrak{p}) = \beta(y_\mathfrak{p}) = \delta_\mathfrak{p}$. Let $q = \alpha \perp -\beta$ and let $v_\mathfrak{p} = (x_\mathfrak{p}, y_\mathfrak{p}) \in F_\mathfrak{p}^{n+m}$ so that $q(v_\mathfrak{p}) = 0$. By the lemma in Exercise 4, there exists $v \in F^{n+m}$ such that $q(v) = 0$ and v is close to $v_\mathfrak{p}$ for every real prime $\mathfrak{p}$. Writing $v = (x, y)$ for some $x \in F^n$ and $y \in F^m$ we define $a = \alpha(x) = \beta(y)$. Then $a \in D_F(\alpha) \cap D_F(\beta)$ and since x is close to $x_\mathfrak{p}$ we know that $a = \alpha(x)$ is close to $\alpha(x_\mathfrak{p}) = \delta_\mathfrak{p}$. Therefore $\mathrm{sgn}_\mathfrak{p}(a) = \delta_\mathfrak{p}$ for every real prime $\mathfrak{p}$. □

The next result settles condition (M1).

11.29 Proposition. *We may assume that $N\theta_i > 0$ at every $\mathfrak{p} \in H(a) - H(b, x)$. (In particular this holds for $\mathfrak{p} \in H(a, b, -x)$.)*

Proof. Given $q \simeq \langle\langle a\rangle\rangle \otimes (\delta_1 \perp \delta_2)$ where $\delta_i \simeq \langle s_i\rangle\langle\langle -c_i\rangle\rangle$ and $c_i = N\theta_i$. Then $\langle s_1\rangle(\delta_1 \perp \delta_2) \simeq \langle 1\rangle \perp -\gamma$ where $\gamma \simeq \langle c_1, -s_1s_2, s_1s_2c_2\rangle$. Then γ represents $c_1 \in D_F(\langle\langle -ab\rangle\rangle)$. Applying (11.28) we see that γ represents some $c \in D_F(\langle\langle -ab\rangle\rangle)$ where $c > 0$ at every $\mathfrak{p}$ where γ is not negative definite. In particular if $\mathfrak{p} \in H(a) - H(b, x)$ then γ is not negative definite and hence $c > 0$ (for at such $\mathfrak{p}$ we have $0 = \mathrm{sgn}_\mathfrak{p}(q) = 2 \cdot \mathrm{sgn}_\mathfrak{p}(\delta_1 \perp \delta_2) = \pm 2(1 - \mathrm{sgn}_\mathfrak{p}(\gamma))$). We express $\delta_1 \perp \delta_2 \simeq \langle s_1\rangle\langle 1, -c\rangle \perp \delta'$ for some binary form δ'. Computing determinants we find that $d\delta' \in D_F(\langle\langle -ab\rangle\rangle)$. Then (11.27) implies that

$$\delta_1 \perp \delta_2 \simeq \langle s_1'\rangle\langle\langle -N\theta_1'\rangle\rangle \perp \langle s_2'\rangle\langle\langle -N\theta_2'\rangle\rangle$$

for some $\theta_i' = r_i' + s_i'\sqrt{ab} \in E$ such that $s_i' \neq 0$ and $N\theta_1' = u^2c$. Hence $N\theta_1' > 0$ at every $\mathfrak{p} \in H(a) - H(b, x)$. Since $\mathrm{sgn}_\mathfrak{p}(q) = 0$ for every such $\mathfrak{p}$ we know that $N\theta_2' > 0$ at $\mathfrak{p}$ as well. Consequently $q \simeq \langle s_1'\rangle\langle\langle a, -N\theta_1'\rangle\rangle \perp \langle s_2'\rangle\langle\langle a, -N\theta_2'\rangle\rangle$ where each term has signature 0 at every $p \notin H(a, b, x)$. □

Now to realize condition (M2) we alter θ_1 by a suitable element ξ of norm 1. The next lemma includes the exact conditions we need for this ξ. Recall that $\ell : E \to F$ is the "trace" function satisfying: $\ell(1) = 0$ and $\ell(\sqrt{ab}) = 1$.

11.30 Lemma. *There exists $\xi \in E^\bullet$ such that*

(1) $N\xi = 1$,

(2) $\ell(\xi\theta_1)$ *and* $\ell(\theta_1)$ *have the same sign at every* $\mathfrak{p} \in H_F(a, b, x)$, *and*

(3) $\xi\theta_1\theta_2 < 0$ *at every* $\mathfrak{P} \in H_E(a, b, -x)$.

Proof. Let $\xi = \beta/\bar{\beta}$ for some $\beta \in E^\bullet$. Then $N\xi = 1$ and we translate the conditions (2) and (3) into restrictions on β. We will determine $u \in F$ such that $\beta = u + \sqrt{ab}$ will work.

Since $\xi = \bar{\beta}^{-2} \cdot N\beta$, condition (3) states that $N\beta$ and $\theta = \theta_1\theta_2$ should have opposite signs at every $\mathfrak{P} \in H_E(a, b, -x)$. This just says that $N\beta$ has certain prescribed signs at the orderings $\mathfrak{p} \in H_F(a, b, -x)$. To verify that statement we must know that there is no inconsistency. Each $\mathfrak{P} \in H_E(a, b, -x)$ induces an ordering $\mathfrak{p} \in H_F(a, b, -x)$. Since $N\beta \in F^\bullet$ its signs are determined by these orderings $\mathfrak{p}$. The difficulty is that $N\beta$ could have inconsistent signs: θ could conceivably have opposite signs at two different orderings $\mathfrak{P}, \mathfrak{P}' \in H_E(a, b, -x)$ which extend the same $\mathfrak{p}$. If this occurs those orderings must be conjugates: $\mathfrak{P}' = \bar{\mathfrak{P}}$. Then the difficulty is that θ and $\bar{\theta}$ might have opposite signs at $\mathfrak{P}$, or equivalently that $\theta \cdot \bar{\theta} = N\theta < 0$ at some $\mathfrak{P} \in H_E(a, b, -x)$. However $N\theta = N\theta_1 \cdot N\theta_2$ is positive at every $\mathfrak{p} \in H_F(a, b, -x)$ by (11.29) so this difficulty cannot arise. Consequently, condition (3) states that

$$N\beta = u^2 - ab \text{ has prescribed signs at } \mathfrak{p} \in H_F(a, b, -x).$$

To analyze condition (2) let $\theta_1 = r + s\sqrt{ab}$. Since $\xi\theta_1 = \frac{1}{N\beta} \cdot \beta^2\theta_1$, condition (2) becomes:

$$\frac{(u^2 + ab)s^2 + 2urs}{u^2 - ab} > 0 \text{ at every } \mathfrak{p} \in H_F(a, b, x).$$

If $\mathfrak{p} \in H(a, b, x)$ then a large enough value of u at $\mathfrak{p}$ will yield a positive value for the rational function displayed above, since the numerator and denominator have positive leading coefficients. Since the two sets of orderings $H(a, b, x)$ and $H(a, b, -x)$ are disjoint, the Weak Approximation Theorem (11.7) implies that an element $u \in F$ can be chosen to fulfill all these conditions. □

11.31 Proposition. *θ_1, θ_2 can be chosen to satisfy* (M1) *and* (M2).

Proof. So far we know know that

$$q \simeq \langle s_1 \rangle \langle\langle a, -N\theta_1 \rangle\rangle \perp \langle s_2 \rangle \langle\langle a, -N\theta_2 \rangle\rangle$$

where $N\theta_i > 0$ at every $\mathfrak{p} \in H(a) - H(b, x)$. Let ξ be the element determined in Lemma 11.30 and define $\theta_1'' = \xi\theta_1 = r_1'' + s_1''\sqrt{ab}$. Note that $N\theta_1'' = N\theta_1$. Allowing the interchange of θ_1 and θ_2, we may assume $s_1'' \neq 0$. (For otherwise both $s_i'' = 0$. But then $\xi\theta_i \in F^\bullet$, so that $N\theta_i \in F^{\bullet 2}$ and $q \simeq 4\mathbb{H}$. In that case Theorem 11.18 is trivial.)

Claim. $s_1 s_1'' \in D_F(\langle\langle a, -N\theta_1\rangle\rangle)$.

It suffices to check this at the real primes. If $\langle\langle a, -N\theta_1\rangle\rangle$ is indefinite at $\mathfrak{p}$ there is nothing to prove, so assume that form is positive definite. Then $a > 0$ and $N\theta_1 < 0$ at $\mathfrak{p}$. Then $\mathfrak{p} \in H(a)$ and therefore $\mathfrak{p} \in H(a, b, x)$ (for otherwise $\mathfrak{p} \in H(a) - H(b, x)$ and $N\theta_1 > 0$ by the choice of θ_i.) Since $s_1 = \ell(\theta_1)$ and $s_1'' = \ell(\xi\theta_1)$, property (2) of Lemma 11.30 implies that $s_1 s_1''$ is positive at $\mathfrak{p}$. This proves the claim.

Therefore we may replace θ_1 by θ_1'' in the representation of q. The condition (3) in the Lemma 11.30 becomes the condition (M2). □

This completes the proof of Theorem 11.18.

Appendix to Chapter 11. Hasse principle for divisibility of forms

To prove the general case of Proposition 11.4 we use a number of results about the quadratic form theory over the complete fields $F_\mathfrak{p}$ in the case $\mathfrak{p}$ is a finite prime. These results are described in more detail in several texts, including Lam (1973) Ch. 6 §1 and Scharlau (1985) Ch. 6 §2.

Suppose $\mathfrak{p}$ lies over the rational prime p. There is a valuation $v_\mathfrak{p} : F_\mathfrak{p} \to \mathbb{Z} \cup \{\infty\}$ extending the usual p-adic valuation on $\mathbb{Q}$. (We use the additive version: $v_\mathfrak{p}(xy) = v_\mathfrak{p}(x) + v_\mathfrak{p}(y)$.)

Here are some of the standard notations:

$\mathcal{O}_\mathfrak{p} = \{a \in F_\mathfrak{p} : v_\mathfrak{p}(a) \geq 0\}$, the valuation ring.

$\mathfrak{m}_\mathfrak{p} = \{a \in F_\mathfrak{p} : v_\mathfrak{p}(a) > 0\} = \mathcal{O}_\mathfrak{p}\pi$, the maximal idea. Here $\pi \in \mathcal{O}_\mathfrak{p}$ and $v_\mathfrak{p}(\pi) = 1$.

$U_\mathfrak{p} = \{a \in F_\mathfrak{p} : v_\mathfrak{p}(a) = 0\}$, the group of units of $\mathcal{O}_\mathfrak{p}$.

$k(\mathfrak{p}) = \mathcal{O}_\mathfrak{p}/\mathfrak{m}_\mathfrak{p}$, the residue field. Then $k(\mathfrak{p})$ is a finite field of characteristic p.

If $u \in U_\mathfrak{p}$ we let $\bar{u} \in k(\mathfrak{p})$ denote its image in the residue field. We assume here that $k(\mathfrak{p})$ has characteristic $\neq 2$ (the "non-dyadic" case).

Any $a \in F_\mathfrak{p}^\bullet$ can be expressed $a = u\pi^n$ where $n = v_\mathfrak{p}(a)$ and $u \in U_\mathfrak{p}$. Therefore any form q over F can be expressed as

$$q = \langle a_1, \ldots, a_m, \pi a_{m+1}, \ldots, \pi a_n\rangle \quad \text{for some } a_j \in U_\mathfrak{p}$$

by diagonalizing and multiplying the entries by suitable even powers of π. Define the first and second residue class forms of q by:

$$\partial_1(q) = \langle \bar{a}_1, \ldots, \bar{a}_m\rangle \quad \text{and} \quad \partial_2(q) = \langle \bar{a}_{m+1}, \ldots, \bar{a}_n\rangle.$$

A.1 Proposition. *Let $\mathfrak{p}$ be a non-dyadic finite prime with uniformizer π as above.*

(1) *If q is a quadratic form over F, then changing the diagonalization of q can change $\partial_1(q)$ and $\partial_2(q)$ only up to Witt equivalence. (Hence $\partial_j : W(K) \to W(k(\mathfrak{p}))$ are well-defined homomorphisms of the Witt groups.)*

(2) *q is anisotropic over F if and only if both $\partial_1(q)$ and $\partial_2(q)$ are anisotropic over $k(\mathfrak{p})$.*

This result is often called "Springer's Theorem". Consequently the isometry class of a form q over the $\mathfrak{p}$-adic field $F_\mathfrak{p}$ is determined by its dimension and by the Witt classes of the forms $\partial_j(q)$ over the finite field $k(\mathfrak{p})$. Forms over finite fields are easy to handle: any quadratic form over $k(\mathfrak{p})$ with dimension ≥ 2 is universal. The next corollary is an immediate consequence of Springer's Theorem (A.1) and this fact about finite fields. A quadratic form α is a "unit form" if it has a diagonalization with unit entries, i.e. if $\partial_2(\alpha) \sim 0$.

A.2 Corollary. *Suppose α is a unit form over $F_\mathfrak{p}$ where $\mathfrak{p}$ is a non-dyadic finite prime.*

(1) *If $\dim \alpha > 1$ then α represents 1.*

(2) *If $\dim \alpha \geq 2$ is even than $\langle u \rangle \alpha \simeq \alpha$ for every $u \in U_\mathfrak{p}$.*

Now we can begin the proof of the Hasse Principle for "divisibility" of forms.

Proof of Proposition 11.4. We use induction on $\dim q$. First let us consider the special case $\dim \varphi = \dim q$ is odd. Then $q_\mathfrak{p} \simeq \langle b_{(\mathfrak{p})} \rangle \varphi_\mathfrak{p}$ for some $b_{(\mathfrak{p})} \in F_\mathfrak{p}^\bullet$. Taking discriminants we find that $\langle b_{(\mathfrak{p})} \rangle \simeq (dq \cdot d\varphi)_\mathfrak{p}$. Therefore $q_\mathfrak{p} \simeq \langle a \rangle \varphi_\mathfrak{p}$ for some $a \in F$ (namely, $\langle a \rangle = dq \cdot d\varphi$). Then Hasse–Minkowski implies that $q \simeq \langle a \rangle \varphi$ over F as hoped. From now on we avoid this special case.

Let

$$S = \{\mathfrak{p} : \text{either } \mathfrak{p} \text{ is infinite, } \mathfrak{p} \text{ is dyadic, or one of } q_\mathfrak{p} \text{ and } \varphi_\mathfrak{p} \text{ is not a unit form}\}.$$

Then S is a finite set of primes of F. We are given forms $\delta_{(\mathfrak{p})}$ over $F_\mathfrak{p}$ such that

$$q_\mathfrak{p} \simeq \varphi_\mathfrak{p} \otimes \delta_{(p)} \quad \text{over } F_\mathfrak{p}.$$

If $\mathfrak{p} \in S$ choose $c_{(\mathfrak{p})} \in D_{F_\mathfrak{p}}(\delta_{(\mathfrak{p})})$. By the Approximation Theorem 11.8 there exists $c \in F^\bullet$ such that $\langle c \rangle_\mathfrak{p} \simeq \langle c_{(\mathfrak{p})} \rangle$ for every $\mathfrak{p} \in S$. We replace q by $\langle c \rangle q$ and $\delta_{(\mathfrak{p})}$ by $\langle c \rangle \delta_{(\mathfrak{p})}$. Therefore we may assume that $\delta_{(\mathfrak{p})}$ represents 1 for every $\mathfrak{p} \in S$.

A.3 Lemma. *If $\mathfrak{p} \notin S$ then $q_\mathfrak{p} \simeq \varphi_\mathfrak{p} \otimes \gamma_{(\mathfrak{p})}$ for some form $\gamma_{(\mathfrak{p})}$ which represents 1 over $F_\mathfrak{p}$.*

Assume this lemma for the moment. Then replacing $\delta_{(\mathfrak{p})}$ by $\gamma_{(\mathfrak{p})}$ for all $\mathfrak{p} \notin S$ we have arranged that $\delta_{(\mathfrak{p})}$ represents 1 for every prime $\mathfrak{p}$. Letting $\delta_{(\mathfrak{p})} \simeq \langle 1 \rangle \perp \omega_{(\mathfrak{p})}$ over $F_\mathfrak{p}$ we find that $q_\mathfrak{p} \simeq \varphi_\mathfrak{p} \perp \varphi_\mathfrak{p} \otimes \omega_{(\mathfrak{p})}$ for every prime $\mathfrak{p}$. Then Hasse–Minkowski implies that $\varphi \subset q$ over F so that $q \simeq \varphi \perp q'$ for some form q' over F. Then $q'_{\mathfrak{p}} \simeq \varphi_{\mathfrak{p}} \otimes \omega_{(\mathfrak{p})}$ over $F_{\mathfrak{p}}$ and by the induction hypothesis we conclude that $\varphi \mid q'$ and therefore $\varphi \mid q$. □

Proof of the lemma. Since $\mathfrak{p} \notin S$ we know that $\mathfrak{p}$ is a non-dyadic finite prime and that $q_{\mathfrak{p}}$ and $\varphi_{\mathfrak{p}}$ are unit forms. Express the given form $\delta_{(\mathfrak{p})}$ as $\delta_{(\mathfrak{p})} \simeq \alpha \perp \langle\pi\rangle\beta$ for some unit forms α, β over $F_{\mathfrak{p}}$. Then $q_{\mathfrak{p}} \simeq \varphi_{\mathfrak{p}} \otimes (\alpha \perp \langle\pi\rangle\beta)$. By Springer's Theorem (A.1) we know that

$$q_{\mathfrak{p}} = \varphi_{\mathfrak{p}} \otimes \alpha \text{ and } 0 = \varphi_{\mathfrak{p}} \otimes \langle\pi\rangle\beta \text{ in the Witt ring } W(F_{\mathfrak{p}}).$$

Therefore $\varphi_{\mathfrak{p}} \otimes \langle\pi\rangle\beta \simeq \varphi_{\mathfrak{p}} \otimes \beta$ since it is a hyperbolic form, and $q_{\mathfrak{p}} \simeq \varphi_{\mathfrak{p}} \otimes (\alpha \perp \beta)$. In other words, we may assume that $\beta = 0$ and $q_{\mathfrak{p}} \simeq \varphi_{\mathfrak{p}} \otimes \alpha$ for a unit form α over $F_{\mathfrak{p}}$.

If $\dim \alpha > 1$ then α represents 1 by (A.2). Otherwise $\dim \alpha = 1$ and $\dim \varphi = \dim q$. Since we settled the special case mentioned at the start, we may assume that $\dim \varphi$ is even. But then (A.2) (2) implies that we may replace α by $\langle 1\rangle$. □

Exercises for Chapter 11

1. Use Hasse–Minkowski to prove the following assertions about a global field F.

(1) "Meyer's Theorem": If q is a quadratic form over F and $\dim q > 4$ then q is isotropic if and only if q is totally indefinite. ("totally indefinite" means that $q_{\mathfrak{p}}$ is indefinite at every real prime $\mathfrak{p}$.)

(2) F is linked.

(3) I^3F is torsion-free in the Witt ring $W(F)$. For every $n \geq 2$, $I^{n+1}F = 2^n \cdot IF$.

2. **Function fields.** Suppose F is an algebraic function field (i.e. a finite extension of $\mathbb{F}_p(t)$ for an indeterminate t). Equivalently, F is a finitely generated extension of $\mathbb{F}_p$ of transcendence degree 1. (As usual, char $F \neq 2$.)

(1) Any valuation v on F extends some $g(t)$-adic valuation on $\mathbb{F}_p(t)$, where $g(t)$ is a monic irreducible polynomial, or v extends the $(\frac{1}{t})$-adic valuation.

(2) Every prime v of F is finite and the completion is $F_v \cong k((x))$, a Laurent series field over some finite field k of characteristic p.

We assume the Hasse–Minkowski Theorem over F.

(3) Every quadratic form of dimension > 4 over F is isotropic. Every 3-fold Pfister form is hyperbolic.

(4) Suppose (σ, τ) is an (s, t)-pair over F. If q is a (σ, τ)- module and q is not hyperbolic then either $\dim q \leq 4$ or (σ, τ) is a special $(2, 2)$-pair and $\dim q = 8$. Knowing Proposition 11.4 for Pfister forms, the only remaining case of the Hasse Principle for (σ, τ) is when (σ, τ) is a special $(2, 2)$-pair.

(5) Theorem 11.18 can be proved for such F, and the Modified Hasse Principle follows.

3. Suppose (σ, τ) is a minimal pair over F where the dimension of an unsplittable is 2^m. Then the unsplittable module is unique up to similarity.

(1) If ψ and ψ' are (σ, τ)-unsplittables which represent 1 then $\psi \simeq \psi'$.

(2) If F is a number field, and $\Omega \subseteq H(\sigma \perp \tau)$ then there exists $(\sigma, \tau) < \mathrm{Sim}(q)$ where $\dim q = 2^{m+1}$ and $H(q) = \Omega$. (Compare (11.11).)

(Hint. (1) If PC(m) holds over F then ψ, ψ' are Pfister forms, hence are round. The statement is unknown over an arbitrary field F. Compare Exercise 9.13 (1).)

4. **Approximations.** The proof of (11.28) uses the following approximation result for isotropic vectors. We follow the method of Cassels (1978), Chapter 6, Lemma 9.1, beginning with a preliminary "transversality" lemma.

(1) **Lemma.** *Let q be an isotropic quadratic form and ℓ a non-zero linear form over $F_{\mathfrak{p}}$. Suppose $v \in F_{\mathfrak{p}}^n$ is a non-zero vector with $q(v) = 0$. If $\mathcal{U}$ is any neighborhood of v in the $\mathfrak{p}$-adic topology on $F_{\mathfrak{p}}^n$ then there exists $w \in \mathcal{U}$ such that $q(w) = 0$ and $\ell(w) \neq 0$.*

(2) **Approximation Lemma.** *Let q be an isotropic quadratic form of dimension $n \geq 3$ over F. Let S be a finite set of primes of F. For each $\mathfrak{p} \in S$ suppose a vector $v_{\mathfrak{p}} \in F_{\mathfrak{p}}^n$ is given such that $q(v_{\mathfrak{p}}) = 0$. For any real $\varepsilon > 0$, there exists $v \in F^n$ such that $q(v) = 0$ and $||v - v_{\mathfrak{p}}||_{\mathfrak{p}} < \varepsilon$ for every $\mathfrak{p} \in S$.*

(3) If α, β, γ are quadratic forms over F which represent common values pairwise (i.e. the forms $\alpha \perp -\beta$, $\beta \perp -\gamma$ and $\gamma \perp -\alpha$ are isotropic), does it follow that the three of them must represent a common value?

(Hint. (1) Compare Exercise 1.15. A proof appears in Cassels, p. 62.

(2) *Proof outline.* (Following Cassels, pp. 89–91.) Choose $0 \neq w \in F^n$ with $q(w) = 0$. We may alter the vectors $v_{\mathfrak{p}}$ if necessary to assume that $b_q(v_{\mathfrak{p}}, w) \neq 0$ for every $\mathfrak{p} \in S$. (For if this fails for some $\mathfrak{p}$ apply (1) using $\ell(x) = b_q(x, w)$.) By (11.9) there exists $u \in F^n$ arbitrarily close to $v_{\mathfrak{p}}$ for each $\mathfrak{p} \in S$. Let $\lambda = -q(u)/2b_q(u, w)$, define $v = u + \lambda w$ and note that $q(v) = 0$. If u is close enough to $v_{\mathfrak{p}}$ in $F_{\mathfrak{p}}^n$ then λ is close to 0 in $F_{\mathfrak{p}}$ and v is close to $v_{\mathfrak{p}}$ in $F_{\mathfrak{p}}^n$. Fill in the details using appropriate estimates.)

5. **Odd Factor Theorem.** *If F is a local field or a global field and $(\sigma, \tau) < \mathrm{Sim}(\alpha \otimes \delta)$ where $\dim \delta$ is odd, then $(\sigma, \tau) < \mathrm{Sim}(\alpha)$.*

Proof outline. (1) If the dimension of an unsplittable is ≤ 4 or if every (σ, τ)-unsplittable is hyperbolic the Odd Factor result holds. (See Exercise 5.22.)

(2) If (σ, τ) is minimal and F is linked the result holds over F. (Compare the proof of (11.5).)

(3) Parts (1) and (2) settle the claim for local fields. In fact, if F is linked, $I^3 F = 0$ and (σ, τ) is not special (e.g. F is $\mathfrak{p}$-adic) the result follows. If F is euclidean (e.g. $F = \mathbb{R}$) the claim also holds.

(5) Assume F is global. Then $(\sigma_{\mathfrak{p}}, \tau_{\mathfrak{p}}) < \mathrm{Sim}(\alpha_{\mathfrak{p}})$ for every $\mathfrak{p}$. Apply the Hasse Principle.

6. **Proposition.** *Let* $\mathcal{M} = \mathcal{M}(x, y)$ *over a global field* F *where* $\langle x \rangle \not\simeq \langle y \rangle$. *Then all* $\mathcal{M}$*-indecomposables have dimension* 2 *or* 4.

Compare the Open Question in Exercise 5.9. Here is an outline of the proof.

(1) **Lemma.** *If* $\langle x \rangle \not\simeq \langle 1 \rangle$ *then* $q \in \mathcal{M}(x)$ *if and only if* $\dim q$ *is even,* $dq \in D_F(\langle\langle -x \rangle\rangle)$ *and* $\operatorname{sgn}_{\mathfrak{p}}(q) = 0$ *for every* $\mathfrak{p} \notin H(x)$.

(2) If α is a form over F with $\dim \alpha$ odd and ≥ 5 then α represents $\det \alpha$. Suppose $q \in \mathcal{M}$ with $\dim q > 4$. We may assume that q represents 1. Then $q \simeq \langle 1, d \rangle \perp \delta$ where $\det q = \langle d \rangle$ and $\det \delta = \langle 1 \rangle$. If $\dim q \equiv 2 \pmod 4$ then $dq = \langle -d \rangle$ and the lemma implies that $\langle 1, d \rangle \in \mathcal{M}$.

(3) Suppose $\dim q \equiv 0 \pmod 4$. By (11.9) there exists a decomposition $q \simeq \langle 1, a, b \rangle \perp \alpha$ where $a, b < 0$ at every $\mathfrak{p} \notin H(x, y)$. Then α represents $c = \det \alpha$ and $q \simeq \langle 1, a, b, c \rangle \perp \alpha_1$ and $\alpha_1 \in \mathcal{M}$.

Open Questions. What are the possible dimensions of $\mathcal{M}(\langle\langle a \rangle\rangle, \langle\langle b \rangle\rangle, \langle x \rangle)$-indecomposables over a general field F? Can there be indecomposables of dimension other than 2, 4 or 8?

(Hint. (1) Let $\omega = \langle\langle -x \rangle\rangle \otimes q$ and compute $d\omega$, $c(\omega)$ and $\operatorname{sgn}_{\mathfrak{p}}(\omega)$. The given conditions hold iff the invariants of ω are all trivial. Apply Hasse–Minkowski.)

7. Suppose F is a global field.

(1) If $\dim q \geq 2$ then $D_F(q)/F^{\bullet 2}$ is infinite.

(2) If $\dim q$ is even and ≥ 2, then $G_F(q) = \{a \in D_F(\langle\langle -dq \rangle\rangle) : a > 0$ at every ordering $\mathfrak{p}$ where $\operatorname{sgn}_{\mathfrak{p}}(q) \neq 0\}$. Consequently, $G_F(q)/F^{\bullet 2}$ is infinite.

(3) The group $G_F(q)$ acts on $D_F(q)$. Either there is one orbit (and q is round), or there are infinitely many orbits.

8. Let $F = k((t))$ be the field of formal Laurent series over k. Springer's Theorem (as in the appendix) holds for F. Characterize Pfister forms and divisibility of forms over F in terms of the residue class forms over k.

9. **Space of orderings.** (1) Let X_F be the set of all orderings of a formally real field F. Recall that for $a \in F^\bullet$, $H(a) = \{\mathfrak{p} \in X_F : a > 0 \text{ at } \mathfrak{p}\}$. Define the "Harrison topology" on X_F by taking the collection of sets $H(a)$ as a subbasis for the topology. Then X_F is compact, totally disconnected and every set $H(\gamma)$ (as in (11.10)) is clopen (i.e. closed and open).

(2) If X_F is finite then every subset is clopen.

Define F to be a "SAP field" if every clopen set in X_F equals $H(a)$ for some $a \in F^\bullet$. Equivalently, if $a_i \in F^\bullet$ are given then there exists $a \in F^\bullet$ such that $H(a_1, \dots, a_n) = H(a)$.

(3) Every algebraic number field is SAP. The iterated Laurent series field $F = \mathbb{R}((x))((y))$ is not SAP.

10. Suppose $v : F \to \mathbb{Z} \cup \{\infty\}$ is a valuation on a field F. Following the notation in the appendix we have $\mathcal{O}$, $\mathfrak{m}$, U, and $k = \mathcal{O}/\mathfrak{m}$. Assume that v is non-dyadic (i.e. $\operatorname{char} k \neq 2$). Choose a uniformizer π for v. Then the residue forms $\partial_1, \partial_2 : W(F) \to W(k)$ are group homomorphisms. Let $G = \{1, g\}$ be the group of 2 elements and define $\partial : W(F) \to W(k)[G]$ by $\partial(q) = \partial_1(q) + \partial_2(q)g$. (Here, $R[G]$ denotes the group ring of G with coefficients in R.)

(1) Then ∂ is a ring homomorphism.

(2) When v is complete (or more generally, "2-henselian") then Springer's Theorem (A.1) holds and ∂ is an isomorphism.

(3) If $v : F \to \Gamma \cup \{\infty\}$ is a Krull valuation into the ordered abelian group Γ there is an analogous ring homomorphism $\partial : W(F) \to W(k)[\Gamma/2\Gamma]$.

Notes on Chapter 11

Discussions of the Hasse–Minkowski Theorem 11.1 appear in Scharlau (1985), Chapter 6, §6, in Lam (1973), Chapter 6, §3 and in Milnor and Husemoller (1973), Appendix 3. These texts assume some knowledge of Class Field Theory in the proof. A self-contained proof in the general case is given in O'Meara (1963).

The Hasse Principle for the compositions of quadratic forms was conjectured in Shapiro (1974). Some special cases were proved (independently) by Ono and Yamaguchi (1979). (A related problem was considered by Ono (1974).) The Hasse Principle and Modified Hasse Principle for (σ, τ) were considered in Shapiro (1978a) but the result for special (2, 2)-pairs was left open. The "trace form" method for (2, 2)-families developed in Chapter 5 and applied here is the new ingredient used to settle these cases.

Proposition 11.4 on the Hasse Principle for division of forms is proved in the appendix. The case $\dim q = \dim \varphi$ (the Hasse Principle for similarity of quadratic forms) was first proved by Ono (1955) and independently by Wadsworth (1972). The general case was proved by Ono and Yamaguchi (1979). We present here a version of an unpublished proof found by Wadsworth in 1977.

Lemma 11.25 was communicated to me by Wadsworth in 1978.

Exercise 4. Lemma 11.28 is valid for n quadratic forms (as proved by Leep). For instance when $n = 3$ it says: Suppose α, β, γ are quadratic forms over F which represent some common value in $F^\bullet$. For every real prime $\mathfrak{p}$ let $\delta_\mathfrak{p} \in \{\pm 1\}$ be a value represented by all three forms over $F_\mathfrak{p}$. Then is there some $a \in D_F(\alpha) \cap D_F(\beta) \cap D_F(\gamma)$ such that $\operatorname{sgn}_\mathfrak{p}(a) = \delta_\mathfrak{p}$. The first step is to reduce to the case of binary forms: $\alpha = \langle 1, a\rangle$, $\beta = \langle 1, b\rangle$, $\gamma = \langle 1, c\rangle$. The process of finding a represented value with prescribed signs involves a careful local argument and an application of Artin Reciprocity.

Exercise 7 was proved by Dieudonné (1954). His method avoids the use of Witt invariants. Also see the appendix of Elman and Lam (1974).

Exercise 9. For further information about SAP and related properties see T. Y. Lam (1983).

Exercise 10. For more about ∂ and Krull valuations see T. Y. Lam (1983), Theorem 4.2.

Part II

Compositions of Size $[r, s, n]$

Introduction

The second part of this book is an exposition of results concerning general composition formulas. The chapters are longer and require more mathematical background than in the first part. We are primarily concerned with algebraic methods and their application to the composition problem. However, at several points in this second part we apply Theorems from other areas of mathematics (algebraic topology, K-theory, differential geometry, etc.). In those cases we attempt to give the reader a description of the situation with suitable references, without getting very deeply into the technicalities.

A composition formula of size $[r, s, n]$ over a field F is a formula of the type:

$$(x_1^2 + x_2^2 + \cdots + x_r^2) \cdot (y_1^2 + y_2^2 + \cdots + y_s^2) = z_1^2 + z_2^2 + \cdots + z_n^2,$$

where $X = (x_1, x_2, \ldots, x_r)$ and $Y = (y_1, y_2, \ldots, y_s)$ are systems of indeterminates and each $z_k = z_k(X, Y)$ is a bilinear form in X and Y with coefficients in F. In this situation we sometimes say that there is an $[r, s, n]$-formula over F or that $[r, s, n]$ is *admissible* over F.

For which r, s, n does there exist an $[r, s, n]$-formula? This question was first asked over a century ago in the seminal paper of Hurwitz (1898). Hurwitz proved that an $[r, s, n]$-formula exists over F if and only if there exist $n \times s$ matrices $A_1, A_2, \ldots, A_r$ over F satisfying

$$\begin{aligned} A_i^\top \cdot A_i &= I_s \quad \text{for } 1 \le i \le r, \\ A_i^\top \cdot A_j + A_j^\top \cdot A_i &= 0 \quad \text{for } 1 \le i, j \le r \text{ and } i \ne j. \end{aligned}$$

In particular $s \le n$. The case $s = n$ was settled in Part I using the Hurwitz–Radon function $\rho(n)$:

There is a composition of size $[r, n, n]$ if and only if $r \le \rho(n)$.

However when $s < n$ those matrices A_i are not square and the Hurwitz–Radon methods do not apply.

Some constructions of composition formulas are easy: we can set some variables equal to zero in a Hurwitz–Radon formula. For example the [8, 8, 8] formula restricts to a $[3, 5, 8]$ formula. But we can do better. Here is one of size $[3, 5, 7]$:

$$\begin{aligned} &(x_1^2 + x_2^2 + x_3^2) \cdot (y_1^2 + y_2^2 + \cdots + y_5^2) \\ &\quad = (x_1^2 + x_2^2 + x_3^2) \cdot (y_1^2 + y_2^2 + y_3^2 + y_4^2) + (x_1y_5)^2 + (y_2y_5)^2 + (y_3y_5)^2. \end{aligned}$$

Since the first term on the right is expressible as a sum of 4 squares (by the 4-square identity), the entire quantity is a sum of 7 squares, as claimed. This formula can be expressed in terms of 7×5 matrices A_1, A_2, A_3 as above, and their entries are all in $\{0, 1, -1\}$. From this example we are quickly led to ask: Is $[3, 5, 6]$ admissible? What about $[3, 6, 7]$ and $[4, 5, 7]$?

Over the field $\mathbb{R}$ of real numbers, these sizes can be eliminated by applying algebraic topology to the problem. This was done in 1940 by Stiefel and Hopf, and those topological connections greatly heightened interest in the study of composition formulas. To apply topological ideas to this composition problem we view it in terms of bilinear mappings. Let $|x|$ denote the euclidean norm of a vector $x \in \mathbb{R}^k$.

Definition. Suppose $f : \mathbb{R}^r \times \mathbb{R}^s \to \mathbb{R}^n$ is a bilinear mapping.

(1) f is *normed*[1] if $|f(x, y)| = |x| \cdot |y|$ whenever $x \in \mathbb{R}^r$ and $y \in \mathbb{R}^s$.

(2) f is *nonsingular* if $f(x, y) = 0$ implies that either $x = 0$ or $y = 0$.

There is a composition of size $[r, s, n]$ over $\mathbb{R}$ if and only if there is a normed bilinear map of size $[r, s, n]$. Certainly every normed map over $\mathbb{R}$ is nonsingular. A nonsingular pairing of size $[n, n, n]$ is exactly an n-dimensional real division algebra (see Chapter 8). Any nonsingular bilinear map f as above induces a map on spheres $S^{r-1} \times S^{s-1} \to S^{n-1}$ and also a map on real projective spaces $\mathbb{P}^{r-1} \times \mathbb{P}^{s-1} \to \mathbb{P}^{n-1}$. These maps lead to the application of geometric methods.

Around 1940 Stiefel applied his theory of characteristic classes of vector bundles to the problem, and Hopf applied his observations about the ring structure of cohomology. They deduced that if there exists a real nonsingular bilinear map of size $[r, s, n]$ then the binomial coefficient $\binom{n}{k}$ is even whenever $n - r < k < s$. As a corollary they concluded that if there is an n-dimensional real division algebra then n must be a power of 2.

In Chapter 12 we outline the proof of this Theorem of Stiefel and Hopf and discuss further applications of topology and K-theory to the study of nonsingular bilinear maps. To help formulate the results we introduce three numerical functions.

Definition.

(1) $r * s = \min\{n :$ there exists a normed bilinear map over $\mathbb{R}$ of size $[r, s, n]\}$. For other base fields F we write $r *_F s$ for this minimum.

(2) $r \# s = \min\{n :$ there exists a nonsingular bilinear map over $\mathbb{R}$ of size $[r, s, n]\}$.

(3) $r \circ s = \min\{n :$ the Stiefel–Hopf criterion holds for $[r, s, n]\}$.

Then $r \circ s \le r \# s \le r * s$. The first inequality is the Stiefel–Hopf Theorem and the second follows since every normed pairing is nonsingular. That "circle function"

---

1 A normed bilinear map is sometimes called an orthogonal multiplication or an orthogonal pairing.

is easily computed and provides a useful lower bound. For instance since $3 \circ 5 = 7$ and there exists a composition of size [3, 5, 7], we know that $3 * 5 = 3 \# 5 = 7$. The Stiefel–Hopf condition is generally not a sharp bound. For example when $r = s = 16$ it yields only the triviality that $16 \# 16 \geq 16$. With Adams' calculation of $KO(\mathbb{P}^n)$ we find that $16 \# 16 \geq 20$. K. Y. Lam constructed a nonsingular bilinear pairing of size [16, 16, 23], and then applied more sophisticated topology to prove that this is best possible: $16 \# 16 = 23$. These ideas are described in more detail in Chapter 12.

Chapter 13 concerns compositions over the integers. This topic is combinatorial in nature, involving matrices with entries in $\{0, 1, -1\}$. We describe several methods for constructing sums of squares formulas. For example, formulas of sizes [10, 10, 16] and [16, 16, 32] are easy to exhibit. Of course it is much harder to show that these sizes are best possible. Non-existence results for such integer pairings were investigated in the 19th century during the search for a 16-square identity. More recently Yuzvinsky (1981) formalized their study and set up the framework of "intercalate" matrices. Yiu has considerably extended that work, investigating the combinatorial aspects of intercalate matrices and their signings. His work in this area culminated with his calculation of $r *_{\mathbb{Z}} s$ for every $r, s \leq 16$. Chapter 13 provides the flavor of Yiu's combinatorial arguments without going deeply into the details.

Chapter 14 deals with compositions of size $[r, s, n]$ over a general field. The topological results of Chapter 12 can be extended to provide some information about compositions over any field of characteristic zero. In particular the Stiefel–Hopf criterion holds for such fields F: $r \circ s \leq r *_F s$. Pfister's theory of multiplicative quadratic forms also relates to the composition problem, but this method again yields results only when the field has characteristic zero. What about fields of other characteristics? Certainly the Hurwitz–Radon Theorem (from Part I) classifies compositions of sizes $[r, n, n]$ over any field F (with characteristic $\neq 2$). Adem used direct matrix methods to reduce the compositions of size $[r, n-1, n]$ over F to the classical Hurwitz–Radon case. An extension of those ideas leads to similar results for codimension 2: sizes $[r, n-2, n]$. It remains unknown whether the Stiefel–Hopf lower bound remains valid over general fields. One result in this direction, due to Szyjewski and Shapiro, provides a somewhat weaker bound valid for arbitrary fields, proved using the machinery of Chow rings.

Chapter 15 describes the application of Hopf maps to the problem of admissibility over $\mathbb{R}$. In the 1930s Hopf introduced a wonderful geometric construction. For any normed bilinear map $f : \mathbb{R}^r \times \mathbb{R}^s \to \mathbb{R}^n$ there is an associated Hopf map on spheres $h_f : S^{r+s-1} \to S^n$. The most familiar example is Hopf's fibration $S^3 \to S^2$ which arises when $r = s = n = 2$. K. Y. Lam (1985) used the geometry of these Hopf maps to uncover certain "hidden" nonsingular pairings associated to the original map f. Lam used these ideas to show that there can be no normed bilinear maps of sizes [10, 11, 17] or [16, 16, 23], providing the first examples where $r \# s < r * s$. Lam and Yiu have exploited these hidden formulas to eliminate further cases for admissibility over $\mathbb{R}$. For example they combined those ideas with arguments from homotopy theory to prove that $16 * 16 \geq 29$.

Finally in Chapter 16 we survey some topics related to compositions of quadratic forms.

- How does the composition theory generalize when higher degree forms are allowed?
- The usual vector product (cross product) in $\mathbb{R}^3$ arose originally from quaternions. Are there more general vector products enjoying similar geometric properties?
- Compositions can also be considered over fields of characteristic 2. Does the Hurwitz–Radon theory work out nicely in that context, or over more general rings?
- Nonsingular bilinear maps lead to linear subspaces of matrices having fixed rank. What is known generally about subspaces of matrices in which all non-zero elements have equal rank?

Chapter 12

$[r, s, n]$-Formulas and Topology

In this chapter we are concerned with compositions over $\mathbb{R}$, the field of real numbers. As mentioned in the introduction, the existence of an $[r, s, n]$-formula over $\mathbb{R}$ is equivalent to the existence of a bilinear map $f : \mathbb{R}^r \times \mathbb{R}^s \to \mathbb{R}^n$ satisfying the *norm property*:

$$|f(x, y)| = |x| \cdot |y| \quad \text{whenever } x \in \mathbb{R}^r \text{ and } y \in \mathbb{R}^s.$$

Such f is called a *normed* bilinear map. It induces $\hat{f} : S^{r-1} \times S^{s-1} \to S^{n-1}$, where S^{k-1} denotes the unit sphere in the space $\mathbb{R}^k$. Consequently f induces a map on real projective spaces $\tilde{f} : \mathbb{P}^{r-1} \times \mathbb{P}^{s-1} \to \mathbb{P}^{n-1}$. H. Hopf (1941) used the ring structure of the cohomology of projective spaces to obtain some necessary conditions for the existence of such a map f. In fact this was the first application of this newly discovered ring structure. These results spurred further interest in the topological side of the problem.

Before describing Hopf's proof we consider the simpler case handled in the Borsuk–Ulam Theorem. All the mappings mentioned here are assumed to be continuous. A map $g : \mathbb{R}^m \to \mathbb{R}^n$ is called *nonsingular* if $g(x) = 0$ implies $x = 0$.[1] A nonsingular map g induces a map on spheres $\hat{g} : S^{m-1} \to S^{n-1}$ defined by $\hat{g}(x) = g(x)/|g(x)|$. Conversely every map between spheres arises this way from a nonsingular map. The map g called *skew* (also called odd, or antipodal) if $g(-x) = -g(x)$ for every $x \in \mathbb{R}^m$.

The Borsuk–Ulam Theorem states that if $m > n \geq 1$ then there is no (continuous) skew map $g : S^m \to S^n$ (see e.g. Spanier (1966), p. 266). That is, the existence of a nonsingular, skew map $\mathbb{R}^m \to \mathbb{R}^n$ implies $m \leq n$. We describe a proof which uses tools motivating Hopf's Theorem. Let $H(X)$ denote the cohomology ring of a topological space X, with coefficients in $\mathbb{F}_2 = \mathbb{Z}/2\mathbb{Z}$. The cohomology ring of real projective space is a truncated polynomial ring:

$$H(\mathbb{P}^{n-1}) \cong \mathbb{F}_2[T]/(T^n),$$

where the class of T represents the class of the fundamental 1-cocycle on $\mathbb{P}^{n-1}$. This ring structure provides a quick proof of the Borsuk–Ulam Theorem as follows: Given a nonsingular skew map $\mathbb{R}^m \to \mathbb{R}^n$ there is an associated skew map on spheres

1 We hope that no confusion arises between this use of the word "nonsingular" and various other meanings familiar to the reader.

$\hat{g} : S^{m-1} \to S^{n-1}$ which induces a map on projective spaces $\tilde{g} : \mathbb{P}^{m-1} \to \mathbb{P}^{n-1}$. This in turn furnishes a map on cohomology $\tilde{g}^* : H(\mathbb{P}^{n-1}) \to H(\mathbb{P}^{m-1})$, which is identified with a ring homomorphism $\tilde{g}^* : \mathbb{F}_2[T]/(T^n) \to \mathbb{F}_2[U]/(U^m)$. Now $\tilde{g}^*(T)$ represents a 1-cocycle, so it must equal 0 or U. But it is non-zero (see the argument in Exercise 2). Therefore $\tilde{g}^*(T) = U$ and consequently $U^n = \tilde{g}^*(T)^n = \tilde{g}^*(T^n) = 0$, which implies $m \le n$.

Hopf discovered an extension of this cohomological argument to the case of nonsingular, bi-skew mappings, a generalization of the bilinear normed maps mentioned above.

12.1 Definition. Suppose $f : \mathbb{R}^r \times \mathbb{R}^s \to \mathbb{R}^n$ is a continuous mapping.

(1) f is *nonsingular* if $f(x, y) = 0$ implies that either $x = 0$ or $y = 0$.

(2) f is *bi-skew* if it is skew in each variable: $f(-x, y) = f(x, -y) = -f(x, y)$.

(3) f is *skew-linear* if it is skew in the first variable and linear in the second. *Linear-skew* maps are defined similarly.

Certainly every normed bilinear map is continuous, nonsingular and bi-skew. Nonsingular bilinear maps of size $[n, n, n]$ were mentioned in Chapter 8 in connection with real division algebras. If there exists a nonsingular bi-skew map of size $[r, s, n]$, the Borsuk–Ulam Theorem implies $r, s \le n$. Hopf generalized the argument above to strengthen this conclusion. We spend some time on this proof since it was the first application of topology to the composition problem, motivating much of the subsequent work.

Hopf's proof uses the cohomology technique above, together with the Künneth formula: $H(X \times Y) \cong H(X) \otimes H(Y)$. These basic results on cohomology are discussed in several texts in algebraic topology, including Spanier (1966) and Greenberg (1967). A more geometric discussion of homology, cohomology and Hopf's Theorem is given by Hirzebruch (1991). He provides an interesting outline of the relevant historical development of homology and cohomology, explaining how the intersection product in the homology of manifolds became identified with the cup product in cohomology.

12.2 Hopf's Theorem. *If there exists a continuous, nonsingular, bi-skew map of size $[r, s, n]$ over $\mathbb{R}$ then the binomial coefficient $\binom{n}{k}$ is even whenever $n - s < k < r$.*

Proof. The given nonsingular bi-skew map $f : \mathbb{R}^r \times \mathbb{R}^s \to \mathbb{R}^n$ induces a map on the real projective spaces

$$\tilde{f} : \mathbb{P}^{r-1} \times \mathbb{P}^{s-1} \to \mathbb{P}^{n-1},$$

and hence a map f^* on the corresponding cohomology rings. The cohomology rings of these spaces can be written using indeterminates R, S, T:

$$H(\mathbb{P}^{r-1}) \cong \mathbb{F}_2[R]/(R^r),$$
$$H(\mathbb{P}^{s-1}) \cong \mathbb{F}_2[S]/(S^s),$$
$$H(\mathbb{P}^{n-1}) \cong \mathbb{F}_2[T]/(T^n).$$

The induced homomorphism on the cohomology rings then becomes

$$f^* : \frac{\mathbb{F}_2[T]}{(T^n)} \to \frac{\mathbb{F}_2[R]}{(R^r)} \otimes \frac{\mathbb{F}_2[S]}{(S^s)}.$$

Since f^* preserves degree we know that $f^*(T) = a \cdot (R \otimes 1) + b \cdot (1 \otimes S)$ for some $a, b \in \mathbb{F}_2$. Since $\tilde{f}$ comes from a bi-skew map $\hat{f}$ on the spheres we find that $f^*(T) = R \otimes 1 + 1 \otimes S$. (To see this, choose basepoints of $\mathbb{P}^{r-1}$ and $\mathbb{P}^{s-1}$ and note that the restriction of $\tilde{f}$ to $\mathbb{P}^{r-1} \vee \mathbb{P}^{s-1} \to \mathbb{P}^{n-1}$ is homotopic to the canonical inclusion on each factor of the wedge. Compare Exercise 2.) Finally since $T^n = 0$,

$$0 = f^*(T^n) = (R \otimes 1 + 1 \otimes S)^n = \sum_{k=0}^{n} \binom{n}{k} R^k \otimes S^{n-k}.$$

Therefore $\binom{n}{k} = 0$ in $\mathbb{F}_2$ whenever $k < r$ and $n - k < s$. □

A weaker version of this Theorem (for nonsingular bilinear maps) was proved by Stiefel (in the same journal issue as Hopf's article in 1941) using certain vector bundle invariants, now called Stiefel–Whitney classes. An algebraic proof valid over real closed fields was found by Behrend (1939) using concepts from real algebraic geometry. Some different approaches to this theorem are described in Chapter 14.

Let $\mathcal{H}(r, s, n)$ be the condition on the binomial coefficients stated in the theorem above. We call this the *Stiefel–Hopf criterion*. For example $\mathcal{H}(3, 5, 6)$ is false since $\binom{6}{2} = 15$ is odd. Consequently $[3, 5, 6]$ is not admissible over $\mathbb{R}$. Similarly $[4, 5, 7]$ and $[3, 6, 7]$ are not admissible over $\mathbb{R}$. Recall that $[3, 5, 7]$ *is* admissible, as mentioned in the introduction above. This criterion $\mathcal{H}(r, s, n)$ is quite interesting and we will spend some effort analyzing its properties.

12.3 Lemma. (1) *$\mathcal{H}(r, s, n)$ implies $r, s \le n$. $\mathcal{H}(r, s, n)$ is true if $n \ge r + s - 1$.*

(2) *$\mathcal{H}(r, s, n)$ is equivalent to $\mathcal{H}(s, r, n)$.*

(3) *$\mathcal{H}(r, s, n)$ implies $\mathcal{H}(r, s, n + 1)$.*

(4) *If $n = 2^m \cdot n_0$ where n_0 is odd, then $\mathcal{H}(r, n, n)$ holds iff $r \le 2^m$.*

Proof. (1) and (2) are easy to check.

(3) Recall that $\mathcal{H}(r, s, n)$ holds if and only if $(R \otimes 1 + 1 \otimes S)^n = 0$.

(4) Consider congruences mod 2:

$$(1 + t)^n \equiv (1 + t^{2^m})^{n_0} \equiv 1 + t^{2^m} + \text{(higher terms)}.$$

Therefore $\binom{n}{k}$ is even for $0 < k < 2^m$ and odd for $k = 2^m$. □

The determination of the possible dimensions of real division algebras was a major open question for many years. The Stiefel–Hopf results of 1941 (Theorem 12.2) provided the first major step toward proving the "1, 2, 4, 8 Theorem" for division algebras.

12.4 Corollary. *Any finite-dimensional real division algebra must have dimension 2^m for some m.*

Proof. There is a real division algebra of dimension n if and only if there is a nonsingular bilinear map of size $[n, n, n]$. Apply (12.2) and (12.3) (4). □

The formulation of the condition $\mathcal{H}(r, s, n)$ using binomial coefficients seems clumsy. Further insights arise by analyzing Hopf's proof directly. We introduce a notation which will help clarify the ideas.

12.5 Definition. $r \circ s = \min\{n : \mathcal{H}(r, s, n) \text{ holds}\}$.

We will spend a few pages discussing properties of $r \circ s$ before returning to our questions about bilinear maps. Lemma 12.3 (1) becomes:

$$\max\{r, s\} \leq r \circ s \leq r + s - 1.$$

To enlarge upon the idea in Hopf's proof suppose c is a nil element in a ring R, and define $\text{ord}(c)$ to be its order of nilpotence: $\text{ord}(c) = \min\{n : c^n = 0\}$. Let $A_r = \mathbb{F}_2[x]/(x^r)$ and $A_{r,s} = A_r \otimes A_s \cong \mathbb{F}_2[x, y]/(x^r, y^s)$ and suppose $a, b \in A_{r,s}$ are the cosets of x and y, respectively. Then $\text{ord}(a) = r$ and $\text{ord}(b) = s$.

12.6 Lemma. (1) *With the notation above:* $r \circ s = \text{ord}(a + b)$.

(2) $r \circ s = \min\{n : (x + y)^n \in (x^r, y^s)$ *in* $\mathbb{F}_2[x, y]\}$.

(3) *If* $i < r$ *and* $j < s$ *then* $(r - i) \circ (s - j) = \min\{n : (x + y)^n \cdot x^i \cdot y^j \in (x^r, y^s)\}$.

Proof. (1) From the proof of (12.2), $\mathcal{H}(r, s, n)$ holds if and only if $(a + b)^n = 0$.

(2) Pull (1) back to the polynomial ring.

(3) $(x + y)^n \cdot x^i \cdot y^j = \sum \binom{n}{k} x^{k+i} y^{n-k+j}$ lies in (x^r, y^s) if and only if $\binom{n}{k}$ is even whenever $k + i < r$ and $n - k + j < s$. This condition is equivalent to $\mathcal{H}(r - i, s - j, n)$, that is: $(r - i) \circ (s - j) \leq n$. □

The formulation using the ring $A_{r,s}$ leads to the observation: $\text{ord}(r^2) = \lceil \frac{\text{ord}(r)}{2} \rceil$. Here $\lceil \alpha \rceil$ denotes the ceiling function of α, that is the smallest integer $\geq \alpha$. If $n \in \mathbb{Z}$ define

$$n^* = \left\lceil \frac{n}{2} \right\rceil.$$

Then $\text{ord}(r^2) = (\text{ord}\, r)^*$.

12.7 Lemma. $r^* \circ s^* = (r \circ s)^*$.

Proof. Let $a, b \in A_{r,s}$ be the usual generators, so that $\operatorname{ord}(a^2) = r^*$, $\operatorname{ord}(b^2) = s^*$ and a^2, b^2 generate an algebra isomorphic to A_{r^*,s^*}. Then

$$r^* \circ s^* = \operatorname{ord}(a^2 + b^2) = \operatorname{ord}((a+b)^2) = (\operatorname{ord}(a+b))^* = (r \circ s)^*.$$

□

Generally $n = 2n^*$ or $2n^* - 1$. Therefore $r \circ s$ can be recovered from the value $r^* \circ s^*$ provided we know exactly what when $r \circ s$ is odd.

12.8 Lemma. *$r \circ s$ is odd if and only if r, s are both odd and $r \circ s = r + s - 1$.*

Proof. Let $a, b \in A_{r,s}$ as above, so that $r \circ s = \operatorname{ord}(a + b)$. The "if" part is clear. "only if": Suppose $r \circ s = 2m + 1$ so that $(a + b)^{2m} \neq 0$ and $(a + b)^{2m+1} = 0$. By the binomial theorem we have:

$$0 \neq (a+b)^{2m} = (a^2 + b^2)^m = \sum a^{2i} b^{2j}, \qquad (*)$$

summed over all $i, j \geq 0$ such that $i + j = m$, $\binom{m}{i}$ is odd, and $2i < r$, $2j < s$. Furthermore,

$$0 = (a+b)^{2m+1} = \sum a^{2i} b^{2j} (a + b) = \sum (a^{2i+1} b^{2j} + a^{2i} b^{2j+1}),$$

summed over the same set of indices. An exponent pair $(2i + 1, 2j)$ cannot equal any other exponent pair in the sum, and similarly for $(2i, 2j + 1)$. Therefore every pair $(2i, 2j)$ appearing in the first sum $(*)$ must satisfy $a^{2i+1}b^{2j} = 0$ and $a^{2i}b^{2j+1} = 0$. Since $a^{2i} \neq 0$ and $b^{2j} \neq 0$, and since $A_{r,s} = \mathbb{F}_2[a] \otimes \mathbb{F}_2[b]$, these conditions imply $a^{2i+1} = 0$ and $b^{2j+1} = 0$. Hence $r = \operatorname{ord}(a) = 2i + 1$ and $s = \operatorname{ord}(b) = 2j + 1$ are both odd and the sum $(*)$ reduces to the single term $a^{r-1}b^{s-1}$. Since $m = i + j$ we also have $r \circ s = 2m + 1 = r + s - 1$. □

12.9 Proposition.

$$r \circ s = \begin{cases} 2(r^* \circ s^*) - 1 & \text{if } r, s \text{ are both odd and } r^* \circ s^* = r^* + s^* - 1, \\ 2(r^* \circ s^*) & \text{otherwise.} \end{cases}$$

Proof. The two cases are distinguished by the parity of $r \circ s$. By the lemma, the first equality holds if and only if r, s are both odd and $r \circ s = r + s - 1$. That occurs if and only if $r = 2r^* - 1$, $s = 2s^* - 1$ and $2(r^* \circ s^*) - 1 = r \circ s = r + s - 1$. The condition on $r^* \circ s^*$ easily follows. □

We can use this recursive method to compute values of $r \circ s$ fairly quickly. For convenience we include a chart of the values of $r \circ s$ when $r, s \leq 17$.

$r \circ s$	1	2	3	4	5	6	7	8	9	10	11	12	13	14	15	16	17
1	1	2	3	4	5	6	7	8	9	10	11	12	13	14	15	16	17
2	2	2	4	4	6	6	8	8	10	10	12	12	14	14	16	16	18
3	3	4	4	4	7	8	8	8	11	12	12	12	15	16	16	16	19
4	4	4	4	4	8	8	8	8	12	12	12	12	16	16	16	16	20
5	5	6	7	8	8	8	8	8	13	14	15	16	16	16	16	16	21
6	6	6	8	8	8	8	8	8	14	14	16	16	16	16	16	16	22
7	7	8	8	8	8	8	8	8	15	16	16	16	16	16	16	16	23
8	8	8	8	8	8	8	8	8	16	16	16	16	16	16	16	16	24
9	9	10	11	12	13	14	15	16	16	16	16	16	16	16	16	16	25
10	10	10	12	12	14	14	16	16	16	16	16	16	16	16	16	16	26
11	11	12	12	12	15	16	16	16	16	16	16	16	16	16	16	16	27
12	12	12	12	12	16	16	16	16	16	16	16	16	16	16	16	16	28
13	13	14	15	16	16	16	16	16	16	16	16	16	16	16	16	16	29
14	14	14	16	16	16	16	16	16	16	16	16	16	16	16	16	16	30
15	15	16	16	16	16	16	16	16	16	16	16	16	16	16	16	16	31
16	16	16	16	16	16	16	16	16	16	16	16	16	16	16	16	16	32
17	17	18	19	20	21	22	23	24	25	26	27	28	29	30	31	32	32

Some interesting patterns can be observed in this table. For example, the rows and columns are non-decreasing; the occurrences of the entries 2, 4, 8, 16 form triangles in the table; the upper left 2×2 square is repeated to the right and down, with added constants and similar patterns hold for the upper left 4×4 and 8×8 squares. These observations can be formulated algebraically as follows.

12.10 Proposition. (1) *If* $r \leq r'$ *then* $r \circ s \leq r' \circ s$.

(2) $r \circ s = 2^m$ *if and only if* $r, s \leq 2^m$ *and* $r + s > 2^m$.

(3) *If* $r \leq 2^m$ *then* $r \circ (s + 2^m) = (r \circ s) + 2^m$.

Proof. (1) $(x + y)^{r' \circ s} \in (x^{r'}, y^s) \subseteq (x^r, y^s)$ and the inequality follows.

(3) $r \circ (s + 2^m) \leq n + 2^m$ if and only if $(x + y)^n y^{2^m} = (x + y)^{n+2^m} \in (x^r, y^{s+2^m})$, and this is equivalent to $r \circ s \leq n$ by (12.6) (3).

(2) "if": Let us use the generators a, b. Since $r, s \leq 2^m$ we have $(a + b)^{2^m} = a^{2^m} + b^{2^m} = 0$ and hence $r \circ s \leq 2^m$. If $r \circ s < 2^m$ then $\mathcal{H}(r, s, 2^m - 1)$ holds, but

$\binom{2^m-1}{k}$ is odd whenever $0 \le k \le 2^m - 1$ (Exercise 6). "only if": We use induction on $r+s$. Certainly $r, s \le r \circ s = 2^m$. If $r, s \le 2^{m-1}$ then $r \circ s \le 2^{m-1}$ contrary to hypothesis. Then we may assume $r \le 2^{m-1} < s$ and express $s = s' + 2^{m-1}$. Then $2^m = r \circ s = r \circ s' + 2^{m-1}$ so that $r \circ s' = 2^{m-1}$. Then by induction $r + s' > 2^{m-1}$ and hence $r + s > 2^m$. □

We introduce analogous notations for nonsingular pairings and normed pairings.

12.11 Definition.
$r * s = \min\{n$: there exists a normed bilinear map over $\mathbb{R}$ of size $[r, s, n]\}$;
$r \# s = \min\{n$: there exists a nonsingular bilinear map over $\mathbb{R}$ of size $[r, s, n]\}$.

12.12 Proposition. (1) $\max\{r, s\} \le r \circ s \le r \# s \le r * s$.
(2) *These operations are sub-distributive:*
$(r + r') \circ s \le r \circ s + r' \circ s;$

$(r + r') * s \le r * s + r' * s;$

$(r + r') \# s \le r \# s + r' \# s.$
(3) $r \# s \le r + s - 1$.
If $2 \mid r, s$ *then* $r \# s \le r + s - 2$.

If $4 \mid r, s$ *then* $r \# s \le r + s - 4$.

If $8 \mid r, s$ *then* $r \# s \le r + s - 8$.

Proof. (1) The middle inequality is a consequence of the Stiefel–Hopf Theorem 12.2.

(2) The inequality for $r \circ s$ can be verified easily using (12.6)(2). Suppose $f : \mathbb{R}^r \times \mathbb{R}^s \to \mathbb{R}^n$ and $g : \mathbb{R}^{r'} \times \mathbb{R}^s \to \mathbb{R}^{n'}$ are bilinear pairings. Define the "direct sum" $h : \mathbb{R}^{r+r'} \times \mathbb{R}^s \to \mathbb{R}^{n+n'}$ by $h((x, x'), y) = (f(x, y), g(x', y))$. If f, g are normed or nonsingular then so is h. We indicate this construction by writing $[r, s, n] \oplus [r', s, n'] = [r + r', s, n + n']$. By the symmetry of r and s we may write $[r, s, n] \oplus [r, s', n'] = [r, s + s', n + n']$ as well. Note that the formula of size $[3, 5, 7]$ mentioned in the introduction to Part II is obtained as a direct sum: $[3, 1, 3] \oplus [3, 4, 4]$.

(3) To get a nonsingular bilinear $f : \mathbb{R}^r \times \mathbb{R}^s \to \mathbb{R}^{r+s-1}$ define the components of $f(x, y)$ to be the coefficients of $1, t, t^2, \ldots, t^{r+s-2}$ in the product

$$\Big(\sum_{i=1}^{r} x_i t^{i-1}\Big)\Big(\sum_{j=1}^{s} y_j t^{j-1}\Big)$$

in the polynomial ring $\mathbb{R}[t]$. If r and s are even, express $r = 2r'$ and $s = 2s'$ and apply the same construction to obtain a nonsingular bilinear map $\mathbb{C}^{r'} \times \mathbb{C}^{s'} \to \mathbb{C}^{r'+s'-1}$. Viewing $\mathbb{C}$ as $\mathbb{R}^2$ this yields a nonsingular map $\mathbb{R}^r \times \mathbb{R}^s \to \mathbb{R}^{r+s-2}$. The other cases are settled by similar arguments using the quaternions and octonions. We call these examples the "Cauchy product" forms. □

Since there exists a normed $[3, 5, 7]$ we know that $3 \circ 5 = 3 \# 5 = 3 * 5 = 7$. Also note that $16 \circ 16 = 16$ but certainly $16 * 16 > 16$ by the Theorem of Hurwitz. From (12.12) (3) we know that $16 \# 16 \leq 24$.

Generally the value of $r \# s$ is larger than the Stiefel–Hopf value $r \circ s$. However they are equal for small values.

12.13 Proposition. *If $r \leq 9$ then $r * s = r \# s = r \circ s$.*

Proof. By the inequality after (12.5), it suffices to show that $r * s \leq r \circ s$, that is, there is a sum-of-squares formula (i.e., a normed, bilinear map) of size $[r, s, r \circ s]$. This can be done explicitly using the n-square identities for $n = 1, 2, 4$ and 8. We work out the case $r = 9$ and omit the smaller cases. If $s \leq 8$ we are done by the symmetry of r and s. Suppose $8 < s \leq 16$. Since there is a formula of size [9, 16, 16] by Hurwitz–Radon, we have $9 * s \leq 16 = 9 \circ s$. Finally suppose $s > 16$ and express $s = 16k + t$ where $0 \leq t < 16$. The sub-distributive property and (12.10) (3) imply: $9 * s \leq 16k + (9 * t) \leq 16k + (9 \circ t) = 9 \circ (16k + t) = 9 \circ s$. □

For which values r, s do there exist normed bilinear pairings of size $[r, s, r \circ s]$? If there is such a pairing we quickly conclude that $r * s = r \# s = r \circ s$. If $r \leq 9$ then we have just noted that there are such pairings. Some further examples are mentioned in Exercise 5, but a general answer remains unclear.

Since $r \circ s \leq r \# s \leq r + s - 1$ we know the exact value of $r \# s$ in the cases where $r \circ s = r + s - 1$. For example $r \# 17 = r \circ 17 = r + 16$ whenever $r \leq 16$. Compare Exercise 3. The exact values of $r \# s$ are quite difficult to find generally and they are known only in a few more cases. The strategy is to derive good upper and lower bounds and hope that they coincide. Lower bounds for $r \# s$ are obtained by explicit constructions. One can construct nonsingular maps by presenting explicit matrices over $\mathbb{R}$, but it is more convenient to use matrices over larger division algebras.

By (12.12) (3) we know $16 \# 16 \leq 24$ using octonion multiplication. K. Y. Lam (1967) improved this bound to 23 by looking more carefully at the octonion algebra $\mathbb{K}$. Let us review the basic properties of the Doubling Process as described in the appendix to Chapter 1. There is a sequence A_n of $\mathbb{R}$-algebras with 1 and having $\dim A_n = 2^n$. These algebras are defined inductively by setting $A_0 = \mathbb{R}$, and $A_{n+1} = A_n \oplus A_n$ as vector spaces, with the multiplication:

$$(a, b) \cdot (c, d) = (ac - \bar{d}b, da + b\bar{c}).$$

Then (1, 0) is the identity element and A_n becomes a subalgebra of A_{n+1} using $a \mapsto (a, 0)$. It follows that $A_1 \cong \mathbb{C}$, $A_2 \cong \mathbb{H}$ (the real quaternions) and $A_3 \cong \mathbb{K}$ (the real octonions).

Each A_n admits a map $a \mapsto \bar{a}$ which is an involution on A_n (i.e. it is an anti-automorphism with $\bar{\bar{a}} = a$) with the property that $a = \bar{a}$ if and only if $a \in \mathbb{R}$. Let $T(a) = a + \bar{a}$ and $N(a) = a \cdot \bar{a} = \bar{a} \cdot a$ be the trace and norm maps $A \to \mathbb{R}$. Then every $a \in A$ satisfies $a^2 - T(a)a + N(a) = 0$. This $N(a)$ is the usual sum-of-squares quadratic form $[a]$ on $A_n \cong \mathbb{R}^{2^n}$. The trace map T is linear and $T(xy) = T(yx)$.

Define $A_n^\circ = \ker(T)$ to be the subspace of "pure" elements, so that $A_n = \mathbb{R} \oplus A_n^\circ$. If $e_1, e_2, \ldots, e_k$ is an orthonormal basis of A_n° (using the norm form) then these elements anti-commute pairwise and $e_j^2 = -1$. For any $x \in A_n$ the subalgebra $\mathbb{R}[x]$ is a field (isomorphic to $\mathbb{R}$ or $\mathbb{C}$).

As we have seen in Chapter 1, the norm form is multiplicative on $A_2 = \mathbb{H}$ and on $A_3 = \mathbb{K}$, so these are division algebras. Moreover $\mathbb{H}$ is associative and $\mathbb{K}$ satisfies the alternative laws: $a \cdot ab = a^2 b$ and $ab^2 = ab \cdot b$. Even though $\mathbb{K}$ is not associative, any two elements $x, y \in \mathbb{K}$ satisfy:

$\mathbb{R}[x, y]$ is an associative subalgebra (isomorphic to $\mathbb{R}$, $\mathbb{C}$ or $\mathbb{H}$);

if $xy = yx$ then $\mathbb{R}[x, y]$ is a field ($\mathbb{R}$ or $\mathbb{C}$).

See Exercise 7. If $n \geq 4$ the algebra A_n does not have a multiplicative norm, is not alternative and is not a division algebra.

We know from (12.12) that there exists a nonsingular bilinear map of size [16, 16, 24]. Here is Lam's improvement:

12.14 Lam's Construction. *Define* $f : \mathbb{K}^2 \times \mathbb{K}^2 \to \mathbb{K}^3$ *by*

$$f((a, b), (c, d)) = (ac - \bar{d}b, da + b\bar{c}, bd - db).$$

Then f is a nonsingular $\mathbb{R}$-bilinear map. This f gives rise to nonsingular bilinear maps of the following sizes:

[16, 16, 23]	[13, 13, 19]	[11, 11, 17]	[10, 10, 16]
[10, 16, 22]	[10, 15, 21]	[10, 14, 20]	[9, 16, 16].

Proof. Note that the usual multiplication on $A_4 = \mathbb{K} \times \mathbb{K}$ provides the first two slots of the formula for f. If $f((a, b), (c, d)) = 0$ then

$$ac = \bar{d}b, \quad da = -b\bar{c}, \quad bd = db. \tag{$*$}$$

Right-multiplying the first equation by $\bar{c}$, left-multiplying the second by $\bar{d}$ and adding the results we obtain

$$(Nc + Nd) \cdot a = ac \cdot \bar{c} + \bar{d} \cdot da = \bar{d}b \cdot \bar{c} - \bar{d} \cdot b\bar{c}. \tag{$**$}$$

Since b, d commute, $\mathbb{R}(b, d)$ is a field which equals $\mathbb{R}(z)$ for some z. Therefore $b, \bar{c}, \bar{d}$ lie in the associative subalgebra $\mathbb{R}(z, \bar{c})$. Hence the right side of $(**)$ vanishes. If $(c, d) \neq (0, 0)$ then $Nc + Nd > 0$ and hence $a = 0$. But then from $(*)$, $\bar{d}b = 0$ and $b\bar{c} = 0$ and therefore $b = 0$ as well. This proves that f is nonsingular.

Since $T(bd - db) = 0$ we see that image$(f) \subseteq \mathbb{K} \times \mathbb{K} \times \ker(T) \cong \mathbb{R}^{23}$ and f furnishes a nonsingular bilinear map of size [16, 16, 23]. To obtain the other sizes listed we restrict f to various subspaces of $\mathbb{K}^2 \times \mathbb{K}^2$. To see how this works let us write out the commutator $bd - db$ when we express $b = (b_1, b_2)$ and $d = (d_1, d_2) \in \mathbb{K} = \mathbb{H} \oplus \mathbb{H}$:

$$bd - db = (b_1 d_1 - d_1 b_1 + \bar{b}_2 d_2 - \bar{d}_2 b_2, d_2(b_1 - \bar{b}_1) - b_2(d_1 - \bar{d}_1)).$$

If b_1 and d_1 are scalars this reduces to $(\bar{b}_2 d_2 - \bar{d}_2 b_2, 0) \in \mathbb{H}^\circ \oplus 0$, a 3-dimensional space. Then $V = \mathbb{R} \oplus \mathbb{H} \subseteq \mathbb{H} \oplus \mathbb{H} = \mathbb{K}$ is a 5-dimensional subspace, and $W = \mathbb{K} \oplus V \subseteq \mathbb{K}^2$ is a 13-dimensional subspace. Restricting f to $W \times W$ then leads to an example of size $[13, 13, 19]$. The other sizes in the list can be obtained similarly by restricting f to suitable subspaces. For example choosing an embedding $\mathbb{C} \subseteq \mathbb{K}$ and restricting f to $(\mathbb{K} \oplus \mathbb{C}) \times (\mathbb{K} \oplus \mathbb{C})$ we get size $[10, 10, 16]$. □

Restriction of Lam's $[16, 16, 23]$ also yields maps of sizes $[11, 11, 17]$, $[11, 15, 21]$ and $[12, 14, 22]$. These can be improved by the following more delicate constructions due to Lam and Adem.

12.15 Proposition. *There exist nonsingular* $\mathbb{R}$*-bilinear maps of sizes* $[12, 12, 17]$ *and* $[12, 15, 21]$.

Proof. We describe the first case, following Lam (1967). Define $g : \mathbb{H}^3 \times \mathbb{H}^3 \to \mathbb{H}^5$ by

$$\begin{aligned} &g((a_1, a_2, a_3), (b_1, b_2, b_3)) \\ &\quad = (a_1 b_1 + \bar{b}_2 a_2 + \bar{b}_3 a_3, a_2 \bar{b}_1 - b_2 a_1, a_3 \bar{b}_1 - b_3 a_1, b_2 \bar{a}_3 + a_2 \bar{b}_3, \bar{b}_3 a_3 + \bar{a}_3 b_3). \end{aligned}$$

Then g is a bilinear map which we prove is nonsingular. If

$$g((a_1, a_2, a_3), (b_1, b_2, b_3)) = 0$$

then

$$a_1 b_1 + \bar{b}_2 a_2 + \bar{b}_3 a_3 = 0,\ b_2 a_1 = a_2 \bar{b}_1,\ b_3 a_1 = a_3 \bar{b}_1,\ b_2 \bar{a}_3 = -a_2 \bar{b}_3,\ \bar{b}_3 a_3 = -\bar{a}_3 b_3.$$

Right-multiplying the first equation by $\bar{b}_1$ and using the other equations to simplify the result, we obtain

$$a_1(b_1 \bar{b}_1 + b_2 \bar{b}_2 + b_3 \bar{b}_3) = 0.$$

Left-multiplying that first equation by b_2 and simplifying similarly yields:

$$a_2(b_1 \bar{b}_1 + b_2 \bar{b}_2 + b_3 \bar{b}_3) = 0.$$

If $(b_1, b_2, b_3) \neq (0, 0, 0)$ then these equations imply $a_1 = a_2 = 0$, and substitution back into the original equations forces $a_3 = 0$ as well. This proves that g is nonsingular. Finally note that $\bar{b}_3 a_3 + \bar{a}_3 b_3$ is a scalar. Hence image(g) is contained in a subspace of dimension $4 + 4 + 4 + 4 + 1 = 17$.

The second formula was constructed by Adem (1971) as a restriction of an explicit bilinear map $g : \mathbb{H}^3 \times \mathbb{H}^4 \to \mathbb{K}^3$. Actually Adem constructs nonsingular bilinear maps of sizes $[12, 15 + 16k, 21 + 16k]$. The details are omitted. □

Further constructions of nonsingular bilinear maps have been given, as in Milgram (1967) and K. Y. Lam (1968a), for example. Sometimes topological results can be used to prove the existence of nonsingular bilinear maps of various sizes. This is the question of whether certain homotopy classes of spheres are "bilinearly representable". Further information appears in K. Y. Lam (1977a, b), (1979) and L. Smith (1978).

The usual application of topology here is to provide "non-existence" results, like Hopf's Theorem 12.2. Deeper topological methods have been used to show that some of the constructions given above are best possible. The first step in this approach is to relate nonsingular bilinear maps to certain vector bundles on projective space. We assume now that the reader has some acquaintance with vector bundles, as described in §§2, 3 of Milnor and Stasheff (1974). Later we will assume further knowledge of K-theory.

Recall that if ξ is a vector bundle given by the projection $\pi : E \to B$, then for each $b \in B$, the fiber $F_b(\xi) = \pi^{-1}(b)$ has the structure of an $\mathbb{R}$-vector space. The bundle ξ has dimension n (or is an *n-plane bundle*) if $\dim F_b(\xi) = n$ for each $b \in B$. For vector bundles ξ and η over the same base space B the Whitney sum $\xi \oplus \eta$ is another vector bundle over B with fibers $F_b(\xi) \oplus F_b(\eta)$. If ε is the trivial line bundle over B then $k \cdot \varepsilon = \varepsilon \oplus \cdots \oplus \varepsilon$ is the trivial k-plane bundle over B.

A *cross-section* of the bundle ξ above is a continuous function $s : B \to E$ which sends each $b \in B$ into the corresponding fiber $F_b(\xi)$. For example a vector field on a smooth manifold M is exactly a cross-section of the tangent bundle of M. Certainly the trivial bundle $k \cdot \varepsilon$ admits k linearly independent cross-sections. Conversely, the bundle ξ admit k linearly independent cross-sections if and only if there is a bundle embedding $k \cdot \varepsilon \to \xi$ over B.

Let $\mathbb{P}^k$ denote real projective space of dimension k and let ξ_k be the canonical line bundle over $\mathbb{P}^k$. (ξ_k is denoted γ_k^1 in Milnor and Stasheff.) For a positive integer n, $n \cdot \xi_k$ denotes the n-fold Whitney sum of ξ_k with itself.

12.16 Proposition. *There is a nonsingular skew-linear map of size $[r, s, n]$ over $\mathbb{R}$ if and only if the bundle $n \cdot \xi_{r-1}$ over $\mathbb{P}^{r-1}$ admits s linearly independent cross-sections.*

Proof. We view $\mathbb{P}^k$ as the quotient S^k/T, where T denotes the antipodal involution of the sphere S^k. The total space of the bundle $n \cdot \xi_k$ over $\mathbb{P}^k$ may be viewed as $E = (S^k \times \mathbb{R}^n)/\tau$, where τ denotes the involution given by $\tau(x, y) = (-x, -y)$. The projection $\pi : E \to \mathbb{P}^k$ for $n \cdot \xi_k$ is induced by projection on the second factor.

Suppose $f : \mathbb{R}^r \times \mathbb{R}^s \to \mathbb{R}^n$ is a nonsingular skew-linear map. Define the related map $\varphi : S^{r-1} \times \mathbb{R}^s \to S^{r-1} \times \mathbb{R}^n$ by: $\varphi(x, v) = (x, f(x, v))$. Since $\varphi(T(x), v) = \tau\varphi(x, v)$ this map induces:

$$\bar{\varphi} : \frac{S^{r-1}}{T} \times \mathbb{R}^s \to \frac{S^{r-1} \times \mathbb{R}^n}{\tau}.$$

This carries the trivial s-plane bundle $s \cdot \varepsilon$ into $n \cdot \xi_{r-1}$, and since f is nonsingular $\bar{\varphi}$ is an injective linear map on each fiber. Then we have the s cross-sections.

Conversely, the cross-sections yield a bundle embedding $s \cdot \varepsilon \to n \cdot \xi_{r-1}$ over $\mathbb{P}^{r-1}$, and we get a fiber preserving map $\bar{\varphi}$ as above. Let $\langle x \rangle$ represent the class of x mod T, and similarly $\langle x, w \rangle$ is the class of (x, w) mod τ.[2] If $(x, v) \in S^{r-1} \times \mathbb{R}^s$ then $\bar{\varphi}(\langle x \rangle, v) = \langle x, w \rangle$, for some $w \in \mathbb{R}^n$. This w is uniquely determined, so that $w = f(x, v)$ for some function $f : S^{r-1} \times \mathbb{R}^s \to \mathbb{R}^n$. Since $\langle x \rangle = \langle -x \rangle$ we have $\bar{\varphi}(\langle -x \rangle, v) = \langle x, w \rangle = \langle -x, -w \rangle$ so that $f(-x, v) = -f(x, v)$. Then f is nonsingular and skew-linear since $\bar{\varphi}$ is injective and linear on fibers. □

Recall the function $\delta(r)$ examined in Exercises 0.6 and 2.3. It was defined as $\delta(r) = \min\{k : r \le \rho(2^k)\}$, where ρ is the Hurwitz–Radon function. Then $r \le \rho(n)$ if and only if $2^{\delta(r)} \mid n$.

12.17 Corollary. *For any $r \ge 1$, the bundle $2^{\delta(r)} \cdot \xi_{r-1}$ is trivial.*

Proof. By the Hurwitz–Radon Theorem there is a normed bilinear map over $\mathbb{R}$ of size $[r, 2^{\delta(r)}, 2^{\delta(r)}]$ and the proposition applies. □

Suppose α, β are vector bundles over a space X. Define α, β to be *stably equivalent* (written $\alpha \sim \beta$) if $\alpha \oplus m \cdot \varepsilon$ is isomorphic to $\beta \oplus n \cdot \varepsilon$ for some integers $m, n \ge 0$. If α is a vector bundle over X define the *geometric dimension*, $\mathrm{gdim}(\alpha)$, to be the smallest integer $k \ge 0$ such that α is stably equivalent to some k-plane bundle. If there is a nonsingular skew-linear map of size $[r, s, n]$ then $\mathrm{gdim}(n \cdot \xi_{r-1}) \le n - s$. (For by (12.16), $n \cdot \xi_{r-1} \cong s \cdot \varepsilon \oplus \eta$ for some $(n - s)$-plane bundle η, and $n \cdot \xi_{r-1} \sim \eta$.)

The total Stiefel–Whitney class $w(\alpha)$ detects this geometric dimension to some extent: $w_i(\alpha) = 0$ whenever $i > \mathrm{gdim}(\alpha)$. (See Exercise 9.) Operations in $\mathrm{KO}(X)$ furnish a finer tool of a similar nature. KO-theory is a generalized cohomology theory classifying real vector bundles up to addition of trivial bundles. We will sketch (without any proofs) the basic idea of K-theory and describe the ring $\widetilde{\mathrm{KO}}(\mathbb{P}^m)$. A more detailed outline of these ideas is presented in Atiyah (1962).

If X is a nice topological space (e.g. $\mathbb{P}^m$) let $\mathrm{Vect}(X)$ be the set of isomorphism classes of real vector bundles over X. This set is a semigroup under Whitney sum, and $\mathrm{KO}(X)$ is the associated Grothendieck group formed as the classes of formal differences of elements of $\mathrm{Vect}(X)$. If $[\alpha]$ denotes the class of the vector bundle α in $\mathrm{KO}(X)$, then $[\alpha] = [\beta]$ if and only if $\alpha \oplus m \cdot \varepsilon \cong \beta \oplus m \cdot \varepsilon$ for some integer $m \ge 0$. The class of the trivial bundle $n \cdot \varepsilon$ is denoted simply by n. The tensor product of vector bundles makes $\mathrm{KO}(X)$ into a commutative ring with 1.

Let $\dim : \mathrm{KO}(X) \to \mathbb{Z}$ be the ring homomorphism induced by the fiber dimension of vector bundles, and define $\widetilde{\mathrm{KO}}(X)$ to be its kernel. For instance, if α is an n-plane bundle then $[\alpha] - n \in \widetilde{\mathrm{KO}}(X)$. As additive groups, $\mathrm{KO}(X) \cong \mathbb{Z} \oplus \widetilde{\mathrm{KO}}(X)$ and the elements of the ideal $\widetilde{\mathrm{KO}}(X)$ may be viewed as the stable equivalence classes of bundles over X. Then the geometric dimension is well-defined on $\widetilde{\mathrm{KO}}(X)$.

---

2 Of course this notation differs from our use of $\langle a \rangle$ and $\langle a, b \rangle$ to represent diagonal quadratic forms or inner products.

The Grothendieck operators $\gamma^i : \mathrm{KO}(X) \to \mathrm{KO}(X)$ are defined using exterior powers. The map $\gamma_t(x) = \sum_{k=0}^{\infty} \gamma^k(x)t^k$ defines a homomorphism $\gamma_t : \mathrm{KO}(X) \to \mathrm{KO}(X)[[t]]$ from the additive group $\mathrm{KO}(X)$ to the multiplicative group of units in the formal power series ring. If $a \in \mathrm{KO}(X)$ then $\gamma^0(a) = 1$ and $\gamma^1(a) = a$. On the other hand, $\gamma_t(1) = (1-t)^{-1}$. The Grothendieck operators are defined in this tricky way in order to obtain the following key property:

$$\text{If } x \in \widetilde{\mathrm{KO}}(X) \text{ then } \gamma^k(x) = 0 \text{ for every } k > \mathrm{gdim}(x).$$

Further information and references appear in Atiyah (1962).

Our application requires knowledge of the ring structure of $\widetilde{\mathrm{KO}}(\mathbb{P}^m)$. Let $\xi = \xi_m$ be the canonical line bundle over $\mathbb{P}^m$. Then $x = [\xi] - 1$ is the corresponding element of $\widetilde{\mathrm{KO}}(\mathbb{P}^m)$. From Exercise 8 we have $[\xi]^2 = 1$ and therefore $x^2 = ([\xi]-1)^2 = -2x$. The topologists define $\varphi(m)$ to be the number of integers j with $0 < j \leq m$ and $j \equiv 0, 1, 2$ or $4 \pmod 8$. This is the same as our function $\delta(m+1)$. From (12.17) we conclude that $2^{\varphi(m)} \cdot x = 0$.

12.18 Theorem. *The additive group $\widetilde{\mathrm{KO}}(\mathbb{P}^m)$ is cyclic of order $2^{\varphi(m)}$ with generator $x = [\xi_m] - 1$. Furthermore $x^2 = -2x$ and $\gamma_t(x) = 1 + xt$.*

This theorem is a major calculation in K-theory first done by Adams (1962) in his work on vector fields on spheres. See Atiyah (1962), p. 130 for further details (but without proofs). For our applications we return to the notation $\delta(r) = \varphi(r-1)$.

12.19 Corollary. *If there is a nonsingular skew-linear map of size $[r, s, n]$ over $\mathbb{R}$, then $\binom{n}{k} \equiv 0 \pmod{2^{\delta(r)-k+1}}$ whenever $n - s < k \leq \delta(r)$.*

Proof. By (12.16) $n \cdot \xi = s \cdot \varepsilon \oplus \eta$ for some $(n-s)$- plane bundle η over $\mathbb{P}^{r-1}$. Then for $x = [\xi] - 1$ we find $\mathrm{gdim}(n \cdot x) \leq n - s$. Therefore $\gamma^k(nx) = 0$ in $\widetilde{\mathrm{KO}}(\mathbb{P}^{r-1})$ whenever $k > n-s$. According to the theorem, $\gamma_t(nx) = (1+xt)^n = \sum_{k=0}^{n} \binom{n}{k} x^k t^k$. If $k > n-s$ then $\binom{n}{k} x^k = \binom{n}{k}(-2)^{k-1}x = 0$ in $\widetilde{\mathrm{KO}}(\mathbb{P}^{r-1})$. Therefore $\binom{n}{k} \cdot 2^{k-1} \equiv 0 \pmod{2^{\delta(r)}}$. □

For convenience let $\mathcal{K}(r, s, n)$ be the number-theoretic condition in the Corollary above. It is not symmetric in r, s so we get some extra information for bilinear maps:

$$\text{If } r \# s \leq n \text{ then } \mathcal{K}(r, s, n) \text{ and } \mathcal{K}(s, r, n).$$

Note that $\mathcal{K}(r, s, n)$ is vacuous if $n \geq s + \delta(r)$ and that $\mathcal{K}(r, s, n)$ implies $\mathcal{K}(r, s, n+1)$.

Since $\delta(16) = 7$ the condition $\mathcal{K}(16, 16, n)$ holds if and only if $2^{8-k} \mid \binom{n}{k}$ whenever $n - 16 < k \leq 7$. Calculation shows that $\mathcal{K}(16, 16, 19)$ is false but $\mathcal{K}(16, 16, 20)$ is true. Consequently $16 \# 16 > 19$. Similar calculations with this criterion yield the following bounds:

$$\begin{array}{llll} 10 \# 11 \geq 16 & 10 \# 16 \geq 18 & 11 \# 16 \geq 20 & 12 \# 12 \geq 16 \\ 12 \# 13 \geq 18 & 12 \# 15 \geq 19 & 13 \# 15 \geq 20. & \end{array}$$

Some of these are improvements on the previous lower bound given by $r \circ s$, but they still do not match the sizes that we can construct. However in the classical case this K-theory does provide a definitive result, generalizing the Hurwitz–Radon Theorem over $\mathbb{R}$.

12.20 Proposition. *If there is a nonsingular skew-linear map of size* $[r, n, n]$ *then* $r \leq \rho(n)$. *Consequently,* $r \# n = n$ *if and only if* $r \leq \rho(n)$.

Proof. If there is such a map (12.19) implies that $\mathcal{K}(r, n, n)$ holds, and consequently $n \equiv 0 \pmod{2^{\delta(r)}}$. When $n = 2^m \cdot n_0$ with n_0 odd, this congruence says that $\delta(r) \leq m$, which is equivalent to $r \leq \rho(2^m) = \rho(n)$. See Exercise 0.6. □

The famous 1, 2, 4, 8 Theorem for real division algebras now follows immediately. For if there is an n-dimensional real division algebra then there is a nonsingular bilinear map of size $[n, n, n]$ over $\mathbb{R}$ and the proposition implies that $n = \rho(n)$ which forces n to be 1, 2, 4 or 8. Compare Exercise 0.8 and the references there. An outline of the proof of this 1, 2, 4, 8 Theorem is given by Hirzebruch (1991). He gives more of the geometric flavor and describes some of the history of these topological methods.

More advanced topological ideas have been applied to determine some of the values $r \# s$. Lam (1972) used the method of modified Postnikov towers to calculate the maximal number of linearly independent cross-sections in $m \cdot \xi_n$ for various small values of m, n. A chart summarizing these maximal values whenever $1 \leq n \leq m \leq 32$ is presented in Lam and Randall (1995). Since we are interested here in $r \# s$, we define

$$\begin{aligned} \sigma(r, s) &= \min\{n : \text{there exists a nonsingular skew-linear map of size } [r, s, n]\} \\ &= \min\{n : n \cdot \xi_{r-1} \text{ has } s \text{ independent cross-sections}\} \end{aligned}$$

From the work above we know that $r \circ s \leq \max\{\sigma(r, s), \sigma(s, r)\} \leq r \# s$. We quote now (without proof) some of the known values for $\sigma(r, s)$.

12.21 Theorem. *Here is a table of values of $\sigma(r,s)$ for $r, s \leq 17$. Moreover $r \# s = \max\{\sigma(r,s), \sigma(s,r)\}$ for these cases, except possibly for those entries which are underlined.*

$\sigma(r,s)$	1	2	3	4	5	6	7	8	9	10	11	12	13	14	15	16	17
1	1	2	3	4	5	6	7	8	9	10	11	12	13	14	15	16	17
2	2	2	4	4	6	6	8	8	10	10	12	12	14	14	16	16	18
3	3	4	4	4	7	8	8	8	11	12	12	12	15	16	16	16	19
4	4	4	4	4	8	8	8	8	12	12	12	12	16	16	16	16	20
5	5	6	7	8	8	8	8	8	13	14	15	16	16	16	16	16	21
6	6	6	8	8	8	8	8	8	14	14	16	16	16	16	16	16	22
7	7	8	8	8	8	8	8	8	15	16	16	16	16	16	16	16	23
8	8	8	8	8	8	8	8	8	16	16	16	16	16	16	16	16	24
9	9	10	11	12	13	14	15	16	16	16	16	16	16	16	16	16	25
10	10	10	12	12	14	14	16	16	16	16	17	17	19	20	$\underline{20}$	22	26
11	11	12	12	12	15	16	16	16	16	17	17	17	19	$\underline{20}$	21	23	27
12	12	12	12	16	16	16	16	16	16	17	17	17	19	$\underline{20}$	21	23	28
13	13	14	15	16	16	16	16	16	16	19	19	19	19	23	23	23	29
14	14	14	16	16	16	16	16	16	16	20	$\underline{20}$	$\underline{20}$	23	23	23	23	30
15	15	16	16	16	16	16	16	16	16	$\underline{20}$	20	20	23	23	23	23	31
16	16	16	16	16	16	16	16	16	16	22	23	23	23	23	23	23	32
17	17	18	19	20	21	22	23	24	25	26	27	28	29	30	31	32	32

Proof. These values come directly from the chart in Lam and Randall (1995). The equality $\sigma(r,s) = r\#s$ occurs whenever there exists a nonsingular bilinear map of size $[r, s, \sigma(r,s)]$. Working through the constructions mentioned in (12.14) and (12.15) above, we see that $r \# s$ is given by the value in the table, except those which are underlined. □

For those underlined values we have the bounds:

$$20 \leq 10 \# 15 \leq 21 \qquad 20 \leq 11 \# 14 \leq 21 \qquad 20 \leq 12 \# 14 \leq 21.$$

The upper bounds here follow from the nonsingular bilinear $[12, 15, 21]$ mentioned in (12.15). The lower bounds are given in the chart above. Determining the exact values seems to be a delicate question. For example, there do exist nonsingular skew-linear *and* linear-skew maps of size $[10, 15, 20]$, but it is unknown whether a bilinear map of that size exists. So far no one has found a topological tool fine enough to distinguish these cases.

The non-symmetry of the chart (12.21) is an interesting phenomenon. For example, $\sigma(11, 15) = 21$ and $\sigma(15, 11) = 20$ and similarly $\sigma(12, 15) \neq \sigma(15, 12)$. These values show that the sizes of nonsingular linear-skew maps differ from the sizes of nonsingular skew-linear maps. This type of behavior was first discovered by Gitler and Lam (1969), who considered the size $[13, 28, 32]$. These examples imply that the bilinear and bi-skew problems are definitely different. However in the important classical case of size $[r, n, n]$ the problems do coincide, as proved in the following result due to Dai, Lam and Milgram (1981).

12.22 Theorem. *If there is a continuous, nonsingular bi-skew map of size* $[r, n, n]$ *over* $\mathbb{R}$ *then* $r \leq \rho(n)$.

The case $[n, n, n]$ was settled by Köhnen (1978). The proof in that case can be done using the non-existence of elements of odd Hopf invariant (due to Adams). The full theorem of Dai, Lam and Milgram uses Adams' deeper work on the J-homomorphism in homotopy theory.

If $r \leq 9$ we know the value of $r \# s$ using (12.13). Further values of $r \# s$ are given in (12.21). For completeness we list (without proof) some further upper bounds for these quantities.

12.23 Proposition. *Here are known upper bounds for $r \# s$ for the range $10 \le r \le 32$ and $17 \le s \le 32$. The underlined entries are known to be the exact value of $r \# s$.*

$r\backslash s$	17	18	19	20	21	22	23	24	25	26	27	28	29	30	31	32
10	$\underline{26}$	$\underline{26}$	$\underline{28}$	$\underline{28}$	$\underline{30}$	$\underline{30}$	$\underline{32}$	$\underline{32}$	$\underline{32}$	$\underline{32}$	$\underline{32}$	$\underline{32}$	$\underline{32}$	$\underline{32}$	$\underline{32}$	$\underline{32}$
11	$\underline{27}$	$\underline{28}$	$\underline{28}$	$\underline{28}$	$\underline{31}$	$\underline{32}$	$\underline{32}$	$\underline{32}$	$\underline{32}$	33	33	35	$\underline{35}$	37	$\underline{37}$	39
12	$\underline{28}$	$\underline{28}$	$\underline{28}$	$\underline{28}$	$\underline{32}$	$\underline{32}$	$\underline{32}$	$\underline{32}$	$\underline{32}$	33	33	35	$\underline{35}$	37	$\underline{37}$	39
13	$\underline{29}$	$\underline{30}$	$\underline{31}$	$\underline{32}$	$\underline{32}$	$\underline{32}$	$\underline{32}$	$\underline{32}$	$\underline{32}$	35	35	35	$\underline{35}$	39	$\underline{39}$	$\underline{39}$
14	$\underline{30}$	$\underline{30}$	$\underline{32}$	$\underline{32}$	$\underline{32}$	$\underline{32}$	$\underline{32}$	$\underline{32}$	$\underline{32}$	36	36	36	39	39	39	$\underline{39}$
15	$\underline{31}$	$\underline{32}$	$\underline{32}$	$\underline{32}$	$\underline{32}$	$\underline{32}$	$\underline{32}$	$\underline{32}$	$\underline{32}$	37	37	39	39	39	39	$\underline{39}$
16	$\underline{32}$	$\underline{32}$	$\underline{32}$	$\underline{32}$	$\underline{32}$	$\underline{32}$	$\underline{32}$	$\underline{32}$	$\underline{32}$	38	39	39	39	39	39	$\underline{39}$
17	$\underline{32}$	$\underline{32}$	$\underline{32}$	$\underline{32}$	$\underline{32}$	$\underline{32}$	$\underline{32}$	$\underline{32}$	$\underline{32}$	39	40	40	40	40	40	40
18		$\underline{32}$	$\underline{33}$	$\underline{34}$	$\underline{35}$	$\underline{36}$	$\underline{37}$	$\underline{38}$	40	$\underline{41}$	42	43	44	45	46	47
19			$\underline{33}$	35	$\underline{35}$	37	37	$\underline{39}$	40	42	42	43	44	46	47	$\underline{47}$
20				35	$\underline{35}$	38	38	$\underline{39}$	40	43	43	43	44	47	47	$\underline{47}$
21					$\underline{35}$	$\underline{39}$	$\underline{39}$	$\underline{39}$	40	44	46	47	47	$\underline{47}$	$\underline{47}$	$\underline{47}$
22						$\underline{39}$	$\underline{39}$	$\underline{39}$	40	45	46	47	$\underline{47}$	$\underline{47}$	$\underline{47}$	$\underline{47}$
23							$\underline{39}$	$\underline{39}$	40	46	46	47	$\underline{47}$	$\underline{47}$	$\underline{47}$	$\underline{47}$
24								$\underline{39}$	40	47	$\underline{47}$	$\underline{47}$	$\underline{47}$	$\underline{47}$	$\underline{47}$	$\underline{47}$
25									47	47	48	48	48	48	48	48
26										48	50	50	52	54	54	54
27											50	50	52	54	54	54
28												50	52	54	54	54
29													52	54	54	54
30														54	54	54
31															54	54
32																$\underline{54}$

Proof. These values were compiled by Yiu (1994c), extending the works of Adem (1968), (1970), (1971), K. Y. Lam (1967), (1972), and Milgram (1967). For example Yiu notes that the values $10 \# s$ are all known, except for the cases $s \equiv 15 \pmod{32}$. In fact, $10 \# (s + 32k) = 10 \# s + 32k$, unless $s = 15$ and $10 \# 15 = 21$. Perhaps this is some hint that there does exist a nonsingular bilinear map of size $[10, 15, 20]$. □

We began this chapter with the question: For which r, s, n does there exist a normed bilinear map of size $[r, s, n]$ over $\mathbb{R}$? This quickly led to similar questions about nonsingular bilinear maps, nonsingular bi-skew maps, etc. We attacked this question by investigating the functions $r * s$ and $r \circ s$ defined in (12.11). That is, fix r and s, then consider pairings of size $[r, s, n]$. The focus changes somewhat if instead we fix r and n.

12.24 Definition. Let $r \le n$ be given.

$$\rho(n, r) = \max\{s : \text{there is a normed bilinear } [r, s, n] \text{ over } \mathbb{R}\}$$
$$= \max\{s : r * s \le n\}.$$
$$\rho_{\#}(n, r) = \max\{s : \text{there is a nonsingular bilinear } [r, s, n] \text{ over } \mathbb{R}\}$$
$$= \max\{s : r \# s \le n\}.$$
$$\rho_{\circ}(n, r) = \max\{s : r \circ s \le n\}.$$

The function $\rho(n, r)$ has been investigated independently by Berger and Friedland (1986) and by K. Y. Lam and Yiu (1987). Using difficult topological methods, Lam and Yiu computed the values of $\rho(n, r)$ in most cases where $n - r \le 5$. Berger and Friedland used more algebraic methods, along with some topology, to compute most of the values of $\rho(n, r)$ when $n - r \le 4$. We will state the results without going into details of the proofs. In order to get a clear description of $\rho(n, r)$ we introduce the basic upper and lower bounds and spend some time on them.

Certainly $\rho(n, n) = \rho(n)$ the standard Hurwitz–Radon function, and (12.20) implies that $\rho_{\#}(n, n) = \rho(n)$ as well. If $\alpha(n, r)$ is any of the functions in (12.24) then the following properties are easily checked:

$$n \le n' \Longrightarrow \alpha(n, r) \le \alpha(n', r), \qquad r \le r' \Longrightarrow \alpha(n, r) \ge \alpha(n, r');$$
$$\alpha(n + n', r) \ge \alpha(n, r) + \alpha(n', r);$$
$$s \le \alpha(n, r) \Longleftrightarrow r \le \alpha(n, s).$$

12.25 Lemma. *Let $r \le n$ and define $\lambda(n, r) = \max\{\rho(r), \rho(r+1), \dots, \rho(n)\}$. Then*

$$\lambda(n, r) \le \rho(n, r) \le \rho_{\#}(n, r) \le \rho_{\circ}(n, r).$$

We call these the "basic bounds" for $\rho(n, r)$.

Proof. If $r \le k \le n$ there exists a normed bilinear $[\rho(k), k, k]$, so there is also a normed $[r, \rho(k), n]$, hence $\rho(k) \le \rho(n, r)$. The other inequalities follow as in (12.12). □

Moreover

$$n - r + 1 \le \rho_{\#}(n, r) \le \rho_{\circ}(n, r) \le n.$$

This lower bound follows from the Cauchy product pairing as in (12.12). For small values it often happens that one of the basic bounds is achieved.

12.26 Lemma. *If $r \leq 9$ or if $\rho_\circ(n, r) \leq 9$ then $\rho(n, r) = \rho_\circ(n, r)$. In particular this occurs whenever $n < 16$.*

Proof. Let $s = \rho_\circ(n, r)$ so that $r \circ s \leq n$. Since either $r \leq 9$ or $s \leq 9$, (12.13) implies that $r * s = r \circ s$. Then $s \leq \rho(n, r)$ and equality follows. When $n < 16$ we see from the table below that if $r > 9$ then $\rho_\circ(n, r) \leq 6$. □

This lemma does not extend much further. For example, $\rho(16, 10) = 10$ (using (12.21)) while $\rho_\circ(16, 10) = 16$. The basic bounds, and this lemma, motivate a closer investigation of the function $\rho_\circ(n, r)$. A table of values is easily constructed from the table for $r \circ s$ given after (12.9).

$r \backslash n$	1	2	3	4	5	6	7	8	9	10	11	12	13	14	15	16	17
1	1	2	3	4	5	6	7	8	9	10	11	12	13	14	15	16	17
2		2	2	4	4	6	6	8	8	10	10	12	12	14	14	16	16
3			1	4	4	4	5	8	8	8	9	12	12	12	13	16	16
4				4	4	4	4	8	8	8	8	12	12	12	12	16	16
5					1	2	3	8	8	8	8	8	9	10	11	16	16
6						2	2	8	8	8	8	8	8	10	10	16	16
7							1	8	8	8	8	8	8	8	9	16	16
8								8	8	8	8	8	8	8	8	16	16
9									1	2	3	4	5	6	7	16	16
10										2	2	4	4	6	6	16	16
11											1	4	4	4	5	16	16
12												4	4	4	4	16	16
13													1	2	3	16	16
14														2	2	16	16
15															1	16	16
16																16	16
17																	1

As done in (12.10), we observe various triangular patterns and codify them algebraically. For consistency, if $r > n$ we set $\rho_\circ(n, r) = 0$.

12.27 Lemma. *Given n define m by $2^m \le n < 2^{m+1}$. Then*

$$\rho_\circ(n,r) = \begin{cases} 2^m + \rho_\circ(n-2^m, r) & \text{if } 1 \le r \le 2^m; \\ \rho_\circ(n-2^m, r-2^m) & \text{if } 2^m < r. \end{cases}$$

The proof of this and the next lemma are left to the interested reader. Note here that if $n - 2^m < r \le 2^m$ the first formula implies that $\rho_\circ(n,r) = 2^m$. This verifies the observed triangles of 2's, 4's, 8's, etc. The patterns of values for small r and for small $n-r$ are easy to guess from the table. To simplify the statements let us define $\rho_\circ(n) = \rho_\circ(n,n)$. Then by (12.3), $\rho_\circ(n)$ is the 2-power in n. That is:

$$\text{If } n = 2^m \cdot (\text{odd}) \text{ then } \rho_\circ(n) = 2^m.$$

12.28 Lemma. $\rho_\circ(n,1) = n;\quad \rho_\circ(n,n) = \rho_\circ(n)$.

If $n = 2a + b$ for $0 \le b < 2$, then:

$$\rho_\circ(n,2) = 2a;\quad \rho_\circ(n,n-1) = \rho_\circ(2a)$$

If $n = 4a + b$ for $0 \le b < 4$, then:

$$\rho_\circ(n,3) = \begin{cases} 4a & \text{if } b = 0,1,2, \\ 4a+1 & \text{if } b = 3; \end{cases} \qquad \rho_\circ(n,n-2) = \begin{cases} \rho_\circ(4a) & \text{if } b = 0,1,2, \\ 3 & \text{if } b = 3; \end{cases}$$

$$\rho_\circ(n,4) = 4a; \qquad \rho_\circ(n,n-3) = \rho_\circ(4a).$$

If $n = 8a + b$ for $0 \le b < 8$ then:

$$\rho_\circ(n,n-4) = \begin{cases} \rho_\circ(8a) & \text{if } b = 0,1,2,3,4, \\ 5 & \text{if } b = 5 \text{ or } 7, \\ 6 & \text{if } b = 6. \end{cases}$$

Now that we are familiar with $\rho_\circ(n,r)$ we return to the analysis of $\rho(n,r)$. In some cases that basic upper bound is achieved.

12.29 Lemma. *If $\lambda(n,r) \le 8$ and $n - r < 8$ then $\rho_\circ(n,r) \le 8$. In this case $\rho(n,r) = \rho_\circ(n,r)$.*

Proof. By hypothesis, $\rho(k) \le 8$ whenever $r \le k \le n$. Then no multiple of 16 lies in the interval from r to n. Suppose $2^m \le n < 2^{m+1}$. If $m \ge 4$ then 2^m is not in that interval, so that $2^m < r \le n$ and $\rho_\circ(n,r) = \rho_\circ(n-2^m, r-2^m)$ by (12.27). If the lemma is false then a counterexample with minimal n must have $m < 4$ so that $n \le 15$. A look at the table of values shows that $\rho_\circ(n,r) \le 8$ in every case where $n - s < 8$. Hence no counterexample can exist. The final equality follows from (12.26). □

The remaining cases, when $\lambda(n,r) > 8$ are more difficult. This inequality holds if and only if $n \equiv 0, 1, \ldots, r \pmod{16}$. The K-theory methods suffice to compute $\rho(n, n-1)$.

12.30 Lemma. (1) $\rho(n, n-1) = \rho_\#(n, n-1) = \max\{\rho(n-1), \rho(n)\}$.
(2) $\rho(n, n-2) = \rho_\#(n, n-2) = \max\{\rho(n-2), \rho(n-1), \rho(n), 3\}$.

Proof. If there is a nonsingular bilinear $[r, n-k, n]$ then (12.16) yields a bundle isomorphism $n \cdot \xi \cong (n-k) \cdot \varepsilon \oplus \eta$, where η is some k-plane bundle. Suppose η happens to be a sum of line bundles. Since ξ and ε are the only line bundles over $\mathbb{P}^{r-1}$ we have $\eta \cong b \cdot \xi \oplus (k-b) \cdot \varepsilon$ for some $0 \le b \le k$. Then $(n-b) \cdot \xi$ is trivial and Theorem 12.18 implies that $2^{\delta(r)}$ divides $n-b$. Then $r \le \rho(n-b)$ for some b, and hence $r \le \lambda(n, n-k)$.

In the case $k = 1$ certainly η is a line bundle. Therefore $\rho_\#(n, n-1) \le \lambda(n, n-1)$ and the basic bounds imply equality. Suppose $k = 2$. A result originally due to Levine (1963) states that if $n > 2$ then every 2-plane bundle over $\mathbb{P}^n$ is a sum of two line bundles. Hence if $\rho_\#(n, n-2) \ge 4$ then it equals $\lambda(n, n-2)$. Otherwise $\lambda(n, n-2) \le \rho_\#(n, n-2) < 4$ which implies $n \equiv 3 \pmod 4$, so that $\lambda(n, n-2) = 2$ and $\rho_\circ(n, n-2) = 3$. Then (12.26) applies. □

An algebraic proof for the calculation of $\rho(n, n-1)$ and $\rho(n, n-2)$, as well as $\rho(n, 2)$ and $\rho(n, 3)$ valid over any base field, is given in Chapter 14. See (14.21).

Finally we are ready to state the results of Lam and Yiu (1987) about pairings of small codimension. The Theorem states that if $n - r$ is small then the basic bound (12.25) is sharp: $\rho(n, r)$ equals either the lower bound or the upper bound.

12.31 Theorem (Lam and Yiu (1987)). *Suppose $n - r \le 4$.*

If $\lambda(n, r) \le 8$ then $\rho_\#(n, r) = \rho(n, r) = \rho_\circ(n, r)$.

If $\lambda(n, r) > 8$ then $\rho(n, r) = \lambda(n, r)$.

The same equalities hold when $n - r = 5$, except possibly for the cases when $\lambda(n, r) > 8$ and $n \equiv 0 \pmod{32}$.

The first part follows from (12.29), but the second part depends on a number of technical details. We state their calculation of $\rho_\#(n, r)$ for the cases when the codimension is at most 4.

12.32 Proposition. *If $n - r \le 4$ and $\lambda(n, r) > 8$ then $\rho_\#(n, r) = \lambda(n, r)$ except possibly in the following cases:*

$n - r = 3$ and $n \equiv 65, 66 \pmod{128}$,

$n - r = 4$ and either $n \equiv 2 \pmod{16}$ or $n \equiv 65, 66 \pmod{128}$.

The proof of this result uses results of Adams concerning the elements of $\widetilde{\mathrm{KO}}(\mathbb{P}^n)$ which have small geometric dimension, and the elements which can be represented by Spin(4)-bundles, or by Spin(5)-bundles. We are not competent to describe further details.

The calculation of $\rho(n, r)$ in Theorem 12.31 is now obtained by using the theory of hidden maps as described below in Chapter 15. The argument is outlined in Exercise 15.11. The omitted cases in (12.32) for $\rho_{\#}(n, r)$ remain unknown.

The methods used by Berger and Friedland (1986) are somewhat simpler. They determine the values of $\rho(n, r)$ when $n - r \leq 3$, and they determine $\rho(n, n-4)$ for odd n. They begin with a purely matrix-theoretic technique, like the methods presented in Chapter 14. When those matrices do not have simple linear expansions they are able to extend the matrix problem to a skew-linear pairing and apply results like (12.20).

12.33 Corollary. (1) $\rho(n, n-3) = \lambda(n, n-3)$.

$$(2)\ \rho(n, n-4) = \begin{cases} \lambda(n, n-3) & \textit{if } n \equiv 0, 1, 2, 3, 4 \\ 5 & \textit{if } n \equiv 5, 7 \\ 6 & \textit{if } n \equiv 6 \end{cases} \pmod 8.$$

Proof. We can verify these formulas by (12.29) when $\lambda(n, r) \leq 8$. For the remaining cases $n \not\equiv 5, 6, 7 \pmod 8$ and Theorem 12.31 applies. □

Theorem 12.31 states essentially that if $n - r \leq 5$ then for every admissible $[r, s, n]$, there is a normed $[r, s, n]$ formula built from the classical Hurwitz–Radon formulas by a process of restrictions and direct sums. This behavior is no longer true for codimension 6. In Chapter 13 we will construct a normed bilinear [10, 10, 16]. Combining this with the non-existence of a nonsingular bilinear [10, 11, 16] noted in (12.21), we find that $\rho(16, 10)$ does not equal either of the basic bounds:

$$\lambda(16, 10) = 9, \qquad \rho(16, 10) = \rho_{\#}(16, 10) = 10, \qquad \rho_{\circ}(16, 10) = 16.$$

The normed [10, 10, 16] is not a direct sum of smaller formulas, and it probably cannot be obtained as a restriction of any Hurwitz–Radon formula. Compare (13.15).

Appendix to Chapter 12. More applications of topology to algebra

This appendix presents some algebraic problems whose solutions involve some of the algebraic topology discussed above. The first two problems concern sums of squares over rings and were posed by R. Baeza. All the rings considered here are commutative with 1.

A.1 Baeza's First Question. If $n = 2^m$ and A is a commutative ring, does the set of units of A which are sums of n squares in A form a group?

Let $D_A(n)$ be the set of sums of n squares in A and let $D_A^\bullet(n) = A^\bullet \cap D_A(n)$. The question is: When is $D_A^\bullet(n)$ closed under multiplication? The classical bilinear identities imply that if $n = 1, 2, 4$ or 8 then $D_A(n)$, and hence $D_A^\bullet(n)$, are always closed under multiplication. When A is a field then Pfister's Theorem (Exercise 0.5) shows that $D_A^\bullet(n)$ is closed under multiplication whenever $n = 2^m$. This was generalized to semilocal rings by Knebusch (1971). The question for general rings was settled by Dai, Lam and Milgram (1981):

A.2 Proposition. *$D_A^\bullet(n)$ is closed under multiplication for every commutative ring A if and only if $n = 1, 2, 4$ or 8.*

Proof. Let r, s, n be positive integers. (We will assume later that $r = s = n$.) Let A be the ring obtained by localizing the polynomial ring $\mathbb{R}[x_1, \dots, x_r, y_1, \dots, y_s]$ at the multiplicative set generated by $u = x_1^2 + \dots + x_r^2$ and $v = y_1^2 + \dots + y_s^2$. Then u and v are units of A. Suppose the product uv is a sum of n squares in A. Then

$$u \cdot v = \left(\frac{f_1}{u^j v^k}\right)^2 + \dots + \left(\frac{f_n}{u^j v^k}\right)^2$$

for some $f_i \in \mathbb{R}[X, Y]$. Clearing the denominators we obtain a polynomial equation:

$$(x_1^2 + \dots + x_r^2)^{2j+1} \cdot (y_1^2 + \dots + y_s^2)^{2k+1} = f_1^2 + \dots + f_n^2$$

in $\mathbb{R}[X, Y]$. Consequently the mapping $(X, Y) \mapsto (f_1(X, Y), \dots, f_n(X, Y))$ provides a nonsingular, bi-skew mapping $\mathbb{R}^r \times \mathbb{R}^s \to \mathbb{R}^n$. Now in the case $r = s = n$, Theorem 12.22 implies that $n \le \rho(n)$ and hence $n = 1, 2, 4$ or 8. □

A.3 Baeza's Second Question. What integers n can occur as the level $s(A)$ of a commutative ring A?

Recall that the *level* (or Stufe) of A is $s(A) = \min\{n : -1 \in D_A(n)\}$. If -1 is not expressible as a sum of squares in A then $s(A) = \infty$. Pfister (1965a) proved that if F is a field with finite level then $s(F)$ must be a power of 2. (See Exercise 5.5.) If A is a Dedekind domain in which 2 is invertible and $s(A)$ is finite, then $s(A) = 2^m$ or $2^m - 1$. (See Baeza (1978), p. 178 and Baeza (1979).) Let $B_n = \mathbb{Z}[x_1, \dots, x_n]/(1 + x_1^2 + \dots + x_n^2)$. Clearly $s(B_n) \le n$ and Baeza noted that if some ring has level n then $s(B_n) = n$. He conjectured that $s(B_n) = n$ for every n. Dai, Lam and Peng (1980) proved this conjecture by a wonderful application of the Borsuk–Ulam Theorem.

A.4 Theorem. *For any $n \ge 1$ there exists an integral domain A with $s(A) = n$.*

Proof. We prove that the ring $B = \mathbb{R}[x_1, \dots, x_n]/(1 + x_1^2 + \dots + x_n^2)$ has level n. Clearly $s(B) \le n$, so let us assume that $s(B) < n$. Then there is an equation

$$-1 = f_1(X)^2 + \dots + f_{n-1}(X)^2 + f_0(X) \cdot (1 + x_1^2 + \dots + x_n^2), \qquad (*)$$

where $X = (x_1, \dots, x_n)$ and $f_j(X) \in \mathbb{R}[X]$. For any real polynomial $f(X)$ we plug in iX for X (where $i = \sqrt{-1}$) and consider the real and imaginary parts: $f(iX) = p(X) + iq(X)$, where p, q are real polynomials and p is even, q is odd. Apply this to each f_j and compare the real parts in the equation $(*)$ above to find

$$-1 = \sum_{j=1}^{n-1}(p_j(X)^2 - q_j(X)^2) + p_0(X) \cdot (1 - x_1^2 - \cdots - x_n^2). \qquad (**)$$

Let $Q : \mathbb{R}^n \to \mathbb{R}^{n-1}$ be the mapping defined by $(q_1, \ldots, q_{n-1})$. This induces a skew map $S^{n-1} \to \mathbb{R}^{n-1}$ and the Borsuk–Ulam Theorem implies that $Q(a) = 0$ for some $a \in S^{n-1}$. Plug this vector a into $(**)$ to obtain $-1 = \sum_{j=1}^{n-1} p_j(a)^2$ in $\mathbb{R}$, a contradiction. □

The ideas used in this proof have been pushed further by Dai and T. Y. Lam (1984) who analyze the "level" of a topological space with involution. An *involution* on a topological space X is a map $x \mapsto \bar{x}$, which is a homeomorphism from X to itself and satisfies $\bar{\bar{x}} = x$. A (continuous) map $f : (X, -) \to (Y, -)$ is *equivariant* if f commutes with the involutions: $f(\bar{x}) = \overline{f(x)}$ for all $x \in X$. For the sphere S^{n-1} we always use the antipodal involution $\bar{x} = -x$.

A.5 Definition. Let $(X, -)$ be a topological space with involution. The *level* and *colevel* are:

$$s(X) = \min\{n : \text{there exists an equivariant map } X \to S^{n-1}\}.$$
$$s'(X) = \max\{m : \text{there exists an equivariant map } S^{m-1} \to X\}.$$

Generally $s'(X) \le s(X)$, for any X. Moreover, $s'(S^{n-1}) = s(S^{n-1}) = n$ for any $n \ge 1$. These assertions follow from Borsuk–Ulam. Determining the level or colevel can be quite difficult, even for the projective spaces. The calculation of $s(\mathbb{P}^{2m-1})$ has been achieved by Stolz, who applied some of the major tools of algebraic topology in his proof. This work is described in the last chapter of the wonderful book of Pfister (1995). He also provides estimates there for the level of complex projective spaces $s(\mathbb{CP}^{2m-1})$.

The level of a Stiefel manifold is of particular interest here. Let $V_{n,m}$ denote the Stiefel manifold of all orthonormal m-frames in $\mathbb{R}^n$. Let δ be the involution $\delta\{v_1, \ldots, v_m\} = \{-v_1, \ldots, -v_m\}$.

A.6 Lemma. *There exists an equivariant map $S^{r-1} \to V_{n,s}$ if and only if there is a nonsingular skew-linear map of size $[r, s, n]$.*

For the proof see Exercise 19. The work on $r \# s$ and $\sigma(r, s)$ mentioned earlier can now be viewed as estimates of the colevel of $V_{n,s}$. Dai and Lam prove that the topological "level" is closely related to the level of a certain ring. We mention some of their results here, without proof. If $(X, -)$ is a space with involution define

$$A_X = \{f : X \to \mathbb{C} : f \text{ is equivariant}\}.$$

Here $\mathbb{C}$ has complex conjugation as the involution. Then A_X is an $\mathbb{R}$-algebra, using the usual addition and multiplication of functions. (It might fail to be a $\mathbb{C}$- algebra.)

A.7 Theorem. $s(X) = s(A_X)$.

Dai and Lam use this correspondence between the topological space X and the ring A_X to provide examples of the behavior of quadratic forms over commutative rings. If α and β are regular quadratic forms over a ring A we write $\alpha \supset \beta$ if α has a subform isometric to β. Then for a ring A the level $s(A)$ is the smallest n such that $n\langle 1\rangle \supset \langle -1\rangle$ over A. (Some care must be taken with the definitions. For our applications, 2 is invertible in A and a "regular quadratic form" is a finitely generated projective A-module P together with a symmetric bilinear form $b : P \times P \to A$ such that the induced map $P \to \mathrm{Hom}_A(P, A)$ is an isomorphism.) If $a \in A^\bullet$ is a unit then $n\langle a\rangle = \langle a, a, \dots, a\rangle$ is a regular quadratic form on the free module A^n.

A.8 Theorem. *$n\langle 1\rangle \supset s\langle -1\rangle$ over A_X if and only if there exists an equivariant $X \to V_{n,s}$.*

The case $s = 1$ is a restatement of (A.7) since $V_{n,1} = S^{n-1}$ and $s(A) \le n$ if and only if $n\langle 1\rangle \supset \langle -1\rangle$ over A. If F is a field (with $2 \ne 0$) Pfister's theory shows that if $2^m\langle 1\rangle$ is isotropic then it must be hyperbolic. Consequently for any n, if $n\langle 1\rangle \supset \langle -1\rangle$ over F and $2^m \le n < 2^{m+1}$, then $n\langle 1\rangle \supset 2^m\langle -1\rangle$. For a general ring A such expansions are not so easy. The existence part of the Hurwitz–Radon Theorem provides a small expansion result of this type for any ring A (see Exercise 20).

A.9 Lemma. *If $a \in A^\bullet$ then $n\langle 1\rangle \supset \langle a\rangle$ over A implies $n\langle 1\rangle \supset \rho(n)\langle a\rangle$ over A.*

Can this expansion result be improved, perhaps in the case $a = -1$? Dai and Lam proved that in general the bound $\rho(n)$ is best possible.

A.10 Proposition. *For any n there exists a ring A_n such that $n\langle 1\rangle \supset \rho(n)\langle -1\rangle$ over A_n but $n\langle 1\rangle \not\supset (\rho(n) + 1)\langle -1\rangle$.*

Proof. Let $A_n = A_{S^{n-1}}$. Then $n\langle 1\rangle \supset s\langle -1\rangle$ over A_n iff there exists an equivariant $S^{n-1} \to V_{n,s}$, iff there is a nonsingular skew-linear map of size $[n, s, n]$, iff $s \le \rho(n)$. This argument uses (A.8), (A.6) and (12.22). □

Using ideas along the same lines (along with some unpublished results on Stiefel manifolds) Dai and Lam provide the following striking examples.

A.11 Theorem. *For any integers $n > r > 0$ there exists a ring $B_{n,r}$ such that $n\langle 1\rangle \supset r\langle 1\rangle \perp \langle -1\rangle$ but $n\langle 1\rangle \not\supset (r + 1)\langle 1\rangle \perp \langle -1\rangle$ over $B_{n,r}$.*

As one corollary they deduce that for any $m > 1$ there exists a ring B for which the Pfister form $2^m\langle 1\rangle$ is isotropic but is not hyperbolic. This result emphasizes the point that much of the quadratic form theory over fields cannot be generalized to arbitrary commutative rings.

One further application of topology to algebra seems to be worth mentioning here. If R is a commutative ring recall that an R-module P is called *stably free* if $P \oplus R^m \cong R^n$ for some integers $m, n \geq 0$. The first examples of stably free modules which are not free were found by Swan (1962) using the ring $R = \mathcal{C}(X)$ of continuous functions $X \to \mathbb{R}$ for a compact Hausdorff space X. Swan established a correspondence between vector bundles over X and finitely generated projective R-modules. (To a vector bundle $E \to X$ associate the $\mathcal{C}(X)$-module $\Gamma(E)$ of all cross sections $X \to E$.) The tangent bundle τ of S^{n-1} satisfies $\tau \oplus \varepsilon \cong n \cdot \varepsilon$. Therefore if $R = \mathcal{C}(S^{n-1})$ the corresponding R-module P satisfies $P \oplus R \cong R^n$. Then if τ is not trivial (i.e. if S^{n-1} is not parallelizable) then P is not free. If fact if n is odd then S^{n-1} admits no non-vanishing tangent vector field, so there is no decomposition $\tau \cong \varepsilon \oplus \eta$ and that module P does not have any free direct summand.

Generally if M is an R-module define $\rho(M)$ to be the supremum of the ranks of the free direct summands of M. For Swan's example above we then have $\rho(P) = \rho(n) - 1$ since Adams proved that $\rho(n) - 1$ is the maximal k such that $\tau \cong k \cdot \varepsilon \oplus \eta$ for some bundle η. Further investigations of this number $\rho(M)$ for various modules M over rings related to spheres and Stiefel manifolds have been carried out by Gabel (1974), Geramita and Pullman (1974) and Allard and Lam (1981).

Exercises for Chapter 12

1. (1) A skew map $h : S^{m-1} \to S^{n-1}$ can be extended to a nonsingular skew map $\mathbb{R}^m \to \mathbb{R}^n$.

(2) There exists a bi-skew map on spheres $g : S^{r-1} \times S^{s-1} \to S^{n-1}$ if and only if there exists a nonsingular, bi-skew map $\mathbb{R}^r \times \mathbb{R}^s \to \mathbb{R}^n$.

(3) A skew-linear $g : S^{r-1} \times \mathbb{R}^s \to \mathbb{R}^n$ is equivalent to a skew map $\theta(g) : S^{r-1} \to \mathbb{M}_{n\times s}(\mathbb{R})$. Define $F_{n,s} \subseteq \mathbb{M}_{n\times s}(\mathbb{R})$ to be the subset of matrices of maximal rank s. There is a nonsingular skew-linear map of size $[r, s, n]$ if and only if there is a skew map $S^{r-1} \to F_{n,s}$.

2. **Liftings.** (1) Suppose $m, n > 1$. Let $h : S^m \to S^n$ be a skew map. Then the induced map $g : \mathbb{P}^m \to \mathbb{P}^n$ must induce an isomorphism of fundamental groups $g_* : \pi(\mathbb{P}^m) \to \pi(\mathbb{P}^n)$.

(2) Conversely if $g : \mathbb{P}^m \to \mathbb{P}^n$ and g_* is non-zero, then g is induced by some skew map on spheres.

(Hint. (1) A half great circle in S^m is carried to a path in S^n connecting a point to its antipode. Such a path induces a non-trivial element in $\pi(\mathbb{P}^n)$. Alternatively the claim follows by the usual proof of Borsuk–Ulam.

(2) There is a lifting of g to $h : S^m \to S^n$. If T is the antipodal map then $h \circ T$ must be either h or $T \circ h$, (since $h \circ T$ is also a lift of g). If $h \circ T = h$ then g is induced by some $g' : \mathbb{P}^m \to S^n$ and $g_* = 0$.)

3. **Observations on $r \circ s$.** Define (r, s) to be *sharp* if $r \circ s = r + s - 1$. In this case, $r \# s = r \circ s = r + s - 1$.

(1) (r, s) is sharp iff (r^*, s^*) is sharp and r, s are not both even.

(2) Define the *bit-sequence* for n to be the reduction mod 2 of the sequence $n, n^*, n^{**}, \ldots$ The number n can be re-built from its bit-sequence. (Look at the dyadic expansion of $n - 1$.)

Lemma. *(r, s) is sharp iff corresponding terms in the bit-sequences for r, s are never both* 0.

(3) If $n = \sum_i n_i 2^i$ is the dyadic expansion of n define $\operatorname{Bit}(n) = \{i : n_i = 1\}$. Define m, n to be *bit-disjoint* if $\operatorname{Bit}(m) \cap \operatorname{Bit}(n) = \emptyset$. Then (r, s) is sharp iff $r - 1$ and $s - 1$ are bit-disjoint.

(4) If $r = 2^k r'$ and $s = 2^k s'$ then $r \circ s \leq r + s - 2^k$, with equality if and only if $r - 2^k$ and $s - 2^k$ are bit-disjoint.

(5) If $n = 2^k \cdot$ (odd) define $\nu_2(n) = k$ (the 2-adic valuation). Suppose $r + s = 2^m$. Then $r \circ s = 2^m - 2^{\nu_2(r)}$. Suppose $r + s = 2^m - 1$. If r is even then $r \circ s = 2^m - 2^{\nu_2(r)}$. If r is odd then $r \circ s = 2^m - 2^{\nu_2(r+1)}$.

4. **New approach to $r \circ s$.** The binary operation $r \circ s$ is the unique operation on positive integers satisfying:

$$r \circ s = s \circ r, \quad 2^m \circ 2^m = 2^m, \quad r \leq r' \Longrightarrow r \circ s \leq r' \circ s$$

and

$$\text{if } r \leq 2^m \text{ then } r \circ (2^m + s) = 2^m + (r \circ s).$$

(1) If $r, s \leq 2^m$ and $r + s > 2^m$, then $r \circ s = 2^m$.

(2) The operation $r \circ s$ is associative. (Note: This becomes clearer using the interpretation in (14.6).)

(3) $p > 1$ is irreducible (relative to $\circ$) if and only if $p = 2^m + 1$ for some $m \geq 0$. Every $n > 1$ can be factored uniquely as a product of distinct irreducibles. In fact if $0 \leq m_1 < m_2 < \cdots < m_k$, then:

$$n - 1 = 2^{m_1} + \cdots + 2^{m_t} \quad \text{if and only if} \quad n = (2^{m_1} + 1) \circ \cdots \circ (2^{m_t} + 1).$$

(4) r, s are $\circ$-coprime if and only if $\operatorname{Bit}(r - 1) \cap \operatorname{Bit}(s - 1) = \emptyset$. (Notation from Exercise 3.) In this case, $r \circ s = r + s - 1$.

(5) 2^m divides $n \Longleftrightarrow 2^m$ $\circ$-divides n.

(6) $2^m + 1$ $\circ$-divides $n \Longleftrightarrow m \in \operatorname{Bit}(n-1)$. Therefore, r $\circ$-divides $n \Longleftrightarrow \operatorname{Bit}(r-1) \subseteq \operatorname{Bit}(n-1)$.

(7) If r, s are not $\circ$-coprime then $r \circ s \le r + s - 2$.

(8) Express $r - 1 = \sum r_i \cdot 2^i$ and $s - 1 = \sum s_i \cdot 2^i$, where $r_i, s_i \in \{0, 1\}$. Define the index $m(r, s) = \max\{j : r_j = s_j = 1\} = \max(\mathrm{Bit}(r-1) \cap \mathrm{Bit}(s-1))$. Then $m(r, s)$ is infinite if and only if r, s are $\circ$-coprime. Deduce Pfister's formula:

$$r \circ s = \begin{cases} \sum\limits_{i \ge m(r,s)} (r_i + s_i) \cdot 2^i & \text{when } m(r,s) \text{ is finite,} \\ r + s - 1 & \text{otherwise.} \end{cases}$$

(Hint. (1) Assume $r \le s$ and induct on m. Either $r \le 2^{m-1} < s$, or $2^{m-1} \le r < s$.

(2) To show $(r \circ s) \circ t = r \circ (s \circ t)$. We may assume $r \le t$.

Case 1: $2^{m-1} < r \le t \le 2^m$. Use subcases $s \le 2^m$ and $2^m < s$.

Case 2: $r \le 2^{m-1} < t \le 2^m$. Consider subcases $s \le 2^{m-1}$; $2^{m-1} \le s < 2^m$ and $2^m < s$.

(5) Part (3) implies $2^m = 2 \circ 3 \circ 5 \circ \cdots \circ (2^{m-1} + 1)$. If $2^m \mid n$ then $n - 1 = 1 + 2 + 2^2 + \cdots + 2^{m-1} +$ (higher terms) and (3) applies. Conversely if $n = 2^m \circ k$ express $k - 1 = 2^{r_1} + \cdots + 2^{r_t}$ where $r_1 < \cdots < r_j < m \le r_{j+1} < \cdots < r_t$. Then $n = 2^m \circ (2^{r_{j+1}} + 1) \circ \cdots \circ (2^{r_t} + 1)$ and (3) implies $n = 2^m + 2^{r_{j+1}} + \cdots + 2^{r_t}$.

(6) Suppose $n = (2^m + 1) \circ k$ and express $k - 1$ as in (5). If m is not one of the r_j apply (3). If $m = r_j$ then $n = 2^{m+1} \circ (2^{r_{j+1}} + 1) \circ \cdots \circ (2^{r_t} + 1)$ and (5) implies $2^{m+1} \mid n$ so that $m \in \mathrm{Bit}(n-1)$.

(7) Suppose $2^m + 1$ is the largest irreducible in common. Express $r = r' \circ (2^m + 1) \circ r''$ where r' involves irreducibles $< 2^m + 1$ and r'' involves the others. Then $r = r' + 2^m + r'' - 1$ by (3). Express s similarly. Then $r \circ s = 2^{m+1} \circ r'' \circ s'' = 2^{m+1} + r'' + s'' - 2 = r + s - (r' + s')$.

(8) Suppose $m = m(r, s)$ is finite. Then $2^m + 1$ is the largest common irreducible. With the notation in (7), $r'' - 1 = \sum_{i>m} r_i 2^i$ and $s'' - 1 = \sum_{i>m} s_i 2^i$.)

5. **When does $r * s = r \circ s$?** (1) **Lemma.** *Suppose $r * s = r \circ s$. If $m \ge \delta(r)$ (or equivalently, if $r \le \rho(2^m)$) then: $r * (2^m + s) = r \circ (2^m + s) = 2^m + (r \circ s)$.*

So if $s \equiv \alpha \pmod{2^{\delta(r)}}$ where $0 \le \alpha < 2^{\delta(r)}$, then $r * \alpha = r \circ \alpha$ implies $r * s = r \circ s$.

(2) If $\alpha \le 9$ or if $2^{\delta(r)} - r < \alpha \le 2^{\delta(r)}$ then $r * \alpha = r \circ \alpha$. For example, $10 * s = 10 \circ s$ whenever $s = 32k + \alpha$ and either $0 \le \alpha \le 10$ or $23 \le \alpha < 32$.

(3) If $\alpha = 2^{\delta(r)} - r$ and $r \not\equiv 0 \pmod 4$ then $r * \alpha = r \circ \alpha$.

If $\alpha = 2^{\delta(r)} - r - 1$ and $r \equiv 1, 2 \pmod 4$ then $r * \alpha = r \circ \alpha = 2^{\delta(r)} - 2$.

(Hint. (1) $r * (2^m + s) \le 2^m + (r * s) = 2^m + (r \circ s) = r \circ (2^m + s) \le r * (2^m + s)$.

(2) Use (12.13). For the second case, $2^{\delta(r)} = r \circ \alpha \le r * \alpha$, and there exists a composition of size $[r, \alpha, 2^{\delta(r)}]$.

(3) The values of $r \circ \alpha$ are given in Exercise 3 (5). Examination of Hurwitz–Radon matrices shows: if $r \le \rho(2^m)$ there exists an $[r, 2^m - r, 2^m - 1]$, and if in addition r is even then there exists an $[r, 2^m - r, 2^m - 2]$. The last statement requires the existence of an $[r, 2^m - r - 1, 2^m - 2]$.)

6. **Binomial coefficients.** (1) Suppose p is a prime. Suppose $n = \sum_i n_i p^i$ and $k = \sum_i k_i p^i$ where $0 \le n_i, k_i < p$. **Lucas' Lemma.** $\binom{n}{k} \equiv \prod_i \binom{n_i}{k_i} \pmod{p}$.

(2) $\binom{n}{k}$ is odd iff $\mathrm{Bit}(k) \subseteq \mathrm{Bit}(n)$. Then n, k are bit-disjoint iff $\binom{n+k}{k}$ is odd.

(3) Write out some lines of Pascal's triangle (mod 2). What patterns are explained by Lucas' Lemma? Which rows are all odd? For what values of n is $\binom{2n-1}{n}$ odd?

(4) How many odd values are there in the nth row of Pascal's triangle?

(5) Compare the table of $r \circ s$ (mod 2) with Pascal's triangle (mod 2).

(Hint. (1) Lucas (1878) proved: if $n = pn' + \nu$ and $k = pk' + \kappa$ then $\binom{n}{k} \equiv \binom{n'}{k'}\binom{\nu}{\kappa}$. Another proof: Compute the coefficient of x^k in $(1 + x)^n$ over $\mathbb{F}_p$ in two ways.

(4) If $c(n)$ is this number, compare $c(2n)$ and $c(n)$. What about $c(2n + 1)$?)

7. **Cayley–Dickson algebras.** Prove the results on A_n stated before (12.14). For $\mathbb{R}[x, y]$, find an orthonormal basis $\{1, e, f, \ldots\}$ with $x, y \in \mathrm{span}\{1, e, f\}$. Then $e^2 = f^2 = -1$, $ef = -fe$ and $\mathbb{R}[x, y] \subseteq \mathbb{R}[e, f]$. Since $\mathbb{K}$ is alternative, $\mathbb{R}[e, f] \cong \mathbb{H}$ is associative.

(Hint. Check the products of elements of $\{1, e, f, ef\}$, using the Moufang identity for $ef \cdot ef$. Compare Exercises 1.24 and 1.25.)

8. (1) Suppose $\xi = \xi_m$ is the canonical line bundle over $\mathbb{P}^m$. Then $\xi \otimes \xi = \varepsilon$, the trivial line bundle. Consequently $[\xi]^2 = 1$ in $\mathrm{KO}(\mathbb{P}^m)$.

Remark. More generally, if ζ is a line bundle over a paracompact space B then $\zeta \otimes \zeta = \varepsilon$.

(2) There is a nonsingular skew-linear $[r, s, n]$ if and only if there is a bundle embedding $s \cdot \xi_{r-1} \to n \cdot \varepsilon$ over $\mathbb{P}^{r-1}$. Prove this in two ways: (i) Tensor (12.16) with ξ. (ii) Imitate the proof of (12.16) noting that the map φ defined there satisfies $\varphi(\tau(x, v)) = (T(x), f(x, v))$.

(Hint. (1) If α, β are vector bundles over B then $\alpha \otimes \beta$ is the bundle over B whose fibers are the tensor products of the fibers of α and β. Any bundle β over B admits a Euclidean metric, hence $\beta \cong \beta^* = \mathrm{Hom}(\beta, \varepsilon)$, the dual bundle. Then $\beta \otimes \beta \cong \mathrm{Hom}(\beta, \beta)$ has a canonical cross-section, so it is trivial if β is a line bundle.)

9. **Stiefel–Whitney classes.** A vector bundle ξ over X has Stiefel–Whitney class $w(\xi) = \sum w_i(\xi) \in H(X) = \bigoplus H^i(X)$, the cohomology ring with $\mathbb{F}_2$-coefficients. These satisfy: $w_i(\xi) = 0$ if $i > \dim \xi$; $w(\xi \oplus \eta) = w(\xi)w(\eta)$; and if ε is a trivial bundle then $w(\varepsilon) = 1$. Recall that $H(\mathbb{P}^{r-1}) \cong \mathbb{F}_2[a]/(a^r)$ where a is a fundamental 1-cocycle. If ξ_{r-1} is the canonical line bundle over $\mathbb{P}^{r-1}$ then $w(\xi_{r-1}) = 1 + a$.

(1) If there is a nonsingular, skew-linear $[r, s, n]$ over $\mathbb{R}$ then $\mathcal{H}(r, s, n)$.

(2) From the same $[r, s, n]$ use the embedding in Exercise 8 to deduce: $\binom{-s}{k}a^k = 0$ whenever $k > n - s$. Equivalently: $\binom{s+k-1}{k}$ is even whenever $n - s < k < r$. The proof shows that this criterion must be equivalent to $\mathcal{H}(r, s, n)$. Is there a direct proof of this equivalence?

(Hint. (1) By (12.16), $n \cdot \xi_{r-1} = s \cdot \varepsilon \oplus \eta$ for some $(n-s)$-plane bundle η. Then $(1+a)^n = w(\eta)$ so that $\binom{n}{j}a^j = 0$ whenever $j > n-s$. This is the idea used by Stiefel (1941).)

10. (1) $\mathcal{K}(r, s, n)$ implies $\mathcal{K}(r, s, n+1)$. $\mathcal{K}(r+1, s, n)$ implies $\mathcal{K}(r, s, n)$. $\mathcal{K}(r, s+1, n)$ implies $\mathcal{K}(r, s, n)$.
$\mathcal{K}(r, n-1, n)$ if and only if $r \le \max\{\rho(n), \rho(n-1)\}$.

(2) Let $r \blacklozenge s = \min\{n : \mathcal{K}(r, s, n) \text{ and } \mathcal{K}(s, r, n)\}$. Then $r \blacklozenge s \le r \# s$. Compute $10 \blacklozenge n$, for $10 \le n \le 17$ and compare the results with the values in the table for $r \# s$.

11. **Vector fields on projective space.** (1) There exist $\rho(n) - 1$ independent (tangent) vector fields on S^{n-1} and on $\mathbb{P}^{n-1}$. (See Exercise 0.7.)

(2) There are k independent vector fields on $\mathbb{P}^{n-1}$ if and only if there are k independent antipodal vector fields on S^{n-1}. Consequently, by Adams' Theorem, $k \le \rho(n) - 1$. (A vector field on S^{n-1} is viewed as a function $f : S^{n-1} \to \mathbb{R}^n$ such that $\langle f(x), x \rangle = 0$. Here $\langle a, b \rangle$ is the usual inner product on $\mathbb{R}^n$, and f is "antipodal" if $f(-x) = -f(x)$.)

12. **Symmetric bilinear maps.** A map $f : \mathbb{R}^r \times \mathbb{R}^r \to \mathbb{R}^n$ is symmetric if $f(x, y) = f(y, x)$.

(1) Hopf (1940) observed: A nonsingular symmetric bilinear $[r, r, n]$ produces an embedding $\mathbb{P}^{r-1}$ into S^{n-1}. If $n > r$ there is an embedding $\mathbb{P}^{r-1}$ into $\mathbb{R}^{n-1}$.

(2) Let $N(r) = \min\{n :$ there is a nonsingular symmetric bilinear $[r, r, n]\}$. Then $r \# r \le N(r) \le 2r - 1$. If r is even then $N(r) \le 2r - 2$. Consequently $\mathbb{P}^n$ embeds in $\mathbb{R}^{2n}$, and if n is odd $\mathbb{P}^n$ embeds in $\mathbb{R}^{2n-1}$.

(3) If $N(r) = r$ then $r \le 2$.

(4) **Hopf Theorem.** *If D is a commutative division algebra over $\mathbb{R}$ then* $\dim D \le 2$. If such D has an identity element then $D \cong \mathbb{R}$ or $\mathbb{C}$. (Note: The fundamental theorem of algebra is one corollary!)

(5) Is there a non-associative example of such D?

(Hint. (1) Given f define $\varphi : S^{r-1} \to \mathbb{R}^n$ by $\varphi(x) = f(x, x)$, inducing $\bar{\varphi} : \mathbb{P}^{r-1} \to S^{n-1}$. This $\bar{\varphi}$ is injective since $\varphi(x) = \varphi(y)$ implies $f(x-y, x+y) = 0$. For the second part use stereographic projection.

(3) If $N(r) = r$ then $\mathbb{P}^{r-1}$ embeds in S^{r-1}, implying they are homeomorphic (by "invariance of domain"), but if $r > 2$ their fundamental groups differ.

(5) For $z, w \in \mathbb{C}$ define $z * w = zw + \ell(zw)$, where $\ell : \mathbb{C} \to \mathbb{R}$ is $\mathbb{R}$-linear.)

13. (1) Prove (12.27) and (12.28). If $n = 8a + b$ for $0 \le b < 8$ then:

$$\rho_\circ(n, n-5) = \begin{cases} \rho_\circ(8a) & \text{if } b = 0, 1, 2, 3, 4, 5, \\ 6 & \text{if } b = 6, 7. \end{cases}$$

(2) $\rho_\circ(n, r) = n$ iff for some m, $r \le 2^m$ and $2^m \mid n$.
$\rho_\circ(n, r) \ge n - r + 1$. When does equality hold?

(3) If $r \le 2^m$ then $\rho_\circ(n + 2^m, r) = \rho_\circ(n, r) + 2^m$.
$\rho_\circ(2^m - 1, r) \le 2^m - r$; $\quad \rho_\circ(2^m - 2, r) \le 2^m - r - 1$.

14. Define $\lambda_\circ(n, r) = \max\{\rho_\circ(r), \rho_\circ(r+1), \ldots, \rho_\circ(n)\}$.
(1) Then $\lambda_\circ(n, r) \le \rho_\circ(n, r)$.
Some of the values in (12.28) can be written more compactly using this function. For example,

$$\rho_\circ(n, n-1) = \lambda_\circ(n, n-1),$$
$$\rho_\circ(n, n-2) = \max\{\lambda_\circ(n, n-1), 3\},$$
$$\rho_\circ(n, n-3) = \lambda_\circ(n, n-3),$$
$$\rho_\circ(n, n-5) = \max\{\lambda_\circ(n, n-5), 6\}.$$

15. In the case $\lambda(n, r) \le 8$ we know that $\rho(n, r) = \rho_\circ(n, r) \le 8$. For which of these values of n, r does it happen that the basic bounds $\lambda(n, r)$ and $\rho_\circ(n, r)$ are not equal?
(*Answer.* This happens when $n - r = 2$ and $n \equiv 3 \pmod 4$; when $n - r = 4$ and $n \equiv 5, 6, 7 \pmod 8$; and when $n - r = 5$ and $n \equiv 6, 7 \pmod 8$.)

16. Define $\rho'_\#(n, r)$ as the skew-linear analog of $\rho_\#(n, r)$.
(1) $n - r + 1 \le \rho_\#(n, r) \le \rho'_\#(n, r) \le \rho_\circ(n, r)$.
$\rho'_\#(n + n', r) \ge \rho'_\#(n, r) + \rho'_\#(n', r)$.
$\rho'_\#(n, r) = \max\{s : n \cdot \xi_{r-1}$ has s independent cross-section $s\}$.
(2) $\alpha(n + 2^{\delta(r)}, r) \ge \alpha(n, r) + 2^{\delta(r)}$ holds when $\alpha = \rho$, $\rho_\#$ or $\rho'_\#$.
If $r > n$ then equality holds. This means: $\alpha(n + 2^{\delta(r)}, r) = 2^{\delta(r)}$.
(3) For $\rho'_\#$ equality holds in all cases: $\rho'_\#(n + 2^{\delta(r)}, r) = \rho'_\#(n, r) + 2^{\delta(r)}$.
(4) **Lemma.** $\mathrm{gdim}(n \cdot \xi_{r-1}) = n - \rho'_\#(n, r)$.

(Hint. (2) The inequality follows by (1) and a normed $[r, 2^{\delta(r)}, 2^{\delta(r)}]$. If $r > n$, or more generally if $\alpha(n, r) = \rho_\circ(n, r)$, note that $r \le 2^{\delta(r)}$ so that $\alpha(n + 2^{\delta(r)}, r) \le \rho_\circ(n + 2^{\delta(r)}, r)$ and equality follows.

(3) If $\rho'_\#(n, r) = \rho_\circ(n, r)$ (e.g. if $r > n$) the reverse inequality follows from (1) and Exercise 13 (3). Suppose $r \le n$. Then $n \cdot \xi_{r-1} \oplus 2^{\delta(r)} \cdot \varepsilon = (n + 2^{\delta(r)})\xi_{r-1} = t \cdot \varepsilon \oplus \nu$ where $t = \rho'_\#(n + 2^{\delta(r)}, r)$ and ν is some bundle. Since $n > \dim(\mathbb{P}^{r-1})$ we may cancel: $n \cdot \xi_{r-1} = (t - 2^{\delta(r)}) \cdot \varepsilon \oplus \nu$. This uses the

Cancellation Theorem. *If* α, β *are vector bundles over* B *and* $\dim \alpha = \dim \beta > \dim B$ *then* $\alpha \oplus \varepsilon \cong \beta \oplus \varepsilon$ *implies* $\alpha \cong \beta$.

(See Sanderson (1964), Lemma 1.2.)

(4) "$\le$" is easy. For "$\ge$" first suppose $n < r$. There is no $[r, s, n]$ and Stiefel–Whitney classes imply $\mathrm{gdim}(n \cdot \xi_{r-1}) = n$. Suppose $n \ge r$. If $s = n - \mathrm{gdim}(n \cdot \xi_{r-1})$, then $n \cdot \xi_{r-1}$ is stably equivalent to some $(n - s)$-plane bundle η. Cancel as in (2).)

17. **Duality.** (1) If $s \le n < 2^k$ then $\rho_\circ(2^k - s, 2^k - n) = \rho_\circ(n, s)$.

(2) Similarly $\lambda(2^k - s, 2^k - n) = \lambda(n, s)$. Does the same equality hold for $\rho(n, s)$?

(3) If k is large compared to n, s, then the same equality holds for $\rho'_{\#}$. More precisely, suppose k is so large that $n \le 2^k$, $r + s \le 2^k$ and $r \le \rho(2^k)$. If there exists a nonsingular skew-linear $[r, s, n]$ then there exists a nonsingular skew-linear $[r, 2^k - n, 2^k - s]$.

(Hint. (1) Equivalently $\mathcal{H}(r, s, n) \implies \mathcal{H}(r, 2^k - n, 2^k - s)$. Given $r \circ s \le n < 2^k$ then $(x + y)^n \in (x^r, y^s)$. Show: $(x + y)^{2^k - s} y^n \in (x^r, y^{2^k})$. Rewrite in terms of x and $z = x + y$.

(2) $\rho(16, 9) = 16$ and $\rho(23, 16) \le \rho_{\#}(23, 16) = 16$. Lam proved that this is a strict inequality (see Chapter 15).

(3) (12.17) implies $2^k \cdot \xi$ is trivial and (12.16) implies $n \cdot \xi \cong s \cdot \varepsilon \oplus \eta$. Deduce that $(2^k - n) \cdot \varepsilon + \eta \otimes \xi = (2^k - s) \cdot \xi$ in $\mathrm{KO}(\mathbb{P}^{r-1})$. This is an isomorphism (cancel as in Exercise 16), hence: $(2^k - s) \cdot \xi$ admits $(2^k - n)$ independent sections. Apply (12.16).)

18. **Borsuk–Ulam and levels.** The algebraic proof of Borsuk–Ulam uses the following real Nullstellensatz. (An elementary proof appears in Pfister (1995), Chapter 4.)

Theorem. *A system of r forms of odd degrees over a real closed field K in $n > r$ variables must have a non-trivial common zero in K.*

(1) **Corollary** (algebraic Borsuk–Ulam). *Suppose K is real closed and $q_1, \dots, q_n \in K[x_0, \dots, x_n]$ are odd polynomials (i.e. $q_i(-X) = -q_i(X)$). Then those polynomials have a common zero $a = (a_0, \dots, a_n) \in K^{n+1}$ with $\sum a_i^2 = 1$.*

(2) Let $B = K[x_1, \dots, x_n]/(1 + x_1^2 + \dots + x_n^2)$ where K is a field of characteristic not 2. Then the level of B is $s(B) = \min\{s(K), n\}$.

(Hint. (1) Each monomial in q_j has odd degree. If the total degree is d_j multiply each monomial by a suitable power of $\sum_{i=0}^n x_i^2$ to bring the degree up to d_j. The result is a form $\bar{q}_j$ of degree d_j. Apply the Nullstellensatz and scale to find a non-trivial common zero a with $\sum a_i^2 = 1$. Note that $\bar{q}_j(a_0, \dots, a_n) = q_j(a_0, \dots, a_n)$.

(2) Compare (A.4).)

19. **Topological level and colevel.** (1) If the involution on X has a fixed point x, then $s(X) = s'(X) = \infty$. If the involution is fixed-point-free and X embeds in $\mathbb{R}^n$, then $s(X) \le n$.

(2) Consider the Stiefel manifold $V_{n,m}$ with involution δ. Prove Lemma A.6.

(3) Use the involution $M \mapsto -M$ on the orthogonal group $\mathrm{O}(n)$. Then $s'(\mathrm{O}(n)) = \rho(n)$, the Hurwitz–Radon function.

(Hint. (2) See Exercise 1. Note that $V_{n,s} \subseteq F_{n,s}$ and the Gram–Schmidt process provides an equivariant map $F_{n,s} \to V_{n,s}$.

(3) $\mathrm{O}(n) = V_{n,n}$ and (2) applies.)

20. (1) Prove Lemma A.9.

(2) For any $m > 1$ there is a commutative ring B such that $2^m\langle 1\rangle$ is isotropic but not hyperbolic.

(Hint. (1) The Hurwitz matrix equations provide $\rho(n)$ maps $f_j : A^n \to A^n$.

(2) Apply (A.11) with $r = 1$.)

21. **Sublevel.** Let q be a regular quadratic form on A^n for a ring A. Define q to be *isotropic* over A if q has a unimodular zero vector, i.e. if there exist $a_1, \dots, a_n \in A$ generating A as an ideal, such that $q(a_1, \dots, a_n) = 0$. Define the *sublevel* $\sigma(A) = \min\{n : (n+1)\langle 1\rangle \text{ is isotropic}\}$. Suppose A is an integral domain with quotient field F.

(1) $s(F) \le \sigma(A) \le s(A)$. Pfister showed that $s(F) = 2^m$ if it is finite. If A is a local ring then $\sigma(A) = s(A)$.

(2) **Lemma.** *If* $2 \in A^\bullet$ *and* q *is isotropic then* $q \supset \langle 1, -1\rangle$ *over* A.

Consequently, $\sigma(A) \le s(A) \le \sigma(A) + 1$.

(3) If $2 \in A^\bullet$ and A is a PID then $\sigma(A) = s(F)$ and $s(A)$ has the form 2^m or $2^m - 1$.

(4) For any ring A, if $s(A) = 1, 2, 4$ or 8 then $\sigma(A) = s(A)$.

(Hint. (4) If $\sigma(A) < s(A)$ then $a_1^2 + \dots + a_s^2 = 0$ and $a_1b_1 + \dots + a_sb_s = 1$. Use an identity $(x_1^2 + \dots + x_s^2) \cdot (y_1^2 + \dots + y_s^2) = (x_1y_1 + \dots + x_sy_s)^2 + f_2^2 + \dots + f_s^2$, where each f_i is a bilinear form over $\mathbb{Z}$.)

22. A nonsingular skew-linear map $f : \mathbb{R}^r \times \mathbb{R}^s \to \mathbb{R}^n$ induces $\gamma(f) \in \pi_{r-1}(V_{n,s})$. $\gamma(f) = 0$ if and only if f extends to a nonsingular skew-linear map of size $[r+1, s, n]$. For example in the case $[10, 10, 16]$ the element $\gamma(f) \in \pi_9(V_{16,10})$ is non-trivial.

(Hint. To construct $\gamma(f)$ use $\theta(f) : S^{r-1} \to F_{n,s}$ composed with the projection $F_{n,s} \to V_{n,s}$ as in Exercises 1 and 19.

Lemma. *If* $h : S^{r-1} \to X$ *is skew, then* h *extends to a skew map* $h' : S^r \to X$ *if and only if* $[h] = 0$ *in* $\pi_{r-1}(X)$.

For the last statement recall that there is no nonsingular skew-linear $[11, 10, 16]$.)

23. Suppose $r \# s = n$ and $f : \mathbb{R}^r \times \mathbb{R}^s \to \mathbb{R}^n$ is nonsingular bilinear. Then f must be surjective. (Hint. If $v \notin \text{image}(f)$, project to $(v)^\perp$.)

24. **Axial maps.** A map $g : \mathbb{P}^{r-1} \times \mathbb{P}^{s-1} \to \mathbb{P}^{n-1}$ is *axial* if $g(x, e) = x$ for every $x \in \mathbb{P}^{r-1}$ and $g(e, y) = y$ for every $y \in \mathbb{P}^{s-1}$. Here e denotes the basepoint of any $\mathbb{P}^k$.

(1) If g is axial then $g^*(T) = R \otimes 1 + 1 \otimes S$, using the notation in the proof of (12.2). Consequently the existence of such an axial map implies $\mathcal{H}(r, s, n)$.

(2) Any axial map g as above is induced by a nonsingular bi-skew map of size $[r, s, n]$.

(Hint. Apply Exercises 1, 2.)

25. **Generalizing $r \circ s$.** For prime p, let

$$\beta_p(r, s) = \min\{n : (x + y)^n \in (x^r, y^s) \text{ in } \mathbb{F}_p[x, y]\}.$$

Then $\beta_2(r, s) = r \circ s$.

(1) $\beta_p(r, s) = \min\{n : \binom{n}{k} \equiv 0 \pmod{p} \text{ whenever } n - s < k < r\}$.

(2) $\max\{r, s\} \leq \beta_p(r, s) \leq r + s - 1$.

(3) Generalize the recursion formulas in (12.10) and the properties in Exercises 3 and 4.

Notes on Chapter 12

Stiefel (1941) was apparently the first to prove that if there is a nonsingular bilinear map of size $[r, s, n]$ then $\mathcal{H}(r, s, n)$. He used vector fields and characteristic classes (compare Exercise 9). Subsequently Hopf (1941) found a different topological proof for this result, yielding the more general result in (12.2). Stiefel and Hopf communicated these results to Behrend, who proved the result for rational bi-skew maps over any real closed field (using methods of real algebraic geometry). Note that Behrend was a student of Hopf at the time and Stiefel was a student of Hopf some years earlier. Some further historical remarks appear in James (1972).

Topologists have been interested in nonsingular bilinear mappings for a number of reasons. In fact much of the work of Stiefel and Hopf was motivated by the problems involving nonsingular maps: embedding projective space into euclidean space, determining the dimensions of real division algebras, finding the maximal number of independent vector fields on S^n and finding non-trivial maps between spheres. Nonsingular mappings are also associated with immersions of projective spaces. Ginsburg (1963) noted that if $r < n$ and there is a nonsingular bilinear map of size $[r, r, n]$ then $\mathbb{P}^{r-1}$ can be immersed in the euclidean space $\mathbb{R}^{n-1}$. Since then many papers on this topic have been published. For example, see the references in Berrick (1980), Lam and Randall (1995), Davis (1998).

Many of the properties of $r \circ s$ given in (12.10) and in Exercise 4 were first established by Pfister (1965a) in another context. Further information on Pfister's work appears in Chapter 14.

The construction of a nonsingular $[r, s, r + s - 1]$ in (12.12) goes back to Hopf (1941).

The evaluation $r \# s = r \circ s$ in (12.13) appears in Behrend (1939) for the cases $r \leq 8$.

The proof of (12.16) follows K. Y. Lam (1967).

The condition (12.19) arising from KO-theory appears implicitly in Atiyah (1962). It was noted explicitly by Yuzvinsky (1981).

The Stiefel–Hopf Theorem says that if there is a nonsingular bilinear $[r, s, n]$ then $\mathcal{H}(r, s, n)$. This condition can be restated as either $r \circ s \leq n$ or $s \leq \rho_\circ(n, r)$. Lam considered the cases when equality holds:

Theorem. *Suppose there exists a real nonsingular bilinear* $[r, s, n]$. *If either* $r \circ s = n$ *or* $s = \rho_\circ(n, r)$, *then* $r + s \le n + \rho(n)$.

The proof appears in K. Y. Lam (1997), Theorems 3.1 and 5.1. In the case $s = n$ this result reduces to Proposition 12.20. Lam noted that the theorem is also true for linear-skew pairings because of the "stable equivalence" of that situation with the bilinear case. (This is stated below in the last note for this Chapter.) It is not clear whether this theorem remains true in the bi-skew case.

Suppose $f : \mathbb{R}^n \times \mathbb{R}^n \to \mathbb{R}^n$ is continuous and nonsingular. If f is also bi-skew then Theorem 12.22 implies that $n = 1, 2, 4$ or 8. The same conclusion holds if instead of bi-skew, we assume there is an identity element $e \ne 0$ (that is $f(e, x) = f(x, e) = x$ for every x). This follows since Adams (1960) proved that S^n is an H-space if and only if $n = 1, 3, 7$. (For given such f, let $\lambda = |e|$ and define $g : S^{n-1} \times S^{n-1} \to S^{n-1}$ by $g(x, y) = \frac{f(\lambda x, \lambda y)}{|f(\lambda x, \lambda y)|}$. Then $\lambda^{-1}e$ is an identity for g, making S^{n-1} into an H-space.)

The unpublished notes by Yiu (1994c) were useful in the preparation of this chapter. They contain further information about the chart in (12.23).

Au-Yeung and Cheng (1993) have found several further cases when $\rho(n, r) = \rho_\circ(n, r)$. This is done by explicit constructions: that equality holds iff there is a normed bilinear $[r, \rho_\circ(n, r), n]$. For example, suppose $m \ge 4$ and $n \equiv -1 \pmod{2^m}$. If $r \le \rho(2^m)$ or $n - r \le \rho(2^m) - 1$ then that equality $\rho = \rho_\circ$ holds. (In those cases $\rho_\circ(n, r) = n - r + 1$.) Similarly the equality $\rho = \rho_\circ$ holds if $n \equiv -2 \pmod{2^m}$ and either $r \le \rho(2^m) - 1$ or $n - r \le \rho(2^m) - 2$.

The calculation of $\rho_\#(n, n-1)$ and $\rho_\#(n, n-2)$ given in (12.30) go back to the work of K. Y. Lam (1966). The evaluation of $\rho(n, n-1)$ and $\rho(n, n-2)$ was also obtained by Hile and Protter (1977). These values are unchanged for compositions over arbitrary fields, as proved below in Chapter 14.

The two problems of Baeza discussed in the appendix appear as problems 12 and 13 in Knebusch (1977b).

Exercise 3 (5) follow results of Anghel (1999).

Exercise 4. This approach to $r \circ s$ using irreducibles was discovered by C. Luhrs and N. Snyder in the 1998 Ross Young Scholars Program. Pfister's formula appears in Pfister (1966). He defined $r \circ s$ in the context described in (14.6) below. Köhnen (1978) was the first to prove that Pfister's $r \circ s$ coincides with the Stiefel–Hopf bound.

Exercise 5. These ideas follow Anghel (1999). The constructions in part (3) can be seen more easily using the language of intercalate matrices described in Chapter 13. Generally if $r < s$ and there exists an $[r, s, n]$ over $\mathbb{Z}$ then there exists an $[r, s-r, n-1]$.

Exercises 8, 9. These ideas and further background appear in Milnor and Stasheff (1974), §§2–4. Compare Exercise 16.D3.

Exercise 11. Stiefel and Hopf pointed out the connection between vector fields on $\mathbb{P}^{n-1}$ and nonsingular skew-linear maps.

Exercise 12 follows Hopf (1940). A related proof (using covering spaces) of Hopf's result appears in Koecher and Remmert (1991), pp. 230–238. This theorem on commutative division algebras is a simply stated algebraic question having a simple answer, but with a proof that involves non-trivial topological methods. This is an early manifestation of the "topological thorn in the flesh of algebra" (as remarked by Koecher and Remmert, p. 223). Of course the 1, 2, 4, 8 Theorem for real division algebras is a much larger "thorn". Springer (1954) proved Hopf's division algebra result using algebraic geometry in place of the topology. Berrick (1980) provides a survey of further results on embeddings of projective spaces.

Exercise 15. This result is stated in Lam and Yiu (1987).

Exercise 16. The functions $s(k, n) = \rho'_{\#}(k, n + 1)$ and $\bar{s}(k, n) = \rho_{\#}(k, n + 1)$ were introduced by K. Y. Lam and studied by him in (1968b) and (1972), and in Gitler and Lam (1969). Also see Lam and Randall (1994). Parts (2) and (3) were proved in K. Y. Lam (1968b).

Exercise 17. (1) is due to Yuzvinsky (1981). (2) Au-Yeung and Cheng (1993) bring up this question at the end of their paper. (3) follows K. Y. Lam (1972), p. 98.

Exercise 18. The calculation of the level in (A.4) works only for that $\mathbb{R}$-algebra. Generalization to other fields requires an algebraic substitute for the Borsuk–Ulam Theorem. Such a substitute was found by Knebusch (1982) and simplified by Arason and Pfister (1982). That real Nullstellensatz was known to topologists, but Behrend (1939) found the first algebraic proof. Lang's proof was somewhat simpler (see Greenberg (1969), p. 158). The Nullstellensatz was generalized to 2-fields, and more general p-fields, by Terjanian and Pfister. An elementary proof of the general result and further historical references appear in Pfister (1995), Chapter 4. The calculation of $s(B)$, generalizing Theorem A.4, was made by Arason and Pfister (1982).

Exercises 19, 20, 21 are due to Dai and Lam (1984), as mentioned in the appendix. What pairs (n, n) and $(n, n+1)$ can be realized as $(\sigma(A), s(A))$ for some ring A? Dai and Lam (1984) showed that all such pairs can be realized, except for the four pairs $(0, 1)$, $(1, 2)$, $(3, 4)$, $(7, 8)$ which are prohibited by Exercise 21 (4). They also relate these values to the colevel $s'(A)$ and the Pythagoras number $P(A)$.

Exercise 22. Compare Yiu (1987).

Exercise 24. Axial maps are defined in James (1963), §3. He mentions in §5 that there is an axial map $g : \mathbb{P}^{r-1} \times \mathbb{P}^{s-1} \to \mathbb{P}^{n-1}$ if and only if there is a nonsingular bi-skew map of size $[r, s, n]$.

Exercise 25. See Eliahou and Kervaire (1998). This $\beta_p(r, s)$ arose in their generalization of Yuzvinsky's Theorem 13.A.1 on sumsets. This function came up independently in Krüskemper's Theorem 14.24.

The main topic of this chapter is the study of nonsingular bilinear maps over $\mathbb{R}$. The topological tools provided the generalizations replacing "linear" by "skew" in various places. This naturally leads to the question: for nonsingular maps of size $[r, s, n]$ are the existence questions for bilinear, linear-skew, skew-linear and bi-skew maps equivalent? We mentioned after (12.21) that the bilinear and bi-skew problems do not coincide in general. All four types do coincide for the classical size [r, n, n], by Theorem 12.22. It is apparently unknown whether the skew-linear and bi-skew questions always coincide. However they are known to be equivalent in the cases (1) $r = s$ and (2) $r \leq 2(n - s)$. (See Adem, Gitler and James (1972) Theorem 1.4 and Gitler (1968) Theorem 3.2.) In fact, in the case $r = s$ there is an equivalence:

There is an immersion $\mathbb{P}^{r-1} \to \mathbb{R}^{n-1}$.

There is a nonsingular skew-linear map of size $[r, r, n]$.

There is an axial map $\mathbb{P}^{r-1} \times \mathbb{P}^{r-1} \to \mathbb{P}^{n-1}$.

There is a nonsingular bi-skew map of size $[r, r, n]$.

See Ginsburg (1963), Adem, Gitler and James (1972), Adem (1978b), and the notes above for Exercise 24. It is remains unknown whether these statements are equivalent to the existence of a nonsingular *bilinear* map of size $[r, r, n]$. See Adem (1968) and James (1972), p. 143.

K. Y. Lam (1968b) proved that the linear-skew and bilinear questions coincide stably: if there is a nonsingular linear-skew pairing of size $[r, s, n]$ then there is a nonsingular bilinear pairing of size $[r, s + q, n + q]$ for some q.

Chapter 13

Integer Composition Formulas

There are several ways to construct sums of squares formulas, and most of them use integer coefficients. In fact the bilinear forms involved have coefficients in $\{0, 1, -1\}$ and the constructions are combinatorial in nature. The most fruitful method for these constructions is to use the terminology of "intercalate matrices" to restate the composition problem, then to apply various ways of gluing such matrices together. This approach to compositions was pioneered by Yuzvinsky (1981) and considerably extended in the works of Yiu.

After the quaternions and octonions were discovered in the 1840s several mathematicians searched for generalizations. Many of them became convinced of the impossibility of a 16-square identity, but no proof was available at that time. In 1848 Kirkman obtained composition formulas of various sizes, including $[10, 10, 16]$ and $[12, 12, 26]$. He was also aware of the simple construction of a $[16, 16, 32]$ formula obtained from the 8-square identity[1]. The work of Kirkman was not widely known, and those formulas were re-discovered and generalized by K. Y. Lam (1966) and others.

To clarify the ideas, we extend the earlier notations and define

$$r *_{\mathbb{Z}} s = \min\{n : \text{there is a composition formula of size } [r, s, n] \text{ over } \mathbb{Z}\}.$$

The values of $r *_{\mathbb{Z}} s$ are already known when $r \leq 9$. In fact, as mentioned in (12.13):

$$\text{if } r \leq 9 \text{ then } r *_{\mathbb{Z}} s = r * s = r \circ s.$$

Lam exhibited several formulas, including a $[10, 10, 16]$, in his 1966 thesis. Subsequently Adem (1975) discovered numerous new formulas derived from the Cayley–Dickson algebras. Based on this experience, Adem conjectured that

$$r *_F r = \begin{cases} 26 & \text{if } r = 11, 12 \\ 28 & \text{if } r = 13 \\ 32 & \text{if } r = 14, 15, 16 \end{cases}$$

for any field F of characteristic not 2. Constructions of formulas of those sizes are described below, but it is unknown whether these sizes are best possible, even if real

---

1 Kirkman attributes this to J. R. Young, who first observed that if $k = 2, 4$, or 8 then there is a $[km, kn, kmn]$ formula. Further historical information is presented in Dickson (1919).

coefficients are used. However using the discrete nature of integer compositions, Yiu has succeeded in proving that Adem's bounds are best possible over $\mathbb{Z}$. In listing his results we may assume $r, s \geq 10$, since we already know the values when $r \leq 9$ or $s \leq 9$.

13.1 Theorem (Yiu). *The values of* $r *_{\mathbb{Z}} s$ *for* $10 \leq r, s \leq 16$ *are listed in the following table:*

$r \backslash s$	10	11	12	13	14	15	16
10	16	26	26	27	27	28	28
11	26	26	26	28	28	30	30
12	26	26	26	28	30	32	32
13	27	28	28	28	32	32	32
14	27	28	30	32	32	32	32
15	28	30	32	32	32	32	32
16	28	30	32	32	32	32	32

Yiu's early work on integer compositions involved a mixture of topological and combinatorial methods to find lower bounds for $r *_{\mathbb{Z}} s$. However the theorem above was proved by elementary (but intricate) combinatorial methods, avoiding the use of topology.

We will present the details for the construction of some of these formulas, but we give only brief hints about Yiu's non-existence proofs. Constructions of composition formulas beyond the range $r, s \leq 16$ have been considered by several authors. Their results appear in Appendix C below.

To begin the analysis let us recall three formulations of the problem of integer compositions.

13.2 Lemma. *The following statements are equivalent.*

(1) *There exists an* $[r, s, n]_{\mathbb{Z}}$ *formula*

$$(x_1^2 + x_2^2 + \cdots + x_r^2) \cdot (y_1^2 + y_2^2 + \cdots + y_s^2) = z_1^2 + z_2^2 + \cdots + z_n^2$$

where each $z_k = z_k(X, Y)$ *is a bilinear form in* X *and* Y *with coefficients in* $\mathbb{Z}$.

(2) *There is a set of* $n \times s$ *matrices* $A_1, \dots, A_r$ *with coefficients in* $\mathbb{Z}$ *and satisfying the Hurwitz equations* $A_i^\top \cdot A_j + A_j^\top \cdot A_i = 2\delta_{ij} I_s$ *for* $1 \leq i, j \leq r$.

(3) *There is a bilinear map* $f : \mathbb{Z}^r \times \mathbb{Z}^s \to \mathbb{Z}^n$ *satisfying the norm condition*

$$|f(x, y)| = |x| \cdot |y|.$$

Proof. This is a special case of Proposition 1.9. We could allow each z_k in (1) to be a polynomial in $\mathbb{Z}[X, Y]$. A degree argument implies that z_k must be a bilinear form. □

Let A, B, C be the standard (orthonormal) bases for $\mathbb{Z}^r, \mathbb{Z}^s, \mathbb{Z}^n$, respectively. Then for instance, every $x \in \mathbb{Z}^r$ can be uniquely expressed as $x = \sum_{a\in A} x_a a$, for $x_a \in \mathbb{Z}$. If $a \in A$ and $b \in B$, then $f(a, b) = \sum_c \gamma_c^{a,b} c$ and $1 = |a|^2 \cdot |b|^2 = |f(a, b)|^2 = \sum_c (\gamma_c^{a,b})^2$. Since these are integers, there is exactly one c for which $\gamma_c^{a,b} = \pm 1$, while all the other terms are 0. That choice of c depends only on a, b and we write it as $c = \varphi(a, b)$. Then φ is a well-defined function on $A \times B$ and $f(a, b) = \pm\varphi(a, b)$. Letting $\varepsilon(a, b)$ be that sign, we obtain functions

$$\varphi : A \times B \to C$$
$$\varepsilon : A \times B \to \{1, -1\}$$

such that $f(a, b) = \varepsilon(a, b) \cdot \varphi(a, b)$ for every $a \in A$ and $b \in B$.

We will translate the norm condition on f to statements about these new functions. As before, $\langle u, v\rangle$ denotes the inner product. Using indeterminates x_a for $a \in A$ and y_b for $b \in B$, we obtain:

$$\sum_{a,b} x_a^2 y_b^2 = \Big|\sum_a x_a a\Big|^2 \cdot \Big|\sum_b y_b b\Big|^2 = \Big|\sum_{a,b} x_a y_b f(a, b)\Big|^2$$
$$= \sum_{a,b}\sum_{a',b'} x_a x_{a'} y_b y_{b'} \langle f(a, b), f(a', b')\rangle.$$

Comparing coefficients of x_a^2 we find that if $b \neq b'$ then $\langle f(a, b), f(a, b')\rangle = 0$. Since C is an orthonormal set, this condition says: if $b \neq b'$ then $\varphi(a, b) \neq \varphi(a, b')$, an injectivity condition on φ.

Similarly the coefficients of y_b^2 show that $a \neq a'$ implies $\varphi(a, b) \neq \varphi(a', b)$. Fixing the indices $a \neq a'$ and $b \neq b'$ and comparing coefficients, we find:

$$0 = \langle f(a, b), f(a', b')\rangle + \langle f(a, b'), f(a', b)\rangle.$$

Therefore: $\varphi(a, b) = \varphi(a', b')$ if and only if $\varphi(a, b') = \varphi(a', b)$. Moreover if these equalities hold for given indices a, b, a', b', then by computing the signs we find:

$$\varepsilon(a, b) \cdot \varepsilon(a', b') = -\varepsilon(a, b') \cdot \varepsilon(a', b).$$

The function φ can be tabulated as an $r \times s$ matrix M (with rows indexed by A and columns indexed by B) with entries in C. Following Yiu's terminology, the entries of M are called *colors* and $n(M)$ denotes the number of distinct colors in M. If $n = n(M)$ we usually take the set of colors to be $\{1, 2, \ldots, n\}$ or $\{0, 1, \ldots, n - 1\}$.

13.3 Definition. Suppose M is an $r \times s$ matrix with entries taken from a set of "colors". Let $M(i, j)$ be the (i, j)-entry of M.

(a) M is an *intercalate*[2] matrix if:

(1) The colors along each row (resp. column) are distinct.

(2) If $M(i, j) = M(i', j')$ then $M(i, j') = M(i', j)$. (intercalacy)

An intercalate matrix M has *type* (r, s, n) if it is an $r \times s$ matrix with at most n colors: $n(M) \leq n$.

(b) An intercalate matrix M is *signed consistently* if there exist $\varepsilon_{ij} = \pm 1$ such that

$$\varepsilon_{ij}\varepsilon_{ij'}\varepsilon_{i'j}\varepsilon_{i'j'} = -1 \text{ whenever } M(i, j) = M(i', j') \text{ and } i \neq i' \text{ and } j \neq j'.$$

The intercalacy condition says that every 2×2 submatrix of M involves an even number of distinct colors. The consistency condition says that every 2×2 submatrix with only two distinct colors must have an odd number of minus signs.

13.4 Lemma. *There is an $[r, s, n]_{\mathbb{Z}}$ formula if and only if there is a consistently signed intercalate matrix of type (r, s, n).*

Proof. This equivalence is explained in the preceding discussion. Note that if $x = \sum_i x_i a_i \in \mathbb{Z}^r$ and $y = \sum_j y_i b_j \in \mathbb{Z}^s$ then $f(x, y) = \sum_k z_k c_k$ where $z_k = \sum \varepsilon_{ij} x_i y_j$ summed over all i, j such that $M(i, j) = k$. Then the terms in z_k correspond to occurrences of the color k in the intercalate matrix. □

These matrices and their signings were first studied by Yuzvinsky (1981) who used the term "monomial pairings". He noted that with this formulation the problem of $[r, s, n]_{\mathbb{Z}}$ formulas separates into two questions:

(1) For which values r, s, n is there an intercalate matrix of type (r, s, n)?

(2) Given an intercalate matrix, does it have a consistent signing?

The reader is invited to verify that the following 3×5 matrix is intercalate, to find a consistent signing and to write out the corresponding composition formula of size [3, 5, 7].

$$\begin{pmatrix} 1 & 2 & 3 & 4 & 5 \\ 2 & 1 & 4 & 3 & 6 \\ 3 & 4 & 1 & 2 & 7 \end{pmatrix}$$

Two intercalate matrices A, B of type (r, s, n) are defined to be *equivalent* if A can be brought to B by permutation of rows, permutation of columns, and relabelling of colors. Up to equivalence $D_1 = \begin{pmatrix} 0 & 1 \\ 1 & 0 \end{pmatrix}$ is the unique intercalate matrix of type $(2, 2, 2)$. One consistent signing of D_1 is $\begin{pmatrix} +0 & +1 \\ +1 & -0 \end{pmatrix}$. Of course these signed values, like $+1$ and -0, should be interpreted formally as a sign and a color, certainly not as

---

2 Pronounced with the accent on the syllable "ter". The word "intercalate" was introduced in this context by Yiu, following some related usage in combinatorics.

a real number. This signed matrix can easily be re-written as a composition formula using the expression for z_k given in the proof of (13.4). With the colors $\{0, 1\}$ here it is convenient to number the rows and columns of D_1 by the indices $\{0, 1\}$ as well. In this case we find $z_0 = +x_0y_0 - x_1y_1$ and $z_1 = +x_0y_1 + x_1y_0$.

If an intercalate matrix is consistently signed then that signing can be carried over to any equivalent matrix. Moreover on a given intercalate matrix M there may be several consistent signings. Starting from one such signing, changing all the signs in any row (or column) yields another consistent signing. Similarly changing the signs of all occurrences of a single color yields another consistent signing. If one signing of M can be transformed to another by some sequence of these three types of changes we say the signings are *equivalent*. Any signing is equivalent to a "standard" signing: all " $+$" signs in the first row and first column.

There are several methods for constructing new intercalate matrices from old ones. In some cases these methods provide consistent signings as well. For example suppose M is an intercalate matrix of type (r, s, n). Then any submatrix M' of M is also intercalate, and if M is consistently signed then so is M'. In this case M' is called a *restriction* of M. On the level of sums of squares formulas this construction is the same as setting a subset of the x's and a subset of the y's equal to zero.

Another construction is the *direct sum*. Suppose A, A' are intercalate matrices of types (r, s, n) and (r, s', n'), respectively. Replace A' by an equivalent matrix if necessary to assume that A and A' involve disjoint sets of colors, and define

$$M = (\,A \quad A'\,)\,.$$

Then M is an intercalate matrix of type $(r, s + s', n + n')$. If A and A' are consistently signed then so is M. On the level of normed mappings this direct sum construction was mentioned in the proof of (12.12). (What is the corresponding construction for composition formulas?)

Of course the construction may be done with the roles of r and s reversed: $(r, s, n) \oplus (r', s, n') \implies (r + r', s, n + n')$. Let us apply these ideas to the standard consistently signed intercalate matrix A of type $(8, 8, 8)$. Define A', A'', A''' to be copies of A, using disjoint sets of 8 colors. Then the matrix

$$M = \begin{pmatrix} A & A' \\ A'' & A''' \end{pmatrix}$$

is the double direct sum of four copies of A. It is a consistently signed intercalate matrix of type $(16, 16, 32)$. The corresponding composition formula was mentioned earlier.

Perhaps the simplest intercalate matrices are of the type (r, s, rs) in which all entries of the matrix are distinct. Every signing of this matrix is consistent and the corresponding sums of squares formula is the trivial one in which all the terms are multiplied out. This example can be built by a sequence of direct sum operations applied to the 1×1 matrix $D_0 = [0]$.

A third construction is the *tensor product* (Kronecker product) of matrices. Suppose $A = (a_{ij})$ and $B = (b_{kl})$ are intercalate matrices of types (r_1, s_1, n_1) and (r_2, s_2, n_2), respectively. Then $A \otimes B = (c_{ik,jl})$ is an intercalate matrix of type (r_1r_2, s_1s_2, n_1n_2). Here the color $c_{ik,jl}$ is the ordered pair (a_{ij}, b_{kl}) and the row-indices (i, k) and the column-indices (j, l) must each be listed in some definite order. The matrix $A \otimes B$ is intercalate if and only if A and B are intercalate. In writing out a tensor product we re-write the colors as integers from 0 to $n-1$. For example starting from $D_1 = \begin{pmatrix} 0 & 1 \\ 1 & 0 \end{pmatrix}$ we obtain

$$D_2 = D_1 \otimes D_1 = \begin{pmatrix} 00 & 01 & 10 & 11 \\ 01 & 00 & 11 & 10 \\ 10 & 11 & 00 & 01 \\ 11 & 10 & 01 & 00 \end{pmatrix} = \begin{pmatrix} 0 & 1 & 2 & 3 \\ 1 & 0 & 3 & 2 \\ 2 & 3 & 0 & 1 \\ 3 & 2 & 1 & 0 \end{pmatrix}.$$

(The translation from bit-strings to integers uses the standard base 2, or dyadic, notation.) This tensoring process can be repeated to obtain intercalate matrices D_t of type $(2^t, 2^t, 2^t)$. These matrices D_t may also be defined inductively, without explicit mention of tensor products, as follows:

$$D_0 = (0) \quad \text{and} \quad D_{t+1} = \begin{pmatrix} D_t & 2^t + D_t \\ 2^t + D_t & D_t \end{pmatrix}.$$

Here D_t is a matrix of integers and $2^t + D_t$ is obtained by adding 2^t to each entry of D_t. Another step of this process yields the 8×8 matrix D_3:

$$\left(\begin{array}{cccc|cccc} 0 & 1 & 2 & 3 & 4 & 5 & 6 & 7 \\ 1 & 0 & 3 & 2 & 5 & 4 & 7 & 6 \\ 2 & 3 & 0 & 1 & 6 & 7 & 4 & 5 \\ 3 & 2 & 1 & 0 & 7 & 6 & 5 & 4 \\ \hline 4 & 5 & 6 & 7 & 0 & 1 & 2 & 3 \\ 5 & 4 & 7 & 6 & 1 & 0 & 3 & 2 \\ 6 & 7 & 4 & 5 & 2 & 3 & 0 & 1 \\ 7 & 6 & 5 & 4 & 3 & 2 & 1 & 0 \end{array}\right)$$

It is not hard to check from this definition that every D_t is intercalate. However, D_t cannot be consistently signed when $t > 3$, by the original 1, 2, 4, 8 Theorem.

This matrix D_t can also be viewed as the table of a binary operation on the interval $[0, 2^t) = \{0, 1, 2, \ldots, 2^t - 1\}$. If $m, n \in [0, 2^t)$ define

$$m \boxplus n = \text{the } (m, n)\text{-entry of } D_t,$$

where the rows and columns of D_t are indexed by the values $0, 1, 2, \ldots, 2^t - 1$. This operation is the well-known "Nim-addition" studied in the analysis of the game of Nim. (For further information on Nim and related games see books on recreational mathematics. A good example is Berlekamp, Conway and Guy (1982).) The Nim-sum $m \boxplus n$ is easily described using the dyadic expansions of m, n: express m, n

as bit-strings of length t, add them as t-tuples in the group $(\mathbb{Z}/2\mathbb{Z})^t$, transform the resulting bit-string back to an integer. For example $3 = (011)$ and $6 = (110)$ in dyadic expansion and $3 \boxplus 6 = (101) = 5$. Certainly the Nim sum makes the non-negative integers into a group, such that $n \boxplus n = 0$ for every n.

Therefore the matrix D_t is just the addition table for the group $(\mathbb{Z}/2\mathbb{Z})^t$, re-written with the labels $0, 1, 2, \ldots, 2^t - 1$ in place of bit-strings. With this interpretation the intercalacy condition is obvious:

$$i \boxplus j = i' \boxplus j' \quad \text{implies} \quad i \boxplus j' = i' \boxplus j.$$

Certainly every submatrix of D_t is intercalate. Define an intercalate matrix to be *dyadic* if it is equivalent to a submatrix of some D_t. The standard dyadic $r \times s$ intercalate matrix is $D_{r,s}$, defined to be the upper left $r \times s$ corner of D_t (where t is chosen so that $r, s \leq 2^t$). For instance the 3×5 matrix mentioned after (13.4) is exactly the matrix $D_{3,5}$ with each entry increased by 1. That matrix $D_{3,5}$ involves 7 of the 8 colors of D_3. How many colors are involved in $D_{r,s}$? Surprisingly the answer is provided by the Stiefel–Hopf function defined in (12.5).

13.5 Lemma. *$D_{r,s}$ involves exactly $r \circ s$ colors.*

Proof. Let $r \bullet s = n(D_{r,s})$, the number of colors in $D_{r,s}$. Certainly $r \bullet s = s \bullet r$; $1 \bullet s = s$; $2^m \bullet 2^m = 2^m$; and if $r \leq r'$ then $r \bullet s \leq r' \bullet s$. Using the inductive definition of D_{t+1} check that $2^m \bullet (2^m + 1) = 2^{m+1}$ and that if $r, s \leq 2^m$ then $r \bullet (s + 2^m) = (r \bullet s) + 2^m$. These properties suffice to determine all values $r \bullet s$, and these match the values $r \circ s$ by (12.10). Another proof is mentioned in Exercise 4. □

This property of $D_{r,s}$ was first noted by Yuzvinsky (1981). He conjectured that every $r \times s$ intercalate matrix contains at least $r \circ s$ colors, and he proved this conjecture for dyadic matrices (that is for submatrices of some D_t). An elegant new proof of this result has been recently discovered by Eliahou and Kervaire, using polynomial methods popularized by Alon and Tarsi. See Appendix A below. Yuzvinsky's conjecture remains open for non-dyadic intercalate matrices, although Yiu has proved the conjecture whenever $r, s \leq 16$.

The classical n-square identities arise from the Cayley–Dickson doubling process, as described in the appendix to Chapter 1. Using a standard basis of the Cayley–Dickson algebra A_t, the multiplication table turns out to be a signed version of the matrix D_t. The signs are not hard to work out (Exercise 5) using the inductive definition of "doubling". For later reference we display here the signing of D_4 which arises from the Cayley–Dickson algebra A_4.

$$\left(\begin{array}{cccccccc|cccccccc}
+0 & +1 & +2 & +3 & +4 & +5 & +6 & +7 & +8 & +9 & +10 & +11 & +12 & +13 & +14 & +15 \\
+1 & -0 & +3 & -2 & +5 & -4 & -7 & +6 & +9 & -8 & -11 & +10 & -13 & +12 & +15 & -14 \\
+2 & -3 & -0 & +1 & +6 & +7 & -4 & -5 & +10 & +11 & -8 & -9 & -14 & -15 & +12 & +13 \\
+3 & +2 & -1 & -0 & +7 & -6 & +5 & -4 & +11 & -10 & +9 & -8 & -15 & +14 & -13 & +12 \\
+4 & -5 & -6 & -7 & -0 & +1 & +2 & +3 & +12 & +13 & +14 & +15 & -8 & -9 & -10 & -11 \\
+5 & +4 & -7 & +6 & -1 & -0 & -3 & +2 & +13 & -12 & +15 & -14 & +9 & -8 & +11 & -10 \\
+6 & +7 & +4 & -5 & -2 & +3 & -0 & -1 & +14 & -15 & -12 & +13 & +10 & -11 & -8 & +9 \\
+7 & -6 & +5 & +4 & -3\bullet & -2 & +1 & -0 & +15 & +14 & -13\bullet & -12 & +11 & +10 & -9 & -8 \\
\hline
+8 & -9 & -10 & -11 & -12 & -13 & -14 & -15 & -0 & +1 & +2 & +3 & +4 & +5 & +6 & +7 \\
+9 & +8 & -11 & +10 & -13\bullet & +12 & +15 & -14 & -1 & -0 & -3\bullet & +2 & -5 & +4 & +7 & -6 \\
+10 & +11 & +8 & -9 & -14 & -15 & +12 & +13 & -2 & +3 & -0 & -1 & -6 & -7 & +4 & +5 \\
+11 & -10 & +9 & +8 & -15 & +14 & -13 & +12 & -3 & -2 & +1 & -0 & -7 & +6 & -5 & +4 \\
+12 & +13 & +14 & +15 & +8 & -9 & -10 & -11 & -4 & +5 & +6 & +7 & -0 & -1 & -2 & -3 \\
+13 & -12 & +15 & -14 & +9 & +8 & +11 & -10 & -5 & -4 & +7 & -6 & +1 & -0 & +3 & -2 \\
+14 & -15 & -12 & +13 & +10 & -11 & +8 & +9 & -6 & -7 & -4 & +5 & +2 & -3 & -0 & +1 \\
+15 & +14 & -13 & -12 & +11 & +10 & -9 & +8 & -7 & +6 & -5 & -4 & +3 & +2 & -1 & -0
\end{array}\right)$$

Observe that this signing is not consistent: for example the signs of colors 3 and 13 in rows 7, 9 and columns 4, 10 do not satisfy the condition for consistent signs. Those entries are marked with bullets "•". However there are some interesting submatrices which are consistently signed. We will analyze $D_{9,16}$ and $D_{10,10}$.

One can verify directly that the signings of these submatrices are consistent. For a more conceptual method, recall that the upper left 8×8 block D_3 is consistently signed since it arises from the standard 8-square identity. Now examine the larger 9×16 block. This provides an example of the following "doubling construction".

13.6 Proposition. *Any consistently signed intercalate matrix of type* (r, s, n) *can be enlarged to one of type* $(r+1, 2s, 2n)$.

Proof. Let A be the given intercalate matrix with sign matrix S. We may assume that the top row of A is $v = (0, 1, 2, \dots, s-1)$ and the top row of S is all " $+$" signs. Let A' be the intercalate matrix obtained from A by replacing every color c by a new color c'. Then the top row of A' is $v' = (0', 1', 2', \dots)$. Define $M = \begin{pmatrix} A & A' \\ v' & v \end{pmatrix}$. Since M is a submatrix of the tensor product $A \otimes D_1$, it is intercalate of type $(r+1, 2s, 2n)$. It remains to show that M can be consistently signed.

Use the given signs $S = (\varepsilon_{ij})$ for the submatrix A, " $+$" signs on the top row of A' and arbitrary signs $(\alpha_0, \alpha_1, \alpha_2, \dots)$ for the v' in the bottom row of M.

Claim. There is a unique way to attach signs to v and to the rest of A' to produce a consistent signing of M.

The sign condition for the top and bottom rows and for columns j and $s+j$ forces the signs attached to the row v to be $(-\alpha_0, -\alpha_1, -\alpha_2, \ldots)$. For given i, j with $0 < i \le r$, we will determine the sign ε'_{ij} attached to the entry $A'(i, j)$. Let $A'(i, j) = k'$ so that $A(i, j) = k$. The intercalacy for the rows 0 and i and for columns j and k shows that $A(i, k) = j$. The sign condition for this rectangle implies that $\varepsilon_{ij}\varepsilon_{ik} = -1$, as well. The following picture of the matrix M may help to clarify this argument.

$$
\begin{array}{r}
\\ \\ i \\ \\ r+1
\end{array}
\begin{array}{c}
\begin{array}{ccccccc|ccccc}
 & & & j & & k & & & & & s+j & \\
\end{array}\\
\left(\begin{array}{ccccccc|ccccc}
+0 & +1 & \ldots & +j & \ldots & +k & \ldots & +0' & +1' & \ldots & +j' & \cdots \\
 & & & & & & & & & & & \\
\ldots & \ldots & \ldots & \varepsilon_{ij}k & \ldots & \varepsilon_{ik}j & \ldots & \ldots & \ldots & \ldots & \varepsilon'_{ij}k' & \ldots \\
\hline
\alpha_0 0' & \alpha_1 1' & \ldots & \alpha_j j' & \ldots & \alpha_k k' & \ldots & -\alpha_0 0 & -\alpha_1 1 & \ldots & -\alpha_j j & \ldots
\end{array}\right)
\end{array}
$$

Now examine the rectangle with opposite corners $M(i, k) = j$ and $M(r+1, s+j) = j$ to see that $\varepsilon_{ik}\varepsilon'_{ij}\alpha_k(-\alpha_j) = -1$. Since $\varepsilon_{ik} = -\varepsilon_{ij}$ we conclude:

$$\varepsilon'_{ij} = -\alpha_k\alpha_j\varepsilon_{ij} \quad \text{where } k = A(i, j).$$

We must verify that this signing is consistent. By construction all the sign conditions involving the bottom row of M are consistent. Since A and A' have no colors in common, it remains only to check the submatrix A'. The signs ε'_{ij} of A' are obtained from the signs S as follows: multiply the jth column by the sign $-\alpha_j$ and multiply every occurrence of the color k by the sign α_k. Therefore the signing of A' is equivalent to the consistent signing of A. □

Now let us return to the multiplication table for A_4 displayed earlier. It is not hard to verify that the first 9 rows are obtained by this doubling construction applied to the standard consistent signing of D_3. Therefore that 9×16 block is consistently signed and we have a sums of squares formula of size [9, 16, 16]. (Of course we already constructed such a formula in the proof of the Hurwitz–Radon Theorem.) Another application of the doubling process, this time with the roles of r and s reversed, yields a formula of size [18, 17, 32], improving on the earlier [16, 16, 32].

Repeated application of the doubling process starting from [8, 8, 8] produces formulas of sizes $[t+5, 2^t, 2^t]_{\mathbb{Z}}$. In fact the corresponding signed intercalate matrix can be found inside the multiplication table of A_t by choosing the columns $0, 1, 2, \ldots, 7$ and 2^k for $k = 3, \ldots, t-1$. On the other hand, Khalil (1993) proved no subset of $t+6$ columns of the multiplication table of A_t is consistently signed. In particular it is not possible to find a [12, 64, 64] formula inside A_6. We can also use that matrix $[9, 16, 16]_{\mathbb{Z}}$ to give another proof of (12.13):

13.7 Corollary. *If $r \leq 9$ then $r *_{\mathbb{Z}} s = r \circ s$.*

Proof. We know generally that $r \circ s \leq r * s \leq r *_{\mathbb{Z}} s$ from the results of Stiefel and Hopf discussed in Chapter 12. Equality holds if there exists an $[r, s, r \circ s]_{\mathbb{Z}}$ formula. Suppose $t \geq 4$ and consider $D_{9,2^t}$. This matrix can be consistently signed by viewing it as the direct sum of 2^{t-4} copies of $D_{9,16}$. Then if $r \leq 9$ and $s \leq 2^t$ the submatrix $D_{r,s}$ is consistently signed and involves exactly $r \circ s$ colors by (13.5). □

The matrices D_t are examples of intercalate matrices of type (n, n, n). Are there any other examples? Consider more generally a square intercalate matrix M of type (r, r, n). A color is called *ubiquitous* in the $r \times r$ matrix M if it appears in every row and every column. If M has a ubiquitous color then M is equivalent to a symmetric matrix, with the ubiquitous color along the diagonal. (This follows from the intercalacy condition.)

13.8 Lemma. *Suppose the intercalate matrix M of type (r, r, n) has two ubiquitous colors. Then r and n are even and M is equivalent to a tensor product $D_1 \otimes M'$.*

Proof. Here $n = n(M)$ and we may assume M is symmetric with one color along the diagonal. Permute the rows and columns to arrange the second ubiquitous color along the principal 2×2 blocks. From this it follows that r is even. Partition M into 2×2 blocks and use the intercalacy condition with the diagonal blocks to deduce that each block is of the form $\begin{pmatrix} a & b \\ b & a \end{pmatrix}$. Then n must be even and the tensor decomposition follows. □

One can now check (as in Exercise 3) that the D_t's are the only intercalate matrices of type (n, n, n). Most of our examples are signings of various submatrices of D_t. However there exist intercalate matrices which are not equivalent to a submatrix of any D_t. (See Exercise 1.)

Suppose M is a symmetric intercalate matrix of type (r, r, n), so that the diagonal of M contains a single (ubiquitous) color. We can enlarge M to a matrix ΣM which is symmetric intercalate of type $(r + 1, r + 1, r + n)$ by appending a new row and column to the bottom and right of M using r new colors (symmetrically) for that row and column, and assigning the diagonal color of M to the lower right corner. For example starting with $L_1 = (0)$ of type $(1, 1, 1)$ we obtain inductively $L_{r+1} = \Sigma L_r$, a symmetric intercalate matrix of type $(r, r, 1 + \binom{r}{2})$. We may choose the colors successively from $\{0, 1, 2, 3, \ldots\}$ to obtain

$$\begin{pmatrix} 0 & 1 & 2 & 4 & 7 & \vdots & \vdots \\ 1 & 0 & 3 & 5 & 8 & \vdots & \vdots \\ 2 & 3 & 0 & 6 & 9 & \vdots & \vdots \\ 4 & 5 & 6 & 0 & 10 & \vdots & \vdots \\ 7 & 8 & 9 & 10 & 0 & \vdots & \vdots \\ \cdots & \cdots & \cdots & \cdots & \cdots & 0 & \vdots \\ \cdots & \cdots & \cdots & \cdots & \cdots & \cdots & 0 \end{pmatrix}$$

Each of these matrices L_r can be consistently signed: endow each color in the upper triangle, including the diagonal, of L_r with " $+$" and each color in the lower triangle with " $-$". The corresponding sums of squares identity is the Lagrange identity:

$$(x_1^2+x_2^2+\cdots+x_r^2)\cdot(y_1^2+y_2^2+\cdots+y_r^2) = (x_1y_1+\cdots+x_ry_r)^2+\sum_{i<j}(x_iy_j-x_jy_i)^2,$$

of type $[r, r, 1+\binom{r}{2}]_{\mathbb{Z}}$. This identity provides one proof of the Cauchy–Schwartz inequality.

Now let us re-examine the 10×10 submatrix of the Cayley–Dickson signing of D_4. That matrix decomposes into 2×2 blocks corresponding to the two ubiquitous colors 0, 1. The basic 8×8 matrix is expanded to the 10×10 using an analog of the Σ construction as follows.

13.9 Lemma. *Suppose M is a consistently signed intercalate matrix of type (r, r, n) with two ubiquitous colors. Then M can be expanded to a consistently signed intercalate matrix M' of type $(r+2, r+2, r+n)$. The same two colors are ubiquitous in M'.*

Proof. Replacing M by an equivalent matrix we may assume that M is decomposed into 2×2 blocks with first diagonal block $\begin{pmatrix} +0 & +1 \\ +1 & -0 \end{pmatrix}$ and subsequent diagonal blocks $\begin{pmatrix} -0 & +1 \\ -1 & -0 \end{pmatrix}$. Construct the matrix M' by appending a new row and column of 2×2 blocks to M. The first $r/2$ blocks in the new column are of the form $\begin{pmatrix} +a & +b \\ +b & -a \end{pmatrix}$, involving r new colors, and the lower right corner block is assigned the diagonal value $\begin{pmatrix} -0 & +1 \\ -1 & -0 \end{pmatrix}$. The entries along the bottom row are determined by the intercalacy and sign conditions, and involve the same r new colors. This matrix M' is intercalate of type $(r+2, r+2, r+n)$ and is consistently signed. □

Applying this construction to the standard (8, 8, 8), we get a consistently signed intercalate matrix of type (10, 10, 16). This matrix appears as the upper left 10×10 submatrix of the signed D_4 displayed earlier. Repeating this construction we obtain a consistently signed intercalate matrix of type (12, 12, 26). Consequently there are integer composition formulas of types [10, 10, 16] and [12, 12, 26]. Another repetition does not yield an interesting result since we already know a formula of type [16, 16, 32].

We saw in Chapters 1 and 2 that for any n there exists a composition formula of size $[\rho(n), n, n]$. In fact we gave an explicit construction for such formulas: First build an $(m+1, m+1)$-family, either by the Construction Lemma (2.7) or by using the trace form on a Clifford algebra (Exercise 3.15). Then apply the Shift Lemma (2.6) and Expansion Lemma (2.5). If the underlying quadratic form is the sum of squares then all entries of the matrices are in $\mathbb{Z}$ and there must be a corresponding signed intercalate matrix. Can it be constructed directly using the combinatorial methods here?

There are two constructions in the literature for explicit signed intercalate matrices which realize the Hurwitz–Radon formulas. There are given in Yiu (1985), and in Yuzvinsky (1984) as corrected by Lam and Smith (1993). Both of these constructions are obtained by consistently signing a suitably chosen $\rho(t) \times 2^t$ submatrix of D_t. These two constructions do not yield equivalent formulas, even though we proved in Chapter 7 that any two formulas of size $[\rho(n), n, n]$ are equivalent, over any field F. The point here is that the notion of equivalence of composition formulas over $\mathbb{Z}$ (i.e. of signed intercalate matrices) is much more restrictive than equivalence over a field.

We will outline (without proofs) some of the underlying ideas involved in the Yuzvinsky–Lam–Smith construction, since that method leads to infinite sequences of new composition formulas. Recall from the discussion before (13.3) that an $[r, s, n]_{\mathbb{Z}}$ formula is determined by two mappings φ and ε where $\varphi : A \times B \to C$ and $\varepsilon : A \times B \to \{1, -1\}$. Here A, B, C are sets of cardinalities r, s, n, respectively. With this notation the three conditions in (13.3) become:

(i) If $a \in A$ the map $\varphi|_{a \times B}$ is injective. If $b \in B$ the map $\varphi|_{A \times b}$ is injective.

(ii) If $a_i \in A$ and $b_i \in B$ and $\varphi(a_1, b_1) = \varphi(a_2, b_2)$ then $\varphi(a_1, b_2) = \varphi(a_2, b_1)$.

(iii) If $a_1 \neq a_2$ and $\varphi(a_1, b_1) = \varphi(a_2, b_2)$ then

$$\varepsilon(a_1, b_1) \cdot \varepsilon(a_1, b_2) \cdot \varepsilon(a_2, b_1) \cdot \varepsilon(a_2, b_2) = -1.$$

To construct maps φ and ε satisfying these conditions consider a normal subgroup H of some finite group G. Left multiplications induces a permutation action $G \times G/H \to G/H$. Choose subsets $A \subseteq G$ and $B \subseteq G/H$, use the map $\varphi : A \times B \to G/H$ given by restriction, and try to find a signing map $\varepsilon : A \times B \to \{\pm 1\}$ so that the three conditions are satisfied. To define ε choose a homomorphism $\chi : H \to \{\pm 1\}$, and a set $\{d_1, \dots, d_n\}$ of coset representatives of H in G. For any d_i and any $g \in G$ then $gd_iH = d_jH$ for some d_j. Define ε by setting $\varepsilon(g, d_iH) = \chi(d_j^{-1} g d_i)$. If φ and ε are constructed this way the three conditions above become the following:

(i′) $g^{-1}g' \notin H$ whenever $g, g' \in A$ and $g \neq g'$.

Suppose $g \neq g'$ in A and there exist $d_i, d_j \in B$ such that $gd_iH = g'd_jH$.
(ii′) $(g^{-1}g')^2 \in H$.
(iii′) $\chi(d_i^{-1}(g^{-1}g')^2 d_i) = -1$.

To apply this criterion we use the group G_r defined as follows by generators and relations:

$$G_r = \langle \epsilon, g_1, \ldots, g_r \mid \epsilon^2 = 1,\ g_i^2 = \epsilon,\ g_i g_j = \epsilon g_j g_i \rangle.$$

This is the group employed by Eckmann (1943b) in his proof of the Hurwitz–Radon Theorem using group representations. In fact this approach was motivated directly by Eckmann's work. If V is an F-vector space and $\pi : G_r \to \mathrm{GL}(V)$ is a group homomorphism with $\pi(\epsilon) = -1$, then the elements $f_i = \pi(g_i)$ generate a Clifford algebra (they anticommute and have squares equal to -1). Now suppose $G = G_r$, H is a normal subgroup containing ϵ, and $\chi : H \to \{\pm 1\}$ is a homomorphism with $\chi(\epsilon) = -1$. Then the three conditions above boil down to one requirement:

$$(g^{-1}g')^2 = \epsilon \text{ whenever } g, g' \in A \text{ and } g \neq g'.$$

For example, these conditions hold if H is a maximal elementary abelian 2-subgroup of G_r, $A = \{1, g_1, g_2, \ldots, g_r\}$ and $B = G/H$. If $|G/H| = 2^m$ this provides an $[r+1, 2^m, 2^m]_{\mathbb{Z}}$ formula. It turns out that this value 2^m is exactly the value needed for a formula of Hurwitz–Radon type; that is, $\rho(2^m) = r + 1$.

Yuzvinsky's idea is to construct new examples by modifying the pairings derived in this way. He found a way to enlarge the set A while decreasing the set B, keeping C the same. He obtained various formulas of size $(2m+2, 2^m - p(m), 2^m)$ where $p(m)$ represents the number of elements in B which must be excluded to accommodate the increase of 1 or 2 elements in A. There are a number of errors and gaps in Yuzvinsky's paper but these have been carefully corrected and clarified in the work of Lam and Smith (1993). Here are the two families of formulas which follow from these methods.

13.10 Proposition. *Suppose* $m > 1$. *Then there exists a* $[2m+2, 2^m - p(m), 2^m]_{\mathbb{Z}}$ *formula in the following two cases:*
(1) $m \equiv 0 \pmod 4$ *and* $p(m) = \binom{m}{m/2}$.
(2) $m \equiv 1 \pmod 4$ *and* $p(m) = 2\binom{m-1}{(m-1)/2}$.

We omit further details. Applying this calculation when $m = 4, 5$ provides $[10, 10, 16]_{\mathbb{Z}}$ and $[12, 20, 32]_{\mathbb{Z}}$ formulas. This last example is important for us since it can be modified to yield some of the values appearing in Theorem 13.1.

13.11 Corollary. *There exist formulas of sizes* $[10, 16, 28]$ *and* $[12, 14, 30]$.

Outline. These formulas arise as restrictions of the explicit $[12, 20, 32]$ constructed by the group-theoretic method above. Signed intercalate matrices of these sizes are displayed in the Appendix to Lam and Smith (1993). These formulas are also mentioned in Smith and Yiu (1992). □

There are several formulas still to construct in order to realize all the values of $r *_{\mathbb{Z}} s$ listed in Theorem 13.1. As in Smith and Yiu (1994) we derive these formulas by explicitly displaying various signed intercalate matrices. The consistently signed intercalate matrix given below, of type (17, 17, 32), is obtained as follows: from the $[18, 17, 32]_{\mathbb{Z}}$ constructed by the doubling process (13.6), delete the bottom row and move the rightmost column to the middle of the matrix. For $12 \le r \le s \le 16$ the $r \times s$ submatrix in the upper left corner contains exactly $24 + (r - 9) \circ (s - 9)$ colors. Therefore this matrix furnishes formulas for all the entries of the table in Theorem 13.1 for the cases $12 \le r \le s$, except for the cases $(r, s) = (12, 12)$ and (12, 14). Since those two sizes were constructed earlier, only the cases $r = 10$ and 11 remain to be verified. In this display we follow the convention of Yiu and use the colors $\{1, 2, \ldots, 32\}$ (rather than $\{0, 1, \ldots, 31\}$).

$$
\left(\begin{array}{cccccccc|c|cccccccc}
+1 & +2 & +3 & +4 & +5 & +6 & +7 & +8 & +17 & +9 & +10 & +11 & +12 & +13 & +14 & +15 & +16 \\
+2 & -1 & +4 & -3 & +6 & -5 & -8 & +7 & +18 & -10 & +9 & -12 & +11 & -14 & +13 & +16 & -15 \\
+3 & -4 & -1 & +2 & +7 & +8 & -5 & -6 & +19 & -11 & +12 & +9 & -10 & -15 & -16 & +13 & +14 \\
+4 & +3 & -2 & -1 & +8 & -7 & +6 & -5 & +20 & -12 & -11 & +10 & +9 & -16 & +15 & -14 & +13 \\
+5 & -6 & -7 & -8 & -1 & -2 & +3 & +4 & +21 & -13 & +14 & +15 & +16 & +9 & -10 & -11 & -12 \\
+6 & +5 & -8 & +7 & -2 & -1 & -4 & +3 & +22 & -14 & -13 & +16 & -15 & +10 & +9 & +12 & -11 \\
+7 & +8 & +5 & -6 & -3 & +4 & -1 & -2 & +23 & -15 & -16 & +13 & +14 & +11 & -12 & +9 & +10 \\
+8 & -7 & +6 & +5 & -4 & -3 & +2 & -1 & +24 & -16 & +15 & -14 & -13 & +12 & +11 & -10 & +9 \\
\hline
+9 & +10 & +11 & +12 & +13 & +14 & +15 & +16 & +25 & -1 & -2 & -3 & -4 & -5 & -6 & -7 & -8 \\
\hline
+17 & +18 & -19 & -20 & -21 & -22 & -23 & -24 & -1 & -25 & -26 & -27 & -28 & -29 & -30 & -31 & -32 \\
+18 & +17 & -20 & +19 & -22 & -21 & +24 & -23 & -2 & +26 & -25 & +28 & -27 & +30 & -29 & -32 & +31 \\
+19 & +20 & +17 & -18 & -23 & -24 & +21 & +22 & -3 & +27 & -28 & -25 & +26 & +31 & +32 & -29 & -30 \\
+20 & -19 & +18 & +17 & -24 & +23 & -22 & +21 & -4 & +28 & +27 & -26 & -25 & +32 & -31 & +30 & -29 \\
+21 & +22 & +23 & +24 & +17 & -18 & -19 & -20 & -5 & +29 & -30 & -31 & -32 & -25 & +26 & +27 & +28 \\
+22 & -21 & +24 & -23 & +18 & +17 & +20 & -19 & -6 & +30 & +29 & -32 & +31 & -26 & -25 & -28 & +27 \\
+23 & -24 & -21 & +22 & +19 & -20 & +17 & +18 & -7 & +31 & +32 & +29 & -30 & -27 & +28 & -25 & -26 \\
+24 & +23 & -22 & -21 & +20 & +19 & -18 & +17 & -8 & +32 & -31 & +30 & +29 & -28 & -27 & +26 & -25
\end{array}\right)
$$

Finally we present below a consistently signed matrix of type (11, 18, 32). It contains a submatrix of type (9, 16, 16) by using the first 9 rows and deleting columns 9, 10. The signing of this (9, 16, 16) matches the first 9 rows of the Cayley–Dickson signing of D_4 listed earlier (renumbering the colors by adding 1). The matrix below also contains a (10, 10, 16) by using the first 10 columns and deleting row 9. Given these two consistently signed parts it is not hard to sign the remaining colors 25, 26, ..., 32 consistently. Now if $11 \le s \le 16$, the first s columns contain exactly $24 + 2 \circ (s - 10)$ colors. This verifies the entries for $11 *_{\mathbb{Z}} s$ in Theorem 3.1.

The verification of the existence of formulas listed in (13.1) is now complete, except for the case (10, 14, 27). We will skip that case, referring the reader to Smith and Yiu (1992).

$$\left(\begin{array}{cccccccc|cc|cccccccc}
+1 & +2 & +3 & +4 & +5 & +6 & +7 & +8 & +17 & +18 & +9 & +10 & +11 & +12 & +13 & +14 & +15 & +16 \\
+2 & -1 & +4 & -3 & +6 & -5 & -8 & +7 & +18 & -17 & +10 & -9 & -12 & +11 & -14 & +13 & +16 & -15 \\
+3 & -4 & -1 & +2 & +7 & +8 & -5 & -6 & +19 & +20 & +11 & +12 & -9 & -10 & -15 & -16 & +13 & +14 \\
+4 & +3 & -2 & -1 & +8 & -7 & +6 & -5 & +20 & -19 & +12 & -11 & +10 & -9 & -16 & +15 & -14 & +13 \\
+5 & -6 & -7 & -8 & -1 & +2 & +3 & +4 & +21 & +22 & +13 & +14 & +15 & +16 & -9 & -10 & -11 & -12 \\
+6 & +5 & -8 & +7 & -2 & -1 & -4 & +3 & +22 & -21 & +14 & -13 & +16 & -15 & +10 & -9 & +12 & -11 \\
+7 & +8 & +5 & -6 & -3 & +4 & -1 & -2 & +23 & -24 & +15 & -16 & -13 & +14 & +11 & -12 & -9 & +10 \\
+8 & -7 & +6 & +5 & -4 & -3 & +2 & -1 & +24 & +23 & +16 & +15 & -14 & -13 & +12 & +11 & -10 & -9 \\
\cdot & \cdot & \cdot & \cdot & \cdot & \cdot & \cdot & \cdot & \cdot & \cdot & \cdot & \cdot & \cdot & \cdot & \cdot & \cdot & \cdot & \cdot \\
+9 & -10 & -11 & -12 & -13 & -14 & -15 & -16 & -25 & -26 & -1 & +2 & +3 & +4 & +5 & +6 & +7 & +8 \\
\cdot & \cdot & \cdot & \cdot & \cdot & \cdot & \cdot & \cdot & \cdot & \cdot & \cdot & \cdot & \cdot & \cdot & \cdot & \cdot & \cdot & \cdot \\
+17 & -18 & -19 & -20 & -21 & -22 & -23 & -24 & -1 & +2 & +25 & +26 & +27 & +28 & +29 & +30 & +31 & +32 \\
+18 & +17 & -20 & +19 & -22 & +21 & +24 & -23 & -2 & -1 & +26 & -25 & +28 & -27 & +30 & -29 & +32 & -31
\end{array}\right)$$

There is one more construction technique of interest for larger matrices. The idea, due to Romero, is to glue together several smaller matrices. An $r \times s$ matrix can be partitioned into five smaller matrices in the following pattern.

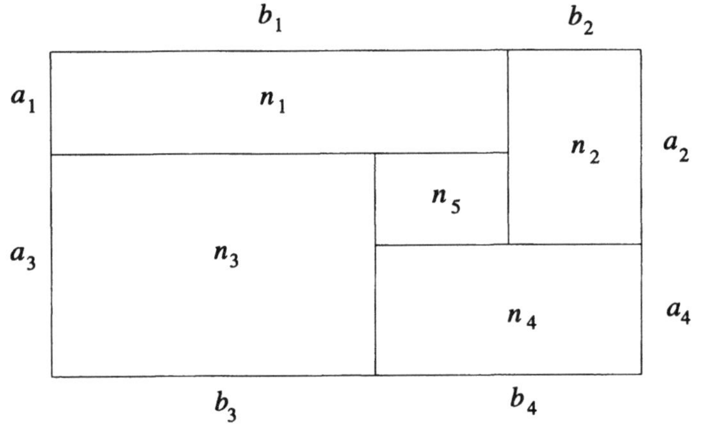

Here we have $r = a_1 + a_3 = a_2 + a_4$ and $s = b_1 + b_2 = b_3 + b_4$, etc. If each subrectangle represents a consistently signed intercalate matrix with dimensions and numbers of colors as indicated, no two of them sharing common colors, then this construction shows that

$$r *_{\mathbb{Z}} s \leq n_1 + n_2 + n_3 + n_4 + n_5.$$

For example using two copies of a $[9, 13, 16]_{\mathbb{Z}}$, two copies of a $[13, 9, 16]$, and one $[4, 4, 4]_{\mathbb{Z}}$ then this construction produces a $[22, 22, 68]_{\mathbb{Z}}$. Therefore $22 *_{\mathbb{Z}} 22 \leq 68$. Using $[9, 16, 16]$'s on the outside yields similarly that $25 *_{\mathbb{Z}} 25 \leq 72$. For further information and extensions of this idea see Romero (1995), Yiu (1996), and Sánches-Flores (1996).

Of course it is far more difficult to prove that the values given in Theorem 13.1 are best possible. Yiu's 1990 paper is devoted to a detailed analysis of small intercalate matrices, culminating in a proof that a $[16, 16, 31]_{\mathbb{Z}}$ formula is impossible. The full result is proved in Yiu (1994a) by modifying and considerably expanding his earlier ideas. The arguments are too intricate to present here, even in outline form. However we will mention one of the simplest tricks that lead toward Yiu's non-existence results.

If M is an intercalate matrix of type (r, s, n), define a *partial signing* of M to be an $r \times s$ matrix S some of whose entries might be undefined, but such that each defined entry is either $+1$ or -1. Each entry of S is viewed as a sign or a blank attached to the corresponding entry of M. A partial signing is *complete* if every entry is defined.

There is a straightforward algorithm to check whether M admits a consistent signing. (Actually it produces all possible consistent signings of M.) First write in " +" signs along the first row and first column. Then attach a " +" to one occurrence of any color which does not appear in the first row or column. Now use the consistency condition to deduce all possible consequences of this partial signing S. More precisely, suppose $M(i, j)$ is an unsigned entry. If it is possible to find indices i', j' such that

$$M(i, j) = M(i', j') \text{ and } S(i', j), S(i, j'), S(i', j') \text{ are all defined,}$$

then endow $M(i, j)$ with the sign $S(i, j) = -S(i', j) \cdot S(i, j') \cdot S(i', j')$. Repeat this procedure as long as possible to obtain a maximal signing matrix S_0. There may be an inconsistency of the signs at this point (a submatrix of type $(2, 2, 2)$ violating the sign condition). If that does not occur then S_0 is consistent. If S_0 is also complete we are done. Otherwise choose an unsigned entry of M, give it an indeterminate sign ε, and repeat the process of deducing all possible consequences. Eventually we will get either an inconsistency or a complete consistent signing. Here is one application of this algorithm.

13.12 Lemma. *The following intercalate matrix M' of type $(7, 7, 15)$ cannot be consistently signed.*

$$\left(\begin{array}{cccc|c|c|c}
1 & 2 & 3 & 4 & 5 & 9 & 13 \\
2 & 1 & 4 & 3 & 6 & 10 & 14 \\
3 & 4 & 1 & 2 & 7 & 11 & 15 \\
\hline
5 & 6 & 7 & 8 & 1 & 13 & 9 \\
6 & 5 & 8 & 7 & 2 & 14 & 10 \\
\hline
9 & 10 & 11 & 12 & 13 & 1 & 5 \\
11 & 12 & 9 & 10 & 15 & 3 & 7
\end{array}\right)$$

Proof. We begin with " +" signs along the first row and column and a " +" for one occurrence of each of the colors 7, 8, 10, 12, 14, 15. Deriving all the consequences we obtain the following partially signed matrix:

$$\begin{pmatrix} +1 & +2 & +3 & +4 & +5 & +9 & +13 \\ +2 & -1 & 4 & 3 & 6 & +10 & +14 \\ +3 & 4 & -1 & 2 & +7 & 11 & +15 \\ +5 & 6 & -7 & +8 & -1 & 13 & 9 \\ +6 & 5 & 8 & 7 & 2 & 14 & 10 \\ +9 & -10 & 11 & +12 & 13 & -1 & 5 \\ +11 & 12 & 9 & 10 & 15 & 3 & 7 \end{pmatrix}$$

Following the algorithm, we next attach indeterminate signs α, β, γ to the unsigned colors 4 in $M(2, 3)$; 6 in $M(2, 5)$; and 3 in $M(7, 6)$. Deducing all the consequences yields:

$$\begin{pmatrix} +1 & +2 & +3 & +4 & +5 & +9 & +13 \\ +2 & -1 & \alpha 4 & -\alpha 3 & \beta 6 & +10 & +14 \\ +3 & -\alpha 4 & -1 & \alpha 2 & +7 & -\gamma 11 & +15 \\ +5 & -\beta 6 & -7 & +8 & -1 & 13 & 9 \\ +6 & \beta 5 & \alpha\beta 8 & \alpha\beta 7 & -\beta 2 & 14 & 10 \\ +9 & -10 & \gamma 11 & +12 & 13 & -1 & 5 \\ +11 & \alpha\gamma 12 & -\gamma 9 & \alpha\gamma 10 & 15 & \gamma 3 & 7 \end{pmatrix}$$

This partial signing is consistent, but now let us consider a sign δ attached to color 5 in $M(6, 7)$. This implies:

$-\delta$ for color 9 in $M(4, 7)$,
$\beta\delta$ for color 10 in $M(5, 7)$,
$\gamma\delta$ for color 7 in $M(7, 7)$,
$-(\alpha\beta)(\beta\delta)(\alpha\gamma) = -\gamma\delta$ also for color 7 in $M(7, 7)$, which is impossible.

This completes the proof. Compare Exercise 7(b). □

The matrix M' above is an example of an intercalate matrix partitioned into blocks in the following way:

$$M = \begin{pmatrix} A_0 & * & * & * \\ * & A_1 & * & * \\ * & * & A_2 & * \\ * & * & * & * \end{pmatrix}$$

such that no colors in A_0 appear in any of the blocks marked with $*$, and every color in A_1 and in A_2 does appear in A_0. We continue to follow Yiu's notation here, using colors $\{1, 2, 3, \ldots\}$.

13.13 Corollary. *Suppose M is an $r \times s$ intercalate matrix, with $r, s \geq 7$, which is partitioned into blocks as above such that:*

$$A_0 = D_{3,4} = \begin{pmatrix} 1 & 2 & 3 & 4 \\ 2 & 1 & 4 & 3 \\ 3 & 4 & 1 & 2 \end{pmatrix}, \quad A_1 = \begin{pmatrix} 1 \\ 2 \end{pmatrix}, \quad A_2 = \begin{pmatrix} 1 \\ 3 \end{pmatrix}.$$

If $n(M) \leq 16$ then M cannot be consistently signed.

Proof. We relate M with the matrix M' of (13.12). Since the four colors in the row directly below A_0 must involve colors not in $\{1, 2, 3, 4\}$ we may number them 5, 6, 7, 8 and use the intercalacy condition to see that the submatrix $\begin{pmatrix} A_0 & * \\ * & A_1 \end{pmatrix}$ must equal

$$\begin{pmatrix} 1 & 2 & 3 & 4 & 5 \\ 2 & 1 & 4 & 3 & 6 \\ 3 & 4 & 1 & 2 & 7 \\ 5 & 6 & 7 & 8 & 1 \\ 6 & 5 & 8 & 7 & 2 \end{pmatrix}.$$

A similar analysis with A_2 yields all of M' except the last column. Since none of the colors in that column can be in $\{1, 2, 3, 4\}$ the intercalacy implies that the top entry must be 13, 14, 15 or 16, and that each of those choices determines the rest of the entries in the column. If that top entry is 13 then $M = M'$ and (13.12) applies. In each of the other cases the proof of (13.12) can be modified to prove that there is no consistent signing. □

Yiu establishes this Corollary and similar results as the first steps toward proving the impossibility of various integer composition formulas, eventually leading to a proof of Theorem 13.1.

We end this chapter with a remark on the interesting structure of a $[10, 10, 16]_{\mathbb{Z}}$ formula.

13.14 Theorem. *Every* $[10, 10, 16]_{\mathbb{Z}}$ *formula is obtained by signing* $D_{10,10}$.

Proof outline. This was proved by Yiu (1987) using topology (namely the homotopy groups of certain Stiefel manifolds, the J-homomorphism and the technique of "hidden formulas" described in Chapter 15), as well as some combinatorial arguments. Yiu reports that this result can also be proved by replacing the topology by the elaborate combinatorics developed in his later papers. □

As mentioned earlier this formula is the smallest one not obviously obtainable as a restriction of one of the Hurwitz–Radon formulas.

13.15 Conjecture. No $[10, 10, 16]_{\mathbb{Z}}$ formula can be a restriction of an $[r, n, n]_{\mathbb{Z}}$ formula. Possibly no $[10.10, 16]$ is a restriction of any $[r, n, n]$ over $\mathbb{R}$ as well.

Yiu reports that he has a proof of the first statement, but I have not seen the details. The idea is to note that any $[10, n, n]_{\mathbb{Z}}$ is an orthogonal sum of $[10, 32, 32]_{\mathbb{Z}}$ formulas. A dimension count should show that the $[10, 10, 16]_{\mathbb{Z}}$ is embedded in some $[10, 32, 32]_{\mathbb{Z}}$, but the corresponding intercalate matrix does not contain a submatrix equivalent to $D_{10,10}$.

Appendix A to Chapter 13. A new proof of Yuzvinsky's theorem

In 1981 Yuzvinsky showed that the number of colors involved in the intercalate matrix $D_{r,s}$ is exactly the Stiefel–Hopf number $r \circ s$. He conjectured that every $r \times s$ intercalate matrix must involve at least $r \circ s$ colors. He proved this conjecture for dyadic intercalate matrices M, that is, for M which are equivalent to a submatrix of some D_t. Replacing M by an equivalent matrix, we may view it as a submatrix of the addition table of an $\mathbb{F}_2$-vector space V. The entries (colors) in M then arise as the set of values obtained by adding elements of certain subsets $A, B \subseteq V$ with $|A| = r$ and $|B| = s$. Yuzvinsky's theorem about dyadic matrices becomes the following counting result.

A.1 Yuzvinsky's Theorem. *If V is an $\mathbb{F}_2$-vector space and $A, B \subseteq V$, then $|A + B| \geq |A| \circ |B|$.*

Of course (13.5) shows that this lower bound cannot be improved. We present here the elegant new proof due to Eliahou and Kervaire (1998).

We work with the polynomial ring $F[x, y]$ over a field F. If g, h are polynomials, then (g, h) is the ideal in $F[x, y]$ generated by g and h. If $A \subseteq F$ is a finite subset, define $g_A(t)$ to be the polynomial in $F[t]$ which vanishes exactly on A. That is, $g_A(t) = \prod_{a \in A}(t - a)$.

A.2 Lemma. *Suppose $A, B \subseteq F$ are finite subsets and $f(x, y) \in F[x, y]$. Then: $f(x, y)$ vanishes on $A \times B$ if and only if $f(x, y) \in (g_A(x), g_B(y))$.*

Proof. Divide $f(x, y)$ by $g_A(x)$ and $g_B(y)$ to determine that $f(x, y) = g_A(x) \cdot u(x, y) + g_B(y) \cdot v(x, y) + h(x, y)$, where $h(x, y)$ vanishes on $A \times B$ and has x-degree $< |A|$ and y-degree $< |B|$. Then for each $a \in A$, the polynomial $h(a, y)$ is identically zero, since it has more zeros than its degree. A similar argument applied to the coefficients of $h(a, y)$ shows that $h(x, y) = 0$. □

The statement of the next lemma uses the idea of the leading form, or top term, of a polynomial. Any $f \in F[x, y]$ can be uniquely expressed $f = f_0 + f_1 + \cdots + f_d$ where f_j is a form (homogeneous polynomial) of degree j. If $f_d \neq 0$ then d is the (total) degree of f and f_d is the *top form* of f. In this case, define $\text{top}(f) = f_d$. (Also define $\text{top}(0) = 0$.) Certainly $\text{top}(g \cdot h) = \text{top}(g) \cdot \text{top}(h)$, but $\text{top}(g + h)$ does not necessarily belong to the ideal $(\text{top}(f), \text{top}(g))$. However in some special cases this property does hold.

A.3 Lemma. *Suppose $g(x), h(y) \in F[x, y]$ are polynomials in one variable, with $\deg(g) = r$ and $\deg(h) = s$. If $f \in (g(x), h(y))$ then $\text{top}(f) \in (x^r, y^s)$.*

Proof outline. Suppose $\text{top}(f) \notin (x^r, y^s)$. Then there exists some monomial $M = c \cdot x^i y^j$ occurring in $\text{top}(f)$ satisfying $i < r$ and $j < s$. Reduce f first modulo $g(x)$, and then modulo $h(y)$. If a monomial $ax^u y^v$ occurs in f and $u \geq r$ or $v \geq s$ then during this reduction that monomial is replaced by a polynomial with smaller total degree. This process cannot produce terms cancelling M, since every monomial in f has total degree $\leq i + j$. Consequently f cannot be reduced to zero by that reduction process. This means that $f \notin (g(x), h(y))$. Contradiction. □

We can now describe how to use these simple polynomial lemmas to prove the result.

Proof of Yuzvinsky's Theorem. Suppose $A, B \subseteq V$ are finite subsets with $|A| = r$ and $|B| = s$, and let $C = A + B$. We may assume V is finite, say with 2^n elements. Identifying V with the field F of 2^n elements, define $f(x, y) = \prod_{c \in C}(x + y - c) \in F[x, y]$. Then f vanishes on $A \times B$ and (A.2) implies $f(x, y) \in (g_A(x), g_B(y))$. Then (A.3) implies $\mathrm{top}(f) = (x + y)^{|C|} \in (x^r, y^s)$ in $F[x, y]$. Choosing an $\mathbb{F}_2$-basis of F and comparing coefficients, we find that this relation holds in $\mathbb{F}_2[x, y]$ as well. Then by (12.6) we conclude that $|C| \geq r \circ s$. □

The Nim sum is also closely related to the "circle function" $r \circ s$. This observation (due to Eliahou and Kervaire) provides yet another aspect of $r \circ s$. Recall that the Nim sum $a \boxplus b$ is defined as the sum in $(\mathbb{Z}/2\mathbb{Z})^t$ of the bit strings determined by the dyadic expansions of a and b. As in Exercise 12.3 let $\mathrm{Bit}(n)$ be the indices of the bits involved in n. For example $10 = 2^1 + 2^3$ so $\mathrm{Bit}(10) = \{1, 3\}$. Integers a, b are "bit-disjoint" if $\mathrm{Bit}(a) \cap \mathrm{Bit}(b) = \emptyset$.

A.4 Lemma. (i) *$a \boxplus b \leq a + b$, with equality iff a, b are bit-disjoint.*
(ii) *If $a, b < 2^m$ then $a \boxplus b < 2^m$.*
(iii) *If $a < 2^m$ then $a \boxplus (2^m + b) = 2^m + (a \boxplus b)$.*
(iv) *If $a \boxplus b = n > 0$ then $n - 1 = a' \boxplus b'$ for some $a' \leq a$ and $b' \leq b$.*

Proof. See Exercise 4. □

A.5 Proposition. $r \circ s = 1 + \max\{a \boxplus b : 0 \leq a < r \text{ and } 0 \leq b < s\}$.

Proof. Let $r \bullet s$ be the quantity on the right. Then certainly $r \bullet s = s \bullet r$ and $1 \bullet s = s$, and also: $s \leq s'$ implies $r \bullet s \leq r \bullet s'$. In particular, $\max\{r, s\} \leq r \bullet s$. By (A.4) (ii) we find: $r, s \leq 2^m$ implies $r \bullet s \leq 2^m$. Consequently, if $r \leq 2^m$ then $r \bullet 2^m = 2^m$. These observations and the following fact suffice to show that $r \bullet s$ and $r \circ s$ coincide, as hoped.

Claim. If $r \leq 2^m$ then $r \bullet (2^m + s) = 2^m + (r \bullet s)$.

Proof. We may assume $s \geq 1$. Suppose $r \bullet s = 1 + (a \boxplus b)$ for some $a < r$ and $b < s$. Then by (A.4) (iii), $2^m + (r \bullet s) = 2^m + 1 + (a \boxplus b) = 1 + a \boxplus (2^m + b) \leq r \bullet (2^m + s)$. Conversely suppose $r \bullet (2^m + s) = 1 + a \boxplus b'$ for some $a < r$ and $b' < 2^m + s$. If $b' < 2^m$ then $a \boxplus b' < 2^m$ and the inequality follows easily. Otherwise $b' \geq 2^m$

so that $b' = 2^m + b$ where $0 \le b < s$. Then $r \bullet (2^m + s) = 1 + a \boxplus (2^m + b) = 1 + 2^m + (a \boxplus b) \le 2^m + r \bullet s$, again using (A.4) (iii). □

With this interpretation of $r \circ s$ we obtain another proof of (13.5). See Exercise 4. Eliahou and Kervaire (1998) generalize all of this to subsets of an $\mathbb{F}_p$-vector space. If $A, B \subseteq V$ are subsets of cardinality r, s, respectively, they prove $|A + B| \ge \beta_p(r, s)$. This $\beta_p(r, s)$ is the p-analog of $r \circ s$ as defined in Exercise 12.25.

As mentioned earlier, Yuzvinsky conjectured that any intercalate matrix of type (r, s, n) must have $n \ge r \circ s$. Theorem A.1 above proves this for dyadic intercalate matrices. For the non-dyadic cases Yiu reports that this conjecture can be proved when $r, s \le 16$ by invoking the complete characterization of small intercalate matrices given in Yiu (1990a) and (1994a).

Appendix B to Chapter 13. Monomial compositions

Let us now consider compositions of more general quadratic forms, not just sums of squares. This generality requires more extensive notations. Suppose α, β, γ are regular quadratic forms over F with dimensions r, s, n, respectively. (Here F is a field in which $2 \ne 0$.) A *composition* for this triple of forms is a formula

$$\alpha(X) \cdot \beta(Y) = \gamma(Z)$$

where each z_k is bilinear in the systems $X = (x_1, \dots, x_r)$ and $Y = (y_1, \dots, y_s)$, with coefficients in F. If (U, α), (V, β), (W, γ) are the corresponding quadratic spaces over F, such a composition becomes a bilinear map $f : U \times V \to W$ satisfying the norm property:

$$\gamma(f(u, v)) = \alpha(u) \cdot \beta(v) \quad \text{for every } u \in U \text{ and } b \in V.$$

Choose orthogonal bases $A = \{u_0, \dots, u_{r-1}\}$ for U and $B = \{v_0, \dots, v_{s-1}\}$ for V. Setting $w_0 = f(u_0, v_0)$, we may choose an orthogonal basis $C = \{w_0, \dots, w_{n-1}\}$. The quadratic forms are then diagonalized: $\alpha \simeq \langle a_0, \dots, a_{r-1}\rangle$, $\beta \simeq \langle b_0, \dots, b_{s-1}\rangle$, $\gamma \simeq \langle c_0, \dots, c_{n-1}\rangle$. By scaling α, β we may assume that $a_0 = 1$ and $b_0 = 1$. Then the norm property implies that $c_0 = a_0 b_0 = 1$ as well.

Each vector $f(u_i, v_j)$ is expressible as a linear combination of $w_0, \dots, w_{n-1}$. Motivated by the integer case above, we restrict attention to pairings such that each $f(u_i, v_j)$ involves only one of the basis vectors w_k.

B.1 Definition. A bilinear pairing $f : U \times V \to W$ is *monomial* (relative to those bases) if for every i, j there exists k such that $f(u_i, v_j) \in F \cdot w_k$.

It quickly follows that a monomial pairing f is determined by two maps

$$\varphi : A \times B \to C \quad \text{and} \quad \varepsilon : A \times B \to F$$

such that $f(u_i, v_j) = \varepsilon(u_i, v_j) \cdot \varphi(u_i, v_j)$. When $a_i = b_j = c_k = 1$ this f is a standard composition over $\mathbb{Z}$ and φ is tabulated by an intercalate matrix M of type (r, s, n), with ε providing a consistent signing.

For a general monomial composition f, consider its extension $\bar{f}$ to a composition over the algebraic closure. This $\bar{f}$ becomes a standard composition over $\mathbb{Z}$, using the basis $\bar{u}_i = \frac{1}{\sqrt{a_i}} u_i$, etc. Therefore f has an associated signed intercalate matrix M. We ask the converse: Given a consistently signed intercalate M, what monomial compositions can be built from it? In particular we are interested in the compositions of indefinite quadratic forms over $\mathbb{R}$.

The associated $r \times s$ matrix M has the property that $M(i, j) = k$ if and only if $f(u_i, v_j) \in F \cdot w_k$, and letting $\varepsilon_{ij} = \varepsilon(u_i, v_j)$ we find that

$$M(i, j) = k \iff f(u_i, v_j) = \varepsilon_{ij} \cdot w_k.$$

Taking the lengths of those vectors we obtain:

$$a_i b_j = \varepsilon_{ij}^2 c_k. \tag{$*$}$$

This condition already puts a restriction on the forms α, β, γ. To keep track of those lengths we label the rows of the matrix M with the scalars a_i and the columns with the scalars b_j, and recall that the "colors" are the indices k corresponding to the scalars c_k. Then these labels satisfy the *square-class consistency* condition:

$$\langle a_i b_j \rangle \simeq \langle c_k \rangle, \quad \text{whenever} \quad M(i, j) = k.$$

For example if M has type $(2, 2, 2)$ the three forms must coincide. To see this we examine the following labeled version of M, recalling the normalization $a_0 = b_0 = 1$:

$$\begin{array}{cc} & \begin{array}{cc} 1 & b_1 \end{array} \\ \begin{array}{c} 1 \\ a_1 \end{array} & \begin{pmatrix} 0 & 1 \\ 1 & 0 \end{pmatrix} \end{array}$$

The occurrences of color 0 then show that $\langle c_0 \rangle \simeq \langle 1 \rangle \simeq \langle a_1 b_1 \rangle$. Similarly, the occurrences of color 1 imply $\langle c_1 \rangle \simeq \langle a_1 \rangle \simeq \langle b_1 \rangle$. Therefore after changing b_j, c_k by squares, we may assume $\alpha = \beta = \gamma$.

Condition $(*)$ above says from the labeled intercalate matrix M alone we know the values ε_{ij}^2 for every i, j. Then determining ε_{ij} is essentially a sign choice. The intercalacy and sign consistency conditions become:

If $M(i, j) = M(i', j') = k$ where $i \neq i'$ and $j \neq j'$, then $M(i, j') = M(i', j) = k'$ for some color k'. In this case:

$$(\varepsilon_{ij} \varepsilon_{i'j'}) \cdot c_k = -(\varepsilon_{i'j} \varepsilon_{ij'}) \cdot c_{k'}.$$

These "signs" are tabulated by writing the value ε_{ij} in parentheses to the left of the entry $M(i, j)$ of the matrix. Then there is a monomial composition for the forms α, β, γ if and only if there exists a consistently signed, labeled intercalate matrix of this type.

As before we may freely change a "signing" of an intercalate matrix M by altering the signs of an entire row, of an entire column, or of all occurrences of a single color. Any sequence of such moves yields equivalent signings.

An example should help clarify these ideas. Starting from the standard intercalate matrix M of type $(4, 4, 4)$, let us analyze all the associated monomial compositions. We first attach the row labels $\{1, a_1, a_2, a_3\}$ and the column labels $\{1, b_1, b_2, b_3\}$ to M. The square-class consistency condition shows that $a_1 = b_1, a_2 = b_2$, and $a_3 = b_3 = a_1a_2$. Therefore we may express $\alpha = \beta = \gamma = \langle 1, a, b, ab\rangle$ for suitable scalars $a, b \in F^\bullet$.

Now we begin inserting the "signs" ε_{ij}. By condition $(*)$ all signs along the first row and column are ± 1, so we may assume they all equal 1. The signs for the occurrences of color 0 there can be calculated using the sign consistency. So far the labeled matrix appears as follows:

$$\begin{array}{c} \\ 1 \\ a \\ b \\ ab \end{array} \begin{array}{c} \begin{array}{cccc} 1 & a & b & ab \end{array} \\ \begin{pmatrix} (1)0 & (1)1 & (1)2 & (1)3 \\ (1)1 & (-a)0 & 3 & 2 \\ (1)2 & 3 & (-b)0 & 1 \\ (1)3 & 2 & 1 & (-ab)0 \end{pmatrix} \end{array}$$

Condition $(*)$ shows that $\varepsilon_{12}^2 = 1$. By changing the sign of every occurrence of color 3 and then changing the signs of the last row and last column, we may assume that $\varepsilon_{12} = 1$. The remaining signs are then determined by the rules above, and we obtain the following signed and labeled matrix:

$$\begin{array}{c} \\ 1 \\ a \\ b \\ ab \end{array} \begin{array}{c} \begin{array}{cccc} 1 & a & b & ab \end{array} \\ \begin{pmatrix} (1)0 & (1)1 & (1)2 & (1)3 \\ (1)1 & (-a)0 & (1)3 & (-a)2 \\ (1)2 & (-1)3 & (-b)0 & (b)1 \\ (1)3 & (a)2 & (-b)1 & (-ab)0 \end{pmatrix} \end{array}$$

This matrix yields the standard composition for $\langle\langle a, b\rangle\rangle$ obtained from multiplication in the quaternion algebra $\left(\frac{-a,-b}{F}\right)$. For example, the formula for z_0 can be read off from the positions and coefficients of the color 0 in the matrix above: $z_0 = x_0y_0 - ax_1y_1 - bx_2y_2 - abx_3y_3$. Moreover we have proved that every monomial composition of type $(4, 4, 4)$ is equivalent to the composition given here, for some scalars $a, b \in F^\bullet$.

The constructions done earlier in this chapter can be generalized to monomial compositions. For example the standard $[8, 8, 8]$ formula for the quadratic form $\langle\langle a, b, c\rangle\rangle$ can be expanded to a monomial $[10, 10, 16]$ formula for the quadratic forms α, β, γ where

$$\alpha = \beta = \langle\langle a, b, c\rangle\rangle \perp d\langle\langle a\rangle\rangle \quad \text{and} \quad \gamma = \langle\langle a, b, c, d\rangle\rangle.$$

Here is the 10×10 matrix which tabulates these formulas. When $a = b = c = d = 1$ this matrix reduces to the standard signed intercalate 10×10 mentioned before (13.6).

$$
\begin{array}{c|cccccccccc}
 & 1 & a & b & ab & c & ac & bc & abc & d & ad \\
\hline
1 & (1)0 & (1)1 & (1)2 & (1)3 & (1)4 & (1)5 & (1)6 & (1)7 & (1)8 & (1)9 \\
a & (1)1 & (-a)0 & (1)3 & (-a)2 & (1)5 & (-a)4 & (-1)7 & (a)6 & (1)9 & (-a)8 \\
b & (1)2 & (-1)3 & (-b)0 & (b)1 & (1)6 & (1)7 & (-b)4 & (-b)5 & (1)10 & (1)11 \\
ab & (1)3 & (a)2 & (-b)1 & (-ab)0 & (1)7 & (-a)6 & (b)5 & (-ab)4 & (1)11 & (-a)10 \\
c & (1)4 & (-1)5 & (-1)6 & (-1)7 & (-c)0 & (c)1 & (c)2 & (c)3 & (1)12 & (1)13 \\
ac & (1)5 & (a)4 & (-1)7 & (a)6 & (-c)1 & (-ac)0 & (-c)3 & (ac)2 & (1)13 & (-a)12 \\
bc & (1)6 & (1)7 & (b)4 & (-b)5 & (-c)2 & (c)3 & (-bc)0 & (-bc)1 & (1)14 & (-1)15 \\
abc & (1)7 & (-a)6 & (b)5 & (ab)4 & (-c)3 & (-ac)2 & (bc)1 & (-abc)0 & (1)15 & (a)14 \\
d & (1)8 & (-1)9 & (-1)10 & (-1)11 & (-1)12 & (-1)13 & (-1)14 & (-1)15 & (-d)0 & (d)1 \\
ad & (1)9 & (a)8 & (-1)11 & (a)10 & (-1)13 & (a)12 & (1)15 & (-a)14 & (-d)1 & (-ad)0
\end{array}
$$

Is every monomial composition of size [10, 10, 16] of this type? That seems to be a difficult question. The construction above over the real field $\mathbb{R}$ provides some new formulas. In addition to the standard positive definite case we obtain some examples where $\gamma = 8\mathbb{H} = 8\langle 1\rangle \perp 8\langle -1\rangle$. After replacing the matrix by an equivalent one, we may assume that $\alpha = \beta$ is a 10-dimensional form with signature ± 6, ± 2, or 0. That is, after scaling to assume the signatures are non-negative we find that $\alpha = \beta$ is one of the forms

$$8\langle 1\rangle \perp 2\langle -1\rangle, \quad 6\langle 1\rangle \perp 4\langle -1\rangle, \quad 5\langle 1\rangle \perp 5\langle -1\rangle.$$

Must every indefinite [10, 10, 16] over $\mathbb{R}$ have γ hyperbolic and (after scaling) $\alpha \simeq \beta$?

It would be interesting to obtain further information about the composition of indefinite quadratic forms over $\mathbb{R}$. Some restrictions of the sizes of such compositions are obtained by lifting to the complex field (see (14.1)). But for an allowable size like [10, 10, 16] it remains unclear what signatures are possible for the three forms.

Appendix C to Chapter 13. Known upper bounds for $r * s$

Upper bounds are provided by constructions. The bound $r * s \le n$ means that there exists a normed bilinear map (over $\mathbb{R}$) of size $[r, s, n]$. All the known constructions can be done with integer coefficients, and hence with intercalate matrices. Much of this chapter was spent describing methods for constructing signed intercalate matrices and showing that the values listed in Theorem 13.1 are upper bounds. (Less space was spent on the much harder task of proving that those values are best possible.)

What about larger values for r, s? In this appendix we list the known upper bounds for $r *_{\mathbb{Z}} s$, following the work of Adem (1975), Yuzvinsky (1984), Lam and Smith (1993), Smith and Yiu (1992), Romero (1995), Yiu (1996), Sánchez-Flores (1996). We list here a table of upper bounds, as presented in Yiu (1996).

To list upper bounds for $r *_{\mathbb{Z}} s$ we may assume $r \le s$. If $r \le 9$ then $r *_{\mathbb{Z}} s$ is known (see(12.13) or (13.7)). If $r, s \le 16$ then Yiu's Theorem 13.1 provides the exact value of $r *_{\mathbb{Z}} s$. Let us consider the next block of values:

$$r \le s, \quad 10 \le r \le 32 \quad \text{and} \quad 17 \le s \le 32.$$

In the following table of upper bounds for $r *_{\mathbb{Z}} s$, each underlined entry is known to be the exact value for $r *_{\mathbb{Z}} s$.

$r\backslash s$	17	18	19	20	21	22	23	24	25	26	27	28	29	30	31	32
10	29	29	30	30	30	30	32	32	32	32	32	32	32	32	32	32
11	32	32	32	32	42	44	44	44	46	48	48	48	48	52	52	52
12	32	32	32	32	42	44	44	44	48	48	48	48	48	52	52	52
13	32	32	43	44	44	44	48	48	48	48	48	58	58	58	58	58
14	32	32	43	44	46	48	48	48	48	48	48	58	58	58	58	58
15	32	32	44	46	48	48	48	48	48	48	48	60	62	63	64	64
16	32	32	44	46	48	48	48	48	48	48	48	60	62	64	64	64
17	32	32	49	50	51	52	53	54	55	56	57	61	64	64	64	64
18		50	50	52	52	54	54	56	56	57	57	64	64	64	64	64
19			56	56	59	60	60	64	64	64	64	64	64	64	64	64
20				56	60	60	60	64	64	64	64	64	64	64	64	64
21					64	64	64	64	72	76	77	80	80	84	84	84
22						68	72	72	72	78	80	80	80	84	84	84
23							72	72	72	78	80	84	88	90	90	90
24								72	72	80	80	88	88	90	90	90
25									72	80	80	88	94	95	96	96
26										80	80	89	94	96	96	96
27											89	89	96	96	96	96
28												96	96	96	96	96
29													96	96	96	96
30														96	96	96
31															116	116
32																116

We conclude with a table of upper bounds for $r * s$ in the range $32 \leq r \leq 64$ and $10 \leq s \leq 16$. (Here we use $s \leq r$ for typographical reasons). Next to that table appear the known upper bounds for $r * r$ in that range.

$r\backslash s$	10	11	12	13	14	15	16
33	42	56	56	63	64	64	64
34	42	56	56	64	64	64	64
35	44	57	58	64	64	64	64
36	44	58	58	64	64	64	64
37	46	58	58	64	64	76	76
38	46	58	58	64	64	78	78
39	48	59	60	64	64	79	80
40	48	59	60	64	64	80	80
41	48	59	60	74	74	80	80
42	48	60	60	78	78	80	80
43	58	60	60	79	80	80	80
44	58	60	60	80	80	80	80
45	59	61	62	80	80	80	80
46	59	61	62	80	80	92	92
47	60	61	62	80	80	94	94
48	60	62	62	80	80	95	96
49	61	62	62	80	80	96	96
50	61	62	62	90	90	96	96
51	62	62	62	92	92	96	96
52	62	62	62	92	94	96	96
53	62	63	64	92	96	96	96
54	62	64	64	96	96	96	96
55	64	64	64	96	96	108	108
56	64	64	64	96	96	108	108
57	64	64	64	96	96	111	112

r	$r * r$
33	127
34	128
35	128
36	128
37	160
38	168
39	168
40	168
41	187
42	188
43	208
44	214
45	216
46	222
47	233
48	240
49	254
50	256
51	273
52	274
53	283
54	304
55	312
56	312
57	320

$r\backslash s$	10	11	12	13	14	15	16
58	64	64	64	96	96	112	112
59	64	64	64	104	106	112	112
60	64	64	64	104	108	112	112
61	64	64	64	104	110	112	112
62	64	64	64	104	112	112	112
63	64	64	64	104	112	112	112
64	64	64	64	104	112	112	112

r	$r * r$
8	320
59	320
60	320
61	360
62	368
63	368
64	368

These two tables and further details of the required constructions were compiled by Yiu (1996). Further tables of upper bounds for all values where $r, s \le 64$ are presented in Sánchez-Flores (1996).

Exercises for Chapter 13

1. The matrix

$$\begin{pmatrix} 1 & 2 & 3 & 4 \\ 2 & 1 & 4 & 3 \\ 5 & 6 & 7 & 8 \\ 6 & 5 & 9 & 10 \end{pmatrix}$$

is a non-dyadic intercalate matrix. That is, it is an intercalate matrix but is not equivalent to a submatrix of any D_t.

2. Let $\mathcal{N}(r, s) = \{n :$ there exists an $r \times s$ intercalate matrix with exactly n colors$\}$. Certainly $r \circ s$ and $rs \in \mathcal{N}(r, s)$, but values in between might not occur. For example:

There exists a $2 \times s$ intercalate with n colors $\iff s \le n \le 2s$ and n is even.

$\mathcal{N}(3, 3) = \{4, 7, 9\}$,

$\mathcal{N}(3, 4) = \{4, 6, 7, 8, 10, 12\}$,

$\mathcal{N}(3, 5) = \{7, 8, 10, 11, 13, 15\}$,

$\mathcal{N}(4, 4) = \{4, 7, 8, 10, 12, 14, 16\}$.

For the dyadic case we ask: If V is an $\mathbb{F}_2$-vector space and $A, B \subseteq V$ with $|A| = r$ and $|B| = s$, then what sizes are possible for $A + B$?

3. **Ubiquitous colors.** Let M be a symmetric intercalate matrix of type (r, r, n). If $r = 2^b \cdot$ (odd) then the number of ubiquitous colors of M is 2^t for some $t \le b$. Let $r = 2^t \cdot r_1$ and $n = 2^t \cdot n_1$. Then M is equivalent to $D_t \otimes N$ where N is symmetric intercalate of type (r_1, r_1, n_1).

Corollary. *If M is intercalate of type (n, n, n) then M is equivalent to D_t for some t.*

(Hint. As in (13.8) we get $M = D_1 \otimes M'$. Each ubiquitous color of M' corresponds to two ubiquitous colors of M.)

4. **Nim sum.** (1) Prove the four parts of (A.4).

(2) Does $2^m(a \boxplus b) = (2^m a) \boxplus (2^m b)$?

(3) Writing $[0, m)$ for the interval, then: $[0, r) \boxplus [0, s) = [0, r \circ s)$. This is a restatement of Lemma 13.5.

(Hint. (1) (ii) Note that $a < 2^m \iff \text{Bit}(a) \subseteq [0, m) = \{0, 1, \dots, m-1\}$.

(iii) Express $b = b_0 + b_1$ where $\text{Bit}(b_0) \subseteq [0, m)$ and $\text{Bit}(b_1) \subseteq [m, \infty)$. Then $a \boxplus (2^m + b) = (a \boxplus b_0) + 2^m + b_1 = 2^m + (a \boxplus b)$.

(iv) Assume a, b are bit-disjoint. If $n = 2^k +$ (higher terms) then none of 1, 2, $2^2, \dots, 2^{k-1}$ occur in a or b. If 2^k occurs in a then $n - 1 = (a-1) \boxplus b$.

(3) Use (A.4) (iv).)

5. **Cayley–Dickson.** Suppose $e_0, e_1, \dots, e_{2^m-1}$ is the standard basis of the Cayley–Dickson algebra A_m as described in Exercise 1.25. The product is given by:

$$e_i \cdot e_j = \varepsilon_{ij} e_k \text{ where } k = i \boxplus j \text{ is the Nim-sum}$$

and the signs $\varepsilon_{ij} = \pm 1$ are determined inductively as follows. Given the signs ε_{ij} for $0 \le i, j < 2^m$, the remaining signs ε_{hk} for $0 \le h, k < 2^{m+1}$ are given by the formulas:

$$\varepsilon_{i,2^m+j} = \begin{cases} +1 & \text{if } i = 0 \text{ or } j = 0, \\ -1 & \text{if } i = j \neq 0, \\ -\varepsilon_{ij} & \text{otherwise;} \end{cases}$$

$$\varepsilon_{2^m+i,j} = \begin{cases} +1 & \text{if } j = 0 \text{ or } i = j, \\ -1 & \text{if } i = 0 \text{ and } j \neq i, \\ -\varepsilon_{ij} & \text{otherwise;} \end{cases}$$

$$\varepsilon_{2^m+i,2^m+j} = \begin{cases} +1 & \text{if } i = 0 \text{ and } j \neq 0, \\ -1 & \text{if } j = 0 \text{ or } i = j, \\ -\varepsilon_{ij} & \text{otherwise.} \end{cases}$$

(Hint. Express $A_{m+1} = A_m \oplus A_m$ and for $0 \le i < 2^m$ identify e_i with $(e_i, 0)$ and e_{i+2^m} with $(0, e_i)$. Work out the products using the "doubling process" stated in (1.A6).)

6. Let D_t be signed according to the Cayley–Dickson process.

(1) If $t \ge 4$, the first 9 rows of D_t are consistently signed. Note that these do *not* form the direct sum of several copies of the upper left 9×16 block.

(2) Examining the displayed signs for D_4 arising from A_4, are the signings of the four 8×8 blocks equivalent?

7. **Equivalence of signs.** (a) All the consistent signings of the intercalate matrix D_4 are equivalent. Similarly for D_8 and D_{10} and for the matrix of type $(12, 12, 26)$ constructed in (13.9).

(b) In the proof of (13.12) we may assume that the new signs α, β, γ are all "+".

(Hint. For example in D_4 assume that the first row and column are all signed with "+", so the three diagonal zeros have sign "−". To alter the signs of the middle 3s, change the signs of all the 3s, and change the sign of the last row and last column. We may assume the 3 in the second row has sign "+". The remaining signs are determined. Similar moves prove the uniqueness for all these examples.)

8. The matrix $D_{10,10}$ can be extended to an intercalate matrix of type $(10, 11, 16)$ in several ways. None of these extensions can be consistently signed. (This follows from $10 \# 11 = 17$, mentioned near the end of Chapter 12. However that proof is difficult.)

(Hint. By Exercise 7 we may assume $D_{10,10}$ has the standard signing coming from A_4 displayed before (13.6). The 11th column of the extension must match one of the remaining columns of A_4. Each case yields a sign inconsistency.)

9. **Hidden formulas.** Let M be an intercalate matrix of type (r, s, n) and suppose the color a has frequency k in M. Permute the rows and columns of M to assume that these k occurrences of a appear along the main diagonal, yielding a partition $M = \begin{pmatrix} A & C \\ B & A'' \end{pmatrix}$, where A is a $k \times k$ matrix with color a along the diagonal, and color a does not appear in C, B or A''.

Lemma. *The matrix* $M_{(a)} = (\, A \; C \; B^\top \,)$ *is also intercalate, of type* $(k, r+s-k, m)$ *for some* $m \le n$. *Furthermore, if* M *is consistently signed so that each occurrence of color* a *has a "+", then* $M_{(a)}$ *is also consistently signed.*

This $M_{(a)}$ is called the intercalate matrix "hidden behind a".

(Hint. Note that $A^\top$ is the same as $-A$, except for the diagonal. Checking the $(\, A \; B^\top \,)$ part is easy. For $(\, C \; B^\top \,)$ suppose color b occurs in C and in B. Permuting rows and columns yields a submatrix of M of the type: $M' = \begin{pmatrix} a & x & -y \\ -x & a & b \\ b & y & x \end{pmatrix}$, where x, y are some other colors. Examine the corresponding parts of $M_{(a)}$.)

10. There exist formulas of the following sizes:

$[17, 17, 32]$ $[18, 18, 50]$ $[20, 20, 56]$ $[21, 21, 64]$ $[22, 22, 68]$

$[25, 25, 72]$ $[26, 26, 80]$ $[27, 27, 89]$ $[30, 30, 96]$

These provide some of the upper bounds listed in the first table of Appendix C.

(Hint. A $[12, 20, 32]_{\mathbb{Z}}$ formula was constructed by Lam and Smith (1993). Use this, earlier formulas, and the techniques of restriction, direct sums and doubling.

Examples: $[18, 19, 50] = [18, 17, 32] \oplus [18, 2, 18]$; $[26, 27, 80] = 3 \cdot [16, 9, 16] \oplus [10, 27, 32]$; and $[27, 27, 89] = [17, 18, 32] \oplus [17, 9, 25] \oplus [10, 27, 32]$. For 22 and 25 recall Romero's construction.)

11. **Generalizing Yuzvinsky.** (1) Generalize (A.2) and (A.3) to n variable polynomials. If V is an $\mathbb{F}_2$-vector space and $A_1, \ldots, A_k \subseteq V$, with $|A_j| = r_j$, what is the minimal value for $|A_1 + \cdots + A_k|$?

(2) Suppose V is an $\mathbb{F}_p$-vector space and $A, B \subseteq V$ with cardinalities $|A| = r$ and $|B| = s$. What is the smallest possibility for $|A + B|$?

12. Generalize the constructions in this chapter to the monomial pairings of Appendix B. For example, what is the analog of the doubling process (13.6) for a composition of α, β, γ? How does (13.9) generalize?

Notes on Chapter 13

In writing this chapter I closely followed the presentations in Lam and Smith (1993), Smith and Yiu (1994), Yiu (1990a) and Yiu (1994a).

Integer composition formulas were analyzed by several of 19th century mathematicians who were seeking to generalize the 8-square identity. Proofs that 16-square identities (over $\mathbb{Z}$) are impossible were given (with various levels of rigor) by several mathematicians, including Young, Cayley, Kirkman and Roberts. For instance, Cayley (1881), using more clumsy terminology, seems to provide a complete list of intercalate matrices of size [16, 16, 16] and shows that none of them has a consistent signing. For further information and references see Dickson (1919).

The work of Kirkman (1848) was tracked down by Yiu, following a reference in Dickson (1919), and reported in Yiu (1990a). Kirkman obtained formulas of types $[2k, 2k, k^2 - 3k + b]$ where $b = 8, 4, 6$ according as $k \equiv 0, 1, 2 \pmod 3$.

Composition formulas of size $[\rho(n), n, n]$ appear implicitly in the works of Hurwitz, Radon and Eckmann. They have been given in more explicit form by a number of authors, including: Wong (1961), K. Y. Lam (1966), Zvengrowski (1968), Geramita and Pullman (1974), Gabel (1974), Shapiro (1977a), Adem (1978b), Yuzvinsky (1981), Bier (1984), K. Y. Lam (1984), Lam and Yiu (1987), Au-Yeung and Cheng (1993). The two methods of constructing signed intercalate matrices of type $(\rho(n), n, n)$ mentioned after (13.9) are also outlined in Smith and Yiu (1992).

This doubling construction of (13.6) is a variation of the one given by Lam and Smith (1993).

Lemma 13.12 and Corollary 13.13 appear in Yiu (1993) in Example 4.10 and Lemma 5.3.

Yuzvinsky (1981), p. 143 mentions Conjecture 13.15 (without proof).

Appendix A. Theorem A.1 is a major result in Yuzvinsky (1981). Our proof closely follows the presentation in Eliahou and Kervaire (1998). They use this polynomial method to prove several related results, including those asked in Exercise 11. For further applications of these ploynomial methods in combinatorics, see Alon (1999). I am grateful to Eliahou and Kervaire for sending me a preliminary version of their paper.

Appendix B. The term "monomial pairing" was used by Yuzvinsky (1981) when he introduced what we call intercalate matrices. The calculations using general quadratic forms here seem to be new.

Exercise 1. See Yiu (1990a), p. 466. Further information on determining whether an intercalate matrix is dyadic see Calvillo, Gitler, Martínez-Bernal (1997a).

Exercise 2. A consistently signed intercalate $r \times s$ matrix with exactly n colors leads to a full composition, as defined in Chapter 14. I believe that these sets $\mathcal{N}(r, s)$ have not been investigated elsewhere.

Exercise 3. See Yiu (1990a) Prop. 2.11. Recall that if $t > 3$ then D_t cannot be consistently signed. It turns out that if a consistently signed intercalate matrix has more than 2 ubiquitous colors then it must have type [4, 4, 4] or [8, 8, 8]. See (15.30) and Exercise 15.16.

Exercise 5. These formulas are also stated in Yiu (1994a), §2.

Exercise 6. See Yiu (1994a), Prop. 2.8.

Exercise 8. Yiu (1987), Prop. 1.3.

Exercise 9. The hidden formulas were first discovered in the general context of quadratic forms between euclidean spheres in Yiu (1986) and Lam and Yiu (1987). See Chapter 15. They were translated into this intercalate matrix version in Lam and Yiu (1989). Also see Yiu (1990a), Theorem 8 and Yiu (1994a), Proposition 14.2. Those hidden formulas play an important role in the proof of Theorem 13.1.

Exercise 10. Smith and Yiu (1992).

Exercise 11. See Eliahou and Kervaire (1998).

Chapter 14

Compositions over General Fields

Methods of algebraic topology were used in Chapter 12 to provide necessary conditions for the existence of a real composition of size $[r, s, n]$. Do these results remain valid over more general base fields? The Lam–Lam Lemma provides a simple way to extend those topological results to any field F of characteristic zero. Another approach to the problem, avoiding the topological machinery, is to apply Pfister's analysis of the set $D_F(n)$ of all sums of n squares in F. He proved the surprising fact that products of these sets are nicely behaved: $D_F(r)D_F(s) = D_F(r \circ s)$. Pfister's work yields another proof of the Stiefel–Hopf Theorem over $\mathbb{R}$ (for normed bilinear pairings). Unfortunately, this approach yields little information when F has positive characteristic. Returning to more elementary methods, Adem pioneered a direct matrix approach valid over any field (at least when $2 \neq 0$). Those techniques apply when the pairings are close to being of the classical Hurwitz–Radon sizes: we obtain results for sizes $[r, s, n]$ when $s \geq n-2$. In the appendix we extend the discussion to compositions of three quadratic forms, not just sums of squares.

The function $r \circ s$ was defined in (12.5) in connection with the following important result, proved by topological methods.

Hopf's Theorem. *If there is a composition of size* $[r, s, n]$ *over* $\mathbb{R}$ *then* $r \circ s \leq n$.

The notation $r \circ s$ was introduced to replace the binomial coefficient conditions in the original statement of Theorem 12.2. We use the term "Hopf's Theorem" here even though separate proofs of stronger results were given by Hopf, Stiefel and Behrend around 1940. Chapter 12 includes Hopf's result (valid for nonsingular bi-skew maps), Stiefel's version (for nonsingular bilinear maps), and several subsequent generalizations. The nonsingular bilinear version was interpreted in (12.12) as the inequality:

$$r \circ s \leq r \# s \leq r * s.$$

Those results are valid for compositions over the field $\mathbb{R}$.

Behrend (1939) generalized Hopf's Theorem to real closed fields (using nonsingular, bi-skew polynomial maps). Behrend used intersection theory in real algebraic geometry, but his result can also be deduced from Hopf's Theorem by using the Tarski Principle from mathematical logic. This transfer principle says roughly that: every "elementary" statement in the theory of real closed fields which is known to be true over $\mathbb{R}$ must also be true over every real closed field.

Suppose F is a field (with characteristic $\neq 2$, as usual). We say that $[r, s, n]$ is *admissible over* F if there exists a normed bilinear pairing (a composition formula) of size $[r, s, n]$ with coefficients in F. For example, we have seen that $[3, 5, 7]$ is admissible over every field. Hopf's Theorem implies that $[3, 5, 6]$ is not admissible over $\mathbb{R}$ since $3 \circ 5 = 7$. In fact $[3, 5, 6]$ is not admissible over any formally real field, since such a field can be embedded in some real closed field where Behrend's Theorem applies. Similarly $[3, 6, 7]$ and $[4, 5, 7]$ are not admissible over any formally real field. But what about more general fields? Could a $[3, 5, 6]$ formula exist if we allow complex coefficients? This possibility is eliminated by a wonderful reduction argument due to K. Y. Lam and T. Y. Lam.

14.1 The Lam–Lam Lemma. *If* $[r, s, n]$ *is admissible over* $\mathbb{C}$ *then there is a nonsingular bilinear pairing of size* $[r, s, n]$ *over* $\mathbb{R}$. *Hence,* $r \# s \leq n$.

Proof. Suppose $(x_1^2 + x_2^2 + \cdots + x_r^2) \cdot (y_1^2 + y_2^2 + \cdots + y_s^2) = z_1^2 + z_2^2 + \cdots + z_n^2$, where each z_k is bilinear in the systems of indeterminates X, Y with coefficients in $\mathbb{C}$. Express $z_k = u_k + i v_k$, where u_k and v_k are bilinear in X, Y with coefficients in $\mathbb{R}$. Compare the real parts in the given formula to find:

$$(x_1^2 + x_2^2 + \cdots + x_r^2) \cdot (y_1^2 + y_2^2 + \cdots + y_s^2) = u_1^2 - v_1^2 + \cdots + u_n^2 - v_n^2.$$

Now consider the map $f : \mathbb{R}^r \times \mathbb{R}^s \to \mathbb{R}^n$ defined by:

$$f(a, b) = (u_1(a, b), \ldots, u_n(a, b)).$$

Then f is bilinear, and the multiplication formula written above implies that f is nonsingular. The definition of $r \# s$ in (12.11) provides the inequality. □

14.2 Theorem. *If* $[r, s, n]$ *is admissible over a field* F *of characteristic zero, then* $r \# s \leq n$.

Proof. One $[r, s, n]$ formula over F involves only finitely many coefficients $\alpha_j \in F$. Then this formula is valid over $\mathbb{Q}(\{\alpha_j\})$. This function field can be embedded into $\mathbb{C}$ as a subfield, so the formula can be viewed over $\mathbb{C}$ and the lemma applies. □

Imitating the notation of Chapter 12 we define:

$$r *_F s = \min\{n : [r, s, n] \text{ is admissible over } F\}$$
$$\rho_F(n, r) = \max\{s : [r, s, n] \text{ is admissible over } F\}.$$

The easy bounds are:

$$\max\{r, s\} \leq r *_F s \leq rs \quad \text{and} \quad \lambda(n, r) \leq \rho_F(n, r) \leq n.$$

Recall that $\lambda(r, s) = \max\{\rho(r), \rho(r+1), \ldots, \rho(n)\}$ as in (12.25). Moreover $\rho_F(n, n) = \rho(n)$ because the Hurwitz–Radon Theorem holds true over F.

We have just proved that if F has characteristic zero then:

$$r \# s \leq r *_F s \quad \text{and} \quad \rho_F(n, r) \leq \rho_\#(n, r).$$

However this algebraic result was proved using non-trivial topology. Is there a truly algebraic proof of the "Hopf Theorem": $r \circ s \leq r *_F s$ for fields of characteristic zero? Does this result remain true if the field has positive characteristic?

One productive idea is to apply Pfister's results on the multiplicative properties of sums of squares. If F is a field (where $2 \neq 0$), define

$$D_F(n) = \{a \in F^\bullet : a \text{ is a sum of } n \text{ squares in } F\}.$$

Recall that if q is a quadratic form over F then $D_F(q)$ is the set of values in $F^\bullet$ represented by q. The notation above is an abbreviated version: $D_F(n) = D_F(n\langle 1\rangle)$. Evaluating one of our bilinear composition formulas at various field elements establishes the following simple result.

14.3 Lemma. *If $[r, s, n]$ is admissible over F then for any field $K \supseteq F$,*

$$D_K(r) \cdot D_K(s) \subseteq D_K(n).$$

Some multiplicative properties of these sets $D_F(n)$ were proved earlier. The classical n-square identities show that $D_F(1)$, $D_F(2)$, $D_F(4)$ and $D_F(8)$ are closed under multiplication. Generally $a \in D_F(n)$ implies $a^{-1} \in D_F(n)$ since $a^{-1} = a \cdot (a^{-1})^2$. Therefore those sets $D_F(n)$ are groups if $n = 1, 2, 4$ or 8. In the 1960s Pfister showed that every $D_F(2^m)$ is a group. This was proved above in Exercise 0.5 and more generally in (5.2). Applying this result to the rational function field $F(X, Y)$ provides some explicit 2^m-square identities. Of course any such identity for $m > 3$ cannot be bilinear (it must involve some denominators).

Here is another proof that $[3, 5, 6]$ is not admissible. We know $[3, 5, 7]$ is admissible over any F and therefore $D_F(3) \cdot D_F(5) \subseteq D_F(7)$. In fact we get equality here. Given any $a = a_1^2 + \cdots + a_7^2 \in D_F(7)$ we may assume that $a_1^2 + a_2^2 + a_3^2 \neq 0$ and factor out that term:

$$a = (a_1^2 + a_2^2 + a_3^2) \cdot \left(1 + \frac{a_4^2 + a_5^2 + a_6^2 + a_7^2}{a_1^2 + a_2^2 + a_3^2}\right).$$

The numerator and denominator of the fraction are in $D_F(4)$ (at least if the numerator is non-zero). Since $D_F(4)$ is a group the quantity in brackets is a sum of 5 squares, so that $a \in D_F(3) \cdot D_F(5)$. When that numerator is zero the conclusion is even easier. Therefore for any field F,

$$D_F(3) \cdot D_F(5) = D_F(7).$$

If [3, 5, 6] is admissible over F then by (14.3):

$$D_K(6) = D_K(7) \quad \text{for every such } K \supseteq F.$$

Of course this equality can happen in some cases. For instance if the form $6\langle 1\rangle$ is isotropic over F then it is isotropic over every $K \supseteq F$ and $D_K(n) = K^\bullet$ for every $n \geq 6$. On the other hand Cassels (1964) proved that in the rational function field $K = \mathbb{R}(x_1, \dots, x_n)$ the element $1 + x_1^2 + \cdots + x_n^2$ cannot be expressed as a sum of n squares. Applied to $n = 6$ this shows that $D_K(6) \neq D_K(7)$ and therefore [3, 5, 6] is not admissible over $\mathbb{R}$.

Cassels' Theorem was the breakthrough which inspired Pfister to develop his theory of multiplicative forms. He observed that a quadratic form φ over F can be viewed in two ways. On one hand φ is a homogeneous quadratic polynomial $\varphi(x_1, \dots, x_n) \in F[X]$. On the other hand it is a quadratic mapping $\varphi : V \to F$ arising from a symmetric bilinear form $b_\varphi : V \times V \to F$, and we speak of subspaces, isometries, etc. We write $\varphi \subset q$ when φ is isometric to a subform of q. The general result we need is the Cassels–Pfister Subform Theorem. It was stated in (9.A.1) and we state it again here.

14.4 Subform Theorem. *Let φ, q be quadratic forms over F such that q is anisotropic. Let $X = (x_1, \dots, x_s)$ be a system of indeterminates where $s = \dim \varphi$. Then $q \otimes F(X)$ represents $\varphi(X)$ over $F(X)$ if and only if $\varphi \subset q$.*

14.5 Corollary. *Suppose s and n are positive integers and $n\langle 1\rangle$ is anisotropic over F. The following statements are equivalent.*

(1) $s \leq n$.

(2) $D_K(s) \subseteq D_K(n)$ *for every field* $K \supseteq F$.

(3) $x_1^2 + \cdots + x_s^2$ *is a sum of n squares in the rational function field* $F(x_1, \dots, x_s)$.

Proof. For (3) $\Longrightarrow$ (1), apply the theorem to the forms $q = n\langle 1\rangle$ and $\varphi = s\langle 1\rangle$. □

To generalize the proof above that [3, 5, 6] is not admissible, we need to express the product $D_F(r) \cdot D_F(s)$ as some $D_F(k)$. This was done by Pfister (1965a). It is surprising that Pfister's function is exactly $r \circ s$, the function arising from the Hopf–Stiefel condition!

14.6 Proposition. $D_F(r) \cdot D_F(s) = D_F(r \circ s)$, *for any field* F.

In the proof we use the fact that $D_F(m + n) = D_F(m) + D_F(n)$. Certainly if $c \in D_F(m+n)$, then $c = a+b$ where a is a sum of m squares, b is a sum of n squares. The Transversality Lemma (proved in Exercise 1.15) shows that this can be done with $a, b \neq 0$. This observation enables us to avoid separate handling of the cases $a = 0$ and $b = 0$.

Proof of 14.6. Since the field F is fixed here we drop that subscript. We will use the characterization of $r \circ s$ given in (12.10). The key property is Pfister's observation, mentioned earlier: $D(2^m) \cdot D(2^m) = D(2^m)$. By symmetry we may assume $r \le s$ and proceed by induction on $r + s$. Choose the smallest m with $2^m < s \le 2^{m+1}$.

We first prove $D(r) \cdot D(s) \subseteq D(r \circ s)$. If $r \ge 2^m$ then $r + s > 2^{m+1}$ and $r \circ s = 2^{m+1}$. Then $D(r) \cdot D(s) \subseteq D(2^{m+1}) = D(r \circ s)$, as hoped. Otherwise $r < 2^m < s$ and we express $s = s' + 2^m$ where $s' > 0$. Then $r \circ s = r \circ (s' + 2^m) = (r \circ s') + 2^m$ by (12.10). Therefore $D(r) \cdot D(s) = D(r) \cdot (D(s') + D(2^m)) \subseteq D(r) \cdot D(s') + D(r) \cdot D(2^m) \subseteq D(r \circ s') + D(2^m) = D((r \circ s') + 2^m) = D(r \circ s)$.

For the equality we begin with a special case.

Claim. $D(2^m) \cdot D(2^m + 1) = D(2^{m+1})$.

For if $c \in D(2^{m+1})$ then $c = a + b$ where $a, b \in D(2^m)$. Then $c = a(1 + b/a)$ and $1 + b/a \in D(2^m + 1)$ since $D(2^m)$ is a group. This proves the claim.

If $r \ge 2^m$ then $r \circ s = 2^{m+1}$ and the claim implies that $D(r \circ s) = D(2^{m+1}) \subseteq D(r) \cdot D(s)$. Otherwise $r < 2^m < s$ and $r \circ s = (r \circ s') + 2^m$ as before. If $c \in D(r \circ s)$ then $c = a + b$ where $a \in D(r \circ s')$ and $b \in D(2^m)$. The induction hypothesis implies that $a = a_1 a_2$ where $a_1 \in D(r)$ and $a_2 \in D(s')$. Then $c = a_1(a_2 + b/a_1)$ and $a_2 + b/a_1 \in D(s') + D(2^m) \cdot D(r) = D(s') + D(2^m) = D(s' + 2^m) = D(s)$. Hence $c \in D(r) \cdot D(s)$. □

14.7 Proposition. *If $(r \circ s)\langle 1 \rangle$ is anisotropic over the field F then $r \circ s \le r *_F s$.*

Proof. Suppose $[r, s, n]$ is admissible over F. By (14.3) and (14.6), $D_K(r \circ s) \subseteq D_K(n)$ for every field $K \supseteq F$. If $n < r \circ s$ then (14.5) provides a contradiction. □

This provides an algebraic proof of Hopf's Theorem over $\mathbb{R}$ (for normed bilinear pairings). Unfortunately these ideas do not apply over $\mathbb{C}$ or over any field of positive characteristic, because $n\langle 1 \rangle$ is isotropic for every $n \ge 3$ in those cases.

Pfister's methods lead naturally to "rational composition formulas", that is, formulas where denominators are allowed.

14.8 Theorem. *For positive integers r, s, n the following two statements are equivalent.*

(1) $r \circ s \le n$.

(2) $D_K(r) \cdot D_K(s) \subseteq D_K(n)$ *for every field K.*

Furthermore if F is a field where $n\langle 1 \rangle$ is anisotropic, then the following statements are also equivalent to (1) *and* (2)*. Here $X = (x_1, \dots, x_r)$ and $Y = (y_1, \dots, y_s)$ are systems of indeterminates.*

(3) $D_K(r) \cdot D_K(s) \subseteq D_K(n)$ *where $K = F(X, Y)$.*

(4) *There is a formula $(x_1^2 + \cdots + x_r^2)(y_1^2 + \cdots + y_s^2) = z_1^2 + \cdots + z_n^2$ where each $z_k \in F(X, Y)$.*

(5) *There is a multiplication formula as above where each z_k is a linear form in Y with coefficients in $F(X)$.*

Proof. The equivalence of (1) and (2) follow from (14.5) and (14.6). Trivially (2) $\Longrightarrow$ (3), (3) $\Longrightarrow$ (4) and (5) $\Longrightarrow$ (4).

Proof that (4) $\Longrightarrow$ (5). Given the formula where $z_k \in F(X, Y)$, let $\alpha = x_1^2 \cdots + x_r^2$. Then $\alpha \cdot (y_1^2 + \cdots + y_s^2)$ is a sum of n squares in $F(X, Y)$. Setting $K = F(X)$ this is the same as saying: $n\langle 1\rangle$ represents $\alpha y_1^2 + \cdots + \alpha y_s^2$ over $K(Y)$. Since $n\langle 1\rangle$ is anisotropic the Subform Theorem 14.4 implies that $s\langle\alpha\rangle \subset n\langle 1\rangle$ over K. Now interpret quadratic forms as inner product spaces to restate this condition as: there is a K-linear map $f : K^s \to K^n$ carrying the form $s\langle\alpha\rangle$ isometrically to a subform of $n\langle 1\rangle$. Equivalently, there is an $n \times s$ matrix A over K such that $A^\top \cdot A = \alpha \cdot 1_s$. Using the column vector $Y = (y_1, \dots, y_s)^\top$ and $Z = AY$ we find

$$(x_1^2 + \cdots + x_r^2)(y_1^2 + \cdots + y_s^2) = \alpha Y^\top \cdot Y = Y^\top (A^\top A) Y = Z^\top \cdot Z = z_1^2 + \cdots + z_n^2.$$

This is a formula of size $[r, s, n]$ where each z_k is a linear form in Y with coefficients in $K = F(X)$, as required.

Proof that (5) $\Longrightarrow$ (1). We start from the formula where each z_k is a linear form in Y. In order to prove $r \circ s \leq n$ it suffices by (14.5) and (14.6) to prove that $D_K(r) \cdot D_K(s) \subseteq D_K(n)$ where $K = F(t_1, \dots, t_{r\circ s})$ is a rational function field. If $\beta \in D_K(s)$, express $\beta = b_1^2 + \cdots + b_s^2$ for $b_j \in K$. Since each z_k in the formula is linear in Y, we may substitute b_k for y_k to obtain:

$$(x_1^2 + \cdots + x_r^2) \cdot \beta = \hat{z}_1^2 + \cdots + \hat{z}_n^2$$

where each $\hat{z}_k \in K(X)$. Equivalently: $n\langle 1\rangle$ represents $\beta x_1^2 + \cdots + \beta x_r^2$ over $K(X)$. Since $n\langle 1\rangle$ is anisotropic over K, the Subform Theorem 14.4 implies that $r\langle\beta\rangle \subset n\langle 1\rangle$ over K. Consequently, $\beta D_K(r) = D_K(r\langle\beta\rangle) \subseteq D_K(n\langle 1\rangle)$. Since $\beta \in D_K(s)$ was arbitrary, we obtain $D_K(r) \cdot D_K(s) \subseteq D_K(n)$, as claimed. □

For a commutative ring A and an element $\alpha \in A$, define its *length*, $\text{length}_A(\alpha)$, to be the smallest integer n such that α is a sum of n squares in A. If no such n exists then define $\text{length}_A(\alpha) = \infty$. The values $r \circ s$ and $r * s$ can be characterized nicely in terms of lengths.

14.9 Corollary. *Let $X = (x_1, \dots, x_r)$ and $Y = (y_1, \dots, y_s)$ be systems of indeterminates. Then*

$$r \circ s = \text{length}_{\mathbb{R}(X,Y)}((x_1^2 + \cdots + x_r^2)(y_1^2 + \cdots + y_s^2)),$$
$$r * s = \text{length}_{\mathbb{R}[X,Y]}((x_1^2 + \cdots + x_r^2)(y_1^2 + \cdots + y_s^2)).$$

Proof. The first formula is the main content of (14.8). The second follows since if $(x_1^2 + \cdots + x_r^2)(y_1^2 + \cdots + y_s^2) = z_1^2 + \cdots + z_n^2$ where each $z_k \in \mathbb{R}[X, Y]$ is a polynomial, then necessarily each z_k is a bilinear form in X, Y. This is seen by computing coefficients and comparing degrees. □

So far in this chapter we have investigated two ideas for generalizing the Hopf Theorem to other fields, but neither applies to fields of positive characteristic. Using the original matrix formulation and linear algebra arguments, J. Adem (1980) was able to prove that $[3, 5, 6]$, $[3, 6, 7]$ and $[4, 5, 7]$ are not admissible over any field (provided $2 \neq 0$). His first two results were subsequently generalized as follows.

14.10 Adem's Theorem. *Let F be any field of characteristic not 2 and suppose $[r, n-1, n]$ is admissible over F.*

(i) *If n is even then $[r, n, n]$ is admissible, so that $r \leq \rho(n)$.*

(ii) *If n is odd then $[r, n-1, n-1]$ is admissible, so that $r \leq \rho(n-1)$.*

Using the function $\rho_F(n, r)$ defined above, Adem's Theorem says

$$\rho_F(n, n-1) = \max\{\rho(n), \rho(n-1)\}$$

for every field F (provided $2 \neq 0$ in F). This matches the value of $\rho(n, n-1)$ over $\mathbb{R}$ determined in Chapter 12.

Following Adem's methods, our proof uses the rectangular matrices directly. To gain some perspective, we will set up the general definitions. Suppose α, β, γ are non-singular quadratic forms over F with dimensions r, s, n, respectively. A *composition* for this triple of forms is a formula

$$\alpha(X) \cdot \beta(Y) = \gamma(Z)$$

where each z_k is bilinear in the systems $X = (x_1, \ldots, x_r)$ and $Y = (y_1, \ldots, y_s)$, with coefficients in F. More geometrically, let (U, α), (V, β), (W, γ) be the corresponding quadratic spaces over F. A composition for α, β, γ becomes a bilinear map $f : U \times V \to W$ satisfying the "norm property":

$$\gamma(f(u, v)) = \alpha(u) \cdot \beta(v) \quad \text{for every } u \in U \text{ and } v \in V.$$

This formulation shows that different bases can be freely chosen for the spaces U, V, W. In particular, if such a composition exists then there are formulas of the type $\alpha(X) \cdot \beta(Y) = \gamma(Z)$ for any choices of diagonalizations for the forms α, β, γ. We will concentrate here on the special case of sums of squares: $\alpha \simeq r\langle 1\rangle$, $\beta \simeq s\langle 1\rangle$ and $\gamma \simeq n\langle 1\rangle$. A composition for these forms over F means that $[r, s, n]$ is admissible over F.

The proofs presented below can be extended to the general case of quadratic forms α, β, γ. We restrict attention to sums of squares only to simplify the exposition. The results in the general case are stated in the appendix.

Given a composition $f : U \times V \to W$, any $u \in U$ provides a map $f_u : V \to W$ defined by $f_u(v) = f(u, v)$. We sometimes blur the distinction between u and f_u and view U as a subset of $\mathrm{Hom}(V, W)$. For any $g \in \mathrm{Hom}(V, W)$ recall that the adjoint $\tilde{g} \in \mathrm{Hom}(W, V)$ is defined by:

$$b_V(v, \tilde{g}(w)) = b_W(g(v), w) \quad \text{for every } v \in V,\ w \in W.$$

Since $(U, \alpha) \simeq r\langle 1\rangle$ there is an orthonormal basis $f_1, \dots, f_r$ of U. These maps $f_i : V \to W$ then satisfy the Hurwitz Equations:

$$\tilde{f}_i f_i = 1_V \quad \text{and} \quad \tilde{f}_i f_j + \tilde{f}_j f_i = 0_V \quad \text{whenever } i \neq j.$$

A choice of orthonormal bases for V and W provides $n \times s$ matrices A_i representing f_i. Then $A_i^{\top}$ is the matrix of $\tilde{f}_i$. Let $X = (x_1, \dots, x_r)$ be a system of indeterminates and define $A = x_1A_1 + \dots + x_rA_r$. As indicated in Chapter 0, the following statements are equivalent:

(1) $[r, s, n]$ is admissible over F.

(2) There exist $n \times s$ matrices $A_1, \dots, A_r$ over F satisfying:

$$A_i^{\top} \cdot A_i = 1_s \quad \text{and} \quad A_i^{\top} \cdot A_j + A_j^{\top} \cdot A_i = 0 \quad \text{whenever } 1 \le i \neq j \le r.$$

(3) There exists an $n \times s$ matrix A over $F(X)$, having entries which are linear forms in X, and satisfying:

$$A^{\top} \cdot A = \alpha(X) \cdot 1_s \quad \text{where } \alpha(X) = x_1^2 + \dots + x_r^2 \in F(X).$$

In the classical case $s = n$ the equations were normalized by arranging $A_1 = 1_n$. To employ a similar normalization here, choose an orthonormal basis $\{v_1, \dots, v_s\}$ for V. Since f_1 is an isometry, the vectors $f_1(v_1), f_1(v_2), \dots, f_1(v_s)$ are orthonormal in W. By Witt Cancellation, these extend to an orthonormal basis $\{f_1(v_1), \dots, f_1(v_s), w_1, \dots, w_{n-s}\}$. Using these bases, the matrix of f_1 is $A_1 = \begin{pmatrix} 1_s \\ 0 \end{pmatrix}$, and the other matrices are $A_i = \begin{pmatrix} B_i \\ C_i \end{pmatrix}$ for $s \times s$ matrices B_i and $(n - s) \times s$ matrices C_i. The Hurwitz Matrix Equations can then be expressed in terms of the B_i's and C_i's. However it turns out to be more convenient to use the version with indeterminates: Let $X' = (x_2, \dots, x_r)$, let $\alpha'(X') = x_2^2 + \dots + x_r^2$ and $A' = x_2A_2 + \dots + x_rA_r$. Then $\alpha(X) = x_1^2 + \alpha'(X')$ and $A = x_1A_1 + A'$. Using $A_1 = \begin{pmatrix} 1_s \\ 0 \end{pmatrix}$ and $A' = \begin{pmatrix} B \\ C \end{pmatrix}$, we obtain a fourth statement equivalent to the admissibility of $[r, s, n]$ over F:

(4) There exist an $s \times s$ matrix B and an $(n - s) \times s$ matrix C over $F(X')$, having entries which are linear forms in X', and satisfying:

$$B^{\top} = -B \quad \text{and} \quad B^2 = -\alpha'(X') \cdot 1_s + C^{\top}C, \ \text{where } \alpha'(X') = x_2^2 + \dots + x_r^2.$$

In the case of Adem's Theorem, $s = n - 1$ and C is a row vector. This leads to the following key result (always assuming $2 \neq 0$). Here we use the dot product notation: If u, v are column vectors then $u \bullet v = u^{\top}v$ is the usual dot product.

14.11 Lemma. *Suppose B is an $s \times s$ matrix over a field K such that $B^\top = -B$ and $B^2 = -d \cdot 1_s + c \cdot uu^\top$ where $u \in K^s$ is a column vector and $c, d \in K^\bullet$. If s is even then $u = 0$. If s is odd then $u \bullet u = c^{-1}d$ and $Bu = 0$.*

Proof. If $u = 0$ then s is even. For in that case $B^2 = -d \cdot 1$ and B has rank s. Since B is skew-symmetric it has even rank, so s must be even. Suppose $u \neq 0$. Since B commutes with $uu^\top = c^{-1}(B^2 + d \cdot 1)$ the matrix $Buu^\top$ is skew symmetric of rank ≤ 1. Then $Buu^\top = 0$ and hence $Bu = 0$ and $\text{rank}(B) < s$. Also $0 = B^2u = -d \cdot u + c \cdot uu^\top u = (-d + c(u \bullet u))u$ and therefore $u \bullet u = c^{-1}d$. Since $s = \text{rank}(d \cdot 1) = \text{rank}(-B^2 + c \cdot uu^\top) \leq \text{rank}(B) + 1 \leq (s-1) + 1 = s$, we find that $\text{rank}(B) = s - 1$. Since B has even rank, s must be odd. □

Proof of Adem's Theorem 14.10. Since $[r, n-1, n]$ is admissible then, as above, the corresponding $n \times (n-1)$ matrix is $A' = \begin{pmatrix} B \\ u^\top \end{pmatrix}$ where B is an $(n-1) \times (n-1)$ matrix and u is a column vector over $F(X')$. The entries of these matrices are linear forms in X' and they satisfy:

$$B^\top = -B \quad \text{and} \quad B^2 = -\alpha'(X') \cdot 1_{n-1} + uu^\top.$$

If s is even we want to "contract" the matrix A' to a skew-symmetric $(n-1) \times (n-1)$ matrix. In that case (14.11) for $K = F(X')$ implies $u = 0$ and $B^2 = -\alpha'(X') \cdot 1_{n-1}$. This says exactly that $[r, n-1, n-1]$ is admissible over F.

If s is odd we want to "expand" A' to a skew-symmetric $n \times n$ matrix. The unique skew-symmetric expansion of A' is $\hat{A}' = \begin{pmatrix} B & -u \\ u^\top & 0 \end{pmatrix}$. Certainly the entries of $\hat{A}'$ are linear forms in X'. The equation $\hat{A}'^\top \cdot \hat{A}' = \alpha'(X')1_n$ follows from (14.11). Consequently $[r, n, n]$ is admissible over F. □

That matrix lemma provides a quick proof, but it hides a basic geometric insight into the problem. View the given $n \times (n-1)$ matrix A as a system of $n-1$ orthogonal vectors of length $\alpha(X)$ in K^n. There is a unique line in K^n orthogonal to those vectors. If n is even, discriminants show that there is a vector on that line of length $\alpha(X)$. Use that vector to expand A to an $n \times n$ matrix $\hat{A}$ which certainly satisfies $\hat{A}^\top \hat{A} = \alpha(X) \cdot 1_n$. The difficulty is to show that the new vector has entries which are linear forms in X. This can be done using an explicit formula for that new vector, found with exterior algebra. If n is odd that line contains a vector with constant entries and we can restrict things to the orthogonal complement. Details for this method appear in Shapiro (1984b). Alternatively, we can use the system of $n \times (n-1)$ matrices $A_1, \ldots, A_r$, perform the expansion on each one and show that those expansions interact nicely. Compare Exercise 6.

This geometric insight into Adem's Theorem depends heavily on the hypothesis of codimension 1. If we have only $n-2$ orthogonal vectors in n-space the expansion of the orthogonal basis is not unique and seems harder to handle. However, Adem (1980) did prove that [4, 5, 7] cannot be admissible, a codimension 2 situation. Yuzvinsky (1983) extracted the geometric idea from Adem's matrix calculations and proved that if $n \equiv 3 \pmod 4$ then $[4, n-2, n]$ cannot be admissible over F. Adem (1986a) simplified Yuzvinsky's proof by returning to the matrix context, and he proved additionally that if $n \equiv 1 \pmod 4$ then any composition of size $[r, n-2, n]$ induces one of size $[r, n-1, n-1]$ and Hurwitz–Radon then implies $r \le \rho(n-1)$. These results are all included in Theorem 14.18 below. The ideas in the proof are clarified using the concept of a "full" pairing.

14.12 Definition. A bilinear map $f : U \times V \to W$ is *full* if image(f) spans W.

Equivalently, the pairing f is full if the associated linear map $f^{\otimes} : U \otimes V \to W$ is surjective. Of course an arbitrary bilinear f has an associated full pairing $f_0 : U \times V \to W_0$, where $W_0 = \text{span}(\text{image}(f))$. However if f is a composition formula for three quadratic spaces over F, this f_0 could fail to be a composition because W_0 might be a singular subspace of W. This problem does not arise if (W, γ) is anisotropic, as in the classical case $\gamma = n\langle 1\rangle$ over $\mathbb{R}$. But even if W_0 is singular we still get a corresponding full composition formula by analyzing the radical $\text{rad}(W_0) = W_0 \cap W_0^{\perp}$. Of course, $\text{rad}(W_0) = (0)$ if and only if W_0 is a regular quadratic space.

14.13 Lemma. *If $f : U \times V \to W$ is a composition of quadratic spaces then there is an associated full composition $\bar{f} : U \times V \to \bar{W}$ where* $\dim \bar{W} \le \dim W$.

Proof. For W_0 as above, let $\bar{W} = W_0/\text{rad}(W_0)$ with induced quadratic form $\bar{\gamma}$ defined $\bar{\gamma}(x + \text{rad}(W_0)) = \gamma(x)$. Then $\bar{\gamma}$ is well defined and $(\bar{W}, \bar{\gamma})$ is a regular space. It is now easy to define $\bar{f}$ and to check that it is a full composition. This $\bar{W}$ can be embedded in W. It is isometric to any subspace of W_0 complementary to $\text{rad}(W_0)$. □

14.14 Lemma. (1) *Suppose a bilinear pairing f is a direct sum of pairings g_1, g_2. Then f is full if and only if g_1 and g_2 are full.*

(2) *If $n = r *_F s$, the minimal size, then every composition of size* $[r, s, n]$ *over F is full.*

(3) *A pairing of size* $[r, s, n]$ *where $n > rs$ cannot be full. The tensor product pairing of size* $[r, s, rs]$ *is full.*

Proof. (1) Suppose $g_j : U \times V_j \to W_j$ are pairings of size $[r, s_j, n_j]$ and the direct sum is $f = g_1 \oplus g_2 : U \times (V_1 \oplus V_2) \to (W_1 \oplus W_2)$, a pairing of size $[r, s_1 + s_2, n_1 + n_2]$. It is defined by: $f(x, (y_1, y_2)) = (g_1(x, y_1), g_2(x, y_2))$. Then $\text{image}(f^{\otimes}) = \text{image}(g_1^{\otimes}) \times 0 + 0 \times \text{image}(g_2^{\otimes}) = \text{image}(g_1^{\otimes}) \times \text{image}(g_2^{\otimes})$, and the statement follows easily.

(2) Suppose $f : U \times V \to W$ is a composition over F of size $[r, s, n]$. If it is not full then apply (14.13) to contradict the minimality of n.

(3) If $f : U \times V \to W$ is a full bilinear pairing of size $[r, s, n]$ then $n = \dim(W) = \dim(\text{image}(f^{\otimes})) \le \dim(U \otimes V) = rs$. The pairing $U \times V \to U \otimes V$ is bilinear and its image contains every decomposable tensor $x \otimes y$. □

With the terminology of full pairings, Adem's Theorem 14.10 can be stated more simply as follows.

14.10^{bis} Adem's Theorem. *Suppose $f : U \times V \to W$ is a full composition of size $[r, n-1, n]$ over F. Then n must be even and f can be extended to a composition $\hat{f} : U \times W \to W$.* □

Here is the matrix version of the condition that f is full.

14.15 Lemma. *Suppose $f : U \times V \to W$ is a composition as above, represented by the $n \times s$ matrices $A_i = \begin{pmatrix} B_i \\ C_i \end{pmatrix}$ where $B_1 = 1_s$ and $C_1 = 0$. View the $(n-s) \times s$ matrix C_i as a linear map $F^s \to F^{n-s}$. Then:*

$$f \textit{ is full if and only if } \text{image}(C_2) + \cdots + \text{image}(C_r) = F^{n-s}.$$

Proof. Recall that A_i is the matrix of the map $f_i = f(u_i, -) : V \to W$. Then $\text{span}(\text{image}(f)) = \text{image}(f_1) + \cdots + \text{image}(f_r)$. The decomposition $W = V_1 \perp V_1^{\perp}$ arises from $V_1 = \text{image}(f_1)$ and provides the maps $B_i : V \to V_1$ and $C_i : V \to V_1^{\perp}$. Then f is full if and only if every $w = v + v' \in W$ can be expressed as $\sum_{i=1}^{r}(B_i v_i + C_i v_i)$ for some $v_i \in V$. Since $B_1 = 1$ and $C_1 = 0$ this is equivalent to saying that every $v' \in V_1^{\perp}$ can be expressed as $\sum_{i=2}^{r} C_i v_i$ for some $v_i \in V$. □

14.16 Lemma. *Suppose B is an $s \times s$ matrix over F and B is similar to $-B$. If $d \in F^{\bullet}$ then:* $\text{rank}(d \cdot 1_s + B^2) \equiv s \pmod 2$.

Proof. We may pass to the algebraic closure and work with Jordan forms. For each $k \times k$ Jordan block J of B, compare $\text{rank}(d \cdot 1_k + J^2)$ to k. If J has eigenvalue λ then J^2 has λ^2 as its only eigenvalue. If $\lambda^2 \ne -d$ then $d \cdot 1_k + J^2$ is nonsingular and $\text{rank}(d \cdot 1_k + J^2) = k$. If $\lambda^2 = -d$, a direct calculation shows that $d \cdot 1_k + J^2$ has rank $k - 1$. Since $B \sim -B$ and $d \ne 0$ there is a matching block J' with eigenvalue $-\lambda$, and the pair $J \oplus J'$ contributes a $2k \times 2k$ block of rank $2k - 2$. Putting these blocks together, we find that $\text{rank}(d \cdot 1_s + B^2)$ differs from s by an even number. □

14.17 Proposition. *There exists a full composition of size* $[2, s, n]$ *if and only if n is even and $s \le n \le 2s$.*

Proof. If there is a composition of that size then certainly $s \le n \le 2s$ by (14.14). With the usual normalizations we get matrices $A_1 = \begin{pmatrix} 1_s \\ 0 \end{pmatrix}$, and $A_2 = \begin{pmatrix} B \\ C \end{pmatrix}$ where B is skew symmetric $s \times s$ and $1 + B^2 = C^\top C$. Since the pairing is full (14.15) implies image$(C) = F^{n-s}$. Then C represents a surjection $F^s \to F^{n-s}$ so that image$(C^\top C) =$ image$(C^\top)$ and rank$(C^\top C) = n - s$. The Lemma now implies n is even. Conversely, full pairings of those sizes are constructed in Exercise 7. □

Now we begin to analyze the compositions of codimension 2, that is, of size $[r, n-2, n]$. Gauchman and Toth (1994) characterized all the full compositions of codimension 2 over $\mathbb{R}$. In (1996) they extended their results to compositions of indefinite forms over $\mathbb{R}$. Here is a new argument which generalizes their results to compositions of codimension 2 over any field F (where $2 \ne 0$, of course).

14.18 Theorem. *Suppose $f : U \times V \to W$ is a full composition over F of size $[r, n-2, n]$.*

(1) *If n is odd then $r = 3$, $n \equiv 3 \pmod 4$ and f is a direct sum of compositions of sizes $[3, n-3, n-3]$ and $[3, 1, 3]$.*

(2) *If n is even then f expands to a composition of size $[r, n, n]$, so that $r \le \rho(n)$.*

Over $\mathbb{R}$ this theorem is stronger than the topological results for those sizes, given in Chapter 12. Those methods eliminate certain sizes, but provide no information about the internal structure of the compositions which do exist. In fact this Theorem works for compositions of arbitrary quadratic forms over F, not just the sums of squares considered here. That version is stated in the appendix.

This theorem quickly yields all the possible sizes for compositions of codimension 2.

14.19 Corollary. *Suppose there is a composition of size $[r, n-2, n]$ over F.*

(1) *If n is odd then: either $r \le \rho(n-1)$ or $r = 3$ and $n \equiv 3 \pmod 4$.*

(2) *If n is even then: either $r \le \rho(n)$ or $r \le \rho(n-2)$.*

Proof. We may assume $r > 1$.

(1) If f is full the theorem applies. Otherwise (14.13) yields a composition of size $[r, n-2, k]$ where $k < n$. Certainly $k \ge n-2$ and equality is impossible since $n-2$ is odd. Then $k = n-1$ and Adem's Theorem (14.10) yields an expansion to $[r, n-1, n-1]$.

(2) If f is not full there is a composition of size $[r, n-2, k]$ as above. If $k = n-1$ Adem's Theorem applies, so in any case there is one of size $[r, n-2, n-2]$. □

The proof of the theorem is fairly long and will be broken into a number of steps. First we will set up the notations, varying slightly from the discussion after (14.10).

Let $s = n - 2$. From the given composition f we obtain $n \times s$ matrices $A_i = \begin{pmatrix} B_i \\ C_i \end{pmatrix}$ over F, where $2 \le i \le r$. Let $X = (x_2, \ldots, x_r)$ be a system of $r - 1$ indeterminates and let K be the rational function field $K = F(X)$. Define $A = \begin{pmatrix} B \\ C \end{pmatrix} = \sum_{i=2}^{r} x_i A_i$. Here is a summary of the given properties:

B is an $s \times s$ matrix; C is a $2 \times s$ matrix;

the entries of B and C are linear forms in $F[X]$;

$B^\top = -B$;

$B^2 = -a1_s + C^\top C$ where $a = x_2^2 + \cdots + x_r^2 \in F[X]$;

the pairing is full: $\text{image}(C_2) + \cdots + \text{image}(C_r) = F^2$.

During this proof we abuse the notations in various ways. For example the square matrix B is sometimes considered as a mapping $K^s \to K^s$ (using column vectors), and other times each B_i is viewed as a mapping $V \to V_1 \subseteq W$ where $V_1 = \text{image}(f_1)$.

Proof of Theorem when n is odd. Since $s = \text{rank}(-B^2 + C^\top C) \le \text{rank}(B) + 2$ we find $s - 2 \le \text{rank}(B) \le s$. Since B is skew symmetric it has even rank. Therefore $\text{rank}(B) = s - 1$. This implies that $\ker(B) = Ku$ is a line generated by some non-zero column vector $u \in K^s$.

Claim 1. $\text{rank}(C) = 1$ and $BC^\top = 0$.

Proof. Note that $BC^\top C = B^3 + aB$ is skew symmetric, hence of even rank ≤ 2. Suppose it is non-zero, so that it has rank 2. Then $C^\top C$ has rank 2, and $\mathcal{S} = \text{image}(C^\top C)$ is a 2-dimensional space. This space is preserved by the map B since B commutes with $C^\top C$. Certainly $u \in \mathcal{S}$ since $0 = B^2 u = -au + C^\top Cu$, and hence B is not injective on $\mathcal{S}$. But $\dim B(\mathcal{S}) = \text{rank}(BC^\top C) = 2$, a contradiction. Therefore $BC^\top C = 0$. We know $C^\top C \ne 0$ because B is singular, and therefore $\text{image}(C^\top C) = \ker(B) = Ku$. If $\text{rank}(C) = 2$ then C represents a surjective map $F^s \to F^2$ and $\text{image}(C^\top C) = \text{image}(C^\top)$ is 2 dimensional, not a line. Then $\text{rank}(C) = 1$ and $\text{image}(C^\top)$ is a line containing $\text{image}(C^\top C) = \ker(B)$. Hence $\text{image}(C^\top) = \ker(B)$ and $BC^\top = 0$, proving the claim.

The vector u is determined up to a scalar multiple in $K^\bullet$. Scale u to assume that its entries are polynomials with no common factor.

Claim 2. $u \in F^s$ is a column vector with constant entries, $u \bullet u \ne 0$, and $C = \begin{pmatrix} \alpha \\ \beta \end{pmatrix} \cdot u^\top$ for some linear forms $\alpha, \beta \in F[X]$.

Proof. Since $\text{image}(C^\top) = Ku$, there exist $\alpha, \beta \in K$ such that $C^\top = (\alpha u, \beta u) = u(\alpha, \beta)$. Then $C^\top C = (\alpha^2 + \beta^2) \cdot uu^\top$. Since $C^\top C \ne 0$ we know $\alpha^2 + \beta^2 \ne 0$. Moreover $0 = B^2 C^\top = (-a1_s + C^\top C)C^\top$ so that $C^\top CC^\top = aC^\top$. If $u \bullet u = 0$ we would have $CC^\top = 0$ and hence $C = 0$, a contradiction. Therefore $u \bullet u \ne 0$.

Express $C^\top = (v_1, v_2)$ for vectors v_i with linear form entries. Switching indices if necessary we may assume $v_1 \ne 0$. Then $v_1 = \alpha u$, and unique factorization implies

$\alpha \in F[X]$. (For if $\alpha = \alpha_1/\alpha_2$ in lowest terms, then α_2 is a common factor of the entries of u.) Therefore $\deg(\alpha) \leq 1$. Suppose $\deg(\alpha) = 0$ so that $\alpha \in F^\bullet$ is a constant. Then the entries of u must be linear forms in X and $v_2 = \beta u$ implies that also $\beta \in F$. Expanding $u = x_2u_2 + \cdots + x_ru_r$ for $u_j \in F^s$ we find that $C_j = \begin{pmatrix} \alpha \\ \beta \end{pmatrix} \cdot u_j^\top$ and $\text{image}(C_j) \subseteq F \cdot \begin{pmatrix} \alpha \\ \beta \end{pmatrix}$ for each j. This contradicts the "full" hypothesis. Therefore $\deg(\alpha) = 1$ and $u \in F^s$ has constant entries, proving the claim.

Now let us undo the identifications and interpret these statements in terms of the original maps $f_j : V \to W$. Recall that $V_1 = \text{image}(f_1)$ was identified with V, and the decomposition $W = V_1 \oplus V_1^\perp$ provided the block matrices. The matrix of f_j was $A_j = \begin{pmatrix} B_j \\ C_j \end{pmatrix}$ where now we view $B_j : V \to V_1$ and $C_j : V \to V_1^\perp$ as linear maps. With this notation, if $y \in V$ then $f_j(y) = B_j(y) + C_j(y)$.

Now $u \in V$ by Claim 2. Define $V_0 = (u)^\perp \subseteq V$, an F-subspace of dimension $s - 1 = n - 3$. If $y \in V_0$ then $u \bullet y = 0$, and computing over K we have: $Cy = \begin{pmatrix} \alpha \\ \beta \end{pmatrix} \cdot u^\top y = 0$, and $u \bullet (By) = (-Bu) \bullet y = 0$. Writing out B and C in terms of the x_j, these equations become:

$$C_j y = 0 \quad \text{and} \quad u \bullet (B_j y) = 0 \quad \text{for every } j \geq 2.$$

Undoing the identification of V and V_1 here, the second condition says: $B_j y$ and $f_1(u)$ are orthogonal in V_1. Let $W_0 = f_1(V_0)$ so that $W_0 = (f_1(u))^\perp$ inside V_1. Then we have proved:

$$\text{If } y \in V_0 \text{ then } f_j(y) = B_j(y) \in W_0.$$

Consequently $f_j : V_0 \to W_0$ for every j, and the original pairing f restricts to a pairing $f' : U \times V_0 \to W_0$ of size $[r, n-3, n-3]$.

Since those maps f_j are isometries they preserve orthogonal complements. The induced composition $f'' : U \times V_0^\top \to W_0^\top$ has size $[r, 1, 3]$, implying $r \leq 3$. Since the original pairing of size $[r, n-2, n]$ is full we know $r \neq 1$ and (14.17) implies $r \neq 2$. Therefore $r = 3$. The pairings f' and f'' provide the direct sum referred to in the statement of the theorem.

Proof of Theorem when n is even. We want to expand the given $n \times s$ matrix $A = \begin{pmatrix} B \\ C \end{pmatrix}$ to an $n \times n$ matrix $\hat{A}$ which has entries which are linear forms, is skew symmetric and satisfies $\hat{A}^2 = -a \cdot 1_n$. This larger matrix must be $\hat{A} = \begin{pmatrix} B & -C^\top \\ C & D \end{pmatrix}$ where $D = \begin{pmatrix} 0 & -d \\ d & 0 \end{pmatrix}$ and d is some linear form in X. The condition on $\hat{A}^2$ becomes:

$$BC^\top = -C^\top D \quad \text{and} \quad CC^\top = (a - d^2)1_2.$$

If we can find a linear form d satisfying these two conditions then the proof is complete.

As before we know that $s-2 \leq \text{rank}(B) \leq s$ and rank(B) is even. Rather than working directly with B we concentrate on C. Note that $C \neq 0$ since the pairing is full.

Claim 1. rank$(C) = 2$.

Proof. Suppose rank$(C) = 1$. Then $C^\top = (\alpha u, \beta u) = u \cdot (\alpha, \beta)$ for some $0 \neq u \in K^s$ and $\alpha, \beta \in K$. Then $C^\top C = (\alpha^2+\beta^2)uu^\top$ has rank ≤ 1 and $B^2 = -a1_s + (\alpha^2+\beta^2)uu^\top$. Since s is even and $u \neq 0$, (14.11) implies that $\alpha^2+\beta^2 = 0$. If $\alpha = \beta = 0$ then $C = 0$, a contradiction. Therefore α, β are non-zero and $(\beta/\alpha)^2 = -1$. (Note: if $\sqrt{-1} \notin F$ then $\sqrt{-1} \notin K$ and this is already impossible.) Then $\beta = \sqrt{-1}\cdot\alpha$, where $\sqrt{-1} \in F$, and $C = \alpha \cdot \begin{pmatrix} 1 \\ \sqrt{-1} \end{pmatrix}$. As in the proof of the odd case we obtain a contradiction to the "full'' hypthesis, proving the claim.

The claim shows that the mapping $C : K^s \to K^2$ is surjective so that $\mathcal{S} = \text{image}(C^\top C) = \text{image}(C^\top)$ is a 2-dimensional subspace of K^s. The map B preserves $\mathcal{S}$ since B commutes with $C^\top C$. Writing $C^\top = (v, w)$ for column vectors $v, w \in K^s$, we have $\mathcal{S} = \text{span}\{v, w\}$ and

$$Bv = \alpha v + \beta w$$
$$Bw = \gamma v + \delta w$$

for some $\alpha, \beta, \gamma, \delta \in K$. These equations say: $BC^\top = C^\top D$ where $D = \begin{pmatrix} \alpha & \gamma \\ \beta & \delta \end{pmatrix}$.

Claim 2. $D = \begin{pmatrix} 0 & -d \\ d & 0 \end{pmatrix}$ for some $d \in K$, and $CC^\top = (a-d^2)1_2$.

Proof. $C^\top DC = BC^\top C = B^3 + aB$ is skew symmetric. Since C is a $2 \times s$ matrix of rank 2 there exists an $s \times 2$ matrix C' satisfying: $CC' = 1_2$. Then $D = C'^\top(B^3+aB)C'$ is skew symmetric so it has the stated form for some $d \in K = F(X)$. The defining equation for D also implies $C^\top D^2 = B^2C^\top = (-a1_s + C^\top C)C^\top = C^\top(-a1_2+CC^\top)$. Multiply by $C'^\top$ to conclude that $D^2 = -a1_2+CC^\top$. The claim follows since $D^2 = -d^2 1_2$.

We know that $d \in K = F(X)$. Since a is a quadratic form and the entries of C are linear forms, the second equation in Claim 2 implies that d^2 is a quadratic form in X. Unique factorization and comparison of highest degree terms implies that d must be a linear form in X. This completes the proof of the theorem. □

We can now determine the admissible sizes $[r, s, n]$ for small values of r. Recall from (12.13) that the admissible sizes over $\mathbb{R}$ are known whenever $r \leq 9$. Using (14.2) this result extends to fields of characteristic 0. It seems far more difficult to prove this when F has positive characteristic.

14.20 Corollary. *Let F be a field of characteristic not 2. If $r \leq 4$ then: $[r, s, n]$ is admissible over F if and only if $r \circ s \leq n$.*

Proof. If $r \circ s \leq n$ then there exists an *integer* composition of size $[r, s, n]$. Such a composition formula is then valid over any field F. This construction of integer formulas works whenever $r \leq 9$, as mentioned in (12.13). Conversely, suppose $r \leq 4$ and $[r, s, n]$ is admissible over F. Since $r \circ s \leq r + s - 1$ we may also assume that $n \leq r + s - 2$. Then:

$$r \circ s \leq n \text{ if and only if } r' \circ s' \leq n \text{ whenever } r' \leq r,\ s' \leq s \text{ and } n = r' + s' - 2.$$

(This reduction, due to Behrend (1939), appears in Exercise 10.) Therefore it suffices to prove the result when $n = r + s - 2$.

The case $r = 1$ is vacuous. If $r = 2$ then $s = n$ and $[r, n, n]$ is admissible. Then $2 \leq \rho(n)$ so that n is even and $2 \circ n = n$. If $r = 3$ then $s = n - 1$ and $[3, n - 1, n]$ is admissible. Adem's Theorem and Hurwitz–Radon imply that $n \equiv 0, 1 \pmod 4$. This is equivalent to the condition $3 \circ (n - 1) \leq n$. Suppose $r = 4$ so that $s = n - 2$ and $[4, n - 2, n]$ is admissible. Theorem 14.18(1) shows that $n \not\equiv 3 \pmod 4$. Check that $4 \circ (n - 2) \leq n$ if and only if $n \not\equiv 3 \pmod 4$. □

The smallest open question here seems to be: Is $[5, 9, 12]$ admissible over some field F? Since $5 \# 9 \geq 5 \circ 9 = 13$, (14.2) implies that $[5, 9, 12]$ is not admissible over any field of characteristic zero. By (14.19) we know that $[5, 10, 12]$ and $[5, 9, 11]$ are not admissible over F. The case $[5, 9, 12]$ can be eliminated by invoking Theorem A.6 below. However the case $[5, 10, 13]$ still remains open.

Theorem 14.18 also provides a calculation of $\rho_F(n, n - 2)$. It matches the values over $\mathbb{R}$ found in (12.30).

14.21 Corollary. *If F is a field with characteristic $\neq 2$.*

If $n - 2 \leq s \leq n$ then $\rho_F(n, s) = \rho(n, s)$.

If $1 \leq r \leq 4$ then $\rho_F(n, r) = \rho_\circ(n, r)$.

Proof. (14.18) implies then first statement. The second follows from (14.20) and the definition of $\rho_\circ$ in (12.24). Values of $\rho_\circ$ are calculated in (12.28). □

These corollaries provide some evidence for a wilder hope:

14.22 Bold Conjecture. If $[r, s, n]$ is admissible over some field F (of characteristic not 2) then it is admissible over $\mathbb{Z}$. Consequently, admissibility is independent of the base field.

This conjecture is true if both r, s are at most 8. It holds true when $s = n$ by the Hurwitz–Radon Theorem. By (14.21) the conjecture is true whenever $r \leq 4$ and whenever $s \geq n - 2$. Every known construction of admissible triples $[r, s, n]$ can be done over $\mathbb{Z}$. But of course not many constructions are known! There really is very little evidence supporting this conjecture, but it certainly would be nice if it could be proved true.

What are the possible sizes of full compositions? Certainly if there exists a full composition of size $[r, s, n]$ over F then $r *_F s \leq n \leq rs$. The case $r = 2$ is settled by (14.17). For monomial compositions (defined in Chapter 13, Appendix B) the answer is fairly easy. A consistently signed intercalate matrix of type (r, s, n) corresponds to a composition of size $[r, s, n]$ over $\mathbb{Z}$ and hence one over F. That composition is full if and only if the matrix involves all n colors. For example, Exercise 13.2 provides consistently signed intercalate 3×3 matrices with exactly 4, 7 and 9 colors. Then we obtain full monomial compositions of sizes $[3, 3, n]$ for $n = 4$, 7 and 9. On the other hand there exist full compositions of size $[3, 3, 8]$, which therefore cannot be monomial. See Exercise 12.

In Chapter 8 we considered the space of all compositions of size $[s, n, n]$. More generally one can investigate the set of all compositions of size $[r, s, n]$ over a field F. Not surprisingly, these are much harder to classify. Let us call two such compositions equivalent if they differ by the action of the group $\mathrm{O}(r) \times \mathrm{O}(s) \times \mathrm{O}(n)$. Yuzvinsky (1981) discussed various versions of this classification problem over $\mathbb{R}$ and gave a complete description of the set of equivalence classes for the sizes $[2, s, n]$. Adem (1986b) worked over an algebraically closed field and determined the set of equivalences classes of pairings of sizes $[2, s, n]$ when $s = n - 1$ and when $s = n - 2$ is even. Using different methods, Toth (1990) noted that for fixed r, s the space of equivalence classes of full compositions of size $[r, s, n]$ over $\mathbb{R}$ can be parametrized by the orbit space of an invariant compact convex body $\mathcal{L}$ in $\mathrm{SO}(r) \otimes \mathrm{SO}(s)$. The compositions of minimum size, the ones with $n = r * s$, form a compact subset of the boundary of $\mathcal{L}$. Good descriptions of this space $\mathcal{L}$ are known in the cases $r = s = 2$ or 3, as studied by Parker (1983). Guo (1996) considers other cases where $r = 2$.

Nonsingular pairings were the central theme of Chapter 12 and the definition makes sense for any base field. Certainly every composition over $\mathbb{R}$ (for sums of squares) is an example of a nonsingular pairing. However over other fields those concepts diverge. Nonsingular pairings are closely related to certain subspaces of matrices.

14.23 Lemma. *For any field F the following are equivalent.*

(1) *There is a full nonsingular bilinear $[r, s, n]$ over F.*

(2) *There is a linear subspace $W \subseteq \mathbb{M}_{r\times s}(F)$ with $\dim W = rs - n$ and such that W contains no matrix of rank 1.*

Proof. Suppose $f : X \times Y \to Z$ is full nonsingular bilinear, where $\dim X = r$, $\dim Y = s$ and $\dim Z = n$. This induces a surjective linear map $f^{\otimes} : X \otimes Y \to Z$, and $U = \ker(f^{\otimes})$ is a subspace of $X \otimes Y$ of dimension $rs - n$ such that: if $x \otimes y \in U$ then $x \otimes y = 0$. The standard identification of $X \otimes Y$ with $\mathrm{Hom}(Y, X) \cong \mathbb{M}_{r\times s}(F)$ sends $x \otimes y$ to $x \cdot y^{\top}$, viewing x, y as column vectors. The pure tensors become matrices of rank ≤ 1 and U becomes the desired subspace W. The converse follows by reversing the process. □

What sorts of nonsingular pairings are there over the complex field $\mathbb{C}$? L. Smith (1978) considered a nonsingular bilinear map $f : \mathbb{C}^r \times \mathbb{C}^s \to \mathbb{C}^n$ and worked out the analog of Hopf's proof by using the induced map on complex projective spaces, and the cohomology over $\mathbb{Z}$. He proved that $n \geq r + s - 1$. We define $r \#_F s$ to help clarify this inequality.

14.24 Definition. $r\#_F s = \min\{n :$ there exists a nonsingular bilinear $[r, s, n]$ over $F\}$.

Then for any field F,

$$\max\{r, s\} \leq r \#_F s \leq r + s - 1.$$

The upper bound follows from the existence of the Cauchy product pairing as defined in (12.12). Smith's result says that the upper bound is achieved if $F = \mathbb{C}$. The matrix ideas above lead to a more algebraic proof. Compare Exercise 13.

14.25 Proposition. *If F is algebraically closed then $r \#_F s = r + s - 1$.*

Proof. Given a nonsingular $[r, s, n]$ over F we will prove $n \geq r + s - 1$. We may assume the map is full (possibly decreasing n) and apply (14.23) to find a subspace $W \subseteq \mathbb{M}_{r\times s}(F)$ with $\dim W = rs - n$ and $W \cap \mathcal{R}_1 = \{0\}$. Here $\mathcal{R}_1$ denotes the set of matrices of rank ≤ 1. This $\mathcal{R}_1$ is an algebraic set (the zero set of all the 2×2 minor determinants). The map $F^r \times F^s \to \mathcal{R}_1$ sending (u, v) to $u \cdot v^\top$ is surjective with 1-dimensional fibers. Since F is algebraically closed the properties of dimension imply that $\dim \mathcal{R}_1 = r + s - 1$. If $n < r + s - 1$ then $\dim W + \dim \mathcal{R}_1 > rs$. Then $W \cap \mathcal{R}_1$ must have positive dimension, contrary to hypothesis. □

At the other extreme, $r \#_F s = \max\{r, s\}$ provided F admits field extensions of every degree. See Exercise 15.

When $F = \mathbb{R}$ the topological methods of Hopf and Stiefel provide the stronger lower bound $r \circ s \leq r \# s$. This lower bound remains valid for a larger class of fields. Behrend (1939) proved it over any real closed field, and his proof has been put into a general context in Fulton (1984). We describe a different generalization here. Recall that for any system of n forms (homogeneous polynomials) in $\mathbb{C}[X]$, involving m variables, if $m > n$ then there exists a common non-trivial zero. This was extended by Behrend to n forms of odd degree in m variables in $\mathbb{R}[X]$. (See the Notes for Exercise 12.18 for references.) We move from $\mathbb{R}$ to a general p-field.

If p is a prime number, a field F is called a *p-field* if $[K : F]$ is a power of p for every finite field extension K/F. Any real closed field is a 2-field, and other examples of p-fields can be constructed in various ways. Pfister (1994) extended the result above to p-fields: If F is a p-field and $f_1, \ldots, f_n \in F[X]$ are forms in m variables with every $\deg(f_i)$ prime to p then $m > n$ implies the existence of a non-trivial common zero in F^m. The proof is elementary and over $\mathbb{R}$ it leads to a proof of the Borsuk–Ulam Theorem.

Krüskemper (1996) generalized Pfister's results to biforms. Suppose X and Y are systems of indeterminates over F. Then $f \in F[X, Y]$ is a *biform* of degree (d, e) if f is homogeneous in X of degree $d \geq 1$ and f is homogeneous in Y of degree $e \geq 1$. For example a biform of degree $(1, 1)$ is exactly a bilinear form. As one corollary to his "Nullstellensatz", Krüskemper deduced the following algebraic version of the Hopf–Stiefel Theorem.

14.26 Krüskemper's Theorem. *Suppose F is a p-field and $f : F^r \times F^s \to F^n$ is a nonsingular biform of degree (d, e) where p does not divide d or e. Then the binomial coefficient $\binom{n}{k} \equiv 0 \pmod p$ whenever $n - s < k < r$.*

The proof of this theorem certainly involves some work, but it is surprisingly elementary. Of course "nonsingular" here means: $f(a, b) = 0$ implies $a = 0$ or $b = 0$. The map f is built from n biforms $f_i : F^r \times F^s \to F$. The hypothesis means that each f_i is a biform of degree (d, e). Actually Krüskemper allows different degrees (d_i, e_i) with the condition that there exist d, e such that for every i: $d_i \equiv d \not\equiv 0$ and $e_i \equiv e \not\equiv 0 \pmod p$. When $F = \mathbb{R}$ (or any 2-field) Krüskemper's Theorem restricts to the Hopf Theorem for nonsingular bi-skew polynomial maps.

With the notation $\beta_p(r, s)$ given in Exercise 12.25 this theorem implies:

If there is a nonsingular bilinear $[r, s, n]$ over some p-field, then $\beta_p(r, s) \leq n$.

When F is algebraically closed this theorem implies (14.25). For in this case, F is a p-field for every p and if there exists k in that interval, then the binomial coefficient would be 0, which is absurd. Therefore the interval is empty and $n - s \geq r - 1$.

Appendix to Chapter 14. Compositions of quadratic forms α, β, γ

We outline here the few results known for general compositions of three quadratic forms over a field F (assuming $2 \neq 0$, as usual). After reviewing the basic notations we state the theorem about compositions of codimension ≤ 2. Next we consider compositions of indefinite forms over the real field, and then we mention the Szyjewski–Shapiro Theorem, which is an analog of the Stiefel–Hopf result valid for general fields F.

Suppose (U, α), (V, β), (W, γ) are (regular) quadratic spaces over F, with dimensions r, s, n respectively. Suppose the bilinear map

$$f : U \times V \to W$$

is a composition for those forms. This means that $\gamma(f(u, v)) = \alpha(u) \cdot \beta(v)$ for every $u \in U$ and $v \in V$, as mentioned after the statement of (14.10). We use the letters α, β, γ to stand for the corresponding bilinear forms as well. Then $2\beta(x, y) = \beta(x + y) - \beta(x) - \beta(y)$ and $\beta(x) = \beta(x, x)$.

The pairing f provides a linear map $\hat{f} : U \to \text{Hom}(V, W)$ given by $\hat{f}(u)(v) = f(u, v)$. If $u \in U$ then $\hat{f}(u) : V \to W$ is a similarity of norm $\alpha(u)$. Then the composition provides a linear subspace of $\text{Sim}(V, W)$, the set of similarities. If $\alpha(u) \neq 0$ then $\hat{f}(u)$ is injective and hence $s \le n$. Of course the case $s = n$ is the classical Hurwitz–Radon situation and we may use all the results of Part I of this book. So let us concentrate here on the cases $r, s < n$. The next lemma is easily proved and shows that we may assume the forms all represent 1.

A.1 Lemma. (1) *Existence of a composition for α, β, γ depends only on the isometry classes of those forms.*

(2) *Suppose $x, y \in F^\bullet$. There exists a composition for α, β, γ if and only if there is a composition for $\langle x\rangle\alpha$, $\langle y\rangle\beta$, $\langle xy\rangle\gamma$.*

The forms β and γ provide an "adjoint" map $\tilde{\ }: \text{Hom}(V, W) \to \text{Hom}(W, V)$ defined in the usual way using the equation: $\beta(v, \tilde{f}(w)) = \gamma(f(v), w)$. Then $g \in \text{Hom}(V, W)$ is a similarity of norm c if and only if $\tilde{g} \circ g = c \cdot 1_V$. If $\alpha \simeq \langle a_1, \dots, a_r\rangle$ there is an orthogonal basis $\{u_1, \dots, u_r\}$ of U with $\alpha(u_i) = a_i$. Letting $f_i = \hat{f}(u_i)$ we obtain the Hurwitz Equations:

$$\begin{aligned} \tilde{f_i} \circ f_i &= a_i 1_V \\ \tilde{f_i} \circ f_j + \tilde{f_j} \circ f_i &= 0 \end{aligned} \qquad \text{whenever } 1 \le i \neq j \le r.$$

Without writing out the details we state the matrix version of the Hurwitz Equations, following the notations used in the proofs of Theorems (14.10) and (14.18).

A basis $\{v_1, \dots, v_s\}$ of V has the Gram matrix $M = (\beta(v_i, v_j))$. A basis $\{f_1(v_1), \dots, f_1(v_s), w_{s+1}, \dots, w_n\}$ of W has Gram matrix of the form $N = \begin{pmatrix} M & 0 \\ 0 & P \end{pmatrix}$ and yields the matrix $\begin{pmatrix} 1_s \\ 0 \end{pmatrix}$ for f_1. Let $X = (x_2, \dots, x_r)$ be indeterminates, let $K = F(X)$ and define $\alpha'(X) = a_2x_2^2 + \cdots + a_rx_r^2$ in K. Then a composition for α, β, γ is provided by matrices B, C over K such that:

B is an $s \times s$ matrix and C is an $(n - s) \times s$ matrix;

the entries of B and C are linear forms in X;

$\tilde{B} = -B$ and $\tilde{B}B + \tilde{C}C = \alpha'(X)1_s$.

This is similar to the previous situation with transposes, but note that $\tilde{B} = M^{-1}B^{\top}M$ and $\tilde{C} = M^{-1}C^{\top}P$.

A.2 Theorem. *Let F be a field (with $2 \neq 0$), and let (U, α), (V, β), (W, γ) be regular quadratic spaces over F, with dimensions r, s, n, respectively. Suppose α represents 1 and $f : U \times V \to W$ is a full composition for α, β, γ.*

(1) *If $s = n - 1$ then f extends to a composition of α, γ, γ.*

(2) *Suppose $s = n - 2$. If n is odd then $r = 3$, $n \equiv 3 \pmod 4$, there are decompositions $\beta \simeq \beta_0 \perp \langle b\rangle$ and $\gamma \simeq \beta_0 \perp \langle b\rangle\alpha$ and f is a direct sum of compositions*

for α, β_0, β_0 *and* α, $\langle b\rangle$, $\langle b\rangle\alpha$. *If* n *is even then* f *expands to a composition of* α, γ, γ.

The proof follows the ideas used in (14.10) and (14.18) above. Further details appear in Shapiro (1997). This Theorem characterizes all compositions of size $[r, s, n]$ where $s \geq n-2$. We can also settle the "dual" situation where $r \leq 2$, generalizing (14.17).

A.3 Proposition. *Suppose there is a full composition for the quadratic forms* α, β, γ *of size* $[2, s, n]$ *over* F. *If* $\alpha = \langle 1, a\rangle$ *then* $\gamma \simeq \langle 1, a\rangle \otimes \varphi$ *for some form* φ.

Proof outline. Given an $n\times s$ matrix $A = \begin{pmatrix} B \\ C \end{pmatrix}$ with $\tilde{B} = -B$ and $B^2 - \tilde{C}C = -a\cdot 1_s$ and $\operatorname{rank}(C) = n-s$. The idea is to find Y so that the expanded matrix $\hat{A} = \begin{pmatrix} B & \tilde{C} \\ C & Y \end{pmatrix}$ satisfies $\tilde{A} = -A$ and $A^2 = -a\cdot 1_n$. Then (1.10) finishes the proof. □

These results help to characterize some of the quadratic forms which occur in various small compositions. Here are some examples. See Exercise 14 for the proofs.

A.4 Corollary. *Suppose there is a composition for* α, β, γ *of size* $[r, s, n]$.

(1) *If* $[r, s, n] = [3, 3, 4]$ *then after scaling (as in* (A.1)*):*

$$\alpha \simeq \beta \simeq \langle 1, a, b\rangle \quad \textit{and} \quad \gamma \simeq \langle\langle a, b\rangle\rangle.$$

(2) *If* $[r, s, n] = [3, 5, 7]$ *then after scaling:*

$$\alpha \simeq \langle 1, a, b\rangle, \quad \beta \simeq \langle\langle a, b\rangle\rangle \perp \langle x\rangle \quad \textit{and} \quad \gamma \simeq \langle\langle a, b\rangle\rangle \perp \langle x\rangle\langle 1, a, b\rangle.$$

There remain many small examples where little is known. For example, if there is a full composition of size [4, 4, 7] then must γ be a subform of a Pfister form? In the monomial examples of size [10, 10, 16] given in Appendix 13.B, γ is a Pfister form and $\alpha \simeq \beta$. Must these conditions hold for any [10, 10, 16]? Possibly the uniqueness result in (13.14) can be extended to monomial pairings and then applied to show that any monomial [10, 10, 16] must involve a Pfister form. But there might exist non-monomial compositions of this size. There might even be a composition of size [10, 10, 15] over some field! This is impossible in characteristic zero since $10 \circ 10 = 16$.

Let us turn now to compositions of quadratic forms over $\mathbb{R}$, the field of real numbers. Chapter 12 involved these compositions in the positive definite case, but we have much less information about indefinite forms. If α is a real quadratic form with $\dim\alpha = r$ then $\alpha \simeq r_+\langle 1\rangle \perp r_-\langle -1\rangle$ where $r_+ + r_- = r$. As a shorthand here we write this simply as $\alpha = (r_+, r_-)$.

A.5 Proposition. *Suppose there is a composition over $\mathbb{R}$ for the quadratic forms α, β, γ of dimensions r, s, n. Then $r \# s \le n$ and*

$$r_+ \# s_+ \le n_+, \quad r_- \# s_- \le n_+, \quad r_+ \# s_- \le n_-, \quad r_- \# s_+ \le n_-.$$

Proof. Use the Lam–Lam Lemma 14.1, as remarked in Exercise 2. □

As a simple example suppose $r = 3$ and $s = 5$. Using (A.4) we obtain compositions of smallest size [3, 5, 7] in three cases:

$$\alpha = (3, 0),\ \beta = (5, 0) \text{ and } \gamma = (7, 0);$$
$$\alpha = (3, 0),\ \beta = (4, 1) \text{ and } \gamma = (4, 3);$$
$$\alpha = (2, 1),\ \beta = (3, 2) \text{ and } \gamma = (4, 3).$$

If $\alpha = (2, 1)$ and $\beta = (4, 1)$ then (A.5) implies that $\gamma \ge (4, 4)$. Compositions with $\gamma = (4, 4)$ can be found by taking suitable subspaces of an octonion algebra. Similarly if $\alpha = (2, 1)$ and $\beta = (5, 0)$ then $\gamma \ge (6, 5)$ and a composition of that size can be found using the monomial constructions in Chapter 13. With slightly larger numbers very little further is known.

There seem to be few elementary techniques for analyzing these general compositions of quadratic forms. (See Exercise 20.) However there is one non-elementary method that has produced results. In 1991 M. Szyjewski remarked that the cohomology ring used by Hopf in his theorem over $\mathbb{R}$ could be replaced by the Chow ring. The same proof would then work over any field F. After he outlined the methods of intersection theory and K-theory, we wrote up the paper (1992). The bare bones of the ideas are sketched here with no attempt made to explain any of the details.

Suppose there is a bilinear composition $f : U \times V \to W$ for the regular quadratic forms α, β, γ over a field F (of characteristic not 2). As usual we suppose the dimensions are r, s, n. The basic strategy is to lift f to a morphism of schemes

$$f^{\#} : \mathbb{P}_F^{r-1} \times \mathbb{P}_F^{s-1} \to \mathbb{P}_F^{n-1}$$

and then pass to the homomorphism induced on the corresponding Chow rings. However that first morphism $f^{\#}$ might fail to exist. The difficulty arises since the quadratic forms vanish at some points over some extension fields of F. For example let Z be the quadric determined by the equation $\gamma = 0$ in $\mathbb{P}_F^{n-1}$, and let C be the open complement of Z in that projective space. Then we really have to work with C in place of the whole projective space.

If Y is a variety, the Chow ring $A^*(Y)$ records information about the intersections of subvarieties of Y. (See Hartshorne (1985) or Fulton (1985).) It acts like a cohomology theory. For example $A^*(\mathbb{P}_F^{n-1}) \cong \mathbb{Z}[T]/(T^n)$, where T is an indeterminate. After some work, and an application of Swan's K- theory calculations, it follows that if C is that open complement where γ doesn't vanish, then

$$A^*(C) \cong \mathbb{Z}[T]/(T^{n-w(\gamma)}, 2T).$$

Here $w(\gamma)$ is the Witt index of γ, that is, $w(\gamma)$ is the dimension of a maximal totally isotropic subspace. This calculation of Chow rings implies a concrete result about the possible sizes, proved in the same style as the original Hopf Theorem.

A.6 Theorem. *Suppose α, β, γ are quadratic forms over F with dimensions r, s, n, respectively. Let $r_0 = r - w(\alpha)$ where $w(\alpha)$ is the Witt index of α. Similarly define s_0 and n_0. If there is a bilinear composition for α, β, γ then the binomial coefficient $\binom{n_0}{k}$ is even whenever $n_0 - s_0 < k < r_0$.*

Further details and references appear in Shapiro and Szyjewski (1992). The conclusion in this theorem says exactly that

$$r_0 \circ s_0 \leq n_0.$$

When α, β, γ are anisotropic then we recover the Stiefel–Hopf criterion. For sums of squares this says: if $n\langle 1\rangle$ is anisotropic then $r \circ s \leq n$. This is the result proved by Pfister's theory in (14.7).

In the weakest case of (A.6), all the forms have maximal Witt index, and $n_0 = n - w(\gamma)$ is $n/2$ or $(n+1)/2$, which is exactly the value n^* defined in Chapter 12. In this case the conclusion of the theorem is:

$$r^* \circ s^* \leq n^*, \quad \text{or equivalently:} \quad r \circ s \leq \begin{cases} n & \text{if } n \text{ is even} \\ n+1 & \text{if } n \text{ is odd.} \end{cases}$$

This equivalence follows from (12.7). For example, $5 \circ 9 = 13$ and $5 \circ 10 = 14$. Therefore no composition of size [5, 9, 12] can exist over any field. However no information is known about compositions of size [5, 10, 13] over fields of positive characteristic, although they are impossible in characteristic zero by (14.2).

Exercises for Chapter 14

1. (1) Suppose A is a commutative ring of characteristic zero. (That is, A has an identity element 1_A and the subring generated by 1_A is isomorphic to $\mathbb{Z}$.) If $[r, s, n]$ is admissible over A then $r \# s \leq n$.

(2) If $[r, s, n]$ is admissible over fields F_j involving infinitely many different characteristics, then $r \# s \leq n$.

(Hint. (1) By (14.2) it suffices to find a prime ideal P where A/P has characteristic 0. Let $S = \{n1_A : n \in \mathbb{Z} - \{0\}\}$, a multiplicatively closed set with $0 \notin S$. Choose P to be an ideal of A maximal such that $P \cap S = \emptyset$.

(2) Apply (1) to an appropriate ring.)

2. (1) In (14.1) we could have used $g(a, b) = (u_1(a, b) - v_1(a, b), \ldots, u_n(a, b) - v_n(a, b))$ in place of $f(a, b)$.

(2) Suppose there is a formula $\left(\sum^r x_i^2\right)\cdot\left(\sum^s y_j^2\right) = u_1^2+\cdots+u_p^2-v_1^2-\cdots-v_k^2$, for some bilinear forms u_i, v_j in $\mathbb{R}[X, Y]$. Then $r \# s \le p$.

Open Question. Must $r * s \le p$ in this situation?

(3) Suppose F is a field (with $2 \ne 0$) and $d \in F$ is not a sum of squares. If there is a composition of size $[r, s, n]$ over $F(\sqrt{d})$, then there is a nonsingular bilinear map $F^r \times F^s \to F^n$.

3. If there exists a composition of size $[r, s, n]$ over a field F of characteristic $p > 0$, then there exists such a composition over a finite field of characteristic p.

(Hint. Let A be the subring generated by the coefficients of that formula and F the field of fractions of A. Let $\mathbb{F}_p^{\text{alg}}$ be the algebraic closure of $\mathbb{F}_p$. Does there exists a place $\lambda : F \to \mathbb{F}_p^{\text{alg}} \cup \{\infty\}$ defined on A ?)

4. **Explicit formulas.** In Exercise 0.5 there are formulas showing that $D_F(2^m)$ is a group. In these formulas each z_k is a linear form in Y with coefficients in $F(X)$. For any given r, s, use those formulas to derive explicit formulas of size $(r, s, r \circ s)$ where each z_k is linear in Y. This provides a proof of (4) $\Longrightarrow$ (5) in Theorem 14.8 avoiding use of the Subform Theorem.

5. Define $r \circ_F s = \text{length}_{F(X,Y)}((x_1^2+\cdots+x_r^2)\cdot(y_1^2+\cdots+y_s^2))$. To avoid notational confusion here, we write level(F) for the level, rather than $s(F)$.

$$r \circ_F s = \begin{cases} r \circ s & \text{if } r \circ s \le \text{level}(F) \\ 1 + \text{level}(F) & \text{if } r \circ s > \text{level}(F). \end{cases}$$

(Hint. Apply (4) $\Longrightarrow$ (1) in (14.8) when $n = r \circ s$ or when $n = \text{level}(F)$.)

6. **Extensions.**

Witt's Theorem. *Suppose W is a regular quadratic space over F (a field where $2 \ne 0$) and $V \subseteq W$ is a subspace. If $\sigma : V \to W$ is an isometry, then there exists $\hat{\sigma} \in \mathrm{O}(W)$ extending σ.*

Suppose $\dim W = 1 + \dim V$.

(1) There is a *unique* extension $\hat{\sigma} \in \mathrm{O}^+(W)$.

(2) If $\dim W$ is even and $f \in \mathrm{Sim}^\bullet(V, W)$ then there is a unique extension $\hat{f} \in \mathrm{Sim}^+(W)$.

(3) Adem's Theorem says that the map $\hat{}: \mathrm{Sim}(V, W) \to \mathrm{Sim}^+(W)$ is linear on every linear subspace.

(Hint. (2) Suppose q is a quadratic form with $\dim q$ even, and $\alpha \subset q$ is a subform of codimension 1. If $\langle c\rangle\alpha \subset q$ then $q \simeq \langle c\rangle q$.)

7. **Full compositions.** (1) If n is even and $s \le n \le 2s$, there exists a full composition of size $[2, s, n]$.

(2) Suppose $f : U \times V \to W$ is a composition of size $[r, s, n]$. Define $f_i : V \to W$ using a basis of U and define $\oplus f : V^r \to W$ by $(\oplus f)(v_1, \dots, v_r) = \sum f_i(v_i)$. If A_i is the matrix of f_i then $\oplus f$ has matrix $\oplus A = (A_1, \dots, A_r)$ of size $n \times rs$. Then f is full $\iff \oplus f$ is surjective. How is this matrix related to the $n \times rs$ matrix of $f^{\otimes} : U \otimes V \to W$?

(Hint. (1) In (14.17) let $C = \begin{pmatrix} 1_{n-s} & 0 \end{pmatrix}$ and B a direct sum of 0's and copies of $\begin{pmatrix} 0 & 1 \\ -1 & 0 \end{pmatrix}$.)

8. If a full composition for forms α, β, γ has size $[r, s, rs]$ then, after scaling: $\gamma \simeq \alpha \otimes \beta$.

9. Suppose $A = v_1 w_1^{\top}$ and $B = v_2 w_2^{\top}$ are two $n \times n$ rank 1 matrices over a field F.
(1) $A = B \iff$ there exists $\lambda \in F^{\bullet}$ such that $v_2 = \lambda v_1$ and $w_1 = \lambda w_2$.
(2) $A + B$ has rank $\leq 1 \iff$ either $\{v_1, v_2\}$ is dependent or $\{w_1, w_2\}$ is dependent.
(3) Restate (2) in terms of decomposable tensors in $V \otimes W$.

10. If $n \leq r + s - 2$ then:

$$r \circ s \leq n \iff r' \circ s' \leq n \text{ whenever } r' \leq r,\ s' \leq s \text{ and } n = r' + s' - 2.$$

(Hint. $\mathcal{H}(r, s, r + s - 2) \iff \binom{r+s-2}{r-1}$ is even. The original definition implies:

$\mathcal{H}(r, s, n) \iff \mathcal{H}(r, n-r+2, n) \& \mathcal{H}(r-1, n-r+3, n) \& \cdots \& \mathcal{H}(n-s+2, s, n)$.)

11. Suppose there is a full composition of size $[r, r, n]$ over F.
(i) If $r = n - 1$ then $n = 4$ or 8.
(ii) If $r = n - 2$ then $n = 4$ or 8.

12. **Full monomial compositions.** If there is a full composition of size $[r, s, n]$ then $r *_F s \leq n \leq rs$. These minimum and maximum values are always realizable. It is harder to decide which values in between are possible.
(1) If a consistently signed intercalate matrix involves exactly n colors, then it yields a full composition over F of size $[r, s, n]$. If a full composition of size $[r, s, n]$ is monomial then the corresponding intercalate matrix must involve n distinct colors.
(2) The intercalate 3×3 matrices in Exercise 13.2 furnish full monomial compositions of sizes $[3, 3, n]$ exactly when $n = 4, 7, 9$. Therefore there cannot exist a full monomial $[3, 3, 8]$.
(3) Construct a full composition over F of size $[3, 3, 8]$.
(4) No full $[3, 3, 5]$ can exist by (14.18). Can a full $[3, 3, 6]$ exist?

(Hint. Monomial compositions appear in Appendix 13.B.
(3) Choose 3 dimensional subspaces U, V of the octonions A_3 such that UV spans A_3. For instance, abusing notations as in the table for A_4 in Chapter 13, let $U = \text{span}\{0, 1, 2\}$ and $V = \text{span}\{0, 4, 2 + 7\}$.

(4) Parker (1983) proved that there is a full $[3, 3, n]$ over $\mathbb{R}$ if and only if $n = 4$, 7, 8 or 9. I have no simpler proof that $[3, 3, 6]$ is impossible.)

13. Suppose $s \leq n$ and let $\mathcal{S}(n, s) \subseteq \mathbb{M}_{n\times s}(F)$ be the set of matrices of rank $< s$. Westwick (1972) showed that $\mathcal{S}(n, s)$ is an irreducible subvariety of $\mathbb{M}_{n\times s}(F)$ of codimension $n - s + 1$. Use this to give another proof of (14.25).

14. (1) Prove the statements in (A.4).

(2) If there is a full composition for α, β, γ of size $[2, 6, 8]$ then must γ be a Pfister form? How about for sizes $[3, 6, 8]$, $[3, 5, 8]$ and $[4, 5, 8]$?

(Hint. (1) These pairings must be full. For $[3, 3, 4]$ Adem yields α, γ, γ so γ is a 2-fold Pfister form and we can arrange $\alpha, \beta \subset \gamma$. Scale to get $\det(\alpha) = \det(\beta)$ and show $\alpha \simeq \beta$. For $[3, 5, 7]$ apply (14.18).

(2) $[3, 6, 8]$ expands to $[3, 8, 8]$, so γ is Pfister by (1.10). For $[2, 6, 8]$ use $\langle 1, a\rangle$, $\langle 1, a, b, c, d\rangle$, $\langle\langle a\rangle\rangle \otimes \langle 1, b, c, d\rangle$ to build a composition. Then γ need not be Pfister. There is a $[3, 5, 8]$ of the same type. I do not whether every full $[4, 5, 8]$ must have γ equal to a Pfister form.)

15. **Nonsingular pairings.** (1) $(ra) \#_F (sb) \leq (r + s - 1) \cdot (a \#_F b)$. For example if there is an n-dimensional F-division algebra then: $n \mid r$ and $n \mid s \Longrightarrow r \#_F s \leq r+s-n$. This generalizes (12.12) (3).

(2) If F has field extensions of every degree then

$$r \#_F s = \max\{r, s\} \quad \text{for every } r, s.$$

(3) Is the converse of (14.25) true?

(4) Suppose F is a 2-field. Then $r \circ s \leq r \#_F s$ by (14.26). If $F = \mathbb{R}$ this is not always an equality. If F is not real closed it has field extensions of every degree 2^m and $r \#_F s = r \circ s$ for every r, s.

(5) There is a full nonsingular $[r, s, n]$ if and only if $r \#_F s \leq n \leq rs$.

(Hint. (3) $2 \#_F s = s + 1 \Longleftrightarrow$ every degree s polynomial in $F[x]$ has a root in F. If this holds for all s then F is algebraically closed.

(4) If F is not real closed then (by Artin–Schreier) there exist extensions of arbitarily large degree 2^t. Galois theory provides extensions of degree 2^m for every m, so there exists a nonsingular $[2^m, 2^m, 2^m]$ for every m. Direct sums of nonsingular pairings are nonsingular. For given r, s, construct a nonsingular $[r, s, r \circ s]$.

(5) There exists a nonsingular $[r, s, r \#_F s]$ which must be full by minimality. By (14.23) there is a corresponding subspace W of dimension $rs - (r \#_F s)$. Choose $W' \subseteq W$ of dimension $rs - n$ and apply (14.23).)

16. **Surjective pairings.** (1) Suppose f is a bilinear pairing of size $[r, s, n]$ over F. If $n = r \#_F s$ then f is surjective.

(2) **Lemma.** *If there is a surjective bilinear $[r, s, n]$ over a field F then $n \leq r+s-1$.*

(3) Let $\mathcal{P}_n$ = {polynomials of degree $< n$}, an F-vector space of dimension n. The Cauchy pairing $c_{(r,s)} : \mathcal{P}_r \times \mathcal{P}_s \to \mathcal{P}_{r+s-1}$, given by multiplication in $F[t]$ is a nonsingular bilinear $[r, s, r+s-1]$. If F is algebraically closed then $c_{(r,s)}$ is surjective. If $F = \mathbb{R}$ then: $c_{(r,s)}$ is surjective if and only if r and s are not both even.

(4) $c_{(r,s)}$ is indecomposable. That is, it is not equivalent to a direct sum of pairings of some sizes $[r, s_1, n_1]$ and $[r, s_2, n_2]$.

(5) Is every surjective $[r, s, r+s-1]$ essentially the same as $c_{(r,s)}$? Must it at least be indecomposable?

(6) If $c_{(2,s)} \oplus \beta$ is surjective over $\mathbb{R}$ where β has size $[2, m, m]$, then s is odd and m is even.

(Hint. (1) If $0 \neq L \subseteq F^n$ is a subspace with $L \cap \text{image}(f) = 0$ consider F^n/L.

(2) Simpler exercise: If $g : F^m \to F^n$ is a surjective polynomial map then $n \leq m$.

Proof. If $n > m$ the components $g_1, \dots, g_n$ in $F[x_1, \dots, x_m]$ are algebraically dependent. Hence there exists a non-zero $G(z_1, \dots, z_n)$ with $G(g_1, \dots, g_n) = 0$. Then $\text{image}(g) \subseteq \mathcal{Z}(G)$. □

Modify this idea. If F is a finite field use a counting argument instead.

(4) Suppose $\mathcal{P}_s = U \oplus W$ such that $\mathcal{P}_{r+s-1} = (\mathcal{P}_r U) \oplus (\mathcal{P}_r W)$. Express $t^j = u_j + w_j$ for $0 \leq j < s$. If $j > 0$ the uniqueness implies $u_j = tu_{j-1}$. Then $U \supseteq \{u_0, tu_0, \dots, t^{s-1}u_0\}$. If $u_0 \neq 0$ then $U = \mathcal{P}_s$. Similarly if $w_0 \neq 0$.

(5) Use a nonsingular $[r, s-1, s-1]$ and the trivial $[r, 1, r]$ over F. The direct sum is an $[r, s, r+s-1]$ which is surjective, nonsingular and decomposable.

(6) $c_{(2,s)}$ and β are surjective, so s is odd. View $\beta : \mathcal{P}_2 \times \mathbb{R}^m \to \mathbb{R}^m$ so that $\beta(x + yt, u) = (xB_0 + yB_1)u$ for some $m \times m$ matrices B_j. If $g \in \mathcal{P}_{s+1}$ and $v \in \mathbb{R}^m$ there exist $x + yt \in \mathcal{P}_2$, $f \in \mathcal{P}_s$ and $u \in \mathbb{R}^m$ such that $(x + yt) \cdot f = g$ and $(xB_0 + yB_1)u = v$. If m is odd, there exists $a + bt \neq 0$ admitting some $v \notin \text{image}(aB_0 + bB_1)$. Choose $g = (a + bt)^s$. The existence of $x + yt$, f, u leads to a contradiction.)

17. **Surjective bilinear maps over the reals.** (1) If r, s are not both even there is a surjective bilinear $[r, s, r+s-1]$ over $\mathbb{R}$. In any case there is a surjective bilinear $[r, s, r+s-2]$ over $\mathbb{R}$.

Conjecture. If r, s are both even there is no surjective bilinear $[r, s, r+s-1]$ over $\mathbb{R}$.

(2) **Proposition.** *The Conjecture is true if* $r = 2$.

That is, if s is even then no real bilinear $[2, s, s+1]$ can be surjective.

(3) **Open Questions.**

- Is there a surjective bilinear $[4, 4, 7]$ over $\mathbb{R}$?
- Is every surjective $[r, s, r+s-1]$ nonsingular?
- If f is a surjective bilinear map over $\mathbb{R}$, is the extension $f_{\mathbb{C}}$ necessarily surjective over $\mathbb{C}$?

(Hint. (1) See Exercise 16 (3).

(2) A bilinear $[2, s, n]$ is essentially a pencil of $n \times s$ matrices $xA + yB$. Kronecker classified such singular pencils in 1890 as follows (see Gantmacher (1959), §12.3). Let x, y be indeterminates.

Theorem. *Suppose $s < n$ and A, B are $n \times s$ matrices over a field F. Then there exists $k > 0$ and square invertible P, Q over F such that $P(xA + yB)Q$ has block-diagonal form $\begin{pmatrix} C & 0 \\ 0 & D \end{pmatrix}$ where D is $(n-k-1) \times (s-k)$ and C is $(k+1) \times k$ of the type* $C = \begin{pmatrix} x & & & \\ y & x & & \\ & \ddots & \ddots & \\ & & y & x \\ & & & y \end{pmatrix}$.

Corollary. *Any bilinear $[2, s, n]$ is a direct sum $c_{(2,k)} \oplus \beta$ for some $k > 0$ and some β of size $[2, s-k, n-k-1]$. In particular the only indecomposable $[2, s, n]$ is the Cauchy pairing $c_{(2,s)}$.*

Now suppose f is a surjective $[2, s, s+1]$. As above, f decomposes with β of size $[2, m, m]$ where $m = s - k$. Surjectivity implies k is odd and m is even by Exercise 16(3), (6). Then $s = k + m$ is odd, contradiction.)

18. **Nonsingular $[n, n, n]$.** (1) There is a nonsingular bilinear $[2, n, n]$ over $F \iff$ there exists $f \in F[x]$ with degree n and no roots in F.

(2) There is a nonsingular $[2, 2, 2]$ over $F \iff$ there is a field $K \supseteq F$ with $[K : F] = 2$.

There is a nonsingular $[3, 3, 3]$ over $F \iff$ there is a field $K \supseteq F$ with $[K : F] = 3$.

(3) The following statements are equivalent:

(a) F admits a quadratic field extension;

(b) There is a nonsingular $[2, 4, 4]$ over F;

(c) There is a nonsingular $[4, 4, 4]$ over F;

(d) F admits either a degree 4 field extension or a quaternion division algebra.

(Hint. (3) Suppose E/F is a quadratic extension but F admits no degree 4 extension. Then E is 2-closed (i.e. $E = E^2$). The Diller–Dress Theorem (see T. Y. Lam (1983), p. 45) implies F is pythagorean. Since F is not 2-closed it is formally real.)

19. Suppose there is a composition of size $[r, s, 2m]$ for forms α, β, and $\gamma = m\mathbb{H}$ over F. If α and β are anisotropic, then $r \#_F s \le m$. If F is a 2-field deduce $r \circ s \le m$.

(Hint. Modify (14.1) to get a nonsingular $[r, s, m]$ over F. Apply (14.26). Compare (A.5).)

20. **Isotropic forms.** Suppose there is a composition for some forms α, β, γ of dimensions r, s, n over F. We may view it as a bilinear pairing $\varphi : U \times V \to W$.

(1) If $u \in U$ and $v \in V$ are non-zero with $\varphi(u, v) = 0$ then $\alpha(u) = \beta(v) = 0$.
(2) If α is isotropic then (W, γ) has a totally isotropic subspace of dimension $\geq s/2$.

Corollary. *If there is a composition as above, where α is isotropic and β is anisotropic then $n \geq 2s$.*

Note: This technique allows another proof of some characteristic zero cases by using function field methods as in the appendix of Chapter 9.

(3) Suppose there is a composition over $\mathbb{R}$ for $\alpha = r\langle 1\rangle$ and $\beta = \gamma = p\langle 1\rangle \perp n\langle -1\rangle$. That is, $r\langle 1\rangle < \mathrm{Sim}(p\langle 1\rangle \perp n\langle -1\rangle)$. Does it follow that $r\langle 1\rangle < \mathrm{Sim}(p\langle 1\rangle)$ and $r\langle 1\rangle < \mathrm{Sim}(n\langle 1\rangle)$?

(Hint. (1) The map $V \to W$ sending $x \mapsto \varphi(u, x)$ is an $\alpha(u)$- similarity with v in the kernel.

(2) Suppose $\beta \simeq s_0\mathbb{H} \perp \beta_1$ where β_1 is anisotropic of dimension s_1. Choose $0 \neq u \in U$ with $\alpha(u) = 0$. Then $f = \varphi(u, -) : V \to W$ has totally isotropic image and kernel. Therefore $\dim\ker(f) \leq s_0$ and $\dim \mathrm{image}(f) = s - \dim\ker(f) \geq s_0 + s_1$.)

(3) Yes. Suppose φ is an unsplittable for $r\langle 1\rangle$ as in Chapter 7. Then $r\langle 1\rangle$ is a minimal form and $\varphi = 2^m\langle 1\rangle$ for some m. Apply (7.11).)

21. Suppose $\left(\sum^r x_i^2\right) \cdot \left(\sum^s y_j^2\right) = z_1^2 + \cdots + z_n^2$ where each z_k is a bilinear in X, Y over $\mathbb{R}$. If $n = r * s$ then $z_1, \ldots, z_n$ are $\mathbb{R}$-linearly independent in $\mathbb{R}[X, Y]$.

(Hint. Suppose $z_n = a_1z_1 + \cdots + a_{n-1}z_{n-1}$ for some $a_j \in \mathbb{R}$. Using variables T_j, let $g(T) = T_1^2 + \cdots + T_{n-1}^2 + (a_1T_1 + \cdots + a_{n-1}T_{n-1})^2$, positive definite in $n - 1$ variables. There exist linear forms L_j in $\mathbb{R}[T]$ with $g(T) = L_1(T)^2 + \cdots + L_{n-1}(T)^2$. Evaluate to get $\left(\sum^r x_i^2\right) \cdot \left(\sum^s y_j^2\right) = g(z_1, \ldots, z_{n-1}) = L_1(Z)^2 + \cdots L_{n-1}(Z)^2$. Each $L_j(Z)$ is bilinear in X, Y. Contradiction.)

Notes on Chapter 14

Behrend's Theorem concerns nonsingular, bi-skew polynomial maps over a real closed field. His proof is related to the later development of real algebraic geometry, as seen in Fulton (1984). A more algebraic proof of Behrend's result is mentioned in (14.26).

The Tarski Principle is proved in textbooks on model theory (or see Prestel (1975)). It is a consequence of the model-completeness of the theory of real closed ordered fields.

The Lam–Lam results (14.1) and (14.2) were first published in Shapiro (1984a).

The observation in (14.6) that Pfister's function $r \circ s$ is the same as the Hopf–Stiefel condition was first made by Köhnen (1978) in his doctoral dissertation (under the direction of Pfister).

The first part of Adem's Theorem 14.10 is also due independently to Yuzvinsky (unpublished).

The idea for this simple proof of (14.11) follows Gauchman and Toth (1996), p. 282.

I first learned about full pairings from Gauchman and Toth (1994). That idea also appears in Parker (1983).

The observation in (14.17) was noted by Guo (1996) over $\mathbb{R}$.

Lemma 14.23 appears in Petrovič (1996).

Proposition 14.25 was also proved in Shapiro and Szyjewski (1992) using Chow rings.

Exercise 1, due to Wadsworth, appears in Shapiro (1984a).

Exercise 2 (2) was noted by T. Y. Lam.

Exercise 5. The definition of $r \circ_F s$ was given by Pfister (1987).

Exercise 6. Witt's Extension Theorem is presented in Scharlau (1985), Theorem 1.5.3.

Exercise 10 as applied in (14.20) was first observed by Behrend (1939).

Exercise 11 was formulated and proved over $\mathbb{R}$ by Gauchman and Toth (1994) (positive definite case) and (1996) (indefinite case).

Exercise 17 (2). The proof for [2, 2, 3] was told to me by A. Leibman. The general $r = 2$ case, with the Kronecker reference, was communicated by I. Zakharevich in 1998.

Exercise 19 yields nothing for other p-fields since all quadratic forms of dim > 1 are isotropic.

Exercise 21 is due to T. Y. Lam.

Chapter 15

Hopf Constructions and Hidden Formulas

When is there a real sum of squares formula of size $[r, s, n]$, or equivalently, a normed bilinear pairing $f : \mathbb{R}^r \times \mathbb{R}^s \to \mathbb{R}^n$? In Chapter 12 we attacked this problem by considering the induced map on spheres $f : S^{r-1} \times S^{s-1} \to S^{n-1}$, or on the associated projective spaces, and applying techniques of algebraic topology. Those methods apply just as well to nonsingular pairings, since any such pairing also induces maps on spheres and projective spaces. Therefore those techniques cannot distinguish between the normed and nonsingular cases.

K. Y. Lam (1985) found a technique that does separate those cases. If f is a normed pairing of size $[r, s, n]$, he began with the well known Hopf map

$$H : \mathbb{R}^r \times \mathbb{R}^s \to \mathbb{R} \times \mathbb{R}^n \quad \text{defined by} \quad H(x, y) = (|x|^2 - |y|^2, 2f(x, y)).$$

This is a quadratic map (i.e. each component is a homogeneous quadratic polynomial in the $r + s$ variables (x, y)) and it restricts to a map on unit spheres

$$H : S^{r+s-1} \to S^n.$$

For example, the normed $[2, 2, 2]$ arising from multiplication of complex numbers provides the map $S^3 \to S^2$ first studied by Hopf (1931).

Lam used the quadratic nature of H to prove that if $q \in S^n$ lies in image(H) then the fiber $H^{-1}(q)$ is a great sphere in S^{r+s-1}, cut out by some linear subspace $W_q \subseteq \mathbb{R}^{r+s}$. The differential dH then induces a nonsingular bilinear pairing $B_{(q)} : W_q \times W_q^{\perp} \to \mathbb{R}^n$ of size $[k, r + s - k, n]$ where $k = \dim W_q$. This is the pairing "hidden" behind the point q. These hidden pairings can be of different sizes as q varies. Knowing that dH has maximal rank at some q, Lam proved that there exists hidden pairings with $k \le r + s - (r \# s)$. As a Corollary he found that no normed bilinear $[16, 16, 23]$ can exist, although there is a nonsingular pairing of that size. Consequently, $24 \le 16 * 16 \le 32$. In subsequent years these ideas were sharpened and refined by Lam and Yiu. For example, using more sophisticated homotopy theory they proved that $16 * 16 \ge 29$.

In the appendix we consider non-constant polynomial maps which restrict to maps of unit spheres $S^m \to S^n$. Which dimensions m, n are possible? A complete answer is provided for quadratic maps.

To begin the chapter, we present the simple geometric arguments developed in Yiu's thesis (1986) to prove general results about quadratic maps of spheres.

If $V = \mathbb{R}^n$ we write $\mathcal{S}(V) = \{v \in V : |v| = 1\}$ for the unit sphere in V. Define a *great k-sphere* of $\mathcal{S}(V)$ to be the intersection of $\mathcal{S}(V)$ with a $(k+1)$-dimensional linear subspace of V. A great 1-sphere is called a *great circle*. If $u, v \in \mathcal{S}(V)$ are distinct and non-antipodal (that is, $u \neq \pm v$), then they lie on a unique great circle.

Suppose $F : V \to V'$ is a quadratic map between two euclidean spaces. This means that each of the components of F (when written out using coordinates) is a homogeneous quadratic polynomial map on V. We may express this, without choosing a basis, as follows:

$$F(av) = a^2F(v) \quad \text{for every } a \in \mathbb{R} \text{ and } v \in V;$$

$$B(u, v) = \frac{1}{2}(F(u+v) - F(u) - F(v)) \text{ is a bilinear map} : V \times V \to V'.$$

In particular, $B(v, v) = F(v)$ for every v. This associated bilinear map also satisfies:

$$F(au + bv) = a^2F(u) + b^2F(v) + 2abB(u, v).$$

Define F to be *spherical* if it preserves the unit spheres, that is: F sends $\mathcal{S}(V)$ to $\mathcal{S}(V')$. Since F is quadratic this amounts to the equation:[1]

$$|F(v)| = |v|^2 \quad \text{for every } v \in V.$$

15.1 Proposition. *Suppose $F : V \to V'$ is a spherical quadratic map and $u, v \in \mathcal{S}(V)$ are orthogonal.*

(a) *If $F(u) = F(v) = q$ then F sends the great circle through u and v to the point $q \in \mathcal{S}(V')$. In this case, $B(u, v) = 0$.*

(b) *If $F(u) \neq F(v)$ then F wraps the great circle through u and v uniformly twice around a circle on $\mathcal{S}(V')$ which has $F(u)$ and $F(v)$ as the endpoints of a diameter.*

(c) *$2B(u, v)$ and $F(u) - F(v)$ are orthogonal vectors of equal length. Consequently, $B(u, v)$ is orthogonal to both $F(u)$ and $F(v)$.*

Proof. The points on that great circle are $u_\theta = (\cos\theta)u + (\sin\theta)v$ for $\theta \in \mathbb{R}$. A short computation shows that

$$F(u_\theta) = \frac{1}{2}\cdot(F(u) + F(v)) + \frac{1}{2}\cos 2\theta \cdot (F(u) - F(v)) + \sin 2\theta \cdot B(u, v). \quad (*)$$

In part (a) this becomes: $F(u_\theta) = q + \sin 2\theta \cdot B(u, v)$ for every $\theta \in \mathbb{R}$. Since this has unit length for every θ, the vector $B(u, v)$ must be zero and $F(u_\theta) = q$ for all θ.

(b) Suppose $F(u) \neq F(v)$.

Claim. $B(u, v)$ and $F(u) - F(v)$ are linearly independent.

If they are dependent the formula (*) shows that $F(u_\theta)$ lies on the line joining $F(u)$ and $F(v)$ as well as on the sphere $\mathcal{S}(V')$. But then $F(u_\theta)$ is in the intersection

1 If $\dim V = m$ and $\dim V' = n$ then the components $F_1, \dots, F_n$ are quadratic forms in variables $x_1, \dots, x_m$ and: $F_1^2 + \cdots + F_n^2 = (x_1^2 + \cdots + x_m^2)^2$.

of that line and sphere, so it is one of the two points $F(u)$, $F(v)$. This contradicts the connectedness of the great circle, proving the claim.

The image of that great circle must lie in the affine plane which passes through the point $\frac{1}{2}(F(u) + F(v))$ and is parallel to the plane spanned by the independent vectors $F(u) - F(v)$ and $B(u, v)$. The intersection of that plane with the sphere $\mathcal{S}(V')$ is a circle. It follows from $(*)$ that the two vectors $F(u) - F(v)$ and $2B(u, v)$ are orthogonal of equal length, the center of the circle is $\frac{1}{2}(F(u) + F(v))$, and $F(u)$ and $F(v)$ are endpoints of a diameter. See Exercise 1.

(c) If $F(u) = F(v)$ then from part (a) we know $B(u, v) = 0$. Suppose $F(u) \neq F(v)$. The proof of (b) settles the first statement. The vector from the center of the sphere to the center of that circle is orthogonal to the plane of the circle. Hence $B(u, v)$, $F(u) - F(v)$, and $F(u) + F(v)$ are pairwise orthogonal. □

In particular if u, v are orthogonal in $\mathcal{S}(V)$ then: $F(u) = F(v)$ if and only if $B(u, v) = 0$.

15.2 Corollary. *Suppose v, w are distinct and non-antipodal in $\mathcal{S}(V)$. The great circle through v and w is either mapped to a single point in $\mathcal{S}(V')$, or it is wrapped uniformly twice around a circle in $\mathcal{S}(V')$ passing through $F(v)$ and $F(w)$.*

Proof. Let u be a point on that great circle with u orthogonal to v. If $F(u) = F(v)$ then (15.1) implies that the great circle goes to this single point. If $F(u) \neq F(v)$ then (15.1) implies that the great circle is wrapped twice around an image circle. □

We avoid extra notation which tells whether F is to be considered as a map on V or on $\mathcal{S}(V)$, hoping that the context will make the interpretation clear. Usually the domain is $\mathcal{S}(V)$.

15.3 Theorem. *If q is in the image of a spherical quadratic map $F : \mathcal{S}(V) \to \mathcal{S}(V')$ then $F^{-1}(q)$ is a great sphere.*

Proof. If $v, w \in F^{-1}(q)$ are distinct non-antipodal points then (15.2) implies that the great circle through v, w lies inside $F^{-1}(q)$. Let $W = \mathbb{R} \cdot F^{-1}(q)$ and check that W is a linear subspace of V. Then $F^{-1}(q) = W \cap \mathcal{S}(V)$ is a great sphere. □

15.4 Definition. Let F be a spherical quadratic map as above. If $q \in \text{image}(F)$ let $W_q = \mathbb{R} \cdot F^{-1}(q)$ be the linear subspace of V such that $F^{-1}(q) = W_q \cap \mathcal{S}(V)$.

These subspaces W_q are closely connected with the bilinear map B. As usual, define the linear maps $B_v : V \to V'$ by $B_v(w) = B(v, w)$.

15.5 Lemma. *Suppose $0 \neq v \in W_q$ and $w \in V$. Then $B(v, w) = 0$ if and only if $w \in W_q$ and w is orthogonal to v. Consequently, $W_q = \mathbb{R} \cdot v \perp \ker(B_v)$.*

Proof. We may assume $|v| = |w| = 1$. Then $F(v) = q$. If $w \in W_q$ is orthogonal to v then $F(w) = q$ and (15.1) implies that $B(v, w) = 0$. Conversely suppose $B(v, w) = 0$. Express $w = c \cdot v + s \cdot u$ for some $c, s \in \mathbb{R}$ and $u \in \mathcal{S}(V)$ orthogonal to v. Then $0 = B(v, w) = c \cdot q + s \cdot B(v, u)$. If $F(u) \neq q$ then (15.1) implies that $0 \neq B(v, u)$ is orthogonal to q, implying $c = s = 0$, impossible. Therefore $F(u) = q$ and (15.1) implies that the great circle maps to q, and $B(v, u) = 0$. Then $c = 0$ so that $w = s \cdot u$ is orthogonal to v, and $F(w) = q$ so that $w \in W_q$. □

The lemma shows that for non-zero vectors $v \in W_q$ and $w \notin W_q$, then $B(v, w) \neq 0$. This provides a nonsingular pairing.

15.6 Proposition. *Suppose $F : S^m \to S^n$ is a spherical quadratic map as above with associated bilinear map $B : V \times V \to V'$. If $q \in \text{image}(F)$ then the restriction of B to*

$$B_{(q)} : W_q \times W_q^{\perp} \to (\mathbb{R}q)^{\perp}$$

is nonsingular. This is the nonsingular pairing "hidden behind q". It has size $[k, m+1-k, n]$, where $k = \dim W_q$.

Proof. The bilinearity is clear and (15.1)(c) implies $B(v, w) \in (\mathbb{R}q)^{\perp}$ whenever $v \in W_q$. Lemma 15.5 proves the pairing is nonsingular. □

These hidden maps are useful because we know restrictions on the existence of nonsingular bilinear maps. For given m, n we will limit the possible values of k. The information so far tells us that $k > 0$ and

$$m + 1 - n \leq k \leq \min\{n, m + 1\}$$

The case $k = m + 1$ can happen if $W_q = V$, or equivalently if F is a constant map. We avoid this triviality by tacitly assuming that our maps F are non-constant.

For future use we observe that B_v is related to the differential of F at v. Viewing F as a mapping on V (rather than on $\mathcal{S}(V)$), at every $v \in V$, there is a differential on the tangent spaces $dF_v : T_v(V) \to T_{F(v)}(V')$. Since the flat spaces V and V' can be identified with their tangent spaces, this becomes $dF_v : V \to V'$. The usual definition $dF_v(w) = \frac{d}{dt}\big|_{t=0} F(v + tw)$ shows that $dF_v(w) = 2B(v, w)$ so that

$$dF_v = 2B_v.$$

Now consider F again as the map on spheres $S^m \to S^n$ (where $\dim V = m + 1$ and $\dim V' = n + 1$). Identifying the tangent space at $v \in \mathcal{S}(V)$ with the orthogonal complement $(v)^{\perp}$, we may restate (15.5) as: $W_q = \mathbb{R} \cdot v \perp \ker(dF_v)$. Therefore

$$\dim W_q = m + 1 - \text{rank}(dF_v) \quad \text{for every } v \in F^{-1}(q).$$

This relation will be exploited later when we consider Hopf maps.

The dimensions of the subspaces W_q can vary with q. For each integer k define

$$Y_k = \{q \in \mathcal{S}(V') : \dim W_q = k\}.$$

If Y_k is nonempty then the restriction of F to $F^{-1}(Y_k)$ provides a great sphere bundle. See Exercise 8.

15.7 Proposition. *If F is a spherical quadratic map and if q and $-q$ are both in the image of F, then W_q and W_{-q} are orthogonal. That is, $W_{-q} \subseteq W_q^{\perp}$.*

Proof. Suppose $F(u) = q$ and $F(v) = -q$. Then u, v are linearly independent and by (15.2) the great circle through them is wrapped uniformly twice around a (great) circle through q and $-q$. If θ is the angle between u and v, then q and $-q$ are separated by the angle 2θ, implying that θ is a right angle. □

The next lemma is an exercize in "polarizing" the equation stating that F is spherical. Here $\langle v, w\rangle$ is the inner product, so that $\langle v, v\rangle = |v|^2$.

15.8 Lemma. *Suppose $F : V \to V'$ is a spherical quadratic map with associated bilinear map B. The following formulas hold true for every $x, y, z, w \in V$.*

(1) $|F(x)| = |x|^2$.

(2) $\langle F(x), B(x, y)\rangle = |x|^2 \cdot \langle x, y\rangle$.
$\langle F(x), F(y)\rangle + 2|B(x, y)|^2 = |x|^2 \cdot |y|^2 + 2\langle x, y\rangle^2$

(3) $\langle F(x), B(y, z)\rangle + 2\langle B(x, y), B(x, z)\rangle = |x|^2 \cdot \langle y, z\rangle + 2\langle x, y\rangle \cdot \langle x, z\rangle$.

(4) $\langle B(x, y), B(z, w)\rangle + \langle B(x, z), B(y, w)\rangle + \langle B(x, w), B(y, z)\rangle$
$= \langle x, y\rangle \cdot \langle z, w\rangle + \langle x, z\rangle \cdot \langle y, w\rangle + \langle x, w\rangle \cdot \langle y, z\rangle$.

Proof. The definition of "spherical" yields (1).

For (2) apply (1) to $x + ty$, expand and equate the coefficients of t and of t^2. In the second equation of (2) substitute $y + z$ for y. Then (3) follows after expanding and canceling.

Similarly for (4) substitute $x + w$ for x in (3), expand and cancel. □

The formulas in (2) above generalize (15.1)(c), showing again that if u, v are orthogonal in $\mathcal{S}(V)$, then $F(u)$ and $B(u, v)$ are orthogonal and

$$4|B(u, v)|^2 = 2 - 2\langle F(u), F(v)\rangle = |F(u) - F(v)|^2.$$

Now suppose $v \in F^{-1}(q)$ so that $|v| = 1$ and $F(v) = q$. By (3) above, with some re-labeling, and moving the terms involving v to the left, we obtain:

$$2\langle B(v, x), B(v, y)\rangle - 2\langle v, x\rangle \cdot \langle v, y\rangle = \langle x, y\rangle - \langle q, B(x, y)\rangle.$$

Writing $B_v(x) = B(v, x)$ as before, the left side can be expressed as

$$2\langle \tilde{B}_v B_v(x), y\rangle - 2\langle \langle v, x\rangle v, y\rangle.$$

Let π_v be the orthogonal projection to the line $\mathbb{R} \cdot v$, that is: $\pi_v(x) = \langle v, x\rangle \cdot v$. This motivates the definition of the map g_q below.

15.9 Corollary. *If* $q \in \mathcal{S}(V')$ *define the map* $g_q : V \to V$ *by*

$$2\langle g_q(x), y\rangle = \langle x, y\rangle - \langle q, B(x, y)\rangle \quad \text{for every } x, y \in V.$$

(1) *For every* $v \in F^{-1}(q)$, $g_q = \tilde{B}_v B_v - \pi_v$. *The projection* π_v *is defined above.*

(2) *Suppose* $u \in \mathcal{S}(V)$. *Then* $g_q(u) = 0$ *if and only if* $F(u) = q$.

(3) $\text{image}(F) = \{q \in \mathcal{S}(V') : \det(g_q) = 0\}$. *Therefore* $\text{image}(F)$ *is an algebraic variety.*

(4) *If* $q \in \text{image}(F)$ *then* $W_q = \ker(g_q)$.

Proof. The work above proves (1). The point here is that this g_q depends only on q and not on the choice of $v \in F^{-1}(q)$.

(2) Suppose $F(u) = q$. If $v \in V$ express $v = \lambda u + u'$ where u' is orthogonal to u. Then (15.1) applied to $u'/|u'|$ implies that $B(u, u')$ is orthogonal to q. Then $\langle q, B(u, v)\rangle = \lambda = \langle u, v\rangle$, and therefore $g_q(u) = 0$. Conversely, suppose $g_q(u) = 0$. Then $0 = \langle g_q(u), u\rangle = 1 - \langle q, F(u)\rangle$, so that $\langle q, F(u)\rangle = 1$. Since both q and $F(u)$ are unit vectors, $q = F(u)$.

Property (2) quickly implies (3) and (4). □

Suppose F is a spherical quadratic map, $q \in \text{image}(F)$, and $B_{(q)} : W_q \times W_q^{\perp} \to (q)^{\perp}$ is the hidden bilinear map of size $[k, m+1-k, n]$. Certainly the space $W_q^{\perp}$ seems harder to understand than W_q. If $-q$ is also in image(F) then $W_{-q} \subseteq W_q^{\perp}$ by (15.7), and we are more familiar with that piece of the hidden map $B_{(q)}$. Can it happen that $W_{-q} = W_q^{\perp}$? This occurs exactly when F is a Hopf map.

15.10 Proposition. *Suppose* $F : \mathcal{S}(V) \to \mathcal{S}(V')$ *is a spherical quadratic map and both* p *and* $-p$ *are in* image(F). *The following statements are equivalent.*

(1) $W_{-p} = W_p^{\perp}$.

(1′) $\dim W_p + \dim W_{-p} = \dim V$.

(2) *There is a decomposition* $V = X \perp Y$ *such that for every* $x \in X$ *and* $y \in Y$,

$$B(x, y) \in (p)^{\perp} \quad \text{and} \quad F(x+y) = (|x|^2 - |y|^2) \cdot p + 2B(x, y).$$

If p *satisfies this property then the restriction of* B *to* $X \times Y \to Z$ *is a normed bilinear map, where* $Z = (p)^{\perp}$. *Such a point* p *is called a pole for the map* F.

Proof. The equivalence of (1) and (1′) is clear.

(1) $\Longrightarrow$ (2). By hypothesis, $V = W_p \perp W_{-p}$. Since $F^{-1}(p)$ is the unit sphere in W_p, we know $F(x) = |x|^2 \cdot p$ for every $x \in W_p$. Similarly if $y \in W_{-p}$ then

$F(y) = -|y|^2 \cdot p$. Then for any $v \in V$ there is a decomposition $v = x + y$ and $F(v) = (|x|^2 - |y|^2) \cdot p + 2B(x, y)$. By (15.1)(c) the vector $B(x, y)$ is orthogonal to $F(x) = p$.

(2) $\Longrightarrow$ (1′). If $x \in X$ and $y \in Y$ the formula implies $F(x) = |x|^2 \cdot p$ and $F(y) = -|y|^2 \cdot p$. Then $X \subseteq W_p$ and $Y \subseteq W_{-p}$. Since $X \perp Y = V$ we find $W_p + W_{-p} = V$ and this is certainly a direct sum.

Finally, suppose these equivalent properties hold. Apply (15.8)(2) to obtain the norm property $|B(x, y)| = |x| \cdot |y|$. □

Now reverse the procedure above, and start from a normed bilinear $f : X \times Y \to Z$. Let p be a new unit vector orthogonal to Z. The *Hopf map* for f with poles $\pm p$ is

$$H_f : \mathcal{S}(X \perp Y) \to \mathcal{S}(\mathbb{R}p \perp Z)$$

defined by $H_f(x, y) = (|x|^2 - |y|^2)p + 2f(x, y)$. If the bilinear map f has size $[r, s, n]$ then $H_f : S^{r+s-1} \to S^n$. These Hopf maps provide important examples of spherical quadratic maps. We will see that every spherical quadratic map is homotopic to some Hopf map.

The bilinear map B_f associated to the Hopf map H_f is easily computed. We record the formula for future reference. If $v = (x, y)$ and $v' = (x', y')$ in $X \perp Y$ then

$$B_f(v, v') = (\langle x, x'\rangle - \langle y, y'\rangle) \cdot p + f(x, y') + f(x', y).$$

The next few results, due to K. Y. Lam (1985), provide examples of sizes r, s where $r \# s < r * s$.

15.11 Lemma. *Suppose $f : X \times Y \to Z$ is a normed bilinear map of size $[r, s, n]$. Then there exists a dense subset $\mathcal{D} \subseteq X \times Y$ such that for every $v \in \mathcal{D}$,* $\operatorname{rank}(df_v) \geq r \# s$.

Proof. As usual, we identify each tangent space of a linear space X with X itself. If $v = (x, y) \in X \times Y$ the differential of f is easily calculated using the bilinearity:

$$df_{(x,y)}(x', y') = f(x, y') + f(x', y).$$

Let $V \subseteq Z$ be a linear subspace maximal with respect to the property: $V \cap \operatorname{image}(f) = \{0\}$. Then the induced map $\bar{f} : X \times Y \to Z/V$ is still nonsingular bilinear, and by the maximality, $\bar{f}$ is surjective. Let $p = \dim(Z/V)$ so that $\bar{f}$ has size $[r, s, p]$. Then $p \geq r \# s$. Since $\bar{f}$ is surjective, Sard's Theorem implies that there is a dense subset of points $(x, y) \in X \times Y$ such that the differential $d\bar{f}_{(x,y)}$ is surjective. (See Exercise 3.) For any such point, $\operatorname{rank}(df_{(x,y)}) \geq \operatorname{rank}(d\bar{f}_{(x,y)}) = p \geq r \# s$. □

Our next step is to compare the ranks of $d(H_f)$ and df. Dropping the subscript, $H(x, y) = (|x|^2 - |y|^2)p + 2f(x, y)$. If $|x| = |y|$ then $H(x, y)$ lies on the equator in $S^n = \mathcal{S}(\mathbb{R}p \perp Z)$. Let S_0 be the set of all $v \in \mathcal{S}(X \perp Y)$ with $H(v)$ on the equator.

Then

$$S_0 = \left\{(x, y) \in X \times Y : |x| = |y| = \frac{1}{\sqrt{2}}\right\},$$

so that S_0 is a torus $S^{r-1} \times S^{s-1}$ of codimension 1 in $\mathcal{S}(X \perp Y)$.

15.12 Lemma. *For every $v \in S_0$, the differentials dH_v and df_v have the same rank.*

Proof. Note that H has domain $\mathcal{S}(X \perp Y)$ while f has domain $X \perp Y$. For $v = (x, y) \in S_0$ we have $H(v) = 2f(x, y)$. Since H and $2f$ coincide on S_0, their differentials coincide on the tangent space:

$$dH_v(w) = 2 \cdot df_v(w) \quad \text{for every } w \text{ tangent to } S_0 \text{ at } v.$$

To complete the proof we need to compute these values when w is normal to S_0 at v.

Since S_0 is the zero set of the two polynomials $g_1 = |x|^2 - \frac{1}{2}$ and $g_2 = |y|^2 - \frac{1}{2}$, the 2-plane in $X \perp Y$ normal to S_0 at v is spanned by the gradient vectors $\nabla g_1 = 2(x, 0)$ and $\nabla g_2 = 2(0, y)$. Certainly $v = (x, y)$ is in that 2-plane and so is $v^* = (-x, y)$. Then v and v^* span the normal 2-plane and v^* is also tangent to $\mathcal{S}(X \perp Y)$ at v, since $\langle v, v^* \rangle = 0$.

From the discussion after (15.6) and the formulas before and after (15.11) we find: for any $v = (x', y') \in X \perp Y$:

$$dH_v(v') = 2B(v, v') = 2(\langle x, x' \rangle - \langle y, y' \rangle)p + 2(f(x, y') + f(x', y))$$
$$df_v(v') = f(x, y') + f(x', y).$$

Therefore

$$dH_v(v^*) = -2p,$$
$$df_v(v^*) = 0 \qquad \text{and} \qquad df_v(v) = 2f(x, y).$$

Hence, $\operatorname{rank}(df_v)$ on the tangent space to $X \perp Y$ equals $\operatorname{rank}(dH_v)$ on the tangent space to $\mathcal{S}(X \perp Y)$. □

15.13 Theorem (Lam (1985)). *Suppose $H : \mathcal{S}(X \perp Y) \to \mathcal{S}(\mathbb{R}p \perp Z)$ is a Hopf map with underlying normed bilinear map $f : X \times Y \to Z$ of size $[r, s, n]$. Then for some $v = (x, y) \in X \perp Y$, the differential dH_v has rank $\geq r+s-r \# s$. Consequently H admits some hidden nonsingular bilinear map of size $[k, r+s-k, n]$ for some $k \leq r+s-r \# s$.*

Proof. By (15.11) there exists $v = (x, y) \in X \times Y$ such that $x, y \neq 0$ and $\operatorname{rank}(df_v) \geq r \# s$. For non-zero scalars α, β, the differentials $df_{(\alpha x, \beta y)}$ and $df_{(x,y)}$ have the same rank, because:

$$df_{(\alpha x, \beta y)}(x', y') = \alpha f(x, y') + \beta f(x', y) = df_{(x,y)}(\beta x', \alpha y').$$

Then by suitably scaling x, y we may assume $v \in S_0$.

If $q = H(v)$ then $W_q = \mathbb{R} \cdot v \perp \ker(dH_v)$ as seen after (15.6). Since the domain of dH_v has dimension $r + s - 1$,

$$\operatorname{rank}(dH_v) = r + s - 1 - \dim\ker(dH_v) = r + s - \dim W_q.$$

By (15.12), $\dim W_q = r+s-\operatorname{rank}(df_v) \le r+s-r\#s$. Finally, using $m = r+s-1$ in (15.6), the nonsingular bilinear map hidden behind this q has size $[k, r+s-k, n]$, where $k = \dim W_q$. □

In the proof of (15.12) we considered the vector $v^* = (-x, y)$ associated to a given vector $v \in S_0$. There is an extension of this "star" operation to all points $v \in \mathcal{S}(X \perp Y)$ with $H(v) \ne \pm p$. This satisfies $H(v^*) = -H(v)$ so that if q lies in the image of a Hopf map, then so does $-q$. This equation also implies $W_q^* \subseteq W_{-q}$ for every $q \ne \pm p$. In fact, this is an equality and "$*$" is a linear map on W_q. Some further details appear in Exercise 5.

In Chapter 12 we observed that $r * s \ge r \# s \ge r \circ s$. Moreover if there exists a normed bilinear pairing of size $[r, s, r \circ s]$ then equalities hold here. As mentioned in (12.13) these equalities do hold for some small cases:

$$r * s = r \# s = r \circ s \quad \text{if } r \le 9 \text{ and if } r = s = 10.$$

The next smallest case is $10 * 11$. Lam proved that $10 \# 11 = 17$ and he constructed a normed [12, 12, 26], described in (13.9). Therefore $17 \le 10 * 11 \le 26$. Using the tools just developed we will show that there is no normed map of size [10, 11, 19]. This example separates the normed and nonsingular cases.

Corollary 15.14. *For r, s between* 10 *and* 17, *the value listed in this table is a lower bound for $r * s$.*

$r \backslash s$	10	11	12	13	14	15	16	17
10	16	20	20	20	20	24	24	26
11		20	20	20	24	24	24	27
12			20	24	24	24	24	28
13				24	24	24	24	29
14					24	24	24	30
15						24	24	31
16							24	32
17								32

Proof. These values follow from Theorem 15.13 together with the lower bounds for $r \# s$, as listed in (12.21). For example suppose there exists a normed [10, 11, 19].

Since $10 \# 11 = 17$ the theorem implies there is a hidden nonsingular bilinear map of size $[k, 21-k, 19]$ for some $k \le 4$. Certainly $21 - k \le 19$ so that $k = 2, 3, 4$. These possibilities are all ruled out by the Stiefel–Hopf condition: $2 \circ 19 = 3 \circ 18 = 4 \circ 17 = 20$. Similarly suppose there exists a normed $[16, 16, 23]$. Since $16 \# 16 = 23$, we find from (15.13) that there is a nonsingular $[9, 23, 23]$ which contradicts Stiefel–Hopf. The other cases are similar. □

In addition to $10 * 10 = 16$, two other values in that table are known to be best possible. The existence of a normed bilinear $[17, 18, 32]$, as mentioned after (13.6), shows that $16 * 17 = 17 * 17 = 32$. The exact values for the other cases remain unknown. The entries for $r *_{\mathbb{Z}} s$ listed in (13.1) are conjectured to equal the values $r * s$. In particular, we suspect that $10 * 11 = 26$ and $16 * 16 = 32$.

Lam's Theorem 15.13 provides the tool needed to complete the calculation of $\rho(n, r)$ when $n - r \le 4$, as stated in (12.31). See Exercise 11 for further details.

The basic Hopf construction for a normed bilinear map $f : \mathbb{R}^r \times \mathbb{R}^s \to \mathbb{R}^n$ provides a quadratic map $H_f : \mathbb{R}^{r+s} \to \mathbb{R}^{n+1}$ which restricts to a map on the unit spheres $H_f : S^{r+s-1} \to S^n$. This construction can also be fruitfully applied if we assume only that f is nonsingular bilinear. In that case it is easy to check that H_f is a quadratic map which restricts to a map into the punctured space $H_f : S^{r+s-1} \to \mathbb{R}^{n+1} - \{0\}$. Radial projection induces a map on spheres $\hat{H}_f : S^{r+s-1} \to S^n$. This map of spheres is certainly smooth, but it might not be quadratic (or even rational). Which homotopy classes in $\pi_{r+s-1}(S^n)$ arise from nonsingular bilinear maps in this way? This question is related to the generalized J-homomorphism and has been investigated by various topologists. For further information see K. Y. Lam (1977a, b), Smith (1978) and Al-Sabti and Bier (1978).

Suppose $F : S^m \to S^n$ is a spherical quadratic map. If $q \in \text{image}(F)$ then hidden behind q is a nonsingular bilinear map $B_{(q)}$ of size $[k, m+1-k, n]$. The Hopf construction for this nonsingular map $B_{(q)}$ yields another map of spheres $\hat{H}_{B(q)} : S^m \to S^n$. How is this map related to the original F? Yiu (1986) proved they are homotopic.

15.15 Proposition. *If $F : \mathcal{S}(V) \to \mathcal{S}(V')$ is a spherical quadratic map, then the Hopf construction of any hidden nonsingular bilinear map $B_{(q)}$ is homotopic to F.*

Proof. If $q \in \text{image}(F)$ then the hidden map $B_{(q)}$ is the restriction of B:

$$2B_{(q)}(u, v) = F(u+v) - F(v) - |u|^2 \cdot q,$$

where $u \in W_q$ and $v \in W_q^{\perp}$. The Hopf construction of $B_{(q)}$ (with poles $\pm q$) is the map $F_{(q)} : \mathcal{S}(V) = \mathcal{S}(W_q \perp W_q^{\perp}) \to V'$ given by

$$\begin{aligned} F_{(q)}(u+v) &= (|u|^2 - |v|^2) \cdot q + 2B_{(q)}(u, v) \\ &= F(u+v) - F(v) - |v|^2 \cdot q. \end{aligned}$$

For $0 \le t \le 1$ define $H_t : \mathcal{S}(V) \to V'$ by $H_t(u, v) = F(u+v) - t \cdot (F(v) + |v|^2 \cdot q)$. This provides a homotopy between $H_0 = F$ and $H_1 = F_{(q)}$. To obtain maps of spheres use the normalized maps $\hat{H}_t(u, v) = \frac{H_t(u,v)}{|H_t(u,v)|}$. This makes sense provided $H_t(u, v)$ is never zero. To prove this, suppose $(u, v) \in \mathcal{S}(V)$ and $H_t(u, v) = 0$ for some t with $0 < t < 1$. Then $0 = 2B(u, v) + (1 - t)F(v) + (|u|^2 - t \cdot |v|^2)q$. The vector $B(u, v)$ is orthogonal to q and to $F(v)$, by (15.1). This dependence relation implies $F(v) \in \mathbb{R}q$ so that $v \in W_q$ and hence $v = 0$. But then $0 = |u|^2 \cdot q$ so that $u = 0$ as well, a contradiction. □

Surprisingly, every hidden nonsingular bilinear map $B_{(q)}$ is homotopic to a normed bilinear map. To establish this homotopy we first prove a lemma. If $f : X \times Y \to Z$ is bilinear and $x \in X$, let $f_x : Y \to Z$ be the induced linear map. Then f is nonsingular if and only if f_x is injective for every $x \in \mathcal{S}(X)$, or equivalently, $\tilde{f}_x f_x$ is injective for every x. The bilinear map f is normed if and only if $\tilde{f}_x f_x = 1_X$ for every $x \in \mathcal{S}(X)$.

15.16 Lemma. *Suppose $f : X \times Y \to Z$ is nonsingular bilinear. If the map $\tilde{f}_x f_x : Y \to Y$ is independent of the choice of $x \in \mathcal{S}(X)$, then f is homotopic to a normed bilinear map, through nonsingular bilinear maps.*

Proof. For any $x \in \mathcal{S}(X)$ the map $\tilde{f}_x f_x$ is symmetric, so it admits a set of eigenvectors $\{\varepsilon_1, \dots, \varepsilon_s\}$ which form an orthonormal basis of Y. Then the vectors $f_x(\varepsilon_i)$ are orthogonal and if λ_i is the eigenvalue for ε_i then $\lambda_i = \langle \varepsilon_i, \tilde{f}_x f_x(\varepsilon_i)\rangle = |f_x(\varepsilon_i)|^2$. Define $L : Y \to Y$ by setting $L(\varepsilon_i) = \lambda_i^{-1/2} \cdot \varepsilon_i$ and extending linearly. Then $f_x \circ L$ is an isometry. Since $\tilde{f}_x f_x$ is independent of x this L works for every choice of x, and hence the map $f'(x, y) = f(x, L(y))$ is a normed bilinear map. Choose a path L_t in $\mathrm{GL}(Y)$ with $L_0 = 1_Y$ and $L_1 = L$. (For instance, set $L_t(\varepsilon_i) = \gamma_i(t) \cdot \varepsilon_i$ for suitable paths γ_i in $\mathbb{R}$.) Then $f_t(x, y) = f(x, L_t(y))$ is a nonsingular bilinear map with $f_0 = f$ and $f_1 = f'$. □

15.17 Proposition. *Suppose F is a spherical quadratic map. Every hidden nonsingular bilinear map of F is homotopic, through nonsingular bilinear maps, to a normed bilinear map.*

Proof. If $x \in \mathcal{S}(W_q)$ then (15.9) implies $\tilde{B}_x B_x = g_q$ on $W_q^{\perp}$, because π_x vanishes there. Therefore $\tilde{B}_x B_x$ is independent of $x \in \mathcal{S}(W_q)$ and the lemma applies. □

When convenient, we will abuse the notations and use $B_{(q)}$ to refer to this hidden normed bilinear map. This extra information in (15.17) helps a bit in the quest for non-existence results. As one application Yiu proved that there is no spherical quadratic map $S^{25} \to S^{23}$. See (A.5) in the appendix below. The machinery of hidden pairings also provides a new proof of the following result originally due to Wood (1968).

15.18 Corollary. *Every spherical quadratic map $F : S^m \to S^n$ is homotopic to a Hopf map $H_f : S^m \to S^n$ for some normed bilinear f.*

Proof. We may assume F is non-constant. Let $g = B_{(q)}$ be the nonsingular bilinear map hidden behind some $q \in \text{image}(F)$. Then by (15.15), F is homotopic to the Hopf construction H_g. Now (15.17) says that g is homotopic, through nonsingular bilinear maps, to some normed bilinear map f, and this induces a homotopy from H_g to H_f.□

Now that we know the hidden maps can be taken to be normed, we can apply Lam's Theorem.

15.19 Corollary. *If there is a non-constant quadratic map $F : S^m \to S^n$ then there exists a normed bilinear map of size $[k, m+1-k, n]$ for some $k \le \rho(m+1-k)$.*

Proof. By (15.17) the hidden maps for F provide normed $[j, m+1-j, n]$, for various values of j. Among all normed maps of such sizes, choose one $[k, m+1-k, n]$ where k is minimal. Theorem 15.13 applied to this pairing yields hidden maps of sizes $[h, m+1-h, n]$ where $h \le m+1-k \# (m+1-k)$. By minimality, $k \le h$, so that $k \# (m+1-k) \le m+1-k$ and the result follows from (12.20). □

A somewhat different proof of this result is given below after (15.30). Any normed bilinear $[r, s, n]$ has Hopf map $S^{r+s-1} \to S^n$ and (15.19) provides some normed $[k, r+s-k, n]$ with $k \le \rho(r+s-k)$. This inequality is usually weaker than the one in (15.13).

The arguments used above have been geometric, based on Yiu's analysis of great circles wrapping twice around, etc. Purely algebraic, polynomial methods lead to the many of the same results, with some variations. We present this alternative approach now, following the ideas of Wood (1968) and Chang (1998). We start again from the beginning, with a spherical quadratic map between unit spheres in euclidean spaces.

15.20 Proposition. *Suppose $F : S^m \to S^n$ is a non-constant quadratic map and $q \in \text{image}(F)$. Then $F^{-1}(q)$ is a great sphere S^{k-1} in S^m, and there is an associated "hidden" nonsingular bilinear map of size $[k, m+1-k, n]$. Moreover, $m-n < k \le \min\{m, n\}$.*

Proof. Suppose $F(p) = q$. Applying isometries to the spheres we may assume $p = (1, 0, \dots, 0) \in S^m$ and $q = (1, 0, \dots, 0) \in S^n$. In terms of coordinates, $F(Z) = (F_0(Z), \dots, F_n(Z))$ where $Z = (z_0, \dots, z_m)$. Since F preserves unit spheres we know that

$$F_0(Z)^2 + \cdots + F_n(Z)^2 = (z_0^2 + \cdots + z_m^2)^2. \tag{1}$$

Since $F(p) = q$ we find that $F_0(Z) = z_0^2 + z_0 L_0 + Q_0$ and $F_j(Z) = z_0 L_j + Q_j$ for $j \ge 1$. Here each L_j is a linear form and each Q_j is a quadratic form in the variables $(z_1, \ldots, z_n)$. Compare coefficients in (1) to obtain: $L_0 = 0$ and $\sum_0^n Q_j^2 = (z_1^2 + \cdots + z_m^2)^2$. By the Spectral Theorem the form Q_0 can be diagonalized by an isometry of $\mathbb{R}^m$. After that change of variables we have $Q_0(Z) = \sum_1^n \mu_i z_i^2$, and the condition on $\sum Q_j^2$ implies $-1 \le \mu_i \le 1$ for each i.

Collect the terms where $\mu_i = 1$ and re-label the variables to obtain $Z = (X, Y)$ where $X = (x_1, \ldots, x_k)$ and $Y = (y_1, \ldots, y_h)$, $k + h = m + 1$, and:

$$F_0(X, Y) = (x_1^2 + \cdots + x_k^2) + (\lambda_1 y_1^2 + \cdots + \lambda_h y_h^2)$$

where $-1 \le \lambda_i < 1$. Since F is non-constant we know $h \ge 1$.

To analyze $F^{-1}(q)$, suppose $Z = (X, Y) \in S^m$ and $F(X, Y) = q$. Then $F_0(X, Y) = 1$ which implies $|X|^2 + \left(\sum_1^h \lambda_i y_i^2\right) = 1 = |X|^2 + |Y|^2$. Then $\sum_1^h (1 - \lambda_i) y_i^2 = 0$ which implies $Y = 0$ since every $\lambda_i < 1$. Therefore $F^{-1}(q) = \{(X, 0) : |X| = 1\} \cong S^{k-1}$, a great sphere in S^m. Then $k \le m$, since $k - 1 = m$ implies F is constant.

The identity (1) implies that no x_i^2 term can occur in $F_j(X, Y)$ for $j \ge 1$. Therefore $F_j(X, Y) = 2b_j(X, Y) + G_j(Y)$ where b_j is a bilinear form and G_j is a quadratic form. This says that $b = (b_1, \ldots, b_n)$ is a bilinear form $\mathbb{R}^k \times \mathbb{R}^h \to \mathbb{R}^n$ and $G = (G_1, \ldots, G_n)$ is a quadratic form $\mathbb{R}^h \to \mathbb{R}^n$. Then

$$F(X, Y) = \Big(|X|^2 + \sum_1^h \lambda_i y_i^2,\ 2b(X, Y) + G(Y)\Big) \in \mathbb{R} \times \mathbb{R}^n.$$

and, after equating like terms, the identity (1) becomes:

$$\begin{aligned} |X|^2 \cdot \Big(\sum_1^h \lambda_i y_i^2\Big) + 2|b(X, Y)|^2 &= |X|^2 \cdot |Y|^2 \\ \langle b(X, Y), G(Y)\rangle &= 0 \\ \Big(\sum_1^h \lambda_i y_i^2\Big)^2 + |G(Y)|^2 &= |Y|^4 \end{aligned} \tag{2}$$

The first equation here can be restated as:

$$2|b(X, Y)|^2 = |X|^2 \cdot \Big(\sum_1^h (1 - \lambda_i) y_i^2\Big). \tag{3}$$

Since $\lambda_j < 1$, this b is a nonsingular bilinear map of size $[k, h, n]$. This immediately implies $k \le n$ and $h = m + 1 - k \le n$. The stated inequalities follow. □

By tracing through the definitions one can check that this b coincides with the hidden nonsingular map $B_{(q)}$ described in (15.6), with $\dim W_q = k$. Moreover equation (3) says that b is almost a normed map. View Y as a column and let D be the diagonal matrix with entries $\left(\frac{2}{1-\lambda_i}\right)^{1/2}$. Then (3) says that $b_D(X, Y) = b(X, DY)$ is a normed bilinear map of same size as b. This leads to another proof that the hidden map b is homotopic to a normed bilinear map, perhaps clearer than the proof in (15.17).

15.21 Corollary. *Let $F : S^m \to S^n$ be a non-constant quadratic form and suppose $q \in \text{image}(F)$ is given with $\dim W_q = n$. Then the hidden b is a normed bilinear map of size $[n, m+1-n, n]$ and F equals the Hopf construction H_b. In this case F is surjective and $m + 1 \leq n + \rho(n)$.*

Proof. Continuing the notations in (15.20), $k = n$ and b is nonsingular bilinear of size $[n, h, n]$ where $h = m + 1 - n$. For $0 \neq Y \in \mathbb{R}^h$ define $b_Y : \mathbb{R}^n \to \mathbb{R}^n$ by $b_Y(X) = b(X, Y)$. Since b is nonsingular each b_Y is bijective. The second equation in (2) above then implies $G(Y) = 0$, and the third equation then yields: $\left(\sum_1^h \lambda_i y_i^2\right)^2 = \left(\sum_1^h y_i^2\right)^2$. Then $\lambda_j^2 = 1$, so that $\lambda_j = -1$ for each j. Consequently b is a normed pairing and $F(X, Y) = (|X|^2 - |Y|^2, 2b(X, Y))$ equals $H_b(X, Y)$. Finally since b is surjective F must also be surjective (see Exercise 12) and the inequality $m + 1 - n \leq \rho(n)$ follows from Hurwitz– Radon. □

The inequality in (15.20) implies $m - n < n$ so that $m \leq 2n - 1$. If this bound is attained, so there is a non-constant quadratic form $F : S^{2n-1} \to S^n$, then (15.21) implies $n = 1, 2, 4$ or 8 and (up to isometry) F equals one of the classical Hopf fibrations.

We now return to the more geometric discussion of the Hopf maps, following Yiu and Lam. These Hopf maps H_f directly generalize the classical Hopf fibrations built from the real composition algebras of dimension $n = 1, 2, 4, 8$. To emphasize the analogy with the classical case, we will write $x \cdot y$ or xy rather than $f(x, y)$ much of the time. For example, the norm property becomes: $|x \cdot y| = |x| \cdot |y|$.

In the composition algebras, if non-zero x and c are given, there exists y with $xy = c$. This y is unique, and in fact, $|x|^2 \cdot y = \bar{x} \cdot xy = \bar{x}c$. This "bar" map on X also acts as an adjoint for the norm form: $\langle xy, c\rangle = \langle y, \bar{x}c\rangle$. Generalizing these properties to pairings of size $[r, s, n]$, we do not obtain a "bar" map on X, but Lam and Yiu (1989) did find a useful analog of the product $\bar{x}c$.

15.22 Definition. Let $X \times Y \to Z$ be a normed bilinear map of size $[r, s, n]$. Define an associated pairing $\varphi : X \times Z \to Y$ by:

$$\langle xy, c\rangle = \langle y, \varphi(x, c)\rangle$$

for $x \in X$, $y \in Y$, $c \in Z$. Also if $c \in Z$ define $\varphi_c : X \to Y$ by $\varphi_c(x) = \varphi(x, c)$.

This map φ is well defined and bilinear. In the classical case, of course, $\varphi(x, c) = \bar{x}c$. This map $c \mapsto \varphi_c$ can be viewed as a dual of the linear map $f^{\otimes} : X \otimes Y \to Z$. See Exercise 17. In the general case the bilinear map φ has size $[r, n, s]$ so we cannot expect it to have the norm property (especially if $s < n$). But it does enjoy some of the properties of its classical analog.

15.23 Lemma. *Suppose* $0 \neq x \in X$ *and* $c \in Z$.

(1) $x \cdot \varphi(x, c)$ *is the orthogonal projection of* $|x|^2 \cdot c$ *to the space* xY.

(2) $|\varphi(x, c)| \leq |x| \cdot |c|$ *with equality if and only if* $c \in xY$.

(3) $c \in xY$ *if and only if* $\tilde{\varphi}_c \varphi_c(x) = |c|^2 \cdot x$.

Proof. (1) That projection is a vector xy such that $\langle |x|^2 \cdot c, xy' \rangle = \langle xy, xy' \rangle$ for every $y' \in Y$. Equivalently, $\langle c, xy' \rangle = \langle y, y' \rangle$ for every y', and $y = \varphi(x, c)$ as claimed.

(2) By (1), $|x \cdot \varphi(x, c)| \leq |\, |x|^2 \cdot c\,|$ with equality if and only if $c \in xY$. The norm property transforms this into the stated inequality.

(3) From (2) we know $\langle \tilde{\varphi}_c \varphi_c(x), x \rangle = \langle \varphi_c(x), \varphi_c(x) \rangle \leq |c|^2 \cdot |x|^2$. Then $\langle (\tilde{\varphi}_c \varphi_c - |c|^2) x, x \rangle \leq 0$ and equality holds if and only if $c \in xY$. If $\tilde{\varphi}_c \varphi_c(x) = |c|^2 \cdot x$ then certainly $c \in xY$. Conversely, suppose $c = xy$ for some $y \in Y$. Then by (1), $x \cdot \varphi_c(x) = |x|^2 \cdot c = x \cdot |x|^2 \cdot y$ and nonsingularity implies: $\varphi_c(x) = |x|^2 \cdot y$. Therefore, for any $x' \in X$, $\langle \varphi_c(x), \varphi_c(x') \rangle = |x|^2 \cdot \langle y, \varphi_c(x') \rangle = |x|^2 \cdot \langle c, x'y \rangle = |x|^2 \cdot \langle xy, x'y \rangle = |x|^2 \cdot |y|^2 \cdot \langle x, x' \rangle = |c|^2 \cdot \langle x, x' \rangle$. Consequently, $\tilde{\varphi}_c(\varphi_c(x)) = |c|^2 \cdot x$. □

In particular if $xy = c$ then $|x|^2 \cdot y = \varphi(x, c)$, just as in the classical case, multiplying by $\bar{x}$.

15.24 Corollary. *Let* $f : X \times Y \to Z$ *be a normed bilinear map. Then* $c \in \text{image}(f)$ *if and only if* $|c|^2$ *is an eigenvalue of* $\tilde{\varphi}_c \varphi_c$. *Therefore,* image$(f)$ *is a real algebraic variety.*

Proof. Apply (15.23)(3). Then image(f) is the zero set of the polynomial $P(c) = \det(\tilde{\varphi}_c \varphi_c - |c|^2)$. □

If $c \in Z$ lies in image(f) then there are expressions $c = xy$ for many different factors $x \in X$ and $y \in Y$. Define the *left-factor set* X_c to be the set of all possible left factors x, as follows:

$$\begin{aligned} X_c &= \{x \in X : c \in xY\} \cup \{0\} \\ &= \{x \in X : \tilde{\varphi}_c \varphi_c(x) = |c|^2 \cdot x\} \\ &= \{x \in X : x \cdot \varphi(x, c) = |x|^2 \cdot c\}. \end{aligned}$$

Then X_c is a linear subspace of X (an eigenspace of $\tilde{\varphi}_c \varphi_c$), a fact that does not seem obvious from the first definition. This space is closely related to $W_c = \mathbb{R} \cdot H^{-1}(c)$ obtained from the Hopf construction $H : \mathcal{S}(X \perp Y) \to \mathcal{S}(\mathbb{R}p \perp Z)$.

15.25 Lemma. *Suppose* $c \in \mathcal{S}(Z)$ *is a point on the equator of* $\mathcal{S}(\mathbb{R}p \perp Z)$. *Then* $W_c \subseteq X \perp Y$ *is the graph of* $\varphi_c : X_c \to Y$.

Proof. Recall that $H(x,y) = (|x|^2 - |y|^2)\cdot p + 2xy$. Then

$$\begin{aligned} W_c &= \mathbb{R}\cdot\left\{(x,y) : |x| = |y| = \frac{1}{\sqrt{2}} \text{ and } 2xy = c\right\} \\ &= \{(x',y') : |x'| = |y'| \text{ and } x'y' = \lambda c \text{ for some } \lambda \geq 0\}. \end{aligned}$$

If $(x',y') \in W_c$ then $x' \in X_c$ so the projection $X \perp Y \to X$ induces an injective linear map $\pi_1 : W_c \to X_c$. If $x \in X_c$ then $x \cdot \varphi(x,c) = |x|^2 \cdot c$, so that $(x, \varphi(x,c)) \in W_c$. Hence π_1 is bijective and the lemma follows. □

This lemma provides some insight into the possible sizes of the hidden bilinear maps. In fact, if $k = \dim W_c$ for $c \in \mathcal{S}(Z)$, then $k = \dim X_c \leq r$. Switching the roles of X and Y throughout, we obtain the right-factor set $Y_c = \{y \in Y : c \in Xy\} \cup \{0\}$ and deduce that $k = \dim Y_c$ as well. In particular, X_c and Y_c have the same dimension k, and $k \leq \min\{r, s\}$.

Since $X_{-c} = X_c$ we find that W_{-c} is the graph of $-\varphi_c : X_c \to Y$, and consequently $\dim W_{-c} = \dim X_c$ as well. What about the spaces W_q when q is not on the equator? If $q \in \mathcal{S}(\mathbb{R}p \perp Z)$ and q is not one of the poles ($\pm p$), then there is a unique great circle through q and p. This great circle is the *meridian* through q. It intersects the equator in some pair of points $\pm c$. Choose $c \in \mathcal{S}(Z)$ so that q and c are on the same half-meridian. Then $q = (\cos\theta)\cdot p + (\sin\theta)\cdot c$ for some $\theta \in (0, \pi)$.

15.26 Proposition. *Let $X \times Y \to Z$ be a normed bilinear map, with Hopf construction $H : \mathcal{S}(X \perp Y) \to \mathcal{S}(\mathbb{R}p \perp Z)$. If $q \in \text{image}(H)$ is not $\pm p$, choose $c \in \mathcal{S}(Z)$ and θ as above. Then W_q is the graph of the map $\tan(\frac{\theta}{2})\cdot\varphi_c : X_c \to Y$.*

Proof. The half-angle identities imply

$$H^{-1}(q) = \{(x,y) \in X \perp Y : |x| = \cos\left(\tfrac{\theta}{2}\right),\ |y| = \sin\left(\tfrac{\theta}{2}\right) \text{ and } 2xy = (\sin\theta)\cdot c\}.$$

If $(u,v) \in W_q$ is non-zero then $(u,v) = (\lambda x, \lambda y)$ for some $\lambda > 0$ and $(x,y) \in H^{-1}(q)$. Since $q \neq \pm p$ we know $u \neq 0$. Then $|u| = \lambda\cdot\cos(\frac{\theta}{2})$ and $|v| = \lambda\cdot\sin(\frac{\theta}{2})$, so that $|v| = \tan(\frac{\theta}{2})\cdot|u|$ and $|u|\cdot|v| = \frac{1}{2}\cdot\lambda^2\sin\theta$. Then $uv = \lambda^2 xy = (\frac{1}{2}\cdot\lambda^2\sin\theta)\cdot c = |u|\cdot|v|\cdot c$. Then $u \in X_c$ and $|u|^2\cdot v = |u|\cdot|v|\cdot\varphi(u,c)$ so that $v = \tan(\frac{\theta}{2})\varphi_c(u)$ as claimed.

Conversely suppose $0 \neq u \in X_c$. Setting $v = \tan(\frac{\theta}{2})\cdot\varphi_c(u)$ we must prove $(u,v) \in W_q$. Then $|v| = \tan(\frac{\theta}{2})\cdot|u|$ and, since $u\cdot\varphi(u,c) = |u|^2\cdot c$, we find $uv = |u|\cdot|v|\cdot c$. Define $x = \lambda u$ and $y = \lambda v$ where $\lambda = \frac{\cos(\theta/2)}{|u|} = \frac{\sin(\theta/2)}{|v|}$. Then $|x| = \cos(\frac{\theta}{2})$ and $|y| = \sin(\frac{\theta}{2})$ and $2xy = 2\lambda^2 uv = (\sin\theta)\cdot c$. Therefore $(x,y) \in H^{-1}(q)$ and $(u,v) \in W_q$, as hoped. □

In fact, the spaces W_q for $q \neq \pm p$ on a meridian are mutually isoclinic. This follows from Exercise 1.22, since $\varphi_c : X_c \to Y$ is an isometry.

15.27 Corollary. *Suppose $H : S^{r+s-1} \to S^n$ is the Hopf construction for some normed bilinear map of size $[r, s, n]$. Then* $\dim W_q$ *is constant on meridians, except possibly at the poles. Moreover,* $\dim W_q \le \min\{r, s\}$.

Proof. Let c be an equatorial point on the given meridian. If q is on that meridian the closest equatorial point is c or $-c$. Then (15.26) implies $\dim W_q = \dim X_{\pm c} = \dim X_c$. As remarked after (15.25), $\dim X_c \le \min\{r, s\}$. □

If $c \in xY$ (or equivalently, $x \in X_c$), then $|\varphi(x, c)| = |x| \cdot |c|$. This looks like the norm property, but to get a composition of quadratic forms we need c to vary within a linear space. To obtain such a space consider the set

$$C = \{c \in Z : c \in xY \text{ whenever } 0 \ne x \in X\}.$$

Lam and Yiu call these elements the "collapse values". Since $C = \bigcap_{x \ne 0} xY$, it is a linear subspace of Z.

15.28 Lemma. *Suppose $X \times Y \to Z$ is a normed pairing of size $[r, s, n]$, and $0 \ne c \in Z$. The following are equivalent.*

(1) *$c \in C$ is a collapse value.*

(2) *$c \in x_i Y$ for some vectors x_i which span X.*

(3) $X_c = X$.

(4) $\dim W_c = r$, *so the hidden map for c has size $[r, s, n]$.*

(5) $x \cdot \varphi(x, c) = |x|^2 \cdot c$ *for every $x \in X$.*

(6) $|\varphi(x, c)| = |x| \cdot |c|$ *for every $x \in X$.*

(7) $\tilde{\varphi}_c \varphi_c = |c|^2 \cdot 1_X$*; that is, $\varphi_c : X_c \to Y$ is a similarity of norm $|c|^2$.*

If these hold then: $\tilde{\varphi}_c \varphi_z + \tilde{\varphi}_z \varphi_c = 2\langle c, z\rangle \cdot 1_X$ for every $z \in Z$.

Proof. Apply (15.23) and (15.25). For the last statement let $x, x' \in X$. Polarize (5) to find $x \cdot \varphi_c(x') + x' \cdot \varphi_c(x) = 2\langle x, x'\rangle c$. Then $\langle \varphi_z(x), \varphi_c(x')\rangle + \langle \varphi_z(x'), \varphi_c(x)\rangle = \langle x \cdot \varphi_c(x') + x' \cdot \varphi_c(x), z\rangle = \langle 2\langle x, x'\rangle c, z\rangle = 2\langle c, z\rangle \cdot \langle x, x'\rangle$. □

15.29 Corollary. *Let $f : X \times Y \to Z$ be a normed bilinear map of size $[r, s, n]$. The set C of collapse values is a linear subspace of Z and $C \subseteq$ image(f). The induced map $\varphi : X \times C \to Z$ is a normed bilinear map of size $[r, \ell, s]$ where $\ell = \dim C$.*

Proof. Apply (15.28). □

Moreover $C +$ image$(f) =$ image(f), so image(f) is a union of cosets of C. Most normed pairings probably have $C = 0$, but there are some important non-zero cases. For example for the Hurwitz–Radon pairings $X \times Z \to Z$ of size $[r, n, n]$, every element of Z is a collapse value. For the integral pairings discussed in Chapter 13, there is a close connection between collapse values and ubiquitous colors. See Exercise 19.

15.30 Proposition. *Suppose* $f : X \times Y \to Z$ *is a normed bilinear map of size* $[r, s, n]$. *Then* $\text{image}(f) = C$ *if and only if every hidden bilinear map for* f *has the same size* $[r, s, n]$. *In this case,* $r \le \rho(s)$ *and* f *restricts to a bilinear map of size* $[r, s, s]$.

Proof. The bilinear maps hidden at the poles always have the size $[r, s, n]$. By (15.27), the sizes of other hidden maps equal the sizes for points on the equator. By (15.28) the map hidden behind c has size $[r, s, n]$ if and only if $c \in C$. This proves the first statement.

If $\text{image}(f) = C$ then f restricts to a surjective normed bilinear map of size $[r, s, \ell]$ where $\ell = \dim C$. For any $0 \ne x \in X$ then $C = xY$ so that $\ell = \dim C = \dim Y = s$. The existence of a normed $[r, s, s]$ implies $r \le \rho(s)$. □

Second proof of 15.19. Given $F : S^m \to S^n$ choose a normed pairing of size $[k, m+1-k, n]$ with k minimal, as before. Let $H : S^m \to S^n$ be its Hopf construction. Any hidden map for H has some size $[h, m+1-h, n]$. By minimality $k \le h$ and by (15.27) $h \le \dim W_q \le k$. Then $h = k$ so that all the hidden bilinear maps for H have the same size. Then (15.30) implies $k \le \rho(m+1-k)$. □

We have been working with left-collapse values, based on the left factors. There is a parallel theory of right-collapse values, based on right factors. Of course if $r < s$ then zero is the only right-collapse value. However if $r = s$ both collapse sets can be non-zero.

15.31 Lemma. *Suppose* $X \times Y \to Z$ *is a normed pairing of size* $[r, s, n]$.

(1) *If* $xy = c$ *is a non-zero left-collapse value then* $Xy \subseteq xY$.

(2) *If* $r = s$ *then left-collapse values are the same as right-collapse values.*

Proof. (1) For any x' there exists $y' \in Y$ with $c = x'y'$. As in the proof of (15.28): $x \cdot \varphi(x', c) + x' \cdot \varphi(x, c) = 2\langle x, x'\rangle c$. Since $|x|^2 y = \varphi(x, c)$ we have $|x|^2 x' y = x' \cdot \varphi(x, c) = 2\langle x, x'\rangle c - x \cdot \varphi(x', c) \in xY$.

(2) If $r = s$ then (1) implies $Xy = xY$, and the two types of collapse values coincide. □

The dimension of C is quite restricted in this case $r = s$. First note that if $\dim C = \ell$ in this case then there is a normed $[r, \ell, r]$ by (15.29) and therefore $\ell \le \rho(r)$. Similarly if $\dim C \ge 2$ then r is even and if $\dim C \ge 3$ then $r \equiv 0 \pmod 4$. Lam and Yiu obtained a much stronger restriction on r. The next lemma provides the tool needed to prove their result.

15.32 Lemma. *Suppose $X \times Y \to Z$ is a normed bilinear map and $x, x' \in X$ and $y, y' \in Y$. Then $\langle xy, x'y'\rangle + \langle xy', x'y\rangle = 2\langle x, x'\rangle \cdot \langle y, y'\rangle$. If x, x', y, y' are unit vectors and either $\langle x, x'\rangle = 0$ or $\langle y, y'\rangle = 0$, then: $xy = x'y'$ implies $xy' = -x'y$.*

Proof. The first identity follows directly from the norm condition. The hypotheses of the second statement imply $\langle xy', x'y\rangle = -1$. The stated equality follows since the two entries are unit vectors. □

This lemma provides a version of the "signed intercalate matrix" condition used in Chapter 13.

15.33 Proposition. *If a normed bilinear map of size* $[r, r, n]$ *has* $\dim C \geq 3$ *then* $r = 4$ *or* 8 *and the map restricts to one of size* $[4, 4, 4]$ *or* $[8, 8, 8]$.

Proof. Let $X \times Y \to Z$ be the given pairing. By hypothesis there is an orthonormal set c_1, c_2, c_3 in C. Choose a unit vector $x_1 \in X$. Since c_i is a collapse value, there exist vectors y_i with $x_1 y_i = c_i$. Then $\{y_1, y_2, y_3\}$ is an orthonormal set in Y. Next we define x_2 and x_3 by: $x_j y_j = c_1$. The Lemma then implies $x_2 y_1 = -x_1 y_2 = -c_2$ and $x_3 y_1 = -x_1 y_3 = -c_3$. Similarly define y_4 and x_4 by: $x_3 y_4 = c_2$ and $x_4 y_4 = c_1$. Finally define $u = x_2 y_3$. Repeated application of the Lemma yields the following multiplication table:

	y_1	y_2	y_3	y_4
x_1	$-c_1$	c_2	c_3	u
x_2	$-c_2$	c_1	u	$-c_3$
x_3	$-c_3$	$-u$	c_1	c_2
x_4	$-u$	c_3	$-c_2$	c_1

The x_i are orthogonal so that $X_4 = \text{span}\{x_1, x_2, x_3, x_4\}$ is 4-dimensional (and $r \geq 4$). Similarly for $Y_4 = \text{span}\{y_1, y_2, y_3, y_4\}$ and $Z_4 = \text{span}\{c_1, c_2, c_3, u\}$. If $r = 4$ this is the whole picture: the original $[4, 4, n]$ restricts to a $[4, 4, 4]$. If $r > 4$ the restriction $X_4^\perp \times Y_4^\perp \to Z$ is a normed pairing of size $[r-4, r-4, n]$ which still has c_1, c_2, c_3 as collapse values. (For if $c = c_i$ then $\varphi_c : X \to Y$ is an isometry carrying X_4 to Y_4, so it restricts to an isometry $\varphi_c : X_4^\perp \to Y_4^\perp$.)

Choose a unit vector x_1' in $X_4^\perp$ and repeat the process above, defining y_i', x_i' and u'. Then the x's and y's form an orthonormal set of 8 vectors (forcing $r \geq 8$). We already know two of the 4×4 blocks of the 8×8 multiplication table.

Claim. $u + u' = 0$.

This is proved by analyzing some of the other entries in the table. Let $v = x_1 y_1'$. The lemma then implies that $x_1' y_1 = -x_1 y_1' = -v$; $x_2' y_2 = x_1 y_1' = v$; and $x_3 y_3' = x_1' y_1 = -v$. Therefore $x_3 y_2 = x_2' y_3'$ implying $-u = u'$ and proving the claim.

Let $X_8 = \text{span}\{x_1, \ldots, x_4'\}$ and $Y_8 = \text{span}\{y_1, \ldots, y_4'\}$. If $r = 8$ this is the whole picture: the original $[8, 8, n]$ restricts to an $[8, 8, 8]$. If $r > 8$ the restriction $X_8^\perp \times Y_8^\perp \to Z$ is a normed pairing of size $[r-8, r-8, n]$ still admitting c_1, c_2, c_3 as collapse values. Choose a unit vector in $X_8^\perp$ and go through the construction of $y_1'', \ldots, x_4'', u''$ as before. The claim above applies to the three different 4×4 blocks to show that $u + u' = u + u'' = u' + u'' = 0$. This is a contradiction. □

Certainly there is a composition of size $[16, 16, 32]$ with integer coefficients. It is conjectured that this 32 cannot be improved. In Chapter 13 we mentioned that Yiu succeeded in proving this in the integer case: $16 *_{\mathbb{Z}} 16 = 32$. If real coefficients are allowed the problem is considerably harder. Lam and Yiu (1989) obtained the best known bound in this case.

15.34 Theorem. $16 * 16 \geq 29$.

The proof uses topological methods that are beyond my competence to describe accurately. A careful outline of the proof appears in Lam and Yiu (1995). We mention here some of the steps they use. Suppose there exists a normed bilinear $f : \mathbb{R}^{16} \times \mathbb{R}^{16} \to \mathbb{R}^{28}$. The hidden normed bilinear maps are of size $[k, 32-k, 28]$ and the Stiefel–Hopf condition implies $k = 4$, 8, 12 or 16. The cases $k = 4$ and 12 are proved impossible by examining the class of f in the stable 3-stem. Therefore $V = \text{image}(f : S^{15} \times S^{15} \to S^{27})$ is a real algebraic variety (by (15.24)) containing only two types of points: the collapse values (with $\dim W_q = 16$) and the generic values (with $\dim W_q = 8$). By (15.33) the collapse values form a linear subspace of dimension $\ell \leq 2$. This structure is simple enough to permit a calculation of the cohomology groups of V. Lam and Yiu then determine the module structure of $H^*(V; \mathbb{Z}_2)$ over the Steenrod algebra and they compute the secondary cohomology operation $\Phi_8 : H^{15}(V) \to H^{23}(V)$. However, a simplicial complex V with such cohomology groups and secondary operations cannot be embedded in S^{27}. Contradiction.

Appendix to Chapter 15. Polynomial maps between spheres

For which m, n do there exist non-constant polynomial maps $S^m \to S^n$? In an elegant paper, Wood (1968) used results of Cassels and Pfister on sums of squares to prove that if there is some t with $n < 2^t \leq m$ then there are no such maps of spheres. It is unknown whether Wood's result is the best possible. However Yiu (1994b) settled the quadratic case, determining exactly when there is a non-constant quadratic form $S^m \to S^n$. This appendix contains proofs of these results of Wood and Yiu.

A.1 Lemma. *If there exist non-constant polynomial maps $S^m \to S^n$ and $S^n \to S^r$ then there is one $S^m \to S^r$.*

Proof. Given $G : S^m \to S^n$ and $F : S^n \to S^r$. For small $\varepsilon > 0$, choose $x, y \in \text{image}(G)$ with $|x - y| = \varepsilon$. Choose points $u, v \in S^n$ with $|u - v| = \varepsilon$ and $F(u) \neq F(v)$. Let φ be a rotation carrying $\{x, y\}$ to $\{u, v\}$ and consider $F\varphi G$. □

A.2 Lemma. *If there exists a non-constant polynomial map $S^m \to S^n$ then there is a non-constant homogeneous polynomial map $S^m \to S^n$.*

Proof. Let G be the Hopf construction of a normed bilinear map of size $[1, m, m]$. Then G is a non-constant quadratic form $S^m \to S^m$. Apply the construction in (A.1) to the given map F and this G to obtain a non-constant polynomial map $S^m \to S^n$ with all monomials of even degree. Multiply each monomial by a suitable power of $|x|^2$. □

A.3 Wood's Theorem. *Suppose there is a non-constant polynomial map $S^m \to S^n$. If $2^t \leq m$ then $2^t \leq n$.*

Proof. We may assume $m \geq 2$. By (A.2) there is a non-constant $h : S^m \to S^n$ homogeneous of degree d. Then $h = (h_0, \dots, h_n)$ where each h_j is a form of degree d in $X = (x_0, \dots, x_m)$. Since h preserves the unit spheres we find: $|h(X)| = |X|^d$. Using $q(X) = x_0^2 + \cdots + x_m^2$ this becomes

$$h_0^2 + \cdots + h_n^2 = q^d.$$

Since $m \geq 2$ this $q(X)$ is irreducible. Therefore $\bar{h}_0^2 + \cdots + \bar{h}_n^2 = 0$ in K, the fraction field of $\mathbb{R}[X]/(q)$. Since h is non-constant, we may assume that some $\bar{h}_j \neq 0$. Therefore K has level $s(K) \leq n$. Apply Pfister's calculation of this level, as in Exercise 9.11 (2). □

A.4 Corollary. *If $m \geq 2^t > n$ then every polynomial map $S^m \to S^n$ is constant. In particular this happens whenever $m \geq 2n$.*

This result leads to an intriguing open question: Which m, n have the property that polynomial maps $S^m \to S^n$ must all be constant? There seem to be no further examples known. However the quadratic case has been settled. Using the machinery of hidden maps and collapse values, Yiu (1994b) has determined the numbers m, n for which all homogeneous quadratic maps on spheres are constant. We will prove his results below.

To warm up, let us do two examples which will be superseded later.

A.5 Proposition. *Suppose F is a spherical quadratic map $S^m \to S^n$.*

(a) *If $(m, n) = (25, 23)$ then F is constant.*

(b) *If $(m, n) = (48, 47)$ then F is constant.*

Proof. (a) If there is a non-constant $F : S^{25} \to S^{23}$ then by (15.6) there is a hidden nonsingular $[k, 26-k, 23]$. The Stiefel–Hopf criterion says $k \circ (26-k) \le 23$, which implies $10 \le k \le 13$. But (15.17) says that there exists a normed bilinear map of that size, which contradicts the values in the table in (15.14).

Similarly for (b), if there exists a non-constant $S^{48} \to S^{47}$ then there is a hidden map of some size $[k, 49-k, 47]$. If $k \le 16$ a calculation shows that $k \circ (49-k) = 48$, a contradiction to Stiefel–Hopf. Therefore $k \ge 17$. By (15.17) there is a normed bilinear map of that size, and (15.13) then provides a hidden map of some size $[p, 49-p, 47]$ where $p \le 49 - k \# (49-k) \le 49 - k \# 32 \le 17$. The cases when $p \le 16$ are impossible as above, so $p = 17$. But then $17 \le 49 - 17 \# 32$, yielding $17 \# 32 \le 32$, a contradiction to (12.20). □

Define the function $q(m)$ by:

$$q(m) = \min\{n : \text{there is a non-constant quadratic } S^m \to S^n\}.$$

Of course "quadratic" here means a spherical (homogeneous) quadratic map. Then for given m, there exists a non-constant quadratic $S^m \to S^{q(m)}$ and if $n < q(m)$ then every quadratic $S^m \to S^n$ is constant.

A.6 Lemma. *$q(m)$ is an increasing function. That is: $m \le m'$ implies $q(m) \le q(m')$.*

Proof. If $F : S^{m'} \to S^{q(m')}$ is a non-constant quadratic form, there exist $a \ne b$ in $S^{m'}$ with $F(a) \ne F(b)$. Choose a linear embedding $i : S^m \to S^{m'}$ whose image contains a and b. Then the composite $F \circ i$ is a non-constant quadratic form $S^m \to S^{q(m')}$. □

Wood's result (A.4) implies that $q(2^t) = 2^t$ for every $t \ge 0$. Moreover, the Hopf construction applied to a formula of Hurwitz–Radon type $[\rho(n), n, n]$ provides a quadratic map $S^{n+\rho(n)-1} \to S^n$. Therefore $q(n + \rho(n) - 1) \le n$. Then (A.6) implies that

$$q(2^t) = q(2^t + 1) = \cdots = q(2^t + \rho(2^t) - 1) = 2^t.$$

For small n the values of $q(n)$ are now easily determined:

$q(1) = 1,$
$q(2) = q(3) = 2,$
$q(4) = q(5) = q(6) = q(7) = 4,$
$q(8) = q(9) = \cdots = q(15) = 8,$
$q(16) = q(17) = \cdots = q(24) = 16.$

By (A.5) there is no quadratic form $S^{25} \to S^{23}$, and the Hopf construction applied to a normed $[2, 24, 24]$ provides a non-constant $S^{25} \to S^{24}$. Therefore $q(25) = 24$. Similarly (A.5) implies that $q(48) = 48$.

A.7 Theorem (Yiu). *The values of* $q(m)$ *are given recursively by*

$$q(2^t + m) = \begin{cases} 2^t & \text{if } 0 \le m < \rho(2^t), \\ 2^t + q(m) & \text{if } \rho(2^t) \le m < 2^t. \end{cases}$$

This formula provides a computation of $q(m)$ for any m. The next proposition provides a key step in the proof. Throughout this proof we will use a shorthand notation, writing "there exists $S^m \to S^n$" to mean that there exists a non-constant homogeneous quadratic map from S^m to S^n.

A.8 Proposition. *Suppose* $\rho(2^t) \le m < 2^t$ *and there exists* $S^{2^t+m} \to S^{2^t+n}$. *Then there exists* $S^m \to S^n$.

Proof. If $t \le 3$ then $\rho(2^t) = 2^t$ and the statement is vacuous. Suppose $t \ge 4$. By (15.19) there is a normed bilinear $[k, 2^t + m + 1 - k, 2^t + n]$ for some $k \le \rho(2^t + m + 1 - k)$. Note that $m + 1 - k \le n$ here. Then $k \le m$, for otherwise $m + 1 - k \le 0$ and $m < k \le \rho(2^t + m + 1 - k) \le \rho(2^t)$, contrary to hypothesis. Since generally $0 < c < 2^t$ implies $\rho(2^t + c) = \rho(c)$, we find $k \le \rho(m + 1 - k)$. From $m + 1 - k \le n$ we obtain a normed $[k, m + 1 - k, n]$ whose Hopf map sends $S^m \to S^n$, as claimed. □

Proof of Theorem A.7. Suppose $\rho(2^t) \le m < 2^t$. We need to prove that $q(2^t + m) = 2^t + q(m)$. Proposition A.8, using $n = q(2^t + m) - 2^t$ implies: $2^t + q(m) \le q(2^t + m)$. The reverse inequality requires the existence of $S^{2^t+m} \to S^{2^t+q(m)}$. To prove this, it suffices to find some normed $[k, 2^t + m + 1 - k, 2^t + q(m)]$. From any $S^m \to S^{q(m)}$ we obtain hidden pairings of size $[k, m + 1 - k, q(m)]$ for some k. If there is such a pairing where $k \le \rho(2^t)$ then we can combine it (direct sum) with the Hurwitz–Radon pairing $[k, 2^t, 2^t]$ to obtain the desired result. The following Lemma proves the existence of such a pairing with an even better bound on k. □

A.9 Lemma. *If* $m < 2^t$ *then there exists a normed* $[k, m + 1 - k, q(m)]$ *for some* $k \le \rho(2^{t-1})$.

Proof. The case $t = 1$ is trivial. Suppose the statement is true for t. In order to prove it for $t + 1$, suppose $2^t \le m' < 2^{t+1}$ and express $m' = 2^t + m$ where $0 \le m < 2^t$. We must produce a normed $[k, 2^t + m + 1 - k, q(2^t + m)]$ for some $k \le \rho(2^t)$.

Case $0 \le m < \rho(2^t)$. Then we know that $q(m') = 2^t$ and $2^t + m + 1 - \rho(2^t) \le 2^t$. Restrict a Hurwitz–Radon pairing to produce a normed $[\rho(2^t), 2^t + m + 1 - \rho(2^t), 2^t]$. This is a pairing of the desired type with $k = \rho(2^t)$.

Case $\rho(2^t) \le m < 2^t$. By induction hypothesis there is a normed $[k, m + 1 - k, q(m)]$ with $k \le \rho(2^{t-1})$. There is a Hurwitz–Radon pairing of size $[k, 2^t, 2^t]$ and a direct sum provides a normed $[k, 2^t + m + 1 - k, 2^t + q(m)]$. Since (A.8) implies that $2^t + q(m) \le q(2^t + m)$, the result follows. □

The recursive description of $q(m)$ can be replaced by an explicit formula involving dyadic expansions.

A.10 Proposition. *Suppose $m > 8$ is written dyadically as $m = \sum_{0\le i<t} m_i 2^i$ where $m_i \in \{0, 1\}$. Define $k(m) = \max\{j : j \le t$ and $(m_0 + m_1 \cdot 2 + \cdots + m_{j-1}2^{j-1}) < \rho(2^j)\}$. Then $q(m) = \sum_{i\ge k(m)} m_i 2^i$.*

Proof. Check that this formula for $q(m)$ satisfies the recursive property in Theorem A.7. If $t > 4$ note that $k(m) < t$. □

This explicit formula does not seem as useful as the recursive definition. For example, when does $q(m) = m$? This is true if m is a 2-power and we know $q(48) = 48$ by (A.5). The formula above implies:

$$q(m) = m \text{ if and only if } 2^{k(m)} \text{ divides } m.$$

But to find examples it seems easier to work recursively.

A.11 Corollary. *Suppose $m = 2^t + m_0$ where $0 < m_0 < 2^t$. Then $q(m) = m$ if and only if $q(m_0) = m_0$ and $\rho(2^t) \le m_0 < 2^t$. The smallest solutions to $q(m) = m$ are:* 1, 2, 4, 8, 16, 32, 48, 64, 80, 96.

Proof. Use (A.7). The small examples for $m > 8$ seem to say that $q(m) = m$ exactly for the multiples of 16. This pattern fails in general. For example, $\rho(2^8) = 17$ so that $q(272) = q(256 + 16) = 256$. See Exercise 22 for a different approach. □

The value $q(m)$ is usually not much smaller than m. In fact, after the first few cases, the fraction $\frac{q(m)}{m}$ becomes close to 1.

A.12 Corollary. (1) $q(m) \le \frac{m+1}{2}$ *if and only if* $m = 1, 3, 7$ *or* 15.

(2) $\lim_{m\to\infty} \frac{q(m)}{m} = 1$.

Proof. (1) Suppose $q(m) \le \frac{m+1}{2}$. Express $m = 2^t + m_0$ where $0 \le m_0 < 2^t$. If $\rho(2^t) \le m_0 < 2^t$ then $2^t + q(m_0) = q(m) \le \frac{m+1}{2}$. This implies $2^t + 2q(m_0) \le m_0 + 1 \le 2^t$ which forces $m_0 = 0$, contrary to hypothesis. Therefore $0 \le m_0 < \rho(2^t)$. Then $2^t = q(m) \le \frac{m+1}{2}$, forcing $m_0 = 2^t - 1$. Now $2^t - 1 \le \rho(2^t)$ implies $t = 0$, 1, 2, 3 and $m = 1, 3, 7, 15$.

(2) See Exercise 23. □

Yiu (1994b) also analyzed the other function related to quadratic maps of spheres:

$$p(n) = \max\{m : \text{there exists } S^m \to S^n\}.$$

It has a similar recursive formula, somewhat more complicated than the one for $q(m)$.

Wood's Theorem produced some examples of integers $m > n$ for which every polynomial map $S^m \to S^n$ is constant. The calculation of $q(m)$ above provides a complete answer for the existence of quadratic maps, but does not address the existence of polynomial maps of higher degree. There does exist a degree 4 map $S^{25} \to S^{23}$ obtained by composing two Hopf maps $S^{25} \to S^{24} \to S^{23}$ (obtained from a normed $[2, 24, 24]$ and $[9, 16, 23]$). Using similar compositions, and Lemma (A.1), we are reduced to asking:

A.13 Open Question. For which m do there exist non-constant polynomial maps $S^m \to S^{m-1}$?

Of course $q(m) \neq m$ if and only if there exists a homogeneous quadratic $S^m \to S^{m-1}$. Wood's Theorem says that if m is a 2-power there is no non-constant polynomial map of that size. Is there a non-constant polynomial map $S^{48} \to S^{47}$? If there is such a map then there must exist one which is homogeneous of some even degree > 2. (See Exercise 21.)

Polynomial maps on spheres seem to be difficult to analyze generally, but we can handle the special case of circles. If $z = x + yi$ is a complex number, let $c_n(x, y)$ and $s_n(x, y)$ be the real and imaginary parts of z^n. Then $f_n = (c_n, s_n)$ maps the circle to itself, wrapping it uniformly n times around. Further examples are found by altering the components modulo $x^2 + y^2 - 1$, and by composing with a rotation of the circle.

A.14 Proposition. *Every polynomial map from S^1 to itself is of this type.*

Proof. If $f, g \in \mathbb{R}[x, y]$ and (f, g) maps S^1 to itself, then $h = f + gi$ can be expressed as a polynomial in $z = x + yi$ and $\bar{z} = x - yi$. Reducing modulo $z\bar{z} - 1$ provides a Laurent polynomial $h(z) \equiv \sum_{k=-m}^{m} b_k z^k$ for some $b_k \in \mathbb{C}$. Let $s(\theta) = h(e^{i\theta})$ so that $|s(\theta)| = 1$ for all $\theta \in \mathbb{R}$. Since the functions $f_n(\theta) = e^{in\theta}$ are linearly independent, the equation $s(\theta) \cdot \overline{s(\theta)} = 1$ implies $b_n \neq 0$ for only one index n. Hence $h(z) \equiv b_n z^n$ and $|b_n| = 1$. Then multiplication by b_n is a rotation and z^n is a uniform wrapping $|n|$ times around. □

Exercises for Chapter 15

1. Suppose p, p' are vectors in euclidean space and $p \neq 0$. If $f(\theta) = (\cos\theta) \cdot p + (\sin\theta) \cdot p'$ traces a circle then that circle has center at the origin, the vectors p, p' are orthogonal and have equal length, and $\pm p$ are the endpoints of a diameter.

2. (1) Suppose $f : S^1 \to S^2$ is a polynomial map. If it is a quadratic form then by (15.1) its image is a circle. Is every circle in S^2 realized as the image of such a quadratic map?

(2) If $f : S^m \to S^n$ is a quadratic form, it maps every great circle to a circle (or point). Does it actually carry every circle in S^m to a circle in S^n?

(3) Suppose the map f in (2) is a quadratic map, not assumed to be homogeneous. Is the image of a great circle necessarily a circle?

(4) If f above is a homogeneous cubic map is its image necessarily a circle?

(5) If $n > 1$ then every polynomial map $S^n \to S^1$ is constant.

(Hint. (1) Rotate the circle to have $(1, 0, 0)$ and $(a, b, 0)$ as endpoints of a diameter. Let $f = (x^2 + ay^2, by^2, cxy)$ for suitable c.

(2) The Hopf map $S^2 \to S^2$ for a normed $[1, 2, 2]$ stretches small circles into other shapes.

(3) Recall spherical coordinates: let $f(x, y) = (x^2, xy, y)$.

(4) Try $f(x, y) = (x^3 + \alpha xy^2, \beta x^2 y, y^3)$ for suitable $\alpha, \beta \in \mathbb{R}$.

(5) Use either (A.3) or (A.14).)

3. **Sard's Theorem.** Suppose $f : M \to N$ is a smooth map of manifolds. A point $x \in M$ is "regular" if the differential $df_x : T_x(M) \to T_{f(x)}(N)$ is surjective, and "critical" otherwise. Let C be the set of critical points.

Sard's Theorem. *Suppose $f : U \to \mathbb{R}^n$ is a smooth map defined on an open set $U \subseteq \mathbb{R}^m$. Then $f(C)$ has Lebesgue measure zero in $\mathbb{R}^n$.*

(A proof is given in Milnor (1965).)

(1) If $f : M \to N$ is surjective, does it follow that the regular points are dense in M?

(2) Suppose M, N are real algebraic varieties and $f : M \to N$ is a surjective polynomial map. Then the set C is an algebraic set, and the set of regular points is open and dense in M.

(Hint. (1) No. Consider $f : \mathbb{R} \to \mathbb{R}$ which is smooth, surjective and constant on some interval.

(2) Say $\dim M = m$, $\dim N = n$. Surjectivity implies $m \geq n$. On some open $U \subseteq M$ find a polynomial function $G(x)$ such that $G(x) = 0$ iff df_x is not surjective. (Use the sum of the squares of the $n \times n$ minors of the matrix df_x.) Deduce that C is a closed algebraic set. Finally $C \neq M$, by Sard's Theorem.)

4. **Classical Hopf maps.** (1) Let D be an n-dimensional real division algebra, so that $n = 1, 2, 4$ or 8. For any u, v with $u \neq 0$ there is a unique $x \in D$ with $xu = v$. Notation: $x = v/u$. Define $\pi : D \times D \to D \cup \{\infty\}$ by $\pi(u, v) = \begin{cases} v/u & \text{if } u \neq 0, \\ \infty & \text{if } u = 0. \end{cases}$ Identifying $D \cup \{\infty\}$ with S^n by stereographic projection, we get an induced map $\pi : S^{2n-1} \to S^n$. Check that $\pi^{-1}(q)$ is a great $(n-1)$-sphere. If $D = \mathbb{C}, \mathbb{H}, \mathbb{O}$ we obtain the classical Hopf fibrations.

(2) Let S^3 be the unit sphere in the quaternions $\mathbb{H}$ and let S^2 be the unit sphere in $\mathbb{H}_0$, the pure quaternions. Fix any $i \in S^2$ and define the quadratic map $h(c) = c \cdot i \cdot \bar{c}$. This $h : S^3 \to S^2$ is essentially the same as the Hopf map. If $h(c) = q$ then

$h^{-1}(q) = \{c \cdot e^{i\theta} : \theta \in \mathbb{R}\}$ is a great circle. In fact if $q \neq q'$ in S^2 then $h^{-1}(q)$ and $h^{-1}(q')$ are linked in S^3. Does h send *every* circle in S^3 to a circle (or point) of S^2?

(3) Suppose $1, i, j$ are orthonormal elements in $\mathbb{O}$, the octonion algebra. Define $h : \mathbb{O} \to \mathbb{O}$ by $h(x) = (i\bar{x})(xj)$. For every x, $h(x) \in \{1, i, j\}^{\perp}$. Then h is a quadratic spherical map $S^7 \to S^4$. This map is essentially the same as the Hopf map arising from a normed $[4, 4, 4]$.

(Hint. (3) Let $\mathbb{H}$ be the quaternion subalgebra generated by i, j, and view $\mathbb{O}$ as $\mathbb{H}^2$. Then $h(a, b) = ((|a|^2 - |b|^2)k, 2bk\bar{a})$.)

5. **Conjugates.** If $V = X \perp Y$ and $v = (x, y) \in V$ with $x \neq 0$ and $y \neq 0$, define $v^* = (-\frac{|y|}{|x|} \cdot x, \frac{|x|}{|y|} \cdot y)$.

(1) $|v^*| = |v|$; $\langle v, v^* \rangle = 0$; $v^{**} = v$.

Then * acts on most of $\mathcal{S}(V)$. Describe this action geometrically in the case $\dim X = \dim Y = 1$.

(2) Let $F : \mathcal{S}(X \perp Y) \to \mathcal{S}(\mathbb{R}p \perp Z)$ be the Hopf map for a normed bilinear $f : X \times Y \to Z$. Then $F(v^*) = -F(v)$. Consequently, if q lies in the image of a Hopf map, then so does $-q$.

(3) The great circle through v and v^* is wrapped uniformly twice around the meridian through $q = F(v)$.

(4) Suppose $c \in \mathcal{S}(Z)$ lies on the equator. If $v = (x, y) \in W_c$ then $v^* = (-x, y)$ and $v \mapsto v^*$ restricts to a linear bijection $W_c \xrightarrow{\cong} W_{-c}$.

(5) Choose an orthonormal basis $v_i = (x_i, y_i)$ of W_c. Then $2f(x_i, y_i) = q$, $|x_i| = 1/\sqrt{2}$, and $x_1, \ldots, x_k$ are orthogonal in X. Similarly for the y_i. Also $B(v_i, v_i^*) = -p$ and if $i \neq j$ then $B(v_i, v_j^*) = f(x_i, y_j) - f(x_j, y_i)$. The hidden map $B_{(c)} : W_c \times W_c^{\perp} \to (q)^{\perp}$ is not easy to determine explicitly, but there is a simple formula for the portion of $B_{(c)}$ of size $[k, k, n]$ on the space $W_c \times W_{-c}$.

(Hint. (3) The great circle is wrapped uniformly twice around a circle which has q and $-q$ as endpoints of a diameter. It passes through p as well.

(5) If $i \neq j$ then $B(v_i, v_j) = 0$ by (15.1) so that $\langle x_i, x_j \rangle - \langle y_i, y_j \rangle = 0$. Since the v_i are orthogonal conclude $\langle x_i, x_j \rangle = \langle y_i, y_j \rangle = 0$.)

6. (1) The antipodal property for image(F) given in Exercise 5(2) does not hold for all spherical quadratic maps.

(2) Is the image of a nonsingular bilinear map always an algebraic variety? (Compare (15.24).)

(Hint. (1) Consider $F : S^1 \to S^2$ defined by $F(x_1, x_2) = (x_1^2, \sqrt{2}x_1x_2, x_2^2)$.

(2) The Cauchy product pairing $c_{(2,2)}$ of size $[2, 2, 3]$ has image$(c_{(2,2)}) = \{(a, b, c) : b^2 \geq 4ac\}$.)

7. **Image (F).** If $F : S^m \to S^n$ is a spherical quadratic map, what can be said about $\Sigma = \text{image}(F)$? This Σ is also the image of the induced map $\bar{F} : \mathbb{P}^m \to S^n$.

(1) Σ is an algebraic subvariety of S^n with the following "2-point property": For any distinct points $a, b \in \Sigma$, there exists a circle C with $a, b \in C \subseteq \Sigma$.

(2) If $m = 1$ then Σ is a circle or point in S^n. After suitable rotation and restriction, F becomes the following map: $F_\theta(x.y) = (x^2 + \cos(2\theta)y^2, 2\sin(\theta)xy, \sin(2\theta)y^2)$.

(3) Suppose $m = 2$. If $\bar{F}$ is not injective on the projective plane, there exist $v \neq \pm w$ in S^2 such that $F(v) = F(w)$. Then the great circle through v, w is mapped to a single point q. If F is non-constant then it is essentially the same as the Hopf map $S^2 \to S^2$ described in Exercise 10 (a) below. In this case $\Sigma \cong S^2$.

(4) If $m = 2$ can it happen that $\bar{F}$ is injective? If so, $\dim W_q = 1$ for every $q \in \text{image}(F)$. Then Σ is an embedded copy of $\mathbb{P}^2$ in S^n and every projective line maps bijectively to a circle in S^n.

8. If $F : S^m \to S^n$ is a spherical quadratic map, let $Y_k = \{q \in S^n : \dim W_q = k\}$. If Y_k is nonempty, the restriction of F to $F^{-1}(Y_k) \to Y_k$ is a great $(k-1)$-sphere bundle projection.

(Hint. Let $G_{k,n}$ denote the Grassmann manifold of k-planes in n-space. Let $W : Y_k \to G_{k,n}$ be the map defined by $W(q) = W_q$. The pullback by W of the canonical k-plane bundle is the restriction of F to $\bigcup_{q \in Y_k} W_q \to Y_k$. This map is a vector bundle projection.)

9. (1) The map g_q of (15.9) satisfies: $\ker(g_q) = W_q$ and $\text{image}(g_q) = W_q^\perp$.

(2) Since g_q is symmetric, V admits an orthonormal basis of eigenvectors. If λ is an eigenvalue for g_q then $0 \leq \lambda \leq 1$.

(3) The 0-eigenspace is W_q and the 1-eigenspace is W_{-q}. Then q is a pole for F if and only if g_q has only 0, 1 as eigenvalues.

(Hint. (2) If $g_q(x) = \lambda x$ then $\langle q, F(x)\rangle = (1 - 2\lambda) \cdot |x|^2$. Apply Cauchy–Schwarz.

(3) Check that $g_q(x) + g_{-q}(x) = x$.)

10. **Degree.** (1) Let $h : S^2 \to S^2$ be the Hopf map for a normed $[1, 2, 2]$. Then $h(x_0, x_1, x_2) = (x_0^2 - x_1^2 - x_2^2, 2x_0x_1, 2x_0x_2)$. Describe the action of h on a typical meridian. If q_s is the south pole then $h^{-1}(q_s)$ is the equator. If $q \neq q_s$ describe $h^{-1}(q)$.

(2) Let $h_n : S^n \to S^n$ be the Hopf map coming from a normed $[1, n, n]$. Describe h_n as in part (1).

(3) A (continuous) map $g : S^n \to S^n$ has a topological *degree* defined as its image in the homotopy group $\pi_n(S^n) \cong \mathbb{Z}$, or in the homology group $H_n(S^n, \mathbb{Z}) \cong \mathbb{Z}$. Alternatively if y is a regular value for f then $\deg(f) = \sum \operatorname{sgn} df_x$, where the sum is over all $x \in f^{-1}(y)$. (See Milnor (1965).) If n is even then $\deg h_n = 0$. If n is odd then $\deg h_n = 2$.

(Hint. (3) Each meridian is wrapped uniformly twice around itself. Little antipodal patches on S^n have opposite orientation iff n is even.)

11. **Computing $\rho(n, r)$.** Assume the formula for $\rho_{\#}(n, n-2)$ given in (12.30).

(1) If $n \not\equiv 0 \pmod{16}$ and $\lambda(n, n-3) > 8$ then $\rho(n, n-3) = \lambda(n, n-3)$.

(2) If $n \not\equiv 0, 1 \pmod{16}$ and $\lambda(n, n-4) > 8$ then $\rho(n, n-4) = \lambda(n, n-4)$.

(3) Assuming (12.32), complete the proof of Theorem 12.31.

(Outline. (1) $\lambda > 8$ implies $n \equiv 0, 1, 2, 3 \pmod{16}$. Suppose there is a normed $[r, n-3, n]$ with $r > \lambda(n, n-3) \geq 9$. There is a hidden $[k, r+n-3-k, n]$ with $k \leq r+n-3-r \# (n-3)$. Use (12.30) to obtain $r \# (n-3) \geq n$. Deduce that $k = r-3$ and $r-3 \leq \rho(n)$ so that $8 \mid n$.

(2) Similarly there is a hidden $[k, r+n-4-k, n]$ with $k \leq r+n-4-r\#(n-4)$. Use (12.30) to obtain $r \# (n-4) \geq n-1$ so that $k = r-4$ or $r-3$. Deduce that either $8 \mid n$ or $8 \mid (n-1)$.

(3) Use (12.32) and (1), (2) to eliminate the remaining cases except for $\rho(n, n-4)$ when $n \equiv 65 \pmod{128}$. In that case $r \# (n-4) = n-1$ is impossible by (12.32) since $n-1 \not\equiv 65, 66 \pmod{128}$.)

12. **Surjectivity.** Suppose $H : \mathcal{S}(X \perp Y) \to \mathcal{S}(\mathbb{R}p \perp Z)$ is the Hopf map for the normed bilinear $f : X \times Y \to Z$ of size $[r, s, n]$.

(1) The following are equivalent:

(a) $f : X \times Y \to Z$ is surjective.

(b) $H : X \perp Y \to \mathbb{R}p \perp Z$ is surjective.

(c) $H : \mathcal{S}(X \perp Y) \to \mathcal{S}(\mathbb{R}p \perp Z)$ is surjective.

(2) If f is surjective then $n \leq r+s-1$.

(3) If $n = r \# s$ then H is surjective.

(4) Let $A, B \subseteq \mathbb{O}$ be subspaces of the octonions, with $\dim A = r$ and $\dim B = s$. Then $A \times B \to \mathbb{O}$ is surjective if and only if $r+s > 8$.

(Hint. (2) Clear from (1) since $H : S^{r+s-1} \to S^n$. Compare Exercise 14.16(2).

(3) Compare Exercise 12.23.

(4) If $r+s > 8$ and $c \neq 0$, then $\bar{A}c \cap B \neq 0$.)

13. **Open Question.** If $m \geq n$ and $F : S^m \to S^n$ is a non-constant spherical quadratic map, then must F is surjective?

(Comment. The answer is yes if $n \leq 2$. By (A.3) the cases $n = 1$, and $n < 4$ with $m \geq 4$ are vacuous. Suppose $n = 2$ and $m \leq 3$. By (15.2), there is a circle $C \subseteq \text{image}(F)$. By (15.9) image$(F)$ is a real algebraic variety, so any circle meeting image(F) in infinitely many points is contained in image(F). Choose $b \in \text{image}(F)$ with $b \notin C$. Consider circles C', C'' in S^2 concentric with C and very close to C, one on each side. For $c \in C$, there is a circle through b and c and entirely in image(F). That circle must meet C' or C''. Since c is arbitrary, either C' or C'' contains infinitely many points of image(F). Suppose $C' \subseteq \text{image}(F)$. The region on S^2 bounded by C and C' lies inside image(F). Deduce F is surjective. For the larger cases, a Hopf map counterexample of size $[r, s, n]$ must have $r \# s < n \leq r+s-1$.)

14. Suppose $F : S^m \to S^n$ is quadratic and $k = \dim W_q$, so that $m - n < k \le \min\{m, n\}$ as in (15.20). The case $k = n \le m$ is mentioned in (15.21).

(1) Analyze the case $k = m < n$.

(2) What can be said when $k = m - n + 1$?

(Hint. (1) F arises from some explicit map $S^m \to S^{m+1} \subseteq S^n$.)

15. (1) Suppose $F : S^m \to S^n$ is a smooth map whose image is a real algebraic variety. Then F is surjective if and only if F has a regular point.

(2) Suppose $f : X \times Y \to Z$ is a normed bilinear map. We suppress the "f" as in (15.22). Then f is surjective if and only if there exist $x \in X$ and $y \in Y$ such that $xY + Xy = Z$.

(3) Check that the [10, 10, 16] described in Chapter 13 is surjective. Is the [12, 12, 26] there also surjective?

(Hint. (1) If F is surjective, regular points exist (Exercise 3). Conversely suppose x is a regular point. By the Inverse Function Theorem, if $q = F(x)$ there is an open ball B in S^n such that $q \in B \subseteq \text{image}(F)$. If C is a circle in S^n through q then $C \cap B$ is infinite and therefore $C \subseteq \text{image}(F)$ (since it is a variety).

(2) f is surjective if and only if the associated Hopf map $H : S^{r+s-1} \to S^n$ is surjective (Exercise 12). By (1) this occurs iff H has a regular point $v = (x, y)$. As in (15.12) deduce that dH_v is surjective iff the map $(x', y') \mapsto f(x, y') + f(x', y)$ is surjective.)

16. **Surjective normed maps.** (1) **Proposition.** *Suppose f is a normed bilinear map of size $[r, s, n]$. If f is surjective then $r + s \le n + \rho(n)$.*

For example any nonsingular $[r, s, r \# s]$ is surjective, by Exercise 12.23. Historically, this proposition provided the earliest examples where $r * s \ne r \# s$. Lam's first proof used a framed cobordism argument.

Is there a normed $[r, s, n]$ with $r + s > n + \rho(n)$? An example would answer the question in Exercise 13.

(2) If $F : S^m \to S^n$ is a spherical quadratic map which is not trivial in $\pi_m(S^n)$, then $n + \rho(n) > m \ge n$.

(3) **Proposition.** *Let 2^m be the smallest 2-power exceeding $k + 1$. Then $2^k * 2^k \ge 2^{k+1} - \rho(2^m)$.*

(Hint. (1) $H : S^{r+s-1} \to S^n$ is surjective and Sard says there is $v \in S^{r+s-1}$ with $\text{rank}(dH_v) = n$. As in (15.13) the associated hidden pairing has size $[n, r+s-n, n]$.

(2) By (15.18) we may assume F is a Hopf map, surjective since non-trivial in homotopy. Apply (1).

(3) James (1963) proved $2^k \# 2^k \ge 2^{k+1} - \rho(2^k)$. Check that $\rho(2^m)$ is the maximal $\rho(j)$ for $1 \le j \le \rho(2^k)$. Given a normed bilinear $[2^k, 2^k, 2^{k+1} - \rho(2^m) - 1]$, apply (15.13) to find a hidden $[t, 2^{k+1} - t, 2^{k+1} - \rho(2^m) - 1]$ where $t \le \rho(2^k)$. The duality in Exercise 12.17 provides a nonsingular skew-linear $[t, \rho(2^m) + 1, t]$, and (12.22) yields a contradiction.)

17. **Duals.** (1) If $f : X \times Y \to Z$ is normed bilinear, we may view is as a linear embedding $X \subseteq \mathrm{Hom}(Y, Z)$. If $c \in Z$ the map $\varphi_c : X \to Y$ defined in (15.22) is then given by $\varphi_c(f) = \tilde{f}(c)$.

(2) If V is an inner product space let $\hat{V} = \mathrm{Hom}(V, \mathbb{R})$ be its dual space. The inner product identifies V with $\hat{V}$, and $V \otimes W$ can be identified with $\mathrm{Hom}(V, W)$. Now suppose X, Y, Z are inner product spaces and a bilinear $f : X \times Y \to Z$ is given. After appropriate identifications, the transpose of $f^{\otimes} : X \otimes Y \to Z$ provides a linear map $\varphi : Z \to \mathrm{Hom}(X, Y)$. This is the same as the map $c \mapsto \varphi_c$.

18. **Poles and collapse values.** By (15.10), a spherical quadratic map is a Hopf map if it admits a pair of poles. Can it admit more than one pair? If F is a classical Hopf map ($S^{2n-1} \to S^n$ for $n = 1, 2, 4, 8$) then every $q \in S^n$ is a pole.

(1) Suppose $F : S^m \to S^n$ is a quadratic map admitting more than one pair of poles. Then $m = 2r - 1$ is odd and F is the Hopf construction of a normed bilinear map of size $[r, r, n]$.

(2) Let $\mathcal{P}$ be the set of poles for F. Then $\mathcal{P}$ is a great sphere and $\dim \mathcal{P} = \dim C$, where C is the linear space of collapse values for F.

(3) If F is not a classical Hopf map then $\dim \mathcal{P} \leq 2$. If $\dim \mathcal{P} = 2$ then r is even.

(Hint. (1) If $\pm p$ are poles let $r = \dim W_p$ and $s = \dim W_{-p}$. Then (15.10) implies F is the Hopf map for a normed $[r, s, n]$. If $\pm q$ is another pair of poles then $r = s$, by (15.27).

(2) $\mathcal{P} = \{q \in S^n : \dim W_q = r\}$. Suppose $p, q \in \mathcal{P}$ and $q \neq \pm p$. The great circle through p, q lies in $\mathcal{P}$, since $\dim W_q$ is constant on meridians. Let c be the corresponding point on the equator. Then q is a pole if and only if $c \in C$. Hence $\mathcal{P}$ is the great sphere with poles $\pm p$ and equator $\mathcal{S}(C)$.

(3) Apply (15.33).)

19. **Integral pairings and collapse values.** Suppose $X \times Y \to Z$ is an integral normed bilinear $[r, s, n]$ corresponding to bases $\{x_1, \ldots, x_r\}$, $\{y_1, \ldots, y_s\}$ and $\{z_1, \ldots, z_n\}$. For the associated $r \times s$ intercalate matrix M, entry m_{ij} is the "color" z_k iff $x_i y_j = \pm z_k$. The *frequency* of a color c is the number of occurrences of c in the matrix M.

(1) **Lemma.** *If c is one of the basis elements z_k then* $\dim W_c$ = *frequency of c.*

(2) The space C of collapse values equals $\mathrm{span}\{z_1, \ldots, z_\ell\}$, where $z_1, \ldots, z_\ell$ are the colors which appear in every row of M.

(3) **Corollary.** *For an integral pairing of size $[r, r, n]$, the space C is the span of the ubiquitous colors. Consequently, if there are more than 2 ubiquitous colors then $r = n = 4$ or 8.*

(4) Every $[5, 5, 8]$ has $\dim C = 1$. The $[10, 10, 16]$ in Chapter 13 has $\dim C = 2$. What about the pairings of sizes $[12, 12, 26]$ and $[16, 16, 32]$?

(Hint. (1) By (15.25) $\dim W_c = \dim X_c$. Color c occurs in row $i \iff c \in x_i Y \iff x_i \in X_c$. To prove these x_i span X_c suppose $0 \neq x \in X_c$ is a linear combination of some $x_i \notin X_c$. The linear independance of the colors leads to a contradiction.

(2) By (1), c is a collapse value iff c has frequency r, iff c occurs in every row.
(3) Ubiquitous colors were defined in (13.8). Apply (15.33).)

20. In the proof of (15.33), complete the 8×8 multiplication table and prove that the three sets of 8 vectors are orthonormal. How is that table related to the octonion multiplication table?

21. Let $f : S^m \to S^n$ be a homogeneous polynomial map of degree d. Then $f = (f_0, f_1, \dots, f_n)$ where each $f_j = f_j(X) \in \mathbb{R}[X]$ is a homogeneous polynomial in $X = (x_0, \dots, x_m)$.
(1) $f_0(X)^2 + \cdots + f_n(X)^2 = (x_0^2 + \cdots + x_m^2)^d$.
(2) If f is constant on S^m then d is even and $f(X) = (x_0^2 + \cdots + x_m^2)^{d/2} \cdot u$, for some $u \in S^n$.
(3) If $m \le n$ then for every d there is a non-constant $f : S^m \to S^n$ of degree d.
(4) Suppose $m > n$ and f is non-constant. Then d must be even.

Open Question. Is there a non-constant $f : S^{48} \to S^{47}$ which is homogeneous of degree 4?

(Hint. (2) If $f(w) = c$ for all $w \in S^m$ then $f(v) = |v|^d \cdot c$ for every $v \in \mathbb{R}^{m+1}$. Use $f(-v)$ to show d is even.
(4) $f_0, \dots, f_n$ is a system of $n+1$ forms of degree d in more than $n+1$ variables. If d is odd, the real Nullstellensatz (see Exercise 12.18) implies that this system has a non-trivial common zero over $\mathbb{R}$.)

22. Here is an alternative approach to (A.11).
(1) The following are equivalent:
(a) $q(m) < m$.
(b) There exists a non-constant quadratic $S^m \to S^{m-1}$.
(c) There exists a normed $[k, m+1-k, n]$ for some k (where $1 < k < m$).
(d) There exists k with $1 < k < m$ and $k \le \rho(m+1-k)$.
(e) There exists k with $1 < k < m$ and $m \equiv k - 1 \pmod{2^{\delta(k)}}$.
(2) Work out (e) when $k \le 9$ to prove:
If $m \not\equiv 0 \pmod{16}$ and $m > 8$ then $q(m) < m$.
If $m \equiv 0 \pmod{16}$ then $q(m) < m$ iff: there exists $k \equiv 1 \pmod{16}$ with $1 < k < m$ and $m \equiv k - 1 \pmod{2^{\delta(k)}}$. Then $m = 272$ is the smallest multiple of 16 for which $q(m) \ne m$.

(Hint. (1) Use (15.19). Recall the properties of $\delta(k)$ given in Exercise 0.6.)

23. (1) Complete the proof of (A.12).
(2) If $t \ge 4$ then $2^t - 8 = q(2^t - 1) = q(2^t - 2) = \cdots = q(2^t - 7)$. What are the next few values?

(Hint (1) Express $m = 2^t + m_0$. If $0 \le m_0 < \rho(2^t)$ then $1 - \frac{q(m)}{m} \to 0$ as t gets large. If $\rho(2^t) \le m_0 < 2^t$ then $\left(1 - \frac{q(m)}{m}\right) \le \frac{1}{2}\left(1 - \frac{q(m_0)}{m_0}\right)$.)

Notes on Chapter 15

We defined $2B(x, y) = F(x + y) - F(x) - F(y)$. In his papers on this subject, Yiu defines the form B without that factor 2. Consequently some of the formulas here differ from those in Yiu's work by various factors of 2. We chose this version to have a notation parallel with the standard inner product, which satisfies $2\langle x, y\rangle = |x + y|^2 - |x|^2 - |y|^2$.

The Hopf construction provides examples of spherical quadratic maps $F : S^{r+s-1} \to S^n$. Hefter (1982) used differential geometry to prove that if $q \in S^n$ is a regular value then $F^{-1}(q)$ is a great sphere (and all these spheres have the same dimension). K. Y. Lam (1984) removed the restriction to regular values by using known facts about the classical Hopf fibration $S^3 \to S^2$. Our presentation follows Yiu's elementary geometric proof of a more general result. This was developed in his thesis (1985), published in Yiu (1986). Chang's algebraic proof is described in (15.20).

Information on the geometric properties of Hopf fibrations is given by Gluck, Warner and Yang (1983) and in Gluck, Warner and Ziller (1986).

Ono (1994) presents some of the basic results on spherical quadratic maps, using different notations. He seems unaware of the work of K. Y. Lam and Yiu. Ono also considers arithmetic properties of Hopf maps defined over the integers.

The conjectures (after (15.14)) that $12*12 = 26$ and $16*16 = 32$ were formulated by Adem (1975).

The polynomial approach described in (15.20) and (15.21) follows Chang (1998). He expands on the methods pioneered by Wood (1968) and further developed by Turisco (1979), (1985) and Ono (1994), §5. Chang also discusses the case $F : S^{2n-2} \to S^n$, proving in this case that $n = 2$, 4 or 8 and F is a restriction of a classical Hopf fibration.

A different version of the map φ defined in (15.22) was considered by Kaplan (1981). If $S \subseteq \mathrm{Sim}(V)$ and $1_V \in S$ let $U = S_1$. The normed pairing $U \times V \to V$ leads to a skew symmetric pairing $V \times V \to U$ which Kaplan uses to make $U \times V$ into a 2-step nilpotent Lie algebra.

The results on collapse values here are all due to Lam and Yiu (1989).

Wood's results, discussed in the appendix, are also presented and extended in Chapter 13 of Bochnak, Coste and Roy (1987).

The idea for the proof of Proposition A.14 was suggested by P. H. Tiep in 1997.

Exercise 4 (1). For any division algebra D this construction yields a smooth $(n-1)$-sphere fibration of S^{2n-1}. (But it is not necessarily a polynomial map.) Isotopic algebras yield smoothly isomorphic fibrations. Conversely starting with a smooth fibration of S^{2n-1} by great $(n-1)$-spheres there is an associated division algebra, unique up to isotopy. For further details of this geometry see Yang (1981) and Gluck, Warner, and Yang (1983), §6. Exercise 4(3) comes from Rigas and Chaves (1998).

Exercise 5. The construction of v^* used by K. Y. Lam in (15.12) is generalized here. This idea was first introduced by Roitberg (1975). With more work the result in (4) can be extended: If $q \neq \pm p$ then * restricts to a *linear* map $W_q \xrightarrow{\cong} W_{-q}$.

Exercises 6, 8, 9 appear in Yiu (1986).

Exercise 6 (2). The Cauchy product form $\mathbb{R}^r \times \mathbb{R}^s \to \mathbb{R}^{r+s-1}$ is nonsingular, as noted in (12.12). Also compare Exercise 14.6. Its Hopf construction $H_{(r,s)} : S^{r+s-1} \to S^{r+s-1}$ has homotopy class determined by its degree (compare Exercise 10). A clever calculation of that degree is given by L. Smith (1978), pp. 727–731.

Exercise 10. Wood (1968) proved that if n is odd then h_n has topological degree 2. He deduced that every $k \in \mathbb{Z} \cong \pi_n(S^n)$ can be represented by a homogeneous polynomial map $S^n \to S^n$ with (algebraic) degree $|k|$. If n is even it apparently remains unknown whether any elements of $\pi_n(S^n)$ other than those corresponding to $0, \pm 1$ can be represented by polynomial maps.

Exercise 11. The result is stated in Lam and Yiu (1987) without full details.

Exercise 14. See Chang (1998). Maps as in part (2) are called "first kind" in Ono (1994), §5.3.4.

Exercise 15. See Lam and Yiu (1989).

Exercise 16 is due to K. Y. Lam (1984), (1985).

Exercise 18. Hopf maps admitting more than one pair of poles are discussed in Yiu (1986). The connection with collapse values is implicit in Lam and Yiu (1989).

Exercise 19 follows Yiu's thesis (1985). The results are also described by Lam and Yiu (1995).

Chapter 16

Related Topics

In this final chapter we mention several topics that are related to compositions of quadratic forms. Most of the proofs are omitted.

Section A. Higher degree forms permitting composition.

Section B. Vector products and composition algebras.

Section C. Compositions over rings and over fields of characteristic 2.

Section D. Linear spaces of matrices of constant rank.

Some of these topics are discussed in greater detail than others, and many deserving topics are omitted altogether. These choices simply reflect the author's interests at the time of writing.

We won't mention the large literature on Gauss's theory of composition of quadratic forms, and its various generalizations. That subject is part of number theory and has been presented in many books and articles.

Section A. Higher degree forms permitting composition

What sorts of compositions are there for forms of degree $d > 2$? Are there restrictions on the dimensions similar to the Hurwitz 1, 2, 4, 8 Theorem? We present here an outline of the ideas from the survey article by R. D. Schafer (1970), and later discuss Becker's conjecture concerning compositions for diagonal forms $x_1^d + x_2^d + \cdots + x_n^d$.

Suppose $\varphi(x_1, \ldots, x_n)$ is a form (homogeneous polynomial) of degree d in n variables with coefficients in a field F. This φ *permits composition* if there is a formula

$$\varphi(X) \cdot \varphi(Y) = \varphi(Z)$$

where X, Y are systems of n indeterminates and each z_k is a bilinear form in X, Y with coefficients in F. In this case the vector space $A = F^n$ admits a bilinear map $A \times A \to A$ which we write as multiplication. Then A is an F-algebra and $\varphi(ab) = \varphi(a) \cdot \varphi(b)$ for every $a, b \in A$. In this case we way that φ permits composition on A.

For example suppose $A = \mathbb{M}_n(F)$ is the matrix algebra of dimension n^2. Then $\det(a)$ is a form of degree n permitting composition. The converse is a beautiful old result.

A.1 Proposition. *Suppose φ is a form of degree $d > 0$ permitting composition on $\mathbb{M}_n(F)$, where F is a field with $|F| > d$. Then for some $s > 0$, $\varphi(a) = (\det a)^s$ for all a.*

Proof outline. Let $K = F(x_{11}, \dots, x_{nn})$ where the x_{ij} are indeterminates. Then φ extends to a form permitting composition on $\mathbb{M}_n(K)$. For the "generic matrix" $X = (x_{ij})$, $\det X$ is an irreducible polynomial in n^2 variables over F. The classical adjoint Z is a matrix over $F[x_{11}, \dots, x_{nn}]$ with $X \cdot Z = (\det X) \cdot 1_n$. Then $\varphi(X)\varphi(Z) = (\det X)^d$ and unique factorization implies $\varphi(X) = (\det X)^s$ for some $s > 0$. □

To avoid trivial examples (like the zero form) we restrict attention to certain "regular" forms. To define the various types of regularity, suppose φ is a form of degree d in n variables. View it geometrically as a map $\varphi : V \to F$, where V is an n-dimensional vector space over F. If $d! \neq 0$ in F (i.e. if the characteristic does not divide d), there is a unique symmetric d-linear map $\theta : V \times \dots \times V \to V$ with the property that $\varphi(v) = \theta(v, v, \dots, v)$ for every v. For example when $d = 3$ we find that

$$\theta(v_1, v_2, v_3) = \frac{1}{3!}\Big(\varphi(v_1 + v_2 + v_3) - \varphi(v_1 + v_2) - \varphi(v_1 + v_3) - \varphi(v_2 + v_3) + \varphi(v_1) + \varphi(v_2) + \varphi(v_3)\Big).$$

This "polarization identity" generalizes the case of a quadratic form and its associated symmetric bilinear form. See Exercise A1.

If k is an integer between 1 and d, define the degree d form φ to be *k-regular* if $v = 0$ is the only vector in V such that

$$\theta(v, v, \dots, v, v_{k+1}, \dots, v_d) = 0 \quad \text{for every } v_{k+1}, \dots, v_d \in V.$$

For example φ is d-regular if it is anisotropic: $\varphi(v) = 0$ implies $v = 0$. If φ is k-regular then it is also $(k-1)$-regular. The regularity of a form φ might decrease under a field extension K/F: φ_K is k-regular over K implies φ is k-regular over F. The converse can fail if $k > 1$. We give special names to two cases.

φ is *regular* if it is 1-regular.
φ is *nonsingular* if it is $(d-1)$-regular over the algebraic closure $\bar{F}$.[1]

For example $\det X$ is a regular form on $\mathbb{M}_n(F)$, but it is singular if $n > 2$. Schafer's work on compositions deals with regular forms.

Generic norms provide further examples of regular forms permitting composition. Jacobson developed that theory for the class of "strictly power associative" algebras.

1 These names are not standard. Some authors use "regular" for what we call nonsingular.

For central simple (associative) algebras, the generic norm coincides with the reduced norm. Further details appear in Schafer (1970) and in Jacobson (1968), pp. 222–226. If A is a finite dimensional F-algebra let N_A be its generic norm. Jacobson proved that if A is alternative then N_A permits composition on A. Moreover if the algebra is also separable then N_A is regular. If A is separable and alternative then it is a direct sum of simple ideals

$$A = A_1 \oplus \cdots \oplus A_r,$$

and the center of each A_i is some separable field extension K_i of F. If A_i is associative, it is a central simple K_i-algebra. If A_i is not associative, it must be an octonion algebra over K_i (as proved by Zorn as mentioned at the end of Chapter 8). Any $a \in A$ is uniquely expressible as $a = a_1 + \cdots + a_r$ and the generic norm is

$$N(a) = N_1(a_1) \dots N_r(a_r),$$

where N_i is the generic norm of the F-algebra A_i. Now if $f_1, \dots, f_r$ are positive integers then

$$\varphi(a) = N_1(a_1)^{f_1} \dots N_r(a_r)^{f_r}$$

is also a regular form on A which permits composition. If N_i had degree d_i then this form φ has degree $d = d_1 f_1 + \cdots + d_r f_r$.

A.2 Schafer's Theorem. *Let A be a finite dimensional F-algebra with 1. Assume $d! \neq 0$ in F. There exists a regular form φ of degree $d > 2$ permitting composition on A if and only if: A is a separable alternative algebra and φ is one of the forms mentioned above, for some positive integers $f_1, \dots, f_r$.*

More details and references appear in Schafer (1970). He also points out that McCrimmon used Jordan algebras and the differential calculus of rational maps to extend this Theorem. McCrimmon proved that there are no infinite dimensional compositions (that is, if A is an algebra with 1 having a regular form which permits composition then dim A is finite), and he removed the restrictions on the characteristic, requiring only that $|F| > d$. That generalization requires a somewhat different definition of "regular" since the associated d-linear form φ is not available when $d! = 0$ in F.

Let's return to the original question about a degree d form φ in n variables such that φ admits a bilinear composition. The bilinear pairing makes $A = F^n$ into an F-algebra, but there might be no identity element. However, if φ is regular we can alter the multiplication to obtain an algebra with 1 so that Schafer's Theorem applies. See Exercise A2. The following restrictions on dimensions are mentioned in Schafer (1970), p. 140.

A.3 Corollary. *Suppose φ is a regular form of degree d in n variables over a field F where $d! \neq 0$. Suppose φ permits composition.*

If $d = 2$ then $n = 1$, 2, 4 or 8.

If $d = 3$ *then* $n = 1, 2, 3, 5$ *or* 9.

If $d = 4$ *then* $n = 1, 2, 3, 4, 5, 6, 8, 9, 12$ *or* 16.

Proof. The case $d = 2$ is the Hurwitz Theorem. Suppose $d = 3$. By Exercise A2 there is an n-dimensional F-algebra A with 1 such that φ is a form on A permitting composition. Schafer's Theorem implies $3 = d_1 f_1 + \dots + d_r f_r$ where d_i is the degree of the generic norm on the simple algebra A_i. If $r = 1$ then A is simple and the degree of its generic norm divides 3. Then A is associative since the octonion algebra has generic norm of degree 2. Then A is a central simple K-algebra where K/F is a separable field extension. Hence either $A = F$ (and $n = 1$), or $A = K$ (and $n = 3$), or A is central simple of degree 3 over F (and $n = 9$). Suppose $r = 2$. If $f_1 > 1$ then $A = F \oplus F$ (and $n = 2$). Otherwise $f_1 = f_2 = 1$ and $A = F \oplus B$ where B is a simple alternative algebra with generic norm of degree 2. Since this norm on B permits composition, Hurwitz implies $n = 1 + \dim B = 2, 3, 5$, or 9. Finally if $r = 3$ then $A = F \oplus F \oplus F$ and $n = 3$.

The cases for $d = 4$ take longer to write out and are omitted. □

The original Hurwitz question involved sums of squares. Rather than generalizing as above to arbitrary forms of degree d we ask the analogous question for sums of dth powers. Every quadratic form can be diagonalized, but for higher degrees these "diagonal" forms are quite special. Define a "degree d diagonal composition of size $[r, s, n]$" to be a formula of the type:

$$(x_1^d + x_2^d + \dots + x_r^d) \cdot (y_1^d + y_2^d + \dots + y_s^d) = z_1^d + z_2^d + \dots + z_n^d$$

where $X = (x_1, x_2, \dots, x_r)$ and $Y = (y_1, y_2, \dots, y_s)$ are systems of indeterminates and each $z_k = z_k(X, Y)$ is a rational function in X and Y. Of course we can simply multiply out the left side and set $z_{ij} = x_i y_j$ to obtain an example of such a composition when $n = rs$. Can there be compositions with smaller n? As in the quadratic case ($d = 2$) we also consider the compositions where each z_k is bilinear in X, Y and the compositions where each z_k in linear in Y and rational in X.

A.4 Becker's Conjecture. *Suppose* $d > 2$ *and* $d! \neq 0$ *in* F. *If there is a degree* d *diagonal composition of size* $[r, s, n]$ *where each* $z_k \in F(X, Y)$, *then* $n \geq rs$.

This conjecture arose from Eberhard Becker's work on higher pythagoras numbers. For instance, see Becker (1982), especially Theorem 2.12. A similar question was raised earlier by Nathanson (1975). Very little seems to be known about this conjecture, but some progress was made by U.-G. Gude (1988). We mention some of his results here, without proofs.

A.5 Proposition. *Becker's Conjecture is true over* $\mathbb{Q}$ *in the following cases:*

$d = 4$ *and* $rs \leq 15$;

$d = 2^{m-2}$ and $rs \le 2^m$ for some $m \ge 5$;

$d = p^{m-1}(p-1)$ and $rs \le p^m$ when p is an odd prime and $m \ge 2$.

For example there is no identity of the type

$$(x_1^4 + x_2^4 + x_3^4) \cdot (y_1^4 + \cdots + y_5^4) = z_1^4 + z_2^4 + \cdots + z_{14}^4$$

where each z_k is a rational function in the x's and y's with coefficients in $\mathbb{Q}$. For the proof, Gude passes to the p-adic field $\mathbb{Q}_p$ (where $p = 2$ in the first two cases), pushes the identity into $\mathbb{Z}_p$ and finally into $\mathbb{Z}/p^m\mathbb{Z}$ where the sums of dth powers are easy to analyze using those values of d.

If $\varphi : V \to F$ is a form of degree d in n variables, its orthogonal group is:

$$\mathrm{O}(\varphi) = \{f \in \mathrm{GL}(V) : \varphi(f(v)) = \varphi(v) \text{ for every } v \in V\}.$$

Suppose $\varphi(X) = x_1^d + x_2^d + \cdots + x_n^d$ where $d > 2$, and let the corresponding $V = F^n$ have basis $\{e_1, \ldots, e_n\}$. Examples of maps $f \in \mathrm{O}(\varphi)$ are given by permuting the basis elements and scaling them by various d^{th} roots of 1. One can show that all the maps in $\mathrm{O}(\varphi)$ are of this type. In particular, $\mathrm{O}(\varphi)$ is finite. This finiteness holds more generally.

A.6 Jordan's Theorem. *Suppose K is an algebraically closed field and $d! \ne 0$ in K. If φ is a nonsingular form of degree $d > 2$ over K then $\mathrm{O}(\varphi)$ is finite.*

This was first proved by C. Jordan over the complex field. For a modern proof see Schneider (1973). Of course the regular forms arising as norm forms of algebras admit many isometries, so they cannot be nonsingular. This finiteness theorem quickly eliminates the possibility of bilinear compositions of size $[r, n, n]$. With more careful work Gude proved the result assuming only linearity in Y.

A.7 Proposition. *Suppose F is a field in which $d! \ne 0$. If $d > 2$ and $r \ge 2$ there is no degree d diagonal composition of size $[r, n, n]$, where each z_k is a linear form in Y with coefficients in $F(X)$.*

Proof idea. Extend F to assume it is infinite. Choose $a \in F^r$ so that $a_1^d + \cdots + a_r^d = 1$ and no denominators in the z_k's become zero when a is substituted for X. For each such a define $L_a : F^n \to F^n$ by $L_a(Y) = (z_1(a, Y), \ldots, z_n(a, Y))$. By hypothesis this is linear, and the composition formula implies $L_a \in \mathrm{O}(\varphi)$ where $\varphi(Y) = y_1^d + \cdots + y_n^d$. There are infinitely many such L_a's, contradicting the finiteness of $\mathrm{O}(\varphi)$. □

Without the linearity hypothesis in this proposition the argument does not work. Gude succeeded in eliminating compositions of size $[2, 2, 2]$ in the general case, assuming no linearity.

A.8 Proposition. *Suppose F is a field of characteristic zero and $d \geq 4$. Then there is no formula of the type*

$$(x_1^d + x_2^d) \cdot (y_1^d + y_2^d) = z_1^d + z_2^d$$

where x_1, x_2, y_1, y_2 are indeterminates and $z_i \in F(x_1, x_2, y_1, y_2)$.

In fact, if $s \geq 2$ there is no composition formula of size $[2, s, 2]$ with degree $d \geq 4$. The proof involves a different finiteness theorem to get the contradiction in this rational case. If V is an algebraic variety of "general type" (also called "hyperbolic" and defined using Kodaira dimension), then the set of dominant rational maps $V \to V$ is finite. This is a generalization of an old theorem of Hurwitz: If C is an irreducible curve of genus ≥ 2 over a field of characteristic zero, then $\mathrm{Aut}(C)$ is finite. (See Hartshorne, Exercise IV.5.2).

There seems to be very little more known about compositions for sums of dth powers.

Section B. Vector products and composition algebras

W. R. Hamilton and his followers in the nineteenth century viewed the algebra of quaternions as a geometric tool, essential for a physical understanding of space and time. In the 1880s the physicists Gibbs and Heaviside (independently) introduced two products for vectors $v, w \in \mathbb{R}^3$ based on the quaternion product. They viewed $\mathbb{R}^3 = \mathbb{H}_0$ as the space of pure quaternions, spanned by i, j and $k = ij$. Then $\mathbb{H} = \mathbb{R} \oplus \mathbb{H}_0$ and the quaternion product vw can be expressed as

$$vw = -\langle v, w\rangle + v \times w,$$

where $\langle v, w\rangle \in \mathbb{R}$ and $v \times w \in \mathbb{H}_0$. It is easy to check that these are the familiar dot product and vector product (cross product) often discussed in basic calculus and physics classes today. The use of i, j and k as the standard unit vectors in $\mathbb{R}^3$ is one remnant of these quaternionic origins.

The vector product enjoys some important geometric properties: it is bilinear; $v \times w$ is orthogonal to v and w; its length $|v \times w|$ equals the area of the parallelogram spanned by v and w. Algebraically this area is $|v| \cdot |w| \cdot |\sin\theta|$ and $|v \times w|^2 = |v|^2|w|^2 - \langle v, w\rangle^2$. Are there similar vector products in other dimensions?

One generalization arises from the following standard algorithm for calculating $v \times w$. Form a 3×3 matrix whose first row is (i, j, k), and with second and third rows given by the coordinates of v and w. The determinant, written as a combination of the basis vectors i, j, k, is $v \times w$. This idea works for any $n - 1$ vectors in $\mathbb{R}^n$. Let A be the $(n-1) \times n$ matrix whose rows are the coordinates of the vectors $v_1, \dots, v_{n-1}$. Define $X(v_1, \dots, v_{n-1})$ to be the vector whose entries are the $(n-1) \times (n-1)$ minor determinants of A, taken with alternating signs. Then "expansion by minors" implies that $X(v_1, \ldots, v_{n-1})$ is a vector orthogonal to each v_i. Further matrix work shows that $|X(v_1, \ldots, v_{n-1})|^2 = \det(A \cdot A^\top)$.

With this motivation we define general vector products, following ideas of Eckmann (1943a) and Brown and Gray (1967).

B.1 Definition. An *r-fold vector product* on the euclidean space $V = \mathbb{R}^n$ is a map $X : V^r \to V$ such that

(1) X is r-linear (that is, $X(v_1, \ldots, v_r)$ is linear in each of its r slots);

(2) $X(v_1, \ldots, v_r)$ is orthogonal to each v_j;

(3) $|X(v_1, \ldots, v_r)|^2 = \det(\langle v_i, v_j \rangle)$.

To avoid trivialities we always assume $r \leq n$.

It is easy to check that a 1-fold vector product exists on $\mathbb{R}^n$ if and only if n is even. The quaternion description of cross products on $\mathbb{R}^3$ leads to an analog with the octonion algebra $\mathbb{O}$. To obtain a 2-fold vector product on $\mathbb{R}^7$ we view $\mathbb{R}^7 = \mathbb{O}_0$ as the space of pure octonions and use the earlier formula to define the product:

$$v \times w = vw - \langle v, w \rangle.$$

It is not hard to check that this is a vector product. (See Exercise B2.) Surprisingly we have already mentioned nearly all of the examples.

B.2 Theorem. *An r-fold vector product exists on $V = \mathbb{R}^n$ if and only if:*

$r = 1$ *and* n *is even.*

$r = n - 1$ *and* n *is arbitrary.*

$r = 2$ *and* $n = 7$.

$r = 3$ *and* $n = 8$.

Exercise 3 describes a 3-fold vector product on $\mathbb{R}^8$.

These vector products were first investigated by Eckmann (1943a) and Whitehead (1963). In fact they proved a much more general theorem, assuming only that X is continuous, not necessarily r-linear. The proof involves algebraic topology, especially the work of Adams (1960). A survey of these ideas is given by Eckmann (1991). Assuming that the product is r-linear, Brown and Gray (1967) provided an algebraic proof of this theorem. They also handled the more general situation when V is a regular quadratic space over any field F (of characteristic $\neq 2$). Different approaches to these results are given by Massey (1983), Dittmer (1994), and Morandi (1999).

The Hurwitz 1, 2, 4, 8 Theorem implies the restrictions on n in the cases $r = 2, 3$. In fact, any 2-fold vector product on V leads to a composition algebra on $\mathbb{R} \perp V$, and any 3-fold vector product on V provides a composition algebra structure on V. See Exercises B2, B3. The connections between 2-fold vector products and composition algebras is also described by Koecher and Remmert (1991), pp. 275–280. More recently Rost (1994) reversed this connection to provide another proof of the Hurwitz

1, 2, 4, 8 Theorem. Rost's ideas lead to a proof using elementary ideas in the theory of graph categories (see Boos (1998)), or they can be performed algebraically within a vector product algebra (see Maurer (1998)).

Changing directions now, let us consider "triple compositions". Suppose (V, q) is a regular quadratic space over a field F.

B.3 Definition. (V, q) permits *triple composition* if there is a trilinear map $\{\,\}: V \times V \times V \to V$ such that

$$q(\{xyz\}) = q(x)q(y)q(z) \quad \text{for every } x, y, z \in V.$$

Certainly if V is a composition algebra then the product $\{xyz\} = x \cdot yz$ provides an example of a triple composition. In the other direction, suppose (V, q) permits triple composition. If $e \in V$ is a unit vector then the product $x \cdot y = \{xey\}$ makes V into a composition algebra (possibly without identity), and consequently $\dim V = 1, 2, 4$ or 8. (What if q does not represent 1 here?) McCrimmon (1983) investigated such triple compositions and found a complete classification of them, up to isotopy.

Such ternary algebras become easier to work with if we add the extra axiom

$$\{xxy\} = \{yxx\} = \langle x, x\rangle y \quad \text{for every } x, y \in V.$$

A triple composition with this property is called a "ternary composition algebra." If V is a (binary) composition algebra then the product $\{xyz\} = x \cdot \bar{y}z$ makes it into a ternary composition algebra. Conversely, given a ternary composition algebra and a unit vector e, the formula $x \cdot y = \{xey\}$ produces a (binary) composition algebra with e as identity element. The advantage of this ternary viewpoint is that the algebra has more symmetries: one unit vector has not been picked out to be the identity element. Ternary compositions are also closely related to 3-fold vector products. These ideas and related topics are explained and extended (for euclidean spaces over $\mathbb{R}$) by Shaw (1987), (1988), (1989), (1990).

Section C. Compositions over rings and over fields of characteristic 2

Suppose σ and q are regular quadratic forms over a field F, where $\dim \sigma = s$ and $\dim q = n$. Then σ and q *admit a composition* if there is a formula

$$\sigma(X) \cdot q(Y) = q(Z)$$

where $X = (x_1, \dots, x_s)$ and $Y = (y_1, \dots, y_n)$ are systems of indeterminates and each z_k is a bilinear form in X and Y, with coefficients in F. Which quadratic forms admit a composition? This is the question studied in Part I of this book, in the case F has characteristic not 2. If the characteristic is 2 the same question can be asked but the methods must be modified.

Suppose $2 = 0$ in F. The norm forms of quadratic extensions of F certainly admit compositions (with themselves). For example, an (inseparable) quadratic extension $F(\sqrt{a})$ has norm form $q_{\text{ins}}(X) = x_1^2 + ax_2^2$. A separable quadratic extension is some $F(\mathcal{P}^{-1}(c))$. Here $\mathcal{P}(x) = x^2 + x$ and $\mathcal{P}^{-1}(c)$ stands for a solution to the equation $\mathcal{P}(x) = c$. The norm form for this extension is $q_{\text{sep}}(X) = x_1^2 + x_1x_2 + ax_2^2$.

A quadratic form $q(X)$ in n variables over F can be viewed geometrically as a map $q : V \to F$ where V is an n-dimensional vector space over F. Such a map q is *quadratic* if

$$q(cv) = c^2q(v) \text{ for } c \in F \text{ and } v \in V, \text{ and}$$
$$b_q(v, w) = q(v + w) - q(v) - q(w) \text{ is bilinear.}$$

Note that $b_q(v, v) = 0$ for every $v \in V$ so that q cannot be recovered from its bilinear form b_q. Define (V, q) to be *nonsingular* if b_q is a nonsingular bilinear form, that is: $V^{\perp} = 0$. The example q_{ins} is singular, with bilinear form $b_{\text{ins}} = 0$, while the example q_{sep} is nonsingular.

Suppose (V, q) is nonsingular with $\dim q = n > 0$. This q cannot be diagonalized, but we can split off binary pieces. To do this, choose $0 \neq v \in V$. By hypothesis $b_q(v, V) \neq 0$ so there exists $w \in V$ with $b_q(v, w) = 1$. Let $U = \text{span}\{v, w\}$. The restriction of q to U is the nonsingular quadratic form $ax^2 + xy + by^2$ where $a = q(v)$ and $b = q(w)$. We denote this binary form by $[a, b]$.[1] Then $q \simeq [a, b] \perp q'$ where q' is the restriction of q to the space $U^{\perp}$. Repeating this process we see that $n = \dim q$ must be even and q is the orthogonal sum of such binary subspaces.

Many of the ideas and results of the classical theory (characteristic not 2) have analogs in characteristic 2. For example, the determinant corresponds to the Arf invariant $\Delta(q)$:

$$\text{If } q \simeq [a_1, b_1] \perp \cdots \perp [a_m, b_m] \text{ define } \Delta(q) = \sum_{j=1}^{m} a_jb_j \text{ in } F/\mathcal{P}(F).$$

One can show that this is well defined (isometric forms have equal Arf invariants). There are similar analogs for the algebras. The quaternion algebra $Q = (a, b]_F$ has generators u, v satisfying

$$u^2 = a, \quad v^2 = v + b, \quad \text{and} \quad uv + vu = u.$$

Octonion algebras can also be defined and analyzed over F. For a nonsingular form q the Clifford algebra $C(q)$ is a central simple algebra, providing an element of the Brauer group $\text{Br}(F)$. There is also a characteristic 2 analog for Pfister forms. Certain "quadratic Pfister forms" $\langle\langle a_1, \ldots, a_n]]$ are defined and these forms are round. We can use these algebraic tools to analyze compositions of quadratic forms.

Two approaches come to mind. The first is to modify the material in Part I of this book, finding the analogs in characteristic 2. Does the same Hurwitz–Radon function work? Is there some sort of Pfister Factor Conjecture that is true at least for small

1 Of course such brackets mean different things in other parts of the book.

dimensions? Etc. The second approach is to develop a single treatment of the theory that handles the questions for all fields (independent of characteristic). Of course the unified treatment could cover compositions over various rings as well.

Parts of this program have been completed. The first work on compositions in characteristic two was probably Albert (1942a). He generalized Hurwitz, proving (for any field F) that if A is a composition algebra over F then either A is one of the familiar four algebras of dimension 1, 2, 4, 8; or else $2 = 0$ and A is a purely inseparable, exponent 2 field extension of F. The general theory of quadratic forms in characteristic 2 has appeared in various texts, including Bourbaki (1959) and Baeza (1978). Baeza discusses compositions for quadratic spaces over a semilocal ring, analyzes the Hurwitz function and proves a 1, 2, 4, 8 Theorem in that context. (See Baeza (1978), pp. 90–93.) Subsequently Baeza's student Junker (1980) studied the analog of the Hurwitz–Radon Theorem over a field of characteristic 2, and proved the Pfister Factor Conjecture for $m \le 4$. Independently, Kaplansky (1979) mentioned that the Clifford algebra approach to Hurwitz–Radon can be extended to characteristic 2. More recently Züger (1995) worked with compositions over a commutative ring (where 2 is not assumed to be a unit). Among other results he obtains some analogs of the Hurwitz–Radon Theorem for compositions of a quadratic form q with another quadratic form, or with a bilinear form, or with a hermitian form.

There has been substantial work recently in presenting characteristic-free versions of the theory of quadratic forms, of central simple algebras, etc. The culmination of many of these efforts appears in the monumental work of Knus, Merkurjev, Rost and Tignol (1998). Perhaps a unified theory of quadratic form compositions can be based on their notion of "quadratic pairs".

Section D. Linear spaces of matrices of constant rank

Nonsingular bilinear maps are related to subspaces of matrices satisfying certain conditions on rank. For example, when is there an r-dimensional subspace U of $\mathrm{GL}_n(\mathbb{R})$? (Of course, this is taken to mean that every *non-zero* element of U is nonsingular.) Such a subspace quickly leads to a nonsingular bilinear $[r, n, n]$, and (12.20) implies that the maximal value for r is the Hurwitz–Radon number $\rho(n)$.

D.1 Definition. A linear subspace of $m \times n$ matrices, $U \subseteq \mathbb{M}_{m,n}(F)$, is said to be a *rank k subspace* if every non-zero element of U has rank k. Define

$$\ell_F(m, n; k) = \text{maximal dimension of a rank } k \text{ subspace of } \mathbb{M}_{m,n}(F).$$

If $F = \mathbb{R}$ we omit the subscript. To avoid trivialities we always assume $1 \le k \le \min\{m, n\}$. Certainly $\ell_F(m, n; k)$ is symmetric in m and n, and we usually arrange the notation so that $m \le n$.

D.2 Lemma. (1) *There is a nonsingular bilinear* $[r, s, n]$ *over* F *if and only if* $s \le \ell_F(r, n; r)$.

(2) $\ell(r, n; r) = \rho_{\#}(n, r)$ *as defined in* (12.24).

Proof. (1) Suppose f is a bilinear $[r, s, n]$. If $u \in F^s$ the induced map $f_u : F^r \to F^n$ corresponds to an $n \times r$ matrix. This provides a linear map $\varphi : F^s \to \mathbb{M}_{n,r}(F)$. If f is nonsingular then φ is injective and every non-zero $\varphi(u)$ is injective, hence of rank r. Then image(φ) is an s-dimensional rank r subspace so that $s \le \ell(n, r; r)$. All the steps are reversible, proving the converse. Part (2) is a restatement of the definition. □

D.3 Lemma. (1) $\ell_F(m, n; k)$ *is an increasing function of* m *and of* n.

(2) *If* $k \le m \le n$ *then* $\ell_F(m, n; m) \le \ell_F(m, n; k)$.

Proof. (1) Enlarge a matrix by adding rows or columns of zeros.

(2) Given a rank m subspace $U \subseteq \mathbb{M}_{m,n}(F)$, every $0 \ne f \in U$ is a surjective map $f : F^n \to F^m$. Choose $g \in \mathbb{M}_m(F)$ of rank k. Then $g \circ f$ has rank k so that $g \circ U$ is a rank k subspace. □

Clearly there exists an n-dimensional subspace of $\mathbb{M}_{m,n}(F)$ consisting of rank 1 matrices. However it takes some work to prove this is best possible: If $m \le n$ then $\ell_F(m, n; 1) = n$. Here is a generalization.

D.4 Proposition. *Suppose* $k \le m \le n$. *If* $|F| > k$ *then:*

$$n - k + 1 \le \ell_F(m, n; k) \le n.$$

Proof comment. The Cauchy product pairing of size $[n - k + 1, k, n]$ shows that $n - k + 1 \le \ell_F(k, n; k) \le \ell_F(m, n; k)$. For the second inequality it suffices to prove $\ell_F(n, n; k) \le n$. Beasley and Laffey (1990) prove this inequality by a linear algebra argument, using standard properties of determinants. Meshulam (1990) proved the real case using topological methods. □

Let us now concentrate on the real case.

D.5 Lemma. *If* $k \le n$ *then*

$$\ell(n, n; k) \ge \max\{\rho(k), \rho(k + 1), \ldots, \rho(n)\}.$$

Proof. By (D.3), if $k \le m \le n$ then $\rho(m) = \ell(m, m; m) \le \ell(m, m; k) \le \ell(n, n; k)$. □

From the topological work in Chapter 12, we already know some values of $\ell(m, n; k)$ when k is large.

D.6 Proposition. $\ell(n, n; n) = \rho(n)$.
$\ell(n-1, n; n-1) = \max\{\rho(n-1), \rho(n)\}$.
$\ell(n-2, n; n-2) = \max\{\rho(n-2), \rho(n-1), \rho(n), 3\}$.

Proof Apply (D.2), (12.20) and (12.30). □

Lam and Yiu (1993) computed the values $\ell(n, n; n-1)$, $\ell(n, n; n-2)$ and $\ell(n-2, n; n-2)$. We outline their first calculation here to give some of the flavor of these topological methods.

Now suppose $r = \ell(m, n; k)$. The given subspace of $\mathbb{M}_{m,n}(\mathbb{R})$ can be regarded as a family of non-zero linear maps $f_x : \mathbb{R}^n \to \mathbb{R}^m$, all of rank k. As x ranges over this r-dimensional space there is an induced map of vector bundles over real projective space $\mathbb{P}^{r-1}$ (see Exercise D3):

$$f : n \cdot \xi_{r-1} \longrightarrow m \cdot \varepsilon.$$

Since each f_x has rank k, image(f) is a k-plane bundle. Therefore:

$$\text{image}(f) \oplus \eta \cong m \cdot \varepsilon$$
$$\zeta \oplus \text{image}(f) \cong n \cdot \xi_{r-1},$$

where η is some $(m-k)$-plane bundle and $\zeta = \ker(f)$ is an $(n-k)$-plane bundle. Combining those equations, we obtain

$$m \cdot \varepsilon \oplus \zeta \cong n \cdot \xi_{r-1} \oplus \eta.$$

Various topological tools can now be applied to deduce restrictions on the numbers k, m, n. For instance, Meshulam (1990) considered Stiefel–Whitney classes for the bundle isomorphisms above to prove (D.4) in the real case. Passing to $\widetilde{\mathrm{KO}}(\mathbb{P}^{r-1})$ as in (12.28) we know that ζ and η become $a \cdot x$ and $b \cdot x$ for some $a, b \in \mathbb{Z}$. Then the last bundle equation above becomes: $(n + b - a) \cdot x = 0$, which implies that $n + b - a \equiv 0 \pmod{2^{\delta(r)}}$, or equivalently: $r \le \rho(n + b - a)$.

D.7 Lemma. $\ell(n, n; n-1) \le \max\{\rho(n-1), \rho(n), \rho(n+1)\}$.

Proof. If $r = \ell(n, n; n-1)$, (D.5) implies $r \ge \max\{\rho(n-1), \rho(n)\} \ge 2$. In the discussion above we have $m = n$ and $k = n-1$ so that ζ and η are line bundles. Every line bundle over $\mathbb{P}^{r-1}$ is either ξ_{r-1} or ε and therefore $a, b \in \{0, 1\}$. Then the inequality $r \le \rho(n + b - a)$ proves the assertion. □

D.8 Proposition. *If* $n \ne 3, 7$ *then* $\ell(n, n; n-1) = \max\{\rho(n-1), \rho(n), \rho(n+1)\}$.

Proof. By (D.5) and (D.7) it suffices to prove: $\rho(n+1) \le \ell(n, n; n-1)$. (This is non-trivial only when $n \equiv 3 \pmod 4$.) This inequality is settled by Exercise D4, replacing n there by $n+1$. □

To complete their analysis of this case, Lam and Yiu also prove that $\ell(3,3;2)=3$ and $\ell(7,7;6)=7$.

Without attempting to provide an accurate survey of the literature, we mention a few more related results. Petrovič (1996) proves that if $2 \le m \le n$ and $n \ne 3$ then

$$\ell(m,n;2)=\begin{cases} n & \text{if } n \text{ even,} \\ n-1 & \text{if } n \text{ is odd.}\end{cases}$$

The hard part here is to prove that if n is odd then $\ell(m,n;2)\ne n$.

Meshulam (1990) shows that if $p>3$ is a prime for which 2 is a generator of $(\mathbb{Z}/p\mathbb{Z})^*$ then $\ell(n,n;k)<n$ for every $k>1$.

Sylvester (1986), working over the complex field $\mathbb{C}$ was the first to use vector bundles in analyzing such problems. Westwick (1987) also discusses $\ell_{\mathbb{C}}(m,n;k)$ over the complex field, and analyzes when the lower bound for $\ell(m,n;k)$ is achieved. Without using topology he proves:

$$\ell_{\mathbb{C}}(m,n;k)=n-k+1 \text{ whenever } n-k+1 \text{ does not divide } \frac{(m-1)!}{(k-1)!}.$$

In particular $\ell_{\mathbb{C}}(m,n;m)=m-n+1$, which provides another proof of (14.25).

Exercises for Chapter 16

A1. **Construction of θ.** Suppose $\theta(X_1,\dots,X_d)$ is a symmetric d-linear form where each X_j is a system of n independent variables. Then $\varphi(X)=\theta(X,X,\dots,X)$ is a degree d form in $X=(x_1,\dots,x_n)$.

(1) For every φ there exists such θ.

(2) Let J range over the subsets of $[1,m]=\{1,\dots,m\}$ with $|J|=\text{card}(J)$. Define $f_d(x_1,\dots,x_m)=\sum_J(-1)^{m-|J|}\left(\sum_{j\in J}x_j\right)^d$. This is a form of degree d in m variables. For example $f_2(x_1,x_2)=(x_1+x_2)^2-x_1^2-x_2^2$.

Lemma. *If $d<m$ then $f_d(x_1,\dots,x_m)=0$. If $d=m$ then $f_d(x_1,\dots,x_d)=n!x_1x_2\dots x_d$.*

(3) The expansion of $\varphi(X+Y)=\theta(X+Y,X+Y,\dots,X+Y)$ acts very much like $(X+Y)^d$, etc. Therefore, $d!\theta(X_1,\dots,X_d)=\sum_J(-1)^{d-|J|}\varphi\left(\sum_{j\in J}X_j\right)$. This "polarization identity" proves that if $d!\ne 0$ in F then the d-linear form θ is uniquely determined by φ.

(4) $\sum_{k=0}^n\binom{n}{k}(-1)^kk^n=(-1)^nn!$.

(Hint. (1) It suffices to check monomials. For example if $d=3$ and $\varphi(X)=x_1^3$ use $\theta(X,Y,Z)=x_1y_1z_1$; if $\varphi(X)=x_1^2x_2$ use $\theta(X,Y,Z)=\frac{1}{3}(x_1y_1z_2+x_1y_2z_1+x_2y_1z_1)$.

(2) Replace J by its characteristic function $\delta:[1,n]\to\{0,1\}$, where $\delta(i)=1$ if and only if $i\in J$. Then $f_d(x_1,\dots,x_m)=\sum_\delta(-1)^{m+\delta(1)+\dots+\delta(m)}\left(\sum_{j=1}^m\delta(j)x_j\right)^d$.

Expand this as $\sum_{(i)} c_{(i)}x_{(i)}$, where $(i) = (i_1, \ldots, i_d) \in [1, n]^d$. Then $c_{(i)} = \sum_\delta (-1)^{m+\delta(1)+\cdots+\delta(m)}\delta(i_1)\ldots\delta(i_d)$.

Claim. If $\{i_1, \ldots, i_d\} \neq [1, n]$ then $c_{(i)} = 0$.

(For $j \notin \{i_1, \ldots, i_d\}$ compare terms where $\delta(j) = 0$ and those where $\delta(j) = 1$.)

(4) Apply (3) when all $X_j = 1$.)

A2. Suppose φ is a regular form of degree d in n variables, and view it as a function on the vector space $V = F^n$.

(1) Suppose $f \in \mathrm{End}(V)$ is a c-similarity for φ, that is: $\varphi(f(v)) = c\varphi(v)$ for every $v \in V$. If $c \neq 0$ then f is bijective.

(2) If φ permits composition then it represents 1 and V can be made into an F-algebra with 1 such that $\varphi(xy) = \varphi(x)\varphi(y)$.

(Hint. (1) If θ is the associated d-linear form, then f is a c-similarity for θ. If $f(v_1) = 0$, regularity implies $v_1 = 0$.

(2) The bilinear composition makes A into an algebra with $\varphi(xy) = \varphi(x)\varphi(y)$. There exists v with $\varphi(v) = 1$. Define a new multiplication as in Exercise 0.8 (2) and check that $\varphi(x \diamond y) = \varphi(x)\varphi(y)$.)

A3. Suppose F is an infinite field, φ is a form over F, and K is an extension field. Let φ_K denote the same form viewed over K.

(1) If φ is regular over F then φ_K is regular over K.

(2) If φ permits composition over F then φ_K permits composition over K.

(3) Do these statements remain valid if F is a finite field?

(Hint. (2) The polynomial $\varphi(XY) - \varphi(X)\varphi(Y)$ vanishes on F^{2n}.)

A4. **Determinant.** Suppose $n! \neq 0$ in F and $n > 2$.

(1) The determinant on $\mathbb{M}_n(F)$ is a form of degree n in n^2 variables. It is 1-regular but not 2-regular.

(2) If $A \in \mathbb{M}_n(F)$ and $\det(A + X) = \det(A) + \det(X)$ for every X, then $A = 0$.

(Hint. Let E_{ij} be the matrix with 1 in the (i, j) position and zeros elsewhere. Let $E = E_{11}$ and express a matrix as $A = \begin{pmatrix} * & * \\ * & A' \end{pmatrix}$ where A' has size $(n-1) \times (n-1)$. Then $\det(E + A) = \det(A) + \det(A')$. If θ_n is the symmetric n-linear form corresponding to det on $\mathbb{M}_n(F)$ then: $\theta_n(E, X_2, \ldots, X_n) = \frac{1}{n}\theta_{n-1}(X_2', \ldots, X_n')$. Choose $X_2 = E$ to see that θ_n is not 2-regular. Prove 1-regularity by induction on n, using various E_{ij} in place of E.)

A5. Suppose $\varphi(X)$ is a form of degree d in n variables over F (and $d! \neq 0$ in F).

(1) Possibly a linear change of variables leads to a form involving fewer than n variables. However: φ is regular if and only if such a reduction in variables cannot occur.

(2) Let $\mathcal{Z}(\varphi)$ be the zero set of φ over the algebraic closure $\bar{F}$. Then φ is a nonsingular form (as defined above) if and only the induced projective hypersurface over $\bar{F}$ is nonsingular.

(Hint. (2) By the Jacobian criterion, that surface is nonsingular if and only if: $\left(\frac{\partial\varphi}{\partial x_1}(a), \dots, \frac{\partial\varphi}{\partial x_n}(a)\right) \neq (0, \dots, 0)$ whenever $0 \neq a \in \mathcal{Z}(\varphi)$.)

B1. **Volumes.** Let $A = (v_1, \dots, v_r)$ be an $n \times r$ matrix formed from columns $v_i \in \mathbb{R}^n$. Let $\mathcal{P}(v_1, \dots, v_r) = \left\{\sum_1^r t_i v_i : 0 \leq t_i \leq 1\right\}$ be the parallelotope spanned by those vectors.

(1) If $r = n$ then the volume is given by the determinant:

$$\operatorname{vol}(\mathcal{P}(v_1, \dots, v_n)) = |\det(A)|.$$

(2) For general $r \leq n$ the volume (as an r-dimensional object) is determined by the $r \times r$ Gram matrix $(\langle v_i, v_j\rangle)$:

$$\operatorname{vol}_r(\mathcal{P}(v_1, \dots, v_r))^2 = \det(A^\top \cdot A) = \det(\langle v_i, v_j\rangle).$$

(3) On the exterior algebra $\Lambda(V) = \sum_{p=0}^n \Lambda^p(V)$ define

$$\langle v_1 \wedge \cdots \wedge v_p, w_1 \wedge \cdots \wedge w_q\rangle = \begin{cases} \det(\langle v_i, w_j\rangle) & \text{if } p = q, \\ 0 & \text{if } p \neq q. \end{cases}$$

This pairing extends linearly to a symmetric bilinear form on $\Lambda(V)$. If $\{e_1, \dots, e_n\}$ is an orthonormal basis for V then the derived basis $\{e^\Delta : \Delta \in \mathbb{F}_2^n\}$ is an orthonormal basis for $\Lambda(V)$. Moreover,

$$\operatorname{vol}_r(\mathcal{P}(v_1, \dots, v_r)) = |v_1 \wedge \cdots \wedge v_r|.$$

(4) These two volume formulas generalize the identity $|v \times w|^2 = |v|^2 \cdot |w|^2 - \langle v, w\rangle^2$ in $\mathbb{R}^3$.

(Hint. (2) One method is to first prove it when the v_i are mutually orthogonal. Then show the formula remains true after applying elementary column operations.)

B2. **2-fold products.** Let $v \times w$ be a 2-fold vector product on euclidean space V. Suppose $\dim V > 1$.

(1) Then $v \times v = 0$ and $v \times w = -w \times v$ for every $v, w \in V$.

(2) Define $A = \mathbb{R} \perp V$ and define a product on A by: $vw = -\langle v, w\rangle + v \times w$. This product makes A into a composition algebra. The Hurwitz Theorem then implies $\dim V = 3$ or 7.

(3) $\langle u \times v, w\rangle = \langle u, v \times w\rangle$, the "interchange rule."

(4) $(u \times v) \times v = \langle u, v\rangle v - \langle v, v\rangle u$.

The identity $(u \times v) \times w = \langle u, w\rangle v - \langle v, w\rangle u$ holds if and only if $\dim V = 3$.

(Hint. (3) $\langle (u + w) \times v, u + w\rangle = 0$. Also note that $\langle uv, w\rangle = \langle u, vw\rangle$, using the product in (2).

(4) Apply (3) to: $\langle u \times v, w \times v\rangle = \langle u, w\rangle\langle v, v\rangle - \langle u, v\rangle\langle w, v\rangle$. Translate the given identity into: $uv \cdot w = \langle u, w\rangle v - \langle v, w\rangle u - \langle u, v\rangle w - \langle uv, w\rangle$. Use "bar" to deduce associativity.)

B3. **3-fold vector products.** (1) If V is a composition algebra with norm form $\langle x, y\rangle$ define $X : V^3 \to V$ by

$$X(a, b, c) = -a \cdot \bar{b}c + \langle a, b\rangle c - \langle c, a\rangle b + \langle b, c\rangle a.$$

Then X is a 3-fold vector product on V.

(2) Conversely suppose X is a 3-fold vector product on a vector space V. Choose a unit vector $e \in V$ and define a multiplication on V by:

$$ac = -X(a, e, c) + \langle a, e\rangle c - \langle c, a\rangle e + \langle e, c\rangle a.$$

This makes V into a composition algebra with e as identity. Consequently dim $V = 4$ or 8.

(Hint. (1) The Flip Law and other basics in Chapter 1, Appendix, show that $\langle a \cdot \bar{b}c, b\rangle = 2\langle a, b\rangle\langle c, b\rangle - \langle b, b\rangle\langle a, c\rangle$.

(2) Calculate $\langle ac, ac\rangle$ and watch most terms cancel.)

C1. Suppose q is a nonsingular quadratic form over F, a field with characteristic 2.

(1) q represents $a \in F \iff q \simeq [a, b] \perp q'$ for some $b \in F$ and nonsingular form q'.

(2) q is isotropic and $\dim q = 2 \iff q \simeq [0, 0]$, corresponding to the form $q(x, y) = xy$. This is the "hyperbolic plane" $\mathbb{H}$.

(3) There is a Witt Decomposition: $q \simeq q_0 \perp k\mathbb{H}$ where q_0 is anisotropic.

(4) If $c \in F^\bullet$ then $\langle c\rangle[a, b] \simeq [ac, bc^{-1}]$.

C2. **Tensor Products.** Let V be a vector space over a field F of characteristic 2. A bilinear form $b : V \times V \to F$ is *alternating* if $b(x, x) = 0$ for every x. If q is a quadratic form on V then b_q is alternating.

(1) Suppose $\{v_1, \dots, v_n\}$ is a basis of V. Given an alternating form b on V and given $a_1, \dots, a_n \in F$, there is a unique quadratic form q on V such that $q(v_i) = a_i$ and $b_q = b$.

(2) Suppose V and W are vector spaces with symmetric bilinear forms b and b'. Then $b \otimes b'$ is a symmetric bilinear form on $V \otimes W$. Both b and b' are nonsingular if and only if $b \otimes b'$ is nonsingular. If b is alternating then $b \otimes b'$ is alternating.

(3) Suppose b is a symmetric bilinear form on V and q is a quadratic form on W. Then there is a unique quadratic form Q on $V \otimes W$ such that $Q(v \otimes w) = b(v, v)q(w)$ and with associated bilinear form $b_Q = b \otimes b_q$.

Remark. Let $W(F)$ be the Witt group of nonsingular symmetric bilinear forms and let $Wq(F)$ be the Witt group of nonsingular quadratic forms over F. Then $W(F)$ is a ring and $Wq(F)$ is a $W(F)$-module.

(Hint. (1) If $a_{ij} = b(v_i, v_j)$ define $q\big(\sum_i x_i v_i\big) = \sum_i a_i x_i^2 + \sum_{i<j} a_{ij} x_i x_j$.)

C3. Suppose $A = (a, b]$ is the quaternion algebra with generators u, v. Then $u^2 = a$, $v^2 = v + b$, $uv + vu = u$, $(uv)^2 = ab$. Define "bar" by: $\bar{1} = 1$, $\bar{u} = u$, $\bar{v} = v + 1$, $\overline{uv} = uv$. Then "bar" is an involution on A and the norm form $N(x) = x \cdot \bar{x}$ provides a composition: $N(xy) = N(x)N(y)$. Check that $N(x_0 + x_1 u + x_2 v + x_3 uv) = x_0^2 + x_0 x_1 + b x_1^2 + a x_2^2 + a x_2 x_3 + ab x_3^2$. Therefore $(A, N) \simeq [1, b] \perp \langle a\rangle[1, b] = \langle 1, a\rangle \otimes [1, b]$, which is the 2-fold quadratic Pfister form $\langle\langle a, b]]$.

D1. (1) Construct your own proof that $\ell_F(n, n; 1) = n$.

(2) Use (D.4) to prove if n is even then $\ell(m, n; 2) = n$ and $\ell(3, 3; 2) = 3$.

D2. We know that $\ell_F(m, n; k) \le n$. When can equality occur?

(1) The following are equivalent statements:

(a) $\ell_F(m, n; k) = n$ whenever $k \le m \le n$.

(b) $\ell_F(k, n; k) = n$.

(c) There exists a nonsingular bilinear $[n, k, n]$ (i.e., $k \le n \#_F n$).

(d) $k \le \ell_F(n, n; n)$ (i.e., there exists a k-dimensional subspace of $\mathrm{GL}_n(F)$.)

(2) If F has field extensions of every degree, property (1) holds for all m, n, k. In fact, if $k \le a, b$ and there exist division algebras of dimensions a and b over F, then there exist nonsingular $[n, k, n]$ for every $n = ax + by$ for $x, y \ge 0$.

D3. **Explicit bundle map.** Recall (12.16) and Exercise 12.8. If $f : S^{r-1} \times \mathbb{R}^s \to \mathbb{R}^n$ is skew-linear, define $\varphi : S^{r-1} \times \mathbb{R}^s \to S^{r-1} \times \mathbb{R}^n$ by $\varphi(x, v) = (x, f_x(v))$. This induces a fiber-preserving map of the total spaces $E(s\xi_{r-1}) \to E(n\varepsilon)$ for bundles over $\mathbb{P}^{r-1}$. This is a bundle morphism provided the images of all the fibers have the same dimension. If $f_x = f(x, -)$ has rank k for every $x \in S^{r-1}$ then this is a bundle morphism $s\xi_{r-1} \to n\varepsilon$ whose image is a k-plane bundle and whose kernel is an $(s - k)$-plane bundle. Compare the discussion after D.6.

D4. **Lemma.** If $n > 2$ and $n \ne 4, 8$ then $\rho(n) \le \ell(n-1, n-1; n-2)$.

Proof outline. (1) If $A = \begin{pmatrix} B & u \\ v^{\top} & 0 \end{pmatrix} \in \mathrm{O}(n)$, with entry 0 in the corner, then $\mathrm{rank}(B) = n - 2$.

(2) For n as in the lemma, then $\rho(n) < n$. Complete the proof.

(Hint. (2) From a normed f of size $[\rho(n), n, n]$ choose $y \in S^{n-1}$. Then there exists $z \in S^{n-1}$ orthogonal to every $f(x, y)$. Choose two bases of $\mathbb{R}^n$ with last elements y and z, respectively. For $x \in S^{\rho(n)-1}$ then $f(x, -)$ has matrix A_x of the type in part (1). Use the space spanned by these B_x's.)

D5. **A "rank < n" question.** What is the largest dimension of a subspace of singular matrices in $\mathbb{M}_n(K)$? There are obvious examples of dimension $n^2 - n$. Is that maximal?

D6. Define $L_F(m, n; k)$ = maximal dimension of a linear subspace of $\mathbb{M}_{m,n}(F)$ in which every non-zero element has rank $\geq k$.

By (14.23) there exists $[r, s, n]/F \iff L_F(r, s; 2) \geq rs - n$. What bounds exist over $L_F(m, n; k)$ generally or over $F = \mathbb{R}$? (Note: See Petrovič (1996) for more information over $\mathbb{R}$.)

Notes on Chapter 16

Conjecture A.4 was told to me by E. Becker in the 1980s, but I have not seen it in print. He might hesitate to assert that it is true, so perhaps we should have called it "Becker's Question".

Grassmann introduced his abstract system for higher dimensional geometry in the 1830s, and Hamilton discovered quaternions in the 1840s. Hamilton and his followers insisted that quaternions provide the best language for discussing anything in geometry and physics. Clifford followed some of the ideas of Grassmann but he also worked with quaternions. In the 1880s the physicists Gibbs and Heaviside rejected the cumbersome quaternion machinery, preferring to work with the dot product and vector product separately. These ideas were also discussed and debated by Peirce, Clifford, Tait, Maxwell, and many others. A careful study of this colorful history is presented by Crowe (1967). Also see Altmann (1989).

Several texts contain information about quadratic forms in characteristic 2, including Bourbaki (1959), Milnor and Husemoller (1973), pp. 110–119, and Baeza (1978). Quadratic Pfister forms were introduced by Baeza. He discussed their basic properties (over semilocal rings) in Baeza (1979).

Paul Yiu told me about the material in Section D, and provided most of the references given there.

Exercise A1. This technique of proving the polarization identity follows Mneimné (1989).

Exercise B2. The interchange rule is part of the definition of vector-product algebras as given by Koecher and Remmert (1991). Further observations on $(u \times v) \times w$ appear in Shaw and Yeadon (1989).

Exercise B3 follows Brown and Gray (1967). The formula for X in part (1) was first noted by Zvengrowski (1966).

Exercise D4 follows Lam and Yiu (1993).

References

Adams, J. F.

1960 On the non-existence of elements of Hopf invariant one, *Ann. of Math.* **72**, 20–104.

1962 Vector fields on spheres, *Ann. of Math.* **75**, 603–632.

Adams, J. F., P. Lax and R. Phillips

1965 On matrices whose real linear combinations are nonsingular, Proc. Amer. Math. Soc. **16**, 318–322; **17** (1966), 945–947.

Adem, J.

1968 Some immersions associated with bilinear maps, *Bol. Soc. Mat. Mexicana* **13**, 95–104.

1970 On nonsingular bilinear maps. In: *The Steenrod Algebra and its Applications*, Lecture Notes in Math. **168**, Springer, 11–24.

1971 On nonsingular bilinear maps II, *Bol. Soc. Mat. Mexicana* **16**, 64–70.

1975 Construction of some normed maps, *Bol. Soc. Mat. Mexicana* **20**, 59–75.

1978a On maximal sets of anticommuting matrices, *Bol. Soc. Mat. Mexicana* **23**, 61–67.

1978b *Algebra Lineal, Campos Vectoriales e Inmersiones*, III ELAM, IMPA, Rio de Janeiro.

1980 On the Hurwitz problem over an arbitrary field I, II, *Bol. Soc. Mat. Mexicana* **25**, 29–51; **26** (1981), 29–41.

1984 On Yuzvinsky's theorem concerning admissible triples over an arbitrary field, *Bol. Soc. Mat. Mexicana* **29**, 65–69.

1986a On admissible triples over an arbitrary field, *Bull. Soc. Math. Belg. Sér. A* **38**, 33–35.

1986b Classification of low dimensional orthogonal pairings, *Bol. Soc. Mat. Mexicana* **31**, 1–28.

Adem, J., S. Gitler and I. M. James

1972 On axial maps of a certain type, *Bol. Soc. Mat. Mexicana* **17**, 59–62.

Adem, J., J. Ławrynowicz and J. Rembieliński

1996 Generalized Hurwitz maps of the type $S \times V \to W$, *Rep. Math. Phys.* **37**, 325–336.

Alarcon, J. I., and P. Yiu

1993 Compositions of hermitian forms, *Linear and Multilinear Algebra* **36**, 141–145.

Albert, A. A.

1931 On the Wedderburn condition for cyclic algebras, *Bull. Amer. Math. Soc.* **37**, 301–312.

1932 Normal division algebras of degree four over an algebraic field, *Trans. Amer. Math. Soc.* **34**, 449–456.

1939 *Structure of Algebras*, Amer. Math. Soc. Colloq. Publ. 24, Amer. Math. Soc., New York. Revised edition 1961.

1942a Quadratic forms permitting composition, *Ann. Math.* **43**, 161–177.

1942b Non-associative algebras, *Ann. Math.* **43**, 685–707.

1963 (ed.), *Studies in Modern Algebra*, vol. 2, Math. Assoc. America; Prentice-Hall, Inc., Englewood Cliffs, N. J.

1972 Tensor products of quaternion algebras, *Proc. Amer. Math. Soc.* **35**, 65–66.

Allard, J., and K. Y. Lam

1981 Freeness of orthogonal modules, *J. Pure Appl. Algebra* **21**, 123–127.

Allen, H. P.

1969 Hermitian forms, I, *Trans. Amer. Math. Soc.* **138**, 199–210.

1968 Hermitian forms, II, *J. Algebra* **10**, 503–515.

Alon, N.

1999, Combinatorial Nullstellensatz, *Combin. Probab. Comput.* **8**, 7–29.

Alpers, B.

1991 Round quadratic forms, *J. Algebra* **18**, 44–55.

Alpers, B., and E. M. Schröder

1991 On mappings preserving orthogonality of non-singular vectors, *J. Geom.* **41**, 3–15.

Al-Sabti, G., and T. Bier

1978 Elements in the stable homotopy groups of spheres which are not bilinearly representable, *Bull. London Math. Soc* **10**, 197–200.

Althoen, S. C., K. D. Hansen and L. D. Kugler

1994 Fused four-dimensional real division algebras, *J. Algebra* **170**, 649–660.

Althoen, S. C., and J. F. Weidner

1978 Real division algebras and Dickson's construction, *Amer. Math. Monthly* **85**, 368–371.

Altmann, S. L.

1986 *Rotations, Quaternions, and Double Groups*, Clarendon Press, Oxford.

1989 Hamilton, Rodrigues and the quaternion scandal, *Math. Mag.* **62**, 291–308.

Amitsur, S. A., L. H. Rowen and J. P. Tignol

1979 Division algebras of degree 4 and 8 with involution, *Israel J. Math.* **33**, 133–148.

Anghel, N.

1999 Clifford matrices and a problem of Hurwitz, preprint.

Arason, J. K., and A. Pfister

1982 Quadratische Formen über affinen Algebren und ein algebraischer Beweis des Satzes von Borsuk–Ulam, *J. reine angew. Math.* **331**, 181–184.

Artin, E.

1957 *Geometric Algebra*, Intersci. Tracts Pure Appl. Mathematics 3, Interscience Publishers, New York.

Atiyah, M.

1962 Immersions and embeddings of manifolds, *Topology* **1**, 125–132.

1967 *K-Theory*, W. A. Benjamin, New York.

Atiyah, M., R. Bott and H. Shapiro

1964 Clifford modules, *Topology* **3**, Suppl. 1, 3–38.

Au-Yeung, Y.-H., and C.-M. Cheng

1993 Two formulas for the generalized Radon–Hurwitz number, *Linear and Multilinear Algebra* **34**, 59–66.

Backelin, J., J. Herzog and H. Sanders

1988 Matrix factorizations of homogeneous polynomials. In: *Algebra – Some Current Trends*, (L. L. Avramov and K. B. Tchakerian, eds.), Lecture Notes in Math. 1352, Springer, Berlin, 1–33.

Baeza, R.

1978 *Quadratic Forms over Semilocal Rings*, Lecture Notes in Math. 655, Springer, Berlin.

1979 Über die Stufe von Dedekind Ringen, *Arch. Math.* **33**, 226–231.

Bayer, E., D. B. Shapiro and J.-P. Tignol

1993 Hyperbolic involutions, *Math. Z.* **214**, 461–476.

Beasley, L. B., and T. J. Laffey

1990 Linear operators on matrices: the invariance of rank-k matrices, *Linear Algebra Appl.* **133**, 175–184.

Becker, E.

1982 The real holomorphy ring and sums of $2n$-th powers. In: *Géometrie Algébrique Réelle et Formes Quadratiques*, (J.-L. Colliot-Thélène, M. Coste, L. Mahé and M.-F. Roy, eds.), Lecture Notes in Math. 959, Springer, Berlin, 139–181.

Behrend, F.

1939 Über Systeme reeller algebraischer Gleichungen, *Compositio Math.* **7**, 1–19.

Benkart, G., D. J. Britten and J. M. Osborn

1982 Real flexible division algebras, *Canad. J. Math.* **34**, 550–588.

Benkart, G., and J. M. Osborn

1981 Real division algebras and other algebras motivated by physics, *Hadronic J.* **4**, 392–443.

Bennett, A. A.

1919 Products of skew-symmetric matrices, *Bull. Amer. Math. Soc.* **25**, 455–458.

Berger, M., and S. Friedland

1986 The generalized Radon–Hurwitz numbers, *Compositio Math.* **59**, 113–146.

Berlekamp, E. R., J. H. Conway and R. K. Guy

1982 *Winning Ways for Your Mathematical Plays*, vol. 1, Academic Press, London.

Berrick, A. J.

1980 Projective space immersions, bilinear maps and stable homotopy groups of spheres. In: *Topology Symposium*, Siegen 1979 (U. Korschorke and W. D. Neumann, eds.), Lecture Notes in Math. 788, Springer, Berlin, 1–22.

Bier, T.
1979 *Geometrische Beiträge zur Homotopietheorie: Gerahmte Mannigfaltigkeiten, normierte und nichtsinguläre Bilinearformen*, doctoral dissertation, Univ. Göttingen.
1983 A remark on the construction of normed and nonsingular bilinear maps, *Proc. Japan Acad.* **56**, 328–330.
1984 Clifford-Gitter, unpublished manuscript (186 pages).

Bier, T., and U. Schwardmann
1982 Räume normierter Bilinearformen und Cliffordstrukturen, *Math. Z.* **180**, 203–215.

Blij, F. van der
1961 History of the octaves, *Simon Stevin* **34**, 106–125.

Blij, F. van der, and T. A. Springer
1960 Octaves and triality, *Nieuw Arch. Wisk.* (3) **8**, 158–169.

Bochnak, J., M. Coste and M.-F. Roy
1987 *Géométrie Algébrique Réelle*, Ergeb. Math. Grenzgeb. (3) 12, Springer, Berlin.

Boos, D.
1998 *Ein tensorkategorieller Zugang zum Satz von Hurwitz*, Diplomarbeit, ETH-Zürich/Universität Regensburg.

Bourbaki, N.
1959 *Algèbre*, Ch. 9, Formes sesquilinéaires et formes quadratiques, Hermann, Paris.

Brauer, R., and H. Weyl
1935 Spinors in *n* dimensions, *Amer. J. Math.* **57**, 425–449.

Brown, R. B., and A. Gray
1967 Vector cross products, *Comment. Math. Helv.* **42**, 222–226.

Bruck, R. H.
1944 Some results in the theory of linear non-associative algebras, *Trans. Amer. Math. Soc.* **56**, 141–199.
1963 What is a loop?. In: [Albert, 1963], 59–99.

Buchanan, T.
1979 Zur Topologie der projektiven Ebenen über reellen Divisionsalgebren, *Geometriae Dedicata* **8**, 383–393.

Buchweitz, R.-O., D. Eisenbud and J. Herzog
1987 Cohen–Macaulay modules on quadrics. In: *Singularities, Representations of Algebras, and Vector Bundles*, Proc. Symp., Lambrecht/Pfalz/FRG 1985 (G.-M. Greuel and G. Trautmann, eds.), Lecture Notes in Math. 1273, Springer, Berlin, 58–95.

Calvillo, G., I. Gitler and J. Martínez-Bernal
1997a Intercalate matrices. I: Recognition of dyadic type. *Bol. Soc. Mat. Mexicana* (3) **3**, 57–67.
1997b Intercalate matrices. II: A characterization of Hurwitz–Radon formulas and an infinite family of forbidden matrices, *Bol. Soc. Mat. Mexicana* (3) **3**, 207–220.

Cartan, E.
1938 *Leçons sur la théorie des spineurs*, Hermann, Paris. English transl.: *The Theory of Spinors*, Dover Publications Inc., 1981.

Cassels, J. W. S.
1964 On the representation of rational functions as sums of squares, *Acta Arith.* **9**, 79–82.
1978 *Rational Quadratic Forms*, Academic Press, London, New York.

Chang, S.
1998 On quadratic forms between spheres, *Geometriae Dedicata* **70**, 111–124.

Chevalley, C.
1946 *The Theory of Lie Groups*, Princeton Univ. Press.
1954 *The Algebraic Theory of Spinors*, Columbia Univ. Press, New York.
1955 *The construction and study of certain important algebras*, Publ. Math. Soc. Japan 1.

Chisolm, J. S. R., and A. K. Common (eds.)
1986 *Clifford Algebras and Their Applications in Mathematical Physics*, Proceedings of the NATO and SERC Workshop, Canterbury, U.K., September 15–27, 1985, D. Reidel, Dordrecht.

Clifford, W. K.
1878 Applications of Grassmann's extensive algebra, *Amer. J. Math* **1**, 350–358. Reprinted in: *Collected Mathematical Papers*, Macmillan, London 1882, 266–276.

Conway, J. H.,
1980 private conversation.

Coxeter, H. S. M.
1946 Quaternions and reflections, *Amer. Math. Monthly* **53**, 136–146.

Crowe, M. J.
1967 *A History of Vector Analysis: The Evolution of the Idea of a Vectorial System*, University of Notre Dame Press, Notre Dame. Reprinted by Dover Publications Inc., 1985.

Crumeyrolle, A.
1990 *Orthogonal and Symplectic Clfford Algebras: Spinor Structures*, Math Appl. **57**, Kluwer Acad. Publ., Dordrecht.

Curtis, C. W.
1963 The four and eight square problem and division algebras. In: [Albert 1963], 100–125.

Dai, Z. D., and T. Y. Lam,
1984 Levels in algebra and topology, *Comment. Math. Helv.* **59**, 376–424.

Dai, Z. D., T. Y. Lam and R. J. Milgram
1981 Application of topology to problems on sums of squares, *Enseign. Math.* **27**, 277–283.

Dai, Z. D., T. Y. Lam and C. K. Peng

1980 Levels in algebra and topology, *Bull. Amer. Math. Soc.* **3**, 845–848.

Davis, D. M.,

1998 Embeddings of real projective spaces, *Bol. Soc. Mat. Mexicana* (3) **4**, 115–122.

Dickson, L. E.

1914 *Linear Algebras*, Cambridge Tracts in Math. 16. (See p. 15.) Reprinted by Hafner Publishing Company, Inc., New York 1960.

1919 On quaternions and their generalizations and the history of the eight square theorem, *Ann. Math.* **20**, 155–171.

Dieudonné, J.

1953 A problem of Hurwitz and Newman, *Duke Math. J.* **20**, 381–389.

1954 Sur les multiplicateurs des similitudes, *Rend. Circ. Mat. Palermo* **3**, 398–408.

Dittmer, A.

1994 Cross product identities in arbitrary dimension, *Amer. Math. Monthly* **101**, 887–891.

Draxl, P. K.

1983 *Skew Fields*, London Math. Soc. Lecture Note Ser. 81, Cambridge Univ. Press.

Drazin, M. P.

1952 A note on skew symmetric matrices, *Math. Gazette* **36**, 253–255.

Dubisch, R.

1946 Composition of quadratic forms, *Ann. of Math.* **47**, 510–527.

Ebbinghaus, H.-D., et al.

1991 *Numbers*, Graduate Texts in Math. 123, Springer, Berlin. A translation into English of the 1988 edition of the book *Zahlen*.

Eckmann, B.

1943a Stetige Lösungen linearer Gleichungssysteme, *Comment. Math. Helv.* **15**, 318–339.

1943b Gruppentheoretischer Beweis des Satzes von Hurwitz–Radon über die Komposition der quadratischen Formen, *Comment. Math. Helv.* **15**, 358–366.

1989 Hurwitz–Radon matrices and periodicity modulo 8, *Enseign. Math.* **35**, 77–91.

1991 Continuous solutions of linear equations – An old problem, its history, and its solution, *Exposition. Math.* **9**, 351–365.

1994 Hurwitz–Radon matrices revisited: from effective solution of the Hurwitz matrix equations to Bott periodicity, *The Hilton Symposium 1993* (G. Mislin, ed.), CRM Proc. Lecture Notes 6, Amer. Math Soc., Providence, 23–35.

Eddington, A.

1932 On sets of anticommuting matrices, *J. London Math. Soc.* **7**, 56–68; **8** (1933), 142–152.

Edwards, B.

1978 On classifying Clifford algebras, *J. Indian Math Soc. (N. S.)* **42**, 339–344.

Eels, J., and P. Yiu

1995 Polynomial harmonic morphisms between euclidean spheres, *Proc. Amer. Math. Soc.* **123**, 2921–2925.

Eichhorn, W.

1969 Funktionalgleichungen in Vektorräumen, Kompositionsalgebren und Systeme partieller Differentialgleichungen, *Aequationes Math.* **2**, 287–303.

1970 Funktionalgleichungen in reellen Vektorräumen und verallgemeinerte Cauchy–Riemannsche Differentialgleichungen, speziell die Weylsche Gleichung des Neutrinos, *Aequationes Math.* **5**, 255–267.

Elduque, A., and H. C. Myung

1993 On flexible composition algebras, *Comm. Algebra* **21**, 2481–2505.

Elduque, A., and J. M. Pérez

1997 Infinite dimensional quadratic forms admitting composition, *Proc. Amer. Math. Soc.* **125**, 2207–2216

Eliahou, S., and M. Kervaire

1998 Sumsets in vector spaces over finite fields, *J. Number Theory* **71**, 12–39.

Elman, R.

1977 Quadratic forms and the u-invariant, III. In: [Orzech 1977], 422–444.

Elman, R., and T. Y. Lam

1973a On the quaternion symbol homomorphism $g_F : k_2F \to B(F)$. In: Algebraic K-Theory 2, Proc. Conf. Battelle Inst. 1972 (H. Bass, ed.), Lecture Notes in Math. 342, Springer, Berlin, 447–463.

1973b Quadratic forms and the u-invariant, I, *Math. Z.* **131**, 283–304.

1973c Quadratic forms and the u-invariant, II, *Invent. Math.* **21**, 125–137.

1974 Classification theorems for quadratic forms over fields, *Comment. Math. Helv.* **49**, 373–381.

1976 Quadratic forms under algebraic extensions, *Math. Ann.* **219**, 21–42.

Elman, R., T. Y. Lam, J.-P. Tignol and A. Wadsworth

1983 Witt rings and Brauer groups under multiquadratic extensions, I, *Amer. J. Math.* **105**, 1119–1170.

Elman, R., T. Y. Lam and A. Wadsworth

1977 Amenable fields and Pfister extensions. In: [Orzech 1977], 445–492.

1979 Pfister ideals in Witt rings, *Math. Ann.* **245**, 219–245.

Faillétaz, J.-M.

1992 *Compositions d'Espaces Bilinéaires par des Espaces Quadratiques*, doctoral dissertation, Univ. de Lausanne.

Freudenthal, H.

1952 Produkte symmetrischer und antisymmetrischer Matrizen, *Nederl. Akad. Wetensch. Proc. Ser. A* **55** = *Indag. Math.* **14**, 193–198.

Frobenius, G.

1910 Über die mit einer Matrix vertauschbaren Matrizen, *Sitzungsber. Preuss. Akad. Wiss.* 1910. Reprinted in *Ges. Abh.*, vol. 3, Springer, Berlin 1968, 415–427.

Fröhlich, A.
1984 *Classgroups and Hermitian Modules*, Progr. Math. 48, Birkhäuser, Basel.

Fröhlich, A., and A. McEvett
1969 Forms over rings with involution, *J. Algebra* **12**, 79–104.

Fulton, W.
1984 *Intersection Theory*, Ergeb. Math. Grenzgeb. (3) 2, Springer, Berlin.

Furuoya, I., S. Kanemake, J. Ławrynowicz and O. Suzuki
1994 Hermitian Hurwitz pairs, In: [Ławrynowicz 1994], 135–154.

Gabel, M.
1974 Generic orthogonal stably free projectives, *J. Algebra* **29**, 477–488.

Gantmacher, F. R.
1959 *The Theory of Matrices* (in 2 volumes), Chelsea, New York.

Gauchman, H., and G. Toth
1994 Real orthogonal multiplications in codimension two, *Nova J. Algebra Geom.* **3**, 41–72.
1996 Normed bilinear pairings for semi-Euclidean spaces near the Hurwitz–Radon range, *Results Math.* **30**, 276–301.

Gentile, E.
1985 A note on the u -invariant of fields, *Arch. Math.* **44**, 249–254

Geramita, A. V., and N. J. Pullman
1974 A theorem of Hurwitz and Radon and orthogonal projective modules, *Proc. Amer. Math. Soc.* **42**, 51–56.

Geramita, A. V., and J. Seberry
1979 *Orthogonal Designs*, Lecture Notes in Pure and Appl. Math. 45, Marcel Dekker, New York.

Gerstenhaber, M.
1964 On semicommuting matrices, *Math. Z.* **83**, 250–260.

de Géry, J. C.
1970 Formes quadratiques dans un corps quelconque, nulles sur un cône donné, *Bull. Soc. Math. France*, 2e série, **94**, 257–279.

Gilkey, P. B.
1987 The eta invariant and non-singular bilinear products in $\mathbf{R}^n$, *Canad. Math. Bull.* **30**, 147–154.

Ginsburg, M.
1963 Some immersions of projective space in Euclidean space, *Topology* **2**, 69–71.

Gitler, S.
1968 The projective Stiefel manifolds II. Applications, *Topology* **7**, 47–53.

Gitler, S., and K. Y. Lam
1969 The generalized vector field problem and bilinear maps, *Bol. Soc. Mat. Mexicana* (2) **14**, 65–69.

Gluck, H., F. Warner and C. T. Yang

1983 Division algebras, fibrations of spheres by great spheres and the topological determination of space by the gross behavior of its geodesics, *Duke Math. J.* **50**, 1041–1076.

Gluck, H., F. Warner and W. Ziller

1986 The geometry of the Hopf fibrations, *Enseign. Math.* **32**, 173–198.

Gordon, N. A., T. M. Jarvix, J. G. Maks and R. Shaw

1994 Composition algebras and $PG(m, 2)$, *J. Geom.* **51**, 50–57.

Gow, R., and T. J. Laffey

1984 Pairs of alternating forms and products of two skew-symmetric matrices, *Linear Algebra Appl.* **63**, 119–132.

Greenberg, M.

1967 *Lectures on Algebraic Topology*, W. A. Benjamin, New York.

1969 *Lectures on Forms in Many Variables*, W. A. Benjamin, New York.

Gude, U.-G.

1988 *Über die Nichtexistenz rationaler Kompositionsformeln bei Formen höheren Grades*, doctoral dissertation, Univ. Dortmund.

Guo Ruizhi

1996 Some remarks on orthogonal multiplication, *Acta Sci. Natur. Univ. Norm. Hunan.* **19**, 7–10.

Hähl, H.

1975 Vierdimensional reelle Divisionsalgebren mit dreidimensionaler Automorphismengruppe, *Geom. Dedicata* **4**, 323–333.

Hahn, A.,

1985 A hermitian Morita theorem for algebras with anti-structure, *J. Algebra* **93**, 215–235.

Halberstam, H., and R. E. Ingham

1967 Four and eight squares theorems, Appendix 3 to *The Math. Papers of Sir William Rowen Hamilton*, Vol. III, 648–656, Cambridge Univ. Press.

Hartshorne, R.

1977 *Algebraic Geometry*, Graduate Texts in Math. 52, Springer, New York.

Hefter, H.

1982 Dehnungsuntersuchungen an Sphärenabbildungen, *Invent. Math.* **66**, 1–10.

Hermann, R.

1974 *Spinors, Clifford and Cayley Algebras*, Interdisciplinary Mathematics, vol. 7, New Brunswick, NJ.

Herstein, I.

1968 *Noncommutative Rings*, Carus Math. Monographs, No. 15, Math. Assoc. America, Washington.

Hile, G. N., and M. H. Protter

1977 Properties of overdetermined first order elliptic systems. *Arch. Rational Mech. Anal.* **66**, 267–293.

Hirzebruch, F.
1991 Division algebras and topology, one chapter in [Ebbinghaus et al. 1991], 281–302.

Hodge, W., and D. Pedoe
1947 *Methods of Algebraic Geometry*, vol. 1, Cambridge Univ. Press.

Hoffmann, D. W., and J.-P. Tignol
1998 On 14-dimensional forms in I^3, 8-dimensional forms in I^2, and the common value property, *Doc. Math.* **3**, 189–214.

Hopf, H.
1931 Über die Abbildungen der dreidimensional Sphäre auf die Kugelfläche, *Math. Ann.* **104**, 637–714.
1935 Über die Abbildungen von Sphären auf Sphären niedrigerer Dimension, *Fund. Math.* **23**, 427–440.
1940 Systeme symmetrischer Bilinearformen und euklidische Modelle der projektiven Räume, *Vierteljahrsschr. Naturforsch. Ges. Zürich* **85**, Beibl. 32, 165–177.
1941 Ein topologischer Beitrag zur reellen Algebra, *Comment. Math. Helv.* **13**, 219–239.

Hornix, E. A. M.
1995 Round quadratic forms, *J. Algebra* **175**, 820–843.

Hughes, D., and F. Piper
1973 *Projective Planes*, Graduate Texts in Math. 6, Springer, New York.

Humphreys, J. E.
1975 *Linear Algebraic Groups*, Graduate Texts in Math. 21, Springer, New York.

Hurwitz, A.
1898 Über die Komposition der quadratischen Formen von beliebig vielen Variabeln, *Nachr. Ges. Wiss. Göttingen*, (Math.-Phys. Kl.), 309–316. Reprinted in *Math. Werke*, Bd. 2, Birkhäuser, Basel 1963, 565–571.
1923 Über die Komposition der quadratischen Formen, *Math. Ann.* **88**, 1–25. Reprinted in *Math. Werke*, Bd. 2, Birkhäuser, Basel 1963, 641–666.

Husemoller, D.
1975 *Fibre Bundles*, McGraw Hill 1966; second edition, Graduate Texts in Math. 20, Springer, New York 1975.

Jacobson, N.
1939 An application of E. H. Moore's determinant of a hermitian matrix, *Bull. Amer. Math. Soc.* **45**, 745–748.
1958 Composition algebras and their automorphisms, *Rend. Circ. Mat. Palermo* (2) **7**, 55–80.
1964 Clifford algebras for algebras with involution of type D, *J. Algebra* **1**, 288–300.
1968 *Structure and Representation of Jordan Algebras*, Amer. Math. Soc. Colloq. Publ. 34, Amer. Math. Soc., Providence. See especially pp. 230–232.
1974 *Basic Algebra I*, W. H. Freeman, San Francisco.
1980 *Basic Algebra II*, W. H. Freeman, San Francisco.

1983 Some applications of Jordan norms to involutorial simple associative algebras, *Adv. Math.* **48**, 149–165. Corrected version appears in: *Coll. Math. Papers*, vol. 3, 251–267. Also see p. 235.

1992 Generic norms. In: *Proc. Internat. Conf. in Algebra*, Novosibirsk 1989 (L. A. Bokut, Yu. L. Ershov and A. I. Kostrikin, eds.), Contemp. Math. 131, part 2, Amer. Math. Soc., Providence, 587–603.

1995 Generic norms II, *Adv. Math.* **114**, 189–196.

1996 *Finite-Dimensional Division Algebras over Fields*, Springer, Berlin.

James, I. M.

1963 On the immersion problem for real projective spaces, *Bull. Amer. Math. Soc.* **69**, 231–238.

1971 Euclidean models of projective spaces, *Bull. London Math. Soc.* **3**, 257–276.

1972 Two problems studied by Heinz Hopf. In: *Lectures on Algebraic and Differential Topology*, by R. Bott, S. Gitler, and I. M. James, Lecture Notes in Math. **279**, Springer, Berlin, 134–160.

Jančevskiĭ, V. I.

1974 Sfields with involution, and symmetric elements (in Russian), *Dokl. Akad. Nauk. BSSR* **18**, 104–107, 186.

Jordan, P., J. von Neumann and E. Wigner

1934 On an algebraic generalization of the quantum mechanical formalism, *Ann. of Math.* **2**, 29–64, particularly pp. 51–54.

Junker, J.

1980 *Das Hurwitz Problem für quadratische Formen über Körper der Charakteristik* 2, Diplom thesis, Univ. Saarbrücken.

Kanemaki, S.

1989 Hurwitz pairs and octonions. In: [Ławrynowicz 1989], 215–223.

Kanemaki, S., and O. Suzuki

1989 Hermitian pre-Hurwitz pairs and the Minkowski space. In: [L awrynowicz 1989], 225–232.

Kaplan, A.

1981 Riemannian nilmanifolds attached to Clifford modules, *Geom. Dedicata* **11**, 127–136.

1984 Composition of quadratic forms in geometry and analysis: some recent applications. In: *Quadratic and Hermitian forms*, Conf. Hamilton/Ont. 1983, CMS Conf. Proc. 4, 193–201.

Kaplansky, I.

1949 Elementary divisors and modules, *Trans. Amer. Math. Soc.* **66**, 464–491.

1953 Infinite-dimensional quadratic forms permitting composition, *Proc. Amer. Math. Soc.* **4**, 956–960.

1969 *Linear Algebra and Geometry*, Allyn and Bacon, Boston. Reprinted by Chelsea, New York, 1974.

1979 Compositions of quadratic and alternate forms, *C. R. Math. Rep. Acad. Sci. Canada* **1**, 87–90.

1983 Products of symmetric and skew-symmetric matrices, unpublished manuscript.

Kasch, F.

1953 Invariante Untermoduln des Endomorphismenringes eines Vektorraums, *Arch. Math.* **4**, 182–190.

Kawada, Y., and N. Iwahori

1950 On the structure and representations of Clifford algebras, *J. Math. Soc. Japan* **2**, 34–43.

Kervaire, M. A.

1958 Non-parallelizability of the n-sphere for $n > 7$, *Proc. Nat. Acad. Sci. USA* **44**, 280–283.

Kestelman, M.

1961 Anticommuting linear transformations, *Canad. J. Math.* **13**, 614–624.

Khalil, S.

1993 *The Cayley–Dickson Algebras*, M.Sc. Thesis, Florida Atlantic Univ.

Khalil, S., and P. Yiu

1997 The Cayley–Dickson algebras, a theorem of A. Hurwitz, and quaternions, *Bull. Soc. Sci. Lett. Łódź Sér. Rech. Déform.* **47**, 117–169.

Kirkman, T.

1848 On pluquaternions, and homoid products of sums of n squares, *Philos. Mag.* (ser. 3) **33**, 447–459; 494–509.

Kleinfeld, E.

1953 Simple alternative rings, *Ann. of Math.* **58**, 544–547.

1963 A characterization of the Cayley numbers. In: [Albert 1963], 126–143.

Knebusch, M.

1970 Grothendieck- und Wittringe von nichtausgeartete symmetrischen Bilinearformen, *Sitzber. Heidelberger Akad. Wiss.*, Math.-Naturwiss. Kl. 1969/70, 3. Abhdl.

1971 Runde formen über semilokalen Ringen, *Math. Ann.* **193**, 21–34.

1976 Generic splitting of quadratic forms, I, *Proc. London Math. Soc.* (3) **33**, 65–93.

1977a Generic splitting of quadratic forms, II, *Proc. London Math. Soc.* (3) **34**, 1–31.

1977b Some open problems. In: [Orzech 1977], 361–370.

1982 An algebraic proof of the Borsuk–Ulam theorem for polynomial mappings, *Proc. Amer. Math. Soc.* **84**, 29–32.

Knebusch, M., and W. Scharlau,

1980 *Algebraic Theory of Quadratic Forms: Generic Methods and Pfister Forms*, DMV Sem. 1, Birkhäuser, Stuttgart.

Kneser, M.

1969 *Lectures on Galois Cohomology of Classical Groups*, Tata notes, No. 47, Tata Institute, Bombay.

Knus, M.-A.

1988 *Quadratic Forms, Clifford Algebras and Spinors*, Seminários de Matemática 1, UNICAMP, Campinas, Brasil.

Knus, M.-A., T. Y. Lam, D. B. Shapiro and J.-P. Tignol

1995 Discriminants of involutions on biquaternion algebras. In: *K-Theory and Algebraic Geometry: Connections with Quadratic Forms and Divisions Algebras,* (B. Jacob and A. Rosenberg, eds.), Proc. Symp. Pure Math. 58.2, Amer. Math. Soc., Providence, 279–303.

Knus, M.-A., A. Merkurjev, M. Rost and J.-P. Tignol

1998 *The Book of Involutions*, Amer. Math. Soc. Colloq. Publ. 44, Amer. Math. Soc., Providence.

Knus, M.-A., and M. Ojanguren

1974 *Théorie de la Descente et Algèbres d'Azumaya*, Lecture Notes in Math. 389, Springer, Berlin.

Knus, M.-A., R. Parimala and R. Sridharan

1989 A classification of rank 6 quadratic spaces via Pfaffians, *J. Reine Angew. Math.* **398**, 187–218.

1991a Pfaffians, central simple algebras and similitudes, *Math. Z.* **206**, 589–604.

1991b Involutions on rank 16 central simple algebras, *J. Indian Math. Soc.* **57**, 143–151.

1991c On the discriminant of an involution, *Bull. Soc. Math. Belgique, Sér. A* **43**, 89–98.

1994 On compositions and triality, *J. reine angew. Math.* **457**, 45–70.

Knuth, D. E.

1968 *The Art of Computer Programming*, vol. 1, Addison-Wesley, Reading, Mass.; revised 1973.

Koecher, M., and R. Remmert

1991 Real division algebras; four chapters in: [Ebbinghaus, et al., 1991] 181–280.

Köhnen, K.

1978 *Definite Systeme und Quadratsummen in der Topologie*, doctoral dissertation, Univ. Mainz.

Krüskemper, M.

1996 On systems of biforms in many variables, unpublished preprint.

Kuz'min, E. N.

1967 Division algebras over the field of real numbers, *Dokl. Akad. Nauk SSSR* **172**, 1014–1017 (in Russian). English translation: *Soviet Math. Dokl.* **8**, 220–223.

Lam, K. Y.

1966 *Non-singular bilinear forms and vector bundles over P^n*, doctoral dissertation, Princeton Univ.

1967 Construction of nonsingular bilinear maps, *Topology* **6**, 423–426.

1968a Construction of some nonsingular bilinear maps, *Bol. Soc. Mat. Mexicana* (2) **13**, 88–94.

1968b On bilinear and skew-linear maps that are nonsingular, *Quart. J. Math. Oxford* (2) **19**, 281–288.

1972 Sectioning vector bundles over real projective spaces, *Quart. J. Math. Oxford* (2) **23**, 97–106.

1977a Some interesting examples of nonsingular bilinear maps, *Topology* **16**, 185–188.

1977b Nonsingular bilinear maps and stable homotopy classes of spheres, *Math. Proc. Cambridge Philos. Soc.* **82**, 419–425.

1979 KO equivalences and the existence of nonsingular bilinear maps, *Pacific J. Math.* **82**, 145–154.

1984 Topological methods for studying the composition of quadratic forms, In: *Quadratic and Hermitian forms*, Conf. Hamilton/Ont. 1983, CMS Conf. Proc. 4, 173–192.

1985 Some new results on composition of quadratic forms, *Invent. Math.* **79**, 467–474.

1997 Borsuk–Ulam type theorems and systems of bilinear equations. In: *Geometry from the Pacific Rim* (A. J. Berrick, B. Loo and H.-Y. Wang, eds.), W. de Gruyter, Berlin, 183–194.

Lam, K. Y., and D. Randall

1995 Geometric dimension of bundles on real projective spaces. In: *Homotopy Theory and Its Applications*, a conference on algebraic topology in honor of Samuel Gitler, August 9-13, 1993, Cocoyoc, Mexico (A. Adem, R. J. Milgram and D. C. Ravenel, eds.), Contemp. Math. 188, Amer. Math. Soc., Providence, 129–152.

Lam, K. Y., and P. Yiu

1987 Sums of squares formulae near the Hurwitz–Radon range. In: *The Lefschetz Centennial Conference. Part II: Proceedings on Algebraic Topology* (S. Gitler, ed.), Contemp. Math. 58, 51–56.

1989 Geometry of normed bilinear maps and the 16-square problem, *Math. Ann.* **284**, 437–447.

1993 Linear spaces of real matrices of constant rank, *Linear Algebra Appl.* **195**, 69–79.

1995 Beyond the impossibility of a 16-square identity. In: *Five Decades as a Mathematician and Educator, On the 80th Birthday of Professor Yung-Chow Wong*, (K. Y. Chan and M. C. Liu, eds.), World Scientific, Singapore, 137–163.

Lam, T. Y.

1973 *The Algebraic Theory of Quadratic Forms*, W. A. Benjamin, New York. Revised Printing: 1980.

1977 Ten lectures on quadratic forms over fields. In: [Orzech 1977], 1–102.

1983 *Orderings, Valuations and Quadratic Forms*, CBMS Regional Conf. Ser. in Math. 52, Amer. Math. Soc., Providence.

1989 Fields of u-invariant 6 after A. Merkurjev. In: *Ring Theory 1989 in Honor of S. Amitsur* (L. Rowen, ed.), Weizmann Science Press, Jerusalem, 12–30.

Lam, T. Y., and T. Smith

1989 On the Clifford–Littlewood–Eckmann groups: a new look at periodicity mod 8, *Rocky Mountain J. Math.* **19**, 749–786.

1993 On Yuzvinsky's monomial pairings, *Quart. J. Math. Oxford (2)* **44**, 215–237.

Ławrynowicz, J.

1989 (ed.) *Deformations of Mathematical Structures. Complex Analysis with Physical Applications*, Kluwer, Dordrecht.

1992 The normed maps $\mathbf{R}^{11} \times \mathbf{R}^{11} \to \mathbf{R}^{26}$ in hypercomplex analysis and in physics. In: [Micali et al. 1992], 447–461.

1994 (ed.) *Deformations of Mathematical Structures II: Hurwitz-Type Structures and Applications to Surface Physics*, Kluwer, Dordrecht.

Ławrynowicz, J., and J. Rembieliński

1985 Hurwitz pairs equipped with complex structures. In: *Seminar on Deformations*, Łódź–Warszawa 1982–84, Proceedings (J. Ławrynowicz, ed.), Lecture Notes in Math. 1165, Springer, Berlin, 184–195.

1986 Pseudo-Euclidean Hurwitz pairs and generalized Fueter equations. In [Chisolm and Commons 1986], 39–48.

1990 On the composition of nondegenerate quadratic forms with an arbitrary index, (a) Inst. of Math. Polish Acad. Sci. Preprint no. 369 (1986); (b) *Ann. Fac. Sci. Toulouse Math.* **11** (1990), 141–168.

Ławrynowicz, J., E. Ramírez de Arellano and J. Rembieliński

1990 The correspondence between type-reversing transformations of pseudo-euclidean Hurwitz pairs and Clifford algebras, I, II, *Bull. Soc. Sci. Lett. Łódź* **40** (1990), 61–97; 99–129.

Lee, H.-C.

1945 On Clifford's algebra, *J. London Math. Soc.* **20**, 27–32.

1948 Sur le théoreme de Hurwitz–Radon pour la composition des formes quadratiques, *Comment. Math. Helv.* **21**, 261–269.

Leep, D. B., D. B. Shapiro and A. R. Wadsworth

1985 Sums of squares in division algebras, *Math. Z.* **190**, 151–162.

Leep, D. B., and A. R. Wadsworth

1989 The transfer ideal of quadratic forms and a Hasse norm theorem mod squares, *Trans. Amer. Math. Soc.* **315**, 415–432.

Lester, J. A.

1977 Cone preserving mappings for quadratic cones over arbitrary fields, *Canad. J. Math.* **29**, 1247–1253.

Lewis, D. W.

1989 New proofs of the structure theorems for Witt rings, *Exposition. Math.* **7**, 83–88.

Lewis, D. W., and J.-P. Tignol

1993 On the signature of an involution, *Arch. Math.* **60**, 128–135.

Levine, J.

1963 Imbedding and immersion of real projective spaces, *Proc. Amer. Math. Soc.* **14**, 801–803.

Lex, W.

1973 Zur Theorie der Divisionsalgebren, *Mitt. Math. Sem. Giessen* **103**, 1–68.

Littlewood, D. E.

1934 Note on the anticommuting matrices of Eddington, *J. London Math. Soc.* **9**, 41–50.

Lucas, E.

1878 Théorie des fonctions numeriques simplement périodiques, *Amer. J. Math.* **1**, 184–240. See especially p. 230.

Mal'cev, A. I.

1973 *Algebraic Systems*, Grundlehren Math. Wiss. 192, Springer, Berlin. See especially pp. 91–94.

Mammone, P., and D. B. Shapiro

1989 The Albert quadratic form for an algebra of degree four, *Proc. Amer. Math. Soc.* **105**, 525–530.

Mammone, P., and J.-P. Tignol

1986 Clifford division algebras and anisotropic quadratic forms: two counterexamples, *Glasgow Math. J.* **29**, 227–228.

Marcus, M.

1975 *Finite Dimensional Multilinear Algebra*, Part II, Marcel Dekker, New York.

Massey, W. S.

1983 Cross products of vectors in higher-dimensional Euclidean spaces, *Amer. Math. Monthly* **90**, 697–701.

Maurer, S.

1998 *Vektorproduktalgebren*, Diplomarbeit, Universität Regensburg.

McEvett, A.

1969 Forms over semisimple algebras with involution, *J. Algebra* **12**, 105–113.

McCrimmon, K.

1983 Quadratic forms permitting triple composition, *Trans. Amer. Math. Soc.* **275**, 107–130.

Merkurjev, A. S.

1981 On the norm residue symbol of degree 2 (in Russian), *Dokl. Akad. Nauk. SSSR* **261**, 542–547. English translation: *Soviet Math. Dokl.* **24**, 546–551.

1991 Simple algebras and quadratic forms (in Russian), *Izv. Akad. Nauk. SSSR Ser. Mat.* **55** , 218–224. English translation: *Math. USSR-Izv.* **38** (1992), 215–221.

Meshulam, R.

1990 On k-spaces of real matrices, *Linear Multilinear Algebra* **26**, 39-41.

Micali, A., R. Boudet and J. Helmstetter (eds.)

1992 *Clifford Algebras and Their Applications in Mathematical Physics*, Proceedings of second workshop held at Montpellier, France, 17-30 September, 1989, Kluwer, Dordrecht.

Micali, A., and P. Revoy

1979 *Modules Quadratiques*, Montpellier 1977. Also appeared in: *Bull. Soc. Math. France*, Mémoire 63.

Milgram, J.

1967 Immersing projective spaces, *Ann. of Math.* **85**, 473–482.

Milnor, J.

1965 *Topology from the Differentiable Viewpoint*, The University Press of Virginia, Charlottesville.

1969 On isometries of inner product spaces, *Invent. Math.* **8**, 83–97.

1970 Algebraic K-theory and quadratic forms, *Invent. Math.* **9**, 318–344.

Milnor, J., and R. Bott

1958 On the parallelizability of the spheres, *Bull. Amer. Math. Soc.* **64**, 87–89.

Milnor, J., and D. Husemoller

1973 *Symmetric Bilinear Forms*, Ergeb. Math. Grenzgeb. 73, Springer, Berlin.

Milnor, J., and J. Stasheff

1974 *Characteristic Classes*, Ann. of Math. Stud. 76, Princeton Univ. Press.

Mneimné, R.

1989 Formule de Taylor pour le déterminant et deux applications, *Linear Algebra Appl.* **112**, 39-47.

Morandi, P. J.

1999 Lie algebras, composition algebras, and the existence of cross products on finite-dimensional vector spaces, *Exposition. Math.* **17**, 63–74.

Morel, F.

1988 Voevodsky's proof of Milnor's conjecture, *Bull. Amer. Math. Soc.* **35**, 123–143.

Moreno, G.

1998 The zero divisors of the Cayley–Dickson algebras over the real numbers, *Bol. Soc. Mat. Mexicana* (3) **4**, 13–28.

Muir, T.

1906 *The Theory of Determinants in the Historical Order of Development*, 2 vols., Macmillan, London.

Mumford, D.

1963 *The Red Book of Varieties and Schemes*, Lecture Notes in Math. **1358**, Springer, Berlin 1988. This is a reprint of the book *Introduction to Algebraic Geometry*, Chapters 1–3, Harvard Univ. 1963.

Myung, H. C.

1986 *Malcev-Admissible Algebras*, Progr. Math. 64, Birkäuser, Basel.

Nathanson, M.

1975 Products of sums of powers, *Math. Mag.* **48** (1975), 112–113.

Neumann, W. D.

1977 *Equivariant Witt Rings*, Bonner Math. Schriften 100, Bonn.

Newman, M. H. A.

1932 Note on an algebraic theorem of Eddington, *J. London Math. Soc.* **7**, 93–99. Correction p. 272.

O'Meara, O. T.

1963 *Introduction to Quadratic Forms*, Grundlehren Math. Wiss. 117, Springer, Berlin.

Ono,T.

1955 Arithmetic of orthogonal groups, *J. Math. Soc. Japan* **7**, 79–91.

1974 Hasse principle for Hopf maps, *J. Reine angew. Math.* **268/269**, 209–212.

1994 *Variations on a Theme of Euler: Quadratic Forms, Elliptic Curves, and Hopf Maps*, Plenum Press, New York. (A translation and enlargement of the Japanese version originally published in 1980.)

Ono, T., and H. Yamaguchi
1979 On Hasse principle for division of quadratic forms, *J. Math. Soc. Japan* **31**, 141–159.

Orzech, G. (ed.)
1977 *Proceedings of the Conference on Quadratic Forms*, August 1–21, 1976 at Queen's University, Kingston, Ontario, Queen's Papers in Pure and Appl. Math. 46, Queen's University, Kingston, Canada.

Palacios, A. R.
1992 One-sided division absolute valued algebras, *Publ. Mat.* **36**, 925–954.

Parker, M.
1983 Orthogonal multiplications in small dimensions, *Bull. London Math. Soc.* **15**, 368–372.

Peng, C. K., and Z. Z. Tang
1997 On representing homotopy classes of spheres by harmonic maps, *Topology* **16**, 867–879.

Petersson, H. P.
1971 Quasi composition algebras, *Abh. Math. Sem. Univ. Hamburg* **35**, 215–222.

Petro, J.
1987 Real division algebras of dimension > 1 contain $\mathbb{C}$, *Amer. Math. Monthly* **94**, 445–449.

Petrović , Z. Z.
1996 *On spaces of matrices satisfying some rank conditions*, doctoral dissertation, Johns Hopkins Univ.

Pfister, A.
1965a Zur Darstellung von -1 als Summe von Quadraten in einem Körper, *J. London Math. Soc.* **40**, 159–165.
1965b Multiplikative quadratische Formen, *Arch. Math.* **16**, 363–370.
1966 Quadratische Formen in beliebigen Körpern, *Invent. Math.* **1**, 116–132.
1979 Systems of quadratic forms, *Bull. Soc. Math. France*, Mémoire **59**, 115–123.
1987 Quadratsummen in Algebra und Topologie, Wiss. Beitr., Martin-Luther-Univ. Halle Wittenberg 1987/88, 195–208.
1994 A new proof of the homogeneous Nullstellensatz for p-fields, and applications to topology. In: *Recent Advances in Real Algebraic Geometry and Quadratic Forms* (W. B. Jacob, T. Y. Lam and R. O. Robson, eds.), Contemp. Math. 155, Amer. Math. Soc., Providence, 221–229.
1995 *Quadratic Forms with Applications to Algebraic Geometry and Topology*, London Math. Soc. Lecture Note Ser. 217, Cambridge Univ. Press.

Pierce, R. S.
1982 *Associative Algebras*, Grad. Texts in Math. 88, Springer, New York.

A. Prestel, A.
1975 *Lectures on Formally Real Fields*, IMPA lecture notes, Rio de Janeiro 1975. Reprinted in: Lecture Notes in Math. 1093, Springer, Berlin 1984.

Prestel, A., and R. Ware
1979 Almost isotropic quadratic forms, *J. London Math. Soc.* (2) **19**, 241–244.

Procesi, C.
1976 The invariant theory of $n \times n$ matrices, *Adv. Math.* **19**, 306–381.

Putter, J.
1967 Maximal sets of anti-commuting skew-symmetric matrices, *J. London Math. Soc.* **42**, 303–308.

Quebbemann, H. G., W. Scharlau and M. Schulte
1979 Quadratic and hermitian forms in additive and abelian categories, *J. Algebra* **59**, 264–289.

Quéguiner, A.
1997 Cohomological invariants of algebras with involution, *J. Algebra* **194**, 299–330.

Racine, M. L.
1974 A simple proof of a theorem of Albert, *Proc. Amer. Math. Soc.* **43**, 487–488.

Radon, J.
1922 Lineare Scharen orthogonaler Matrizen, *Abh. Math. Sem. Univ. Hamburg* **1**, 1–14. Reprinted in *Collected Works*, vol. 1, Birkhäuser, Basel 1987, 307–320.

Rajwade, A. R.
1993 *Squares*, London Math. Soc. Lecture Note Ser. 171, Cambridge Univ. Press.

Riehm, C.
1982 The automorphism group of a composition of quadratic forms, *Trans. Amer. Math. Soc.* **269**, 403–414.
1984 Explicit spin representation and Lie algebras of Heisenberg type, *J. London Math. Soc.* (2) **29**, 49–62.

Rigas, A., and L. M. Chaves
1998 Hopf maps and triality. *Math. J. Okayama Univ.* **38**, 197–208.

Robert, E.
1912 *Composition des formes quadratiques de quatre et de huit variables indépendantes*, doctoral dissertation, ETH Zürich.

Roitberg, J.
1975 Dilatation phenomena in the homotopy groups of spheres, *Adv. Math.* **15**, 198–206.

Romero, D.
1995 New constructions for integral sums of squares formulae, *Bol. Soc. Mat. Mexicana* (3) **1**, 87–90.

Rosenblatt, J., and D. B. Shapiro
1989 Homometric elements in semisimple rings, *Comm. Algebra* **17**, 3043–3051.

Rost, M.
1994 On 14-dimensional quadratic forms, their spinors and the difference of two octonion algebras, preprint.
1996 On the dimension of a composition algebra, *Doc. Math.* **1**, 209–214.

Rothaus, O. S.
1978 Choosing simple bases for subspaces of tensor products, *J. Pure Appl. Algebra* **12**, 129–135.

Rowen, L. H.
1978 Central simple algebras, *Israel J. Math.* **29**, 285–301.
1980 *Polynomial Identities in Ring Theory*, Pure Appl. Math. 84. Academic Press, New York.

Rowen, L. H., and D. Saltman
1991 The discriminant of an involution, unpublished manuscript.

Saltman, D.
1978 Azumaya algebras with involution, *J. Algebra* **52**, 526–539.

Samuel, P.
1967 Qu'est-ce qu'une quadrique?, *Enseign. Math.* (2) **13**, 129–130.

Sánchez-Flores, A.
1996 A method to generate upper bounds for the sums of squares formulae problem, *Bol. Soc. Mat. Mexicana* (3) **2**, 79–92.

Sanderson, B. J.
1964 Immersions and embeddings of projective spaces, *Proc. London Math. Soc.* (3) **14**, 137–153.

Schafer, R. D.
1954 On the algebras formed by the Cayley–Dickson process, *Amer. J. Math.* **76**, 435–446.
1966 *An Introduction to Nonassociative Algebras*, Pure Appl. Math. 22, Academic Press, New York.
1970 Forms permitting composition, *Adv. Math.* **4**, 127–148.

Scharlau, W.
1984 Involutions on simple algebras and orders. In: *Quadratic and Hermitian forms*, Conf. Hamilton/Ont. 1983, CMS Conf. Proc. 4, 141–158.
1985 *Quadratic and Hermitian Forms*, Grundlehren Math. Wiss. 270, Springer, Berlin.

Scheiderer, C.
1992 Real algebra and its applications to geometry in the last ten years: some major developments and results. In: *Real Algebraic Geometry*, Proceedings Conf., Rennes 1991 (M. Coste, L. Mahé and M. F. Roy, eds.), Lecture Notes in Math. **1524**, Springer, Berlin, 75–96.

Schneider, J. E.
1973 Orthogonal groups of nonsingular forms of higher degree, *J. Algebra* **27**, 112–116.

Shapiro, D. B.
1974 *Similarities, quadratic forms, and Clifford algebras*, doctoral dissertation, University of California, Berkeley.
1976 Cancellation of semisimple hermitian pairings, *J. Algebra* **41**, 212–223.
1977a Spaces of similarities I, II, *J. Algebra* **46**, 148–170; 171–181.
1977b Spaces of similarities IV, *Pacific J. Math.* **69**, 223–244.

1978a Hasse principle for the Hurwitz problem, *J. reine angew. Math.* **301**, 179–190.
1978b Similarities and isoclinic planes, *Geom. Dedicata* **7**, 479–497.
1984a Products of sums of squares, *Exposition. Math.* **2**, 235–261.
1984b On the Hurwitz problem over an arbitrary field, *Bol. Soc. Math. Mexicana* (2) **29**, 1–4.
1992 Symmetric and skew-symmetric linear transformations, unpublished notes.
1997 Compositions of codimension 2, in preparation.

Shapiro, D. B., and M. Szyjewski
1992 Product formulas for quadratic forms, *Bol. Soc. Mat. Mexicana* (2) **37**, 463–474.

Shaw, R.
1987 Ternary vector cross products, *J. Phys. A* **20**, L689–L694.
1988 Ternary composition algebras and Hurwitz's theorem, *J. Math. Phys.* **29** (11), 2329–2333.
1989 Ternary composition algebras: 8 dimensions out of 4?, *Nuovo Cimento B* **103**, 161–183.
1990 Ternary composition algebras. I, II, *Proc. Royal Soc. London Ser. A* **431**, 1–19; 21–36

Shaw, R., and F. J. Yeadon
1989 On $(a \times b) \times c$, *Amer. Math. Monthly* **96**, 623–629.

Smith, L.
1978 Nonsingular bilinear forms, generalized J homomorphisms, and the homotopy of spheres I, *Indiana Univ. Math. J.* **27**, 697–737.

Smith, T. L., and P. Yiu
1992 Construction of sums of squares formulae with integer coefficients, *Bol. Soc. Mat. Mexicana* (2) **37**, 479–495.

Spanier, E.
1966 *Algebraic Topology*, McGraw-Hill, New York. Corrected edition: Springer, New York 1981.

Springer, T. A.
1954 An algebraic proof of a theorem of H. Hopf, *Indag. Math.* **16**, 33–35.

Stampfli-Rollier, C.
1983 4-dimensionale quasikompositions Algebren, *Arch. Math.* **40**, 516–525.

Stenzel, H.
1922 Über die Darstellbarkeit einer Matrix als Produkt von zwei symmetrischen Matrizen, als Produkt von zwei alternierenden Matrizen und als Produkt von einer symmetrischen und einer alternierenden Matrix, *Math. Z.* **15**, 1–25.

Stiefel, E.
1941 Über Richtungsfelder in den projektiven Räumen und einen Satz aus der reellen Algebra, *Comment. Math. Helv.* **13**, 201–218.

Storer, T.
1971 Hurwitz on Hadamard designs, *Bull. Austral. Math. Soc.* **4**, 109–112.

Swan, R.
1962 Vector bundles and projective modules, *Trans. Amer. Math. Soc.* **105**, 264–277.

1987 Vector bundles, projective modules and the K-theory of spheres. In: *Algebraic Topology and Algebraic K-theory* (W. Browder, ed.), Proc. of 1983 John Moore Princeton Conference, Ann. of Math. Stud. 113, Princeton Univ. Press

Sylvester, J. J.
1884 On quaternions, nonions, sedenions, etc., Johns Hopkins Univ. Circ. 3 (1884), 7–9. Reprinted in: *The Collected Mathematical papers of J. J. Sylvester*, vol. IV, 1912, (reprinted by Chelsea, New York, 1973), 122–132.

Sylvester, J.
1986 On the dimension of spaces of linear transformations satisfying rank conditions, *Linear Algebra Appl.* **78**, 1–10.

Szymiczek, K.
1977 Quadratic forms over fields, *Dissertationes Math. (Rozprawy Mat.)* 152, 63 pp.

Tamagawa, T.
1977 On quadratic forms and pfaffians, *J. Fac. Sci. Tokyo Sect. IA Math.*, **24**, 213–219.

Taussky, O.
1939 An algebraic property of Laplace's differential equation, *Quart. J. Math. Oxford* **10**, 99–103.
1966 A determinantal identity for quaternions and a new eight square identity, *J. Math. Anal. Appl.* **15**, 162–164.
1970 Sums of squares, *Amer. Math. Monthly* **77**, 805–830.
1971 $(1, 2, 4, 8)$-sums of squares and Hadamard matrices. In: *Combinatorics*, Proc. Symp. Pure Math. 19, UCLA 1968, Amer. Math. Soc., 229–233.
1981 History of sums of squares in algebra. In: *American Mathematical Heritage: Algebra and Applied Math.*, Texas Tech. Univ. Math. Ser. 13, 73–90.
1982 The many aspects of the Pythagorean triangles, *Linear Algebra Appl.* **43**, 285–295.
1987 Radon's contributions to matrix theory. In: *J. Radon, Collected Works*, vol. 1, Birkhäuser, Basel, 302–306.

Taussky, O., and H. Zassenhaus
1959 On the similarity transformation between a matrix and its transpose, *Pacific J. Math.* **9**, 893–896.

Thong, N.-Q.-D.
1975 Structure de groupe de similitudes, *Linear Algebra Appl.* **12**, 241–246.

Tignol, J.-P.
1977 Sur les corps à involution de degré 8, *C. R. Acad. Sci. Paris Sér. A* **284**, 1349–1352.
1978 Sur les classes de similitude de corps à involution de degré 8, *C. R. Acad. Sc. Paris Sér. A***286**, 875–876.
1991 Private communication.

Tits, J.
1968 Formes quadratiques, groupes orthogonaux et algèbres de Clifford, *Invent. Math.* **5**, 19–41.

Toth, G.
1990 *Harmonic Maps and Minimal Immersions Through Representation Theory*, Perspect. Math. 12, Academic Press, Boston.

Turisco,J.
1979 Quadratic mappings of spheres, *Linear Algebra Appl.* **23**, 261–274.
1985 A family of quadratic forms associated to quadratic mappings of spheres, *Linear Algebra Appl.* **65**, 249–260.

Tyrrell, J. A., and J. G. Semple
1971 *Generalized Clifford Parallelism*, Cambridge Univ. Press.

Urbanik, K., and F. B. Wright
1960 Absolute valued algebras, *Bull. Acad. Polon. Sci. Sér. Sci. Math. Astronom. Phys.* **8**, 285–286.

Veldkamp, F. D.
1991 Freudenthal and the octonions, *Nieuw. Arch. Wisk.* **9**, 145–162.

Voss, A.
1896 Symmetrische und alternierende Lösungen der Gleichung $SX = XS'$, *Sitzungsber. Math.-Phys. Cl. Akad. Wiss. München* **26**, 273–281.

Wadsworth, A.
1972 *Noetherian pairs and function fields of quadratic forms*, doctoral dissertation, University of Chicago.
1975 Similarity of quadratic forms and isomorphism of their function fields, *Trans. Amer. Math. Soc.* **208**, 352–358.
1990 16-dimensional algebras with involution, manuscript.

Wadsworth, A., and D. B. Shapiro
1977a Spaces of similarities III, *J. Algebra* **46**, 182–188.
1977b On multiples of round and Pfister forms, *Math. Z.* **157**, 53–62.

van der Waerden, B. L.
1976 Hamilton's discovery of the quaternions, *Math. Mag.* **49**, 227–234.
1985 *A History of Algebra: From al-Khwārizmī to Emmy Noether*, Springer, Berlin.

Waterhouse, W. C.
1987 Nonassociative quaternion algebras, *Algebras, Groups and Geoms.* **4**, 365–378.

Westwick, R.
1972 Spaces of linear transformations of equal rank, *Linear Algebra Appl.* **5**, 49–64.
1987 Spaces of matrices of fixed rank, *Linear and Multilinear Algebra* **20**, 171–174.

Whitehead, G. W.
1963 Note on cross-sections in Stiefel manifolds, *Comment. Math. Helv.* **37**, 239–240.

Witt, E.
1937 Theorie der quadratischer Formen in beliebigen Körpern, *J. reine angew. Math.* **176**, 31–44.

Wolf, J. A.
1963a Geodesic spheres in Grassmann manifolds, *Illinois. J. Math.* **7**, 425–446.
1963b Elliptic spaces in Grassmann manifolds, *Illinois. J. Math.* **7**, 447–462.

Wolfe, W.

1976 Amicable orthogonal designs – existence, *Canad. J. Math.* **28**, 1006–1020.

1977 Rational quadratic forms and orthogonal designs. In: *Number Theory and Algebra* (H. Zassenhaus, ed.), 339–348, Academic Press, New York.

Wonenburger, M. J.

1962a The Clifford algebra and the group of similitudes, *Canad. J. Math.* **14**, 45–59.

1962b Study of certain similitudes, *Canad. J. Math.* **14**, 60–68.

Wong, Y. C.

1961 Isoclinic n-Planes in Euclidean $2n$-space, Clifford Parallels in Elliptic $(2n-1)$-Space, and the Hurwitz Matrix Equations, *Mem. Amer. Math. Soc.* 41. Second printing with corrections and minor changes, 1971.

Wong, Y. C., and K.-P. Mok

1988 Isoclinic n-planes in $\mathbb{R}^{2n}$ and the Hopf–Steenrod sphere bundles $S^{2n-1} \to S^n$, $n = 2, 4, 8$, *Enseign. Math.* **34**, 167–204.

1990 Normally related n-planes and isoclinic n-planes in $\mathbb{R}^{2n}$ and strongly linearly independent matrices of order n, *Linear Algebra Appl.* **139**, 31–52.

Wood, R.

1968 Polynomial maps from spheres to spheres, *Invent. Math.* **5**, 163–168.

Yabe, T.

1976 On a certain quadratic form, *Mem. Defense Acad. Japan* **16**, 71–74.

Yanchevskiĭ, V. I., see V. I. Jančevskiĭ.

Yang, C. T.

1981 Division algebras and fibrations of spheres by great spheres, *J. Differential Geom.* **16**, 577–593.

Yiu, P.

1985 *Topological and combinatoric methods for studying sums of squares*, doctoral dissertation, Univ. of British Columbia.

1986 Quadratic forms between spheres and the non-existence of sums of squares formulae, *Math. Proc. Cambridge Philos. Soc.* **100**, 493–504.

1987 Sums of squares formulae with integer coefficients, *Canad. Math. Bull.* **30**, 318–324.

1990a On the product of two sums of 16 squares as a sum of squares of integral bilinear forms, *Quart. J. Math. Oxford* (2) **41**, 463–500.

1990b Strongly regular graphs and Hurwitz–Radon numbers, *Graphs Combin.* **6**, 61–69.

1993 Maximal normal sets of n-planes in $\mathbf{R}^{2n}$, *Linear Algebra Appl.* **181**, 241–249.

1994a Composition of sums of squares with integer coefficients. In: [Ławrynowicz 1994], 7–100.

1994b Quadratic forms between euclidean spheres, *Manuscripta Math.* **83**, 171–181.

1994c Tables of nonsingular bilinear maps, unpublished notes.

1996 Some upper bounds for composition numbers, *Bol. Soc. Mat. Mexicana* (3) **2**, 65–78.

Yuzvinsky, S.

1981 Orthogonal pairings of Euclidean spaces, *Michigan Math. J.* **28**, 131–145.

1983 On the Hopf condition over an arbitrary field, *Bol. Soc. Mat. Mexicana* (2) **28**, 1–7.

1984 A series of monomial pairings, *Linear and Multilinear Algebra* **15**, 109–119.

1985 Composition of quadratic forms and tensor product of quaternion algebras, *J. Algebra* **96**, 347–367.

Zassenhaus, H., and W. Eichhorn

1966 Herleitung von Acht-und Sechzehn Quadrate-Identitäten mit Hilfe von Eigenschaften der verallgemeinerten Quaternionen und der Cayley-Dicksonschen Zahlen, *Arch. Math.* **17**, 492–496.

Zhevlakov, K. A., A. M. Slin'ko, I. P. Shestakov and A. I. Shirshov

1982 *Rings that are Nearly Associative*, Pure Appl. Math. 104. Academic Press, New York.

Zorn, M. A.

1930 Theorie der alternative Ringe, *Abh. Math. Sem. Hamburg Univ.* **8**, 123–147.

1933 Alternativkörper und quadratische Systeme, *Abh. Math. Sem. Hamburg* **9**, 395–402.

1935 The automorphisms of Cayley's non-associative algebra, *Proc. Nat. Acad. Sci.* **21**, 355–358.

Züger, R.

1995 *Kompositionen von quadratischen Formen mit quadratischen, bilinearen oder hermiteschen Formen*, doctoral dissertation, Univ. Bern.

Zvengrowski, P.

1966 A 3-fold vector product in $\mathbf{R}^8$, *Comment. Math. Helv.* **40**, 149–152.

1968 Canonical vector fields on spheres, *Comment. Math. Helv.* **43**, 341–348.

List of Symbols

Each symbol includes page references to definitions and to a few later occurrences.

Index

Each term includes a page reference to its first use and (sometimes) to a few later occurrences. A person's name is listed when it occurs in the text as more than a direct reference to a publication.